Molecular Mechanisms of Bioactive Nutrients Promoting Health through Gut Microbiota 2.0

Molecular Mechanisms of Bioactive Nutrients Promoting Health through Gut Microbiota 2.0

Guest Editors

Baojun Xu
Matteo Bordiga

MDPI

Basel • Beijing • Wuhan • Barcelona • Belgrade • Novi Sad • Cluj • Manchester

Guest Editors

Baojun Xu
Department of Life Sciences
BNU-HKBU United
International College
Zhuhai
China

Matteo Bordiga
Department of
Pharmaceutical Sciences
Università degli Studi del
Piemonte Orientale "A. Avogadro"
Novara
Italy

Editorial Office

MDPI AG
Grosspeteranlage 5
4052 Basel, Switzerland

This is a reprint of the Special Issue, published open access by the journal *International Journal of Molecular Sciences* (ISSN 1422-0067), freely accessible at: www.mdpi.com/journal/ijms/special_issues/Gut_Microbiota_Bioactive_Nutrients.

For citation purposes, cite each article independently as indicated on the article page online and using the guide below:

Lastname, A.A.; Lastname, B.B. Article Title. *Journal Name* **Year**, *Volume Number*, Page Range.

ISBN 978-3-7258-3254-5 (Hbk)
ISBN 978-3-7258-3253-8 (PDF)
https://doi.org/10.3390/books978-3-7258-3253-8

Contents

About the Editors

Baojun Xu

Dr. Xu is a Chair Professor at the Beijing Normal University-Hong Kong Baptist University United International College (UIC, a full English teaching college in China), a Fellow of the Royal Society of Chemistry, a Zhuhai Scholar Distinguished Professor, Department Head of the Department of Life Sciences, the Program Director of the Food Science and Technology Program, and the author of over 370 peer-reviewed papers. Dr. Xu received his Ph.D. in Food Science from Chungnam National University, South Korea. He conducted his postdoctoral research work at North Dakota State University (NDSU), Purdue University, and the Gerald P. Murphy Cancer Foundation in the USA during 2005–2009. He did short-term visiting research at NDSU in 2012 and the University of Georgia in 2014, followed by visiting research during his sabbatical leave (7 months) at Pennsylvania State University in the USA in 2016. Dr. Xu is serving as the Associate Editor-in-Chief of *Food Science and Human Wellness*, the Associate Editor of *Food Research International*, the Associate Editor of *Food Frontiers*, and an editorial board member of around 10 international journals. He received the inaugural President's Award for Outstanding Research of UIC in 2016 and the President's Award for Outstanding Service of UIC in 2020. Dr. Xu has been listed in the world's top 2% of scientists by Stanford University for the past 5 consecutive years and has been listed in the Best Scientist in the World ranking in the field of Biology and Biochemistry at Research.com in 2023 and 2024. Prof. Xu was named an inaugural Highly Ranked Scholar (top 0.05%) by ScholarGPS in 2024 and 2025.

Matteo Bordiga

Dr. Bordiga is currently an Assistant Professor of Food Chemistry at Università del Piemonte Orientale (UPO), Novara, Italy. He earned his PhD in Food Science and MS in Chemistry and Pharmaceutical Technologies from the same university. His main research activity concerns food chemistry, investigating the different classes of polyphenols from an analytical, technological, and nutritional point of view. More recently, he moved his research interests to wine chemistry, focusing his attention on the entire production process. His research activity mainly concerns the chemical characterization and study of the antioxidant properties of food matrices of vegetable origin (cocoa and chocolate, grapes, wine, cheeses, peppers, hazelnuts, and cereals such as rice, barley, and wheat) and by-products of the agro-food industry (hazelnut cuticle, cocoa peel, and pomace), with particular interest in the impact that the technological transformation processes (fermentation, roasting, grinding and hulling, and food cooking) have on the composition and activity of products. His research activities also concern the optimization and validation of chemical–analytical methods (in particular, spectrophotometric methods and chromatographic methods) applied in the food and nutraceutical fields.

He has published more than 60 research papers in peer-reviewed international and national journals. He is currently the Editor of *LWT - Food Science and Technology* (Q1; Elsevier) and is the Associate Editor of *Food Science and Nutrition* (Q1; Wiley) and *Quality Assurance and Safety of Crops & Foods* (Q1; Codon Publications).

International Journal of *Molecular Sciences*

MDPI

Editorial

Special Issue "Molecular Mechanisms of Bioactive Nutrients Promoting Health Through Gut Microbiota 2.0"

Matteo Bordiga [1] and Baojun Xu [2,*]

1 Department of Pharmaceutical Sciences, Università degli Studi del Piemonte Orientale "A. Avogadro", Largo Donegani 2, 28100 Novara, Italy; matteo.bordiga@uniupo.it

2 Food Science and Technology Program, Department of Life Sciences, BNU-HKBU United International College, Zhuhai 519087, China

* Correspondence: baojunxu@uic.edu.cn; Tel.: +86-756-3620636

This Special Issue of the *International Journal of Molecular Sciences* focuses on the highly relevant and rapidly developing topic of gut microbiota research. As the articles within this collection highlight, the human gut microbiota plays a critical role in human health. Dietary intake directly influences gut microbiota composition, and the resulting changes in gut metabolites can have profound effects on the host. This Special Issue brings together the latest findings on the intricate relationship between bioactive food components, gut microbiota, colon health, and chronic metabolic diseases.

The gut microbiota is a complex ecosystem composed of trillions of microorganisms, including bacteria, archaea, fungi, and viruses. These microbes play essential roles in digestion, nutrient absorption, immune function, and even mental health. Importantly, the composition of the gut microbiota is not static; it is constantly changing in response to diet, lifestyle factors, and medications.

A growing body of research suggests that bioactive components derived from dietary sources can be harnessed to modify gut microbiota composition and promote health. These bioactive components include phytochemicals (naturally occurring plant chemicals) and complex carbohydrates. However, the exact mechanisms by which these dietary components influence gut microbiota metabolism and how these changes affect human health remain unclear.

This Special Issue features a selection of articles that address these critical gaps in our knowledge. For instance, the article by Velderrain-Armenta et al. investigates the combined effects of *Bifidobacterium longum* and *Chlorella sorokiniana* on the antiviral cellular immune response [1]. Their findings suggest that this combination may be effective in boosting the immune system and protecting against rotavirus infection.

In another article by Mollace et al. explores the potential of bergamot polyphenolic extract combined with albedo and pulp fibers to counteract changes in gut microbiota associated with a high-fat diet [2]. Their study suggests that this combination may help to improve gut health and metabolic profiles in individuals with high-fat diet-induced dyslipidemia.

The article by Kato et al. takes a more holistic approach, examining the integrated multi-omics effects of fructo-oligosaccharide (FOS) supplementation on the human gut ecosystem [3]. Their findings highlight the significant inter-individual variability in response to FOS supplementation, suggesting the need for personalized approaches to prebiotic consumption.

Received: 19 January 2025
Accepted: 24 January 2025
Published: 26 January 2025

Citation: Bordiga, M.; Xu, B. Special Issue "Molecular Mechanisms of Bioactive Nutrients Promoting Health Through Gut Microbiota 2.0". *Int. J. Mol. Sci.* **2025**, *26*, 1081. https://doi.org/10.3390/ijms26031081

Moving beyond the gut, Liu et al. explore the potential of egg-derived peptides to prevent obesity in a mouse model [4]. Their study suggests that these peptides may help to reduce lipid deposition and reprogram gut microbiota, ultimately leading to weight loss and improved metabolic health.

The complex interplay between diet, gut microbiota, and mental health is the focus of the article by Randeni and Xu [5], in which they review the latest evidence on how dietary components can influence gut microbiota composition and function, thereby impacting mood and reducing the risk of depression.

Randeni et al. also contribute another article to this Special Issue, this time exploring the triangular relationship between diet, gut microbiota, and inflammation [6]. Chronic inflammation is a key factor in many diseases, and this review highlights the potential of dietary interventions to modulate gut microbiota composition and reduce inflammation.

The article by Rodríguez-Daza and de Vos focuses on the role of dietary polyphenols in promoting the growth of *Akkermansia muciniphila*, a beneficial gut bacterium associated with improved gut health and metabolic function [7]. Their review explores the mechanisms by which polyphenols may exert these effects and the potential therapeutic implications.

Finally, Santhiravel et al. discuss the impact of plant phytochemicals on the gut microbiota [8], highlighting the potential of these bioactive compounds to modify gut microbiota composition and promote human health.

In conclusion, this Special Issue of the *International Journal of Molecular Sciences* provides a comprehensive overview of the latest research on the interplay between bioactive nutrients, gut microbiota, inflammation [9], and human health [10]. The research contributions to this Special Issue highlight the vast potential of dietary interventions to modulate gut microbiota composition and promote health. As our understanding of this complex ecosystem continues to grow, we can expect to see the development of novel therapeutic strategies for a wide range of chronic diseases.

Conflicts of Interest: The authors declare no conflicts of interest.

References

1. Velderrain-Armenta, F.; González-Ochoa, G.; Tamez-Guerra, P.; Romero-Arguelles, R.; Romo-Sáenz, C.I.; Gomez-Flores, R.; Flores-Mendoza, L.; Icedo-García, R.; Soñanez-Organis, J.G. *Bifidobacterium longum* and *Chlorella sorokiniana* combination modulates IFN-γ, IL-10, and SOCS3 in rotavirus-infected cells. *Int. J. Mol. Sci.* **2024**, *25*, 5514. [CrossRef] [PubMed]
2. Mollace, R.; Macrì, R.; Nicita, M.; Musolino, V.; Gliozzi, M.; Carresi, C.; Bava, I.; Maiuolo, J.; Tavernese, A.; Cardamone, A.; et al. Bergamot polyphenolic extract combined with albedo and pulp fibres counteracts changes in gut microbiota associated with high-fat diet: Implications for lipoprotein size re-arrangement. *Int. J. Mol. Sci.* **2023**, *24*, 12967. [CrossRef] [PubMed]
3. Kato, T.; Kagawa, M.; Suda, W.; Tsuboi, Y.; Inoue-Suzuki, S.; Kikuchi, J.; Hattori, M.; Ohta, T.; Ohno, H. Integrated multi-omics analysis reveals differential effects of fructo-oligosaccharides (FOS) supplementation on the human gut ecosystem. *Int. J. Mol. Sci.* **2022**, *23*, 11728. [CrossRef] [PubMed]
4. Liu, Z.; Ding, S.; Jiang, H.; Fang, J. Egg protein transferrin-derived peptides Irw (Lle-Arg-Trp) and Iqw (Lle-Gln-Trp) prevent obesity mouse model induced by a high-fat diet via reducing lipid deposition and reprogramming gut microbiota. *Int. J. Mol. Sci.* **2022**, *23*, 11227. [CrossRef] [PubMed]
5. Randeni, N.; Xu, B. Critical review of the cross-links between dietary components, the gut microbiome, and depression. *Int. J. Mol. Sci.* **2025**, *26*, 614. [CrossRef] [PubMed]
6. Randeni, N.; Bordiga, M.; Xu, B. A comprehensive review of the triangular relationship among diet-gut microbiota-inflammation. *Int. J. Mol. Sci.* **2024**, *25*, 9366. [CrossRef] [PubMed]
7. Rodríguez-Daza, M.C.; de Vos, W. Polyphenols as drivers of a homeostatic gut microecology and immuno-metabolic traits of *Akkermansia muciniphila*: From mouse to man. *Int. J. Mol. Sci.* **2023**, *24*, 45. [CrossRef] [PubMed]
8. Santhiravel, S.; Bekhit, A.E.-D.A.; Mendis, E.; Jacobs, J.L.; Dunshea, F.R.; Rajapakse, N.; Ponnampalam, E.N. The impact of plant phytochemicals on the gut microbiota of humans for a balanced life. *Int. J. Mol. Sci.* **2022**, *23*, 8124. [CrossRef] [PubMed]

9. Ma, Z.; Du, B.; Li, J.; Yang, Y.; Zhu, F. An insight into anti-inflammatory activities and inflammation related diseases of anthocyanins: A review of both in vivo and in vitro investigations. *Int. J. Mol. Sci.* **2021**, *22*, 11076. [CrossRef] [PubMed]
10. Megur, A.; Daliri, E.B.-M.; Baltriukienė, D.; Burokas, A. Prebiotics as a tool for the prevention and treatment of obesity and diabetes: Classification and ability to modulate the gut microbiota. *Int. J. Mol. Sci.* **2022**, *23*, 6097. [CrossRef] [PubMed]

International Journal of
Molecular Sciences

MDPI

Review

Critical Review of the Cross-Links Between Dietary Components, the Gut Microbiome, and Depression

Nidesha Randeni and Baojun Xu *

Food Science and Technology Program, Department of Life Sciences, BNU-HKBU United International College, Zhuhai 519087, China; nidesha.randeni96@gmail.com
* Correspondence: baojunxu@uic.edu.cn; Tel.: +86-756-3620636

Abstract: The complex relationship between diet, the gut microbiota, and mental health, particularly depression, has become a focal point of contemporary research. This critical review examines how specific dietary components, such as fiber, proteins, fats, vitamins, minerals, and bioactive compounds, shape the gut microbiome and influence microbial metabolism in order to regulate depressive outcomes. These dietary-induced changes in the gut microbiota can modulate the production of microbial metabolites, which play vital roles in gut–brain communication. The gut–brain axis facilitates this communication through neural, immune, and endocrine pathways. Alterations in microbial metabolites can influence central nervous system (CNS) functions by impacting neuroplasticity, inflammatory responses, and neurotransmitter levels—all of which are linked to the onset and course of depression. This review highlights recent findings linking dietary components with beneficial changes in gut microbiota composition and reduced depressive symptoms. We also explore the challenges of individual variability in responses to dietary interventions and the long-term sustainability of these strategies. The review underscores the necessity for further longitudinal and mechanistic studies to elucidate the precise mechanisms through which diet and gut microbiota interactions can be leveraged to mitigate depression, paving the way for personalized nutritional therapies.

Keywords: diet; gut health; dietary patterns; mental health; neurotransmitters

Academic Editor: Hirokazu Fukui

Received: 19 December 2024
Revised: 6 January 2025
Accepted: 11 January 2025
Published: 13 January 2025

Citation: Randeni, N.; Xu, B. Critical Review of the Cross-Links Between Dietary Components, the Gut Microbiome, and Depression. *Int. J. Mol. Sci.* **2025**, *26*, 614. https://doi.org/10.3390/ijms26020614

1. Introduction

The intricate relationship between diet, the gut microbiome, and mental health has emerged as a focal point in contemporary biomedical research. This triad of components is the basis of the gut–brain axis—the bidirectional communication network between the enteric nervous system (ENS) and the CNS that "talks" back and forth and integrates information from the gut and the brain to regulate bodily functions and mental health. The gut–brain axis encompasses several different pathways (immunological, hormonal, and neurological), all contributing to GI functions and the mental state of individuals [1].

Dietary influences play a major role in shaping the gut microbiome—the ecological community of billions of microorganisms residing within the gastrointestinal (GI) tract [2]. These microbes are not only present passively but also play active roles in several physiological processes, like the modulation of immune response, production of neurotransmitters, and even synthesis of some essential metabolites. There are several kinds of dietary intervention that cause changes in the diversity and composition of the gut microbiota [3], with some foods being more beneficial for enhancing certain bacterial groups while others cause an imbalance called dysbiosis, which is known to be associated with several diseases [4].

Research on depression, which is a widespread and debilitating mental disorder, has revealed that changes in the gut–brain axis may be implicated in the condition. There is a growing body of evidence indicating that gut microbiota may influence functions of the brain and behaviors through the synthesis of neuroactive substances, changes in the immune response, and the modification of the integrity of the gut barrier. All these factors are important and contribute to the understanding of the disease mechanisms underlying depression, as well as providing new approaches for treatment by manipulating the gut microbiome [5].

The objective of this review is to conduct an integral and objective survey of all literature available on the links connecting depression, gut microbiota, and diet. In particular, we will elucidate the mechanisms within the gut–brain axis—particularly focusing on neurotransmitter production, immune response, and microbial byproduct composition—that we hypothesize can explain how dietary interventions can modify gut microbiota to improve mental health. This review will be framed within the context of recent advances in the field, highlighting key studies and identifying gaps in the knowledge that warrant further investigation.

2. Gut Microbiome and Depression

2.1. Composition and Function of the Gut Microbiome

The gut microbiome comprises numerous microorganisms that play critical roles within the host. The gut microbiota is dominated by bacteria, with the majority belonging to four dominant phyla: *Firmicutes*, *Bacteroidetes*, *Actinobacteria*, and *Proteobacteria* [6]. These phyla are interconnected to form a microbial community, which is considered unique for each species in the community because of the effects of the host's genetics, diet, age, geography, and lifestyle [7]. *Firmicutes* is the major phylum of the gut microbiome and contain several families, such as *Lactobacillaceae*, *Clostridiaceae*, and *Ruminococcaceae*. Microorganisms belonging to *Firmicutes* are critical for the fermentation of complex plant materials into SCFAs such as butyrate, which has been shown to possess anti-inflammatory properties while also supporting the health of intestinal tissues [8]. In particular, *Bacteroidetes*, including species from the genus *Bacteroides*, is involved in the breakdown of protein and polysaccharide complexes in the gut into smaller units for utilization by the host and other microbes. These organisms are important for the fermentation of carbohydrates and the production of SCFAs important for metabolic health, such as acetate and propionate [9]. While members of the *Actinobacteria* phylum, such as *Bifidobacterium*, are quite common in the gut microbiome of young children, this group is also significant in adult microbiomes. *Bifidobacterium* species include patented probiotic products that have effects on the immune system and prevent the colonization of the gut by pathogenic organisms [10]. Although not as abundant as the other major phyla, *Proteobacteria* comprises *Escherichia*, *Salmonella*, and *Helicobacter* species that play a role in the gut. Whereas some *Proteobacteria* are non-pathogenic members of the skin microbiota, others are opportunistic pathogens that cause dysbiosis at elevated population densities. *Proteobacteria* species promote inflammation and can be upregulated in diseases such as IBD [11].

In addition to bacteria, the gut microbiome also includes archaea, viruses, fungi, and protozoa [12]. One of the archaea species is *Methanobrevibacter smithii*, involved in methane production, and some others are involved in specific digestive processes. Many of the components of gut virome are bacteriophages, viruses that infect bacteria and that function as population control agents. Although they are fewer in number, fungi are also involved in maintaining gut health, but when they start to proliferate, problems occur, as seen in Candida dysbiosis [13].

Every organism in an individual microbiome is connected with the other organisms to form associations that are considered significant for the host organism's health status. The proportion of these microbial communities is essential because the disruption of the balance of the microbiome, known as dysbiosis, is associated with many diseases, for instance, metabolic, autoimmune, and psychiatric disorders.

2.2. Gut Microbiome Dysbiosis and Depression

The connection between depression and gut microbiome dysbiosis is a topic that has garnered progressively increasing scientific attention concerning its usefulness in explaining how imbalances in the microbial populations in the human gut can cause mental health disorders. Dysbiosis is defined as a reduction in beneficial bacteria (for example, *Bifidobacterium* and *Lactobacillus*) and an increase in potentially pathological bacteria (some strains of *Proteobacteria* and *Clostridia*) [14]. This imbalance can disrupt the integrity of the intestinal barrier, promote immune activation, and create a state of low-grade inflammation, which is believed to affect depressive-like behavior via bidirectional communication between the gut and the brain (Figure 1).

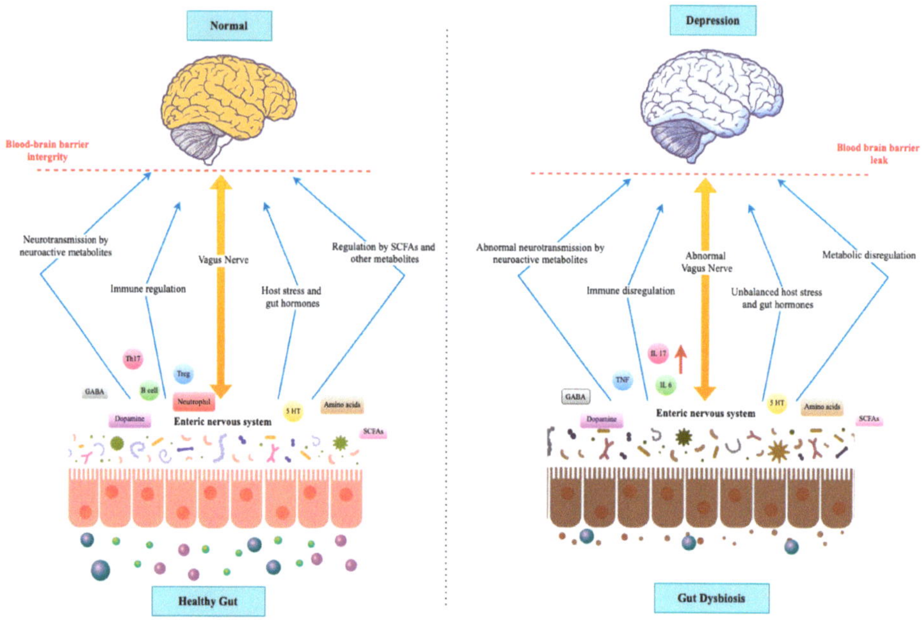

Figure 1. Gut dysbiosis in depression.

This is supported by one of the effects of dysbiosis, in which dysbiosis triggers the hypothalamic–pituitary–adrenal (HPA) axis, which is very crucial in the stress response pathway. An imbalance in microbiota results in enhanced permeability of the intestinal barrier, or "leaky gut", and causes the liberation of bacterial endotoxins such as lipopolysaccharide (LPS). These endotoxins act as immunostimulants, thus increasing the production of pro-inflammatory cytokines. These cytokines appear to modulate neurotransmitter function and neural plasticity, both factors related to mood regulation [15]. The dysbiotic gut also impacts the levels of neuroactive metabolites that function in the brain. A healthy gut microbiota synthesizes hormones, including serotonin precursors and SCFAs such as butyrate, necessary for modulating brain health. An ideal microbiome makes up to 90% of the body's serotonin, which is responsible for mood. These metabolites include

SCFAs, the production of which decreases in dysbiosis—a state that interferes with these neurotransmitter pathways and influences the development and severity of depression [16]. In addition, dysbiosis is linked with the overproduction of reactive oxygen species (ROS), which influence inflammatory processes in the gut and within the brain. Furthermore, an imbalance in the gut microbiota may affect neurogenesis, causing a decline in the generation of neurons in the hippocampal part of the brain [17]. In particular, the data also imply that dysbiosis affects the endocannabinoid system, regulating such processes as emotions, stress, and anxiety. Some of the gut microbiota directly modulate endocannabinoid tone, affecting the brain and mood. This homeostasis is imbalanced in dysbiosis, with possible dysregulation of mood and increased levels of depressive symptoms [18]. With depressive-like behavior, experiments on in vivo serotonin, dopamine, norepinephrine, 5-HIAA, 3-MT, HVA, and MHPG metabolic reduction have revealed a gut microbiota shift [19]. Overall, gut microbiota imbalance plays a role in the disruption of the anti-inflammatory and anti-neurotoxic milieu; it participates in mood regulation through numerous processes. Knowing this connection means there are new options for the use of dietary and probiotic interventions targeting the restoration of normal gut microbiota, possibly for the prevention and treatment of depression.

Diet is the main factor associated with dysbiosis. It is worth noting that the human gut microbiome is very dynamic and sensitive to the types of foods eaten, where various diets and types of foods support the growth of particular microbial communities. A diet rich in fiber helps in the growth of friendly bacteria that produce SCFAs, which are important in gut health and immune defense [20]. On the other hand, the Western diet, associated with the high consumption of saturated fats and sugars, can have negative health effects that translate to an increase in inflammation-promoting bacteria and a decrease in human bacterial counts, otherwise known as dysbiosis [21]. This imbalance can disrupt the gut barrier function, allowing for the translocation of microbial products into the bloodstream and triggering systemic inflammation. Further, low nutrient density, as found in foods with few vitamins and minerals, affects microbial metabolism and decreases the synthesis of important metabolites that are beneficial for host physiology [2]. In contrast, the oral intake of red ginseng lowered the levels of depression in mice and stabilized changes in the gut microbiota during anxiety and depression [22]. In this regard, the type and contents of food consumed decide the nature of the gut microbiota, and a substandard diet tends to create dysbiosis and result in various health complications, including metabolic disease, inflammation, and other mental health disorders such as depression.

2.3. The Gut–Brain Axis

The gut–brain axis is a two-way communication system connecting the CNS with the ENS through neural, hormonal, and immune system messengers. This bidirectional system is capable of coordinating messaging between the brain and GI system and, therefore, has a significant effect on the physical and psychological well-being of the body [23]. Specifically, it has been found that the gut microbiota and its bidirectional communication with the brain affect mood [24], mental health, decision-making [24], and even preference for sweetness [25]. Factors such as the ability of the gut microbiota to use the host and influence it to act in a manner favorable for the microbiota via reliance or local manipulation seem probable. In this case, the ways the host perceives and activates their brain and body may be modulated by gut flora "by force".

The vagus nerve is believed to be the primary neural pathway in the gut–brain axis, mediating communications between the brain and the gut. It plays an important role in digestion, mood, and controlling inflammation [26]. The gut sensory neurons communicate through enteric nerves about the state of the gut to the brain, while motor neurons carry signals from the brain to the gut for the purpose of motility and secretion processes [5]. Furthermore, the ENS, also known as the "second brain", consists of a massive network of neurons operating largely independently from the CNS, though modulated by it via the vagus nerve and other autonomic pathways [27] (Figure 2). Other important components of the gut–brain axis are hormones and neuropeptides. These gut hormones—ghrelin, leptin, and peptide YY, among others—influence appetite, metabolism, and even mood [28]. For example, the "hunger hormone" ghrelin, besides stimulating appetite, influences mood and cognitive functions, acting via the hypothalamus and other encephalic regions [29]. Furthermore, the involvement of the HPA axis, specifically stress hormones such as cortisol, interferes with gut function and microbiome composition, thus having impacts on mental health [30] (Figure 2). The gut and the brain interrelate through the immune system, wherein GALT plays a major role [31]. Moreover, inflammatory cytokines produced within the gut may cross the blood–brain barrier or signal through the vagus nerve to affect brain functions, influencing the development of mood disorders such as depression [32]. Dysbiosis leads to increased intestinal permeability known as "leaky gut", which allows bacterial endotoxins to enter the bloodstream and induce systemic inflammation. This inflammation can then impact brain function and behavior [33].

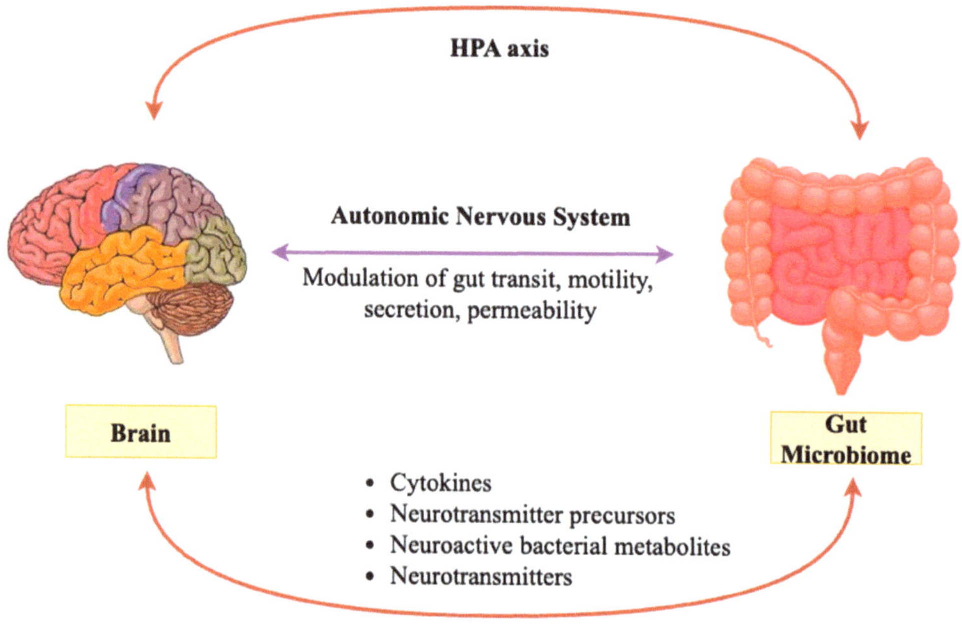

Figure 2. The gut–brain axis.

The gut microbiota produces metabolites such as SCFAs and neurotransmitters (e.g., serotonin, GABA) that can modulate brain function. Serotonin, for instance, is predominantly produced in the gut, and its levels are influenced by the gut microbiome composition. Microbial metabolites can affect the integrity of the blood–brain barrier and modulate neuroinflammation, influencing mental health outcomes [34]. Serotonin, or 5-hydroxytryptamine (5-HT), is a key neurotransmitter involved in regulating mood, cognition, and GI function. Approximately 90% of the body's serotonin is synthesized in the gut, predominantly by enterochromaffin cells, with a smaller proportion produced by neurons in the ENS [35]. Certain bacteria, such as *Enterococcus*, *Streptococcus*, and *Escherichia* species, can produce metabolites that influence enterochromaffin cells to increase serotonin production. GABA is the primary inhibitory neurotransmitter in the CNS, reducing neuronal excitability and playing a role in stress responses and mood regulation [36]. Certain gut bacteria, including *Lactobacillus* and *Bifidobacterium* species, can produce GABA. These bacteria convert glutamate, an excitatory neurotransmitter, into GABA through the action of glutamate decarboxylase [34]. GABAergic signaling in the brain is essential for regulating anxiety, stress, and mood. The dysregulation of serotonin levels in the gut and alterations in gut microbiota composition that affect GABA production may contribute to psychiatric conditions such as depression and anxiety disorders [34].

The gut–brain axis is significantly influenced by the immune system through mechanisms involving cytokine modulation, intestinal barrier integrity, and microglial activation. The gut microbiota interacts with the immune system, affecting both gut and brain health primarily by influencing cytokine production [37]. Cytokines are small proteins released by immune cells that play a key role in cell signaling, particularly in immune response. Dysbiosis can lead to the overproduction of pro-inflammatory cytokines, which can cross the blood–brain barrier or signal through the vagus nerve, promoting neuroinflammation and contributing to mood disorders such as depression [38]. Conversely, beneficial bacteria like *Bifidobacterium* and *Lactobacillus* promote the production of anti-inflammatory cytokines such as IL-10, supporting immune homeostasis [39]. Gut bacteria can also influence the expression and function of tight junction proteins that regulate intestinal permeability. Dysbiosis also compromises intestinal barrier integrity, resulting in a "leaky gut", allowing endotoxins like LPS to enter the bloodstream and trigger systemic inflammation. This inflammation can activate microglia, the brain's immune cells, leading to neuroinflammation, a hallmark of neuropsychiatric disorders. These interactions underscore the potential of targeting the gut microbiota for therapeutic interventions in improving gut and mental health [40].

2.4. Metabolites Produced by Microorganisms and Their Physiological Roles

The term microbial-derived metabolites refers to the biochemical compounds that are formed by the gut microbiome as it ferments fibers, proteins, and other substances in food. These compounds, which include SCFAs, bile acid derivatives, indoles, and neurotransmitter precursors, among others, have many physiological functions and are utilized outside of the gastrointestinal tract; they affect immunity, metabolism, and inflammation, as well as mental health. They act as crucial mediators integrating the gut with all other systems, including the brain, via the gut–brain axis. In addition, it has been established that the gastrointestinal tract has numerous endogenous gut hormones such as leptin, ghrelin, cholecystokinin, and glucagon-like peptide-1 (GLP-1). The mentioned hormones also include neuropeptide Y, connected to mood, anxiety, and the immunological system, as well as peptide YY, originating from two pancreatic polypeptides that promote satiety and enhance glucose balance and behavior [41].

SCFAs that are produced as a result of the bacterial fermentation of dietary fiber include acetate, propionate, and butyrate. These metabolites are important in the gut, as they enhance barrier function by promoting tight junctions and the secretion of mucus. SCFAs exert an anti-inflammatory effect and immunomodulatory function by inducing T regulatory cells (Tregs), which curb excessive immune responses and thereby help in controlling inflammation throughout the body [42]. Moreover, SCFAs upregulate the levels of the enzyme tyrosine hydroxylase, which regulates the rate at which the biosynthesis of catecholamines, including dopamine, adrenaline, and noradrenaline, occurs and also upregulates tryptophan 5-hydroxylase 1, which is a critical enzyme involved in producing the neurotransmitter serotonin [43]. Tryptophan, an indispensable dietary amino acid, is a source of numerous important metabolites, including serotonin, that are crucial for neuroendocrine signaling. The use of tryptophan to produce various metabolites, including tryptamine, kynurenine, and indoles, in particular, is attributable to a specific gut microbiome [44]. Tryptophan metabolites reach the CNS through the circulation or the vagal afferent pathway and, therefore, play an important role in neuroendocrine and neuroimmune activity [45].

The primary bile acids synthesized in the liver are converted by gut microorganisms into secondary bile acids that play a key role in lipid catabolism and the maintenance of cholesterol levels. These bile acids altered by microorganisms are also involved in cellular signaling and bind to receptors such as farnesoid X receptor (FXR) and G-protein coupled bile acid receptor 1 (TGR5), thereby modulating glucose metabolism, energy balance, and immune responses. In addition, bile acid metabolites help regulate inflammation, as certain bile acid derivatives can inhibit pro-inflammatory pathways, reducing the risk of chronic inflammatory diseases [46].

Kynurenine is another tryptophan metabolite, and *Lactobacillus* taxa regulate its production [47]. In order to decrease host kynurenine metabolism, *Lactobacilli* create hydrogen peroxide, an ROS that inhibits the production of the enzyme indoleamine-2,3-dioxygenase (IDO1). In the GI tract, IDO1 contributes to the synthesis of kynurenine from tryptophan [48]. The reduction in *Lactobacillus* brought on by stress lessened the inhibition of IDO1 mediated by hydrogen peroxide in a rat model of chronic varied stress, which increased the synthesis of kynurenine from tryptophan [49]. It has been demonstrated that kynurenine, which crosses the blood–brain barrier, causes neuroinflammation and neurodegeneration, which are also linked to depression and Alzheimer's disease [47].

Derived from the bacterial metabolism of the amino acid tryptophan, indoles play a critical role in modulating gut barrier integrity, reducing gut permeability, and influencing immune responses. While the generation of kynurenine and serotonin from tryptophan is mostly dependent on gut microorganisms, indole synthesis is entirely microbe dependent because only specific microbes produce the tryptophanase enzyme needed to produce these chemicals from tryptophan [50]. Indole derivatives like indole-3-propionic acid (IPA) act as antioxidants and have neuroprotective effects. These metabolites also engage with the immune cell aryl hydrocarbon receptor (AhR), aiding in achieving immune homeostasis and reducing tissue inflammation. Indoles' role in immune modulation, gut health, and barrier function underscores their significance in ensuring there is neither "leaky gut" nor systemic inflammation, which contribute to mood dysregulation and other disorders [51].

As shown in Figure 3, tryptophan is essential for the synthesis of serotonin, kynurenine, and indoles. Ingesting foodstuffs containing tryptophan, such as turkey, eggs, and dairy products, may serve to enhance serotonin levels. In addition, it is thought that dietary carbohydrates may further increase the level of tryptophan absorption by elevating insulin levels, which may suppress other amino acid competition [52]. This amino acid can be found in meat and dairy products, nuts, and soy products [53]. The proper levels of tyrosine are critical for the production and regulation of dopamine levels. It has also been argued that the consumption of many antioxidant-rich foods, including fruits and vegetables, can help shield neurons that generate dopamine from oxidative damage and stress [54]. GABA is produced when sufficient amounts of dietary glutamine are present. Foods such as nuts, seeds, and a few other fermented foods also help in GABA synthesis [55].

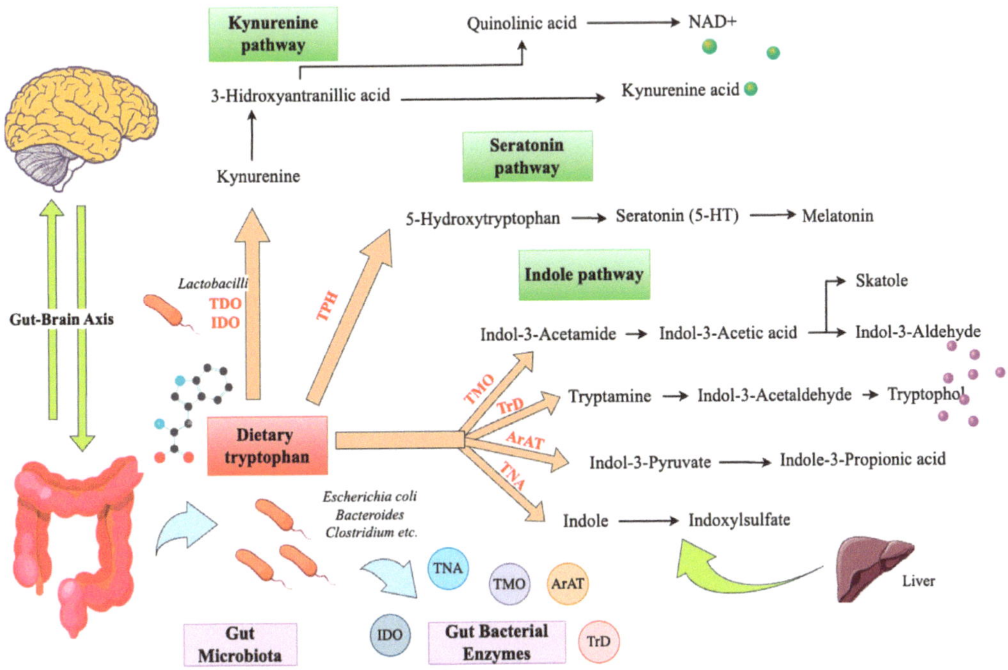

Figure 3. Neuroactive metabolites produced by the gut microbiome.

3. The Impact of Dietary Components on the Gut Microbiota

3.1. Carbohydrates

Carbohydrates are a major component of the human diet and significantly alter the activity and composition of the gut microbiota. The types of carbohydrates consumed, their fermentation by gut bacteria, and their impact on microbiome diversity are crucial factors in maintaining gut and overall health. Considering their chemical composition and digestibility, carbohydrates form two categories: simple and complex [56]. Unlike simple carbohydrates, many complex carbohydrates and fibers reach the colon, where gut bacteria ferment them, since they cannot be broken down in the small intestine. The breakdown of fibers depends on the breakdown of carbohydrates by the intestinal microbiota. Fibers are composed of resistant starches; non-digestible oligosaccharides, including raffinose, stachyose, oligofructose, and inulin; and non-starch polysaccharides such cellulose, hemicellulose, glucans, gums, and pectins [57]. This fermentation process has significant implications for gut microbiome diversity and function [58]. Gut bacteria produce monosaccharides, specific gases (such as carbon dioxide and methane), and short-chain fatty acids (SCFAs) like butyrate, propionate, and acetate through the saccharolytic fermentation of food fibers and resistant starches. SCFAs, with their anti-inflammatory properties, provide colonocytes with the energy needed, enhance gut barrier function, and influence host metabolism [59]. As illustrated in Figure 4, SCFAs like acetate and propionate are absorbed by the blood and move to the liver through the portal vein. While a significant portion of SCFAs is utilized by colonocytes and the liver, some unmetabolized SCFAs may enter the systemic circulation and exert effects on other tissues and organs like the brain, influencing brain function, neurotransmitter production, and brain inflammation [60].

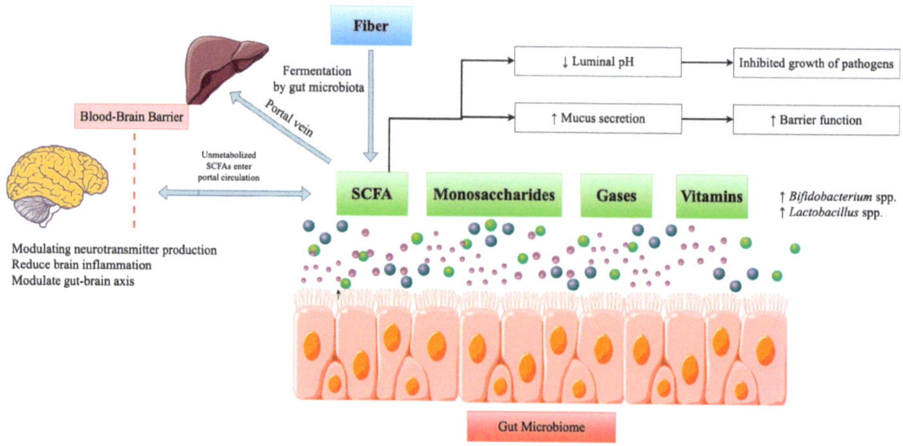

Figure 4. The function of dietary fiber in connection with brain function and gut microbiota. ↑, increase; ↓, decrease.

The types and amounts of carbohydrates consumed can significantly influence the gut microbiome's diversity and composition [61]. Diets that have high fiber contents encourage the development of good bacteria, like *Bifidobacterium* and *Lactobacillus* species, which have been acknowledged for their ability to enhance health. These bacteria utilize dietary fiber to produce SCFAs, contributing to a healthy gut environment and supporting immune function [62]. By encouraging the growth of harmful bacteria and decreasing the number of helpful microorganisms, diets heavy in refined carbohydrates—such as sugars and processed grains—can have a detrimental effect on microbiome diversity. This shift can lead to dysbiosis, associated with various health issues, including obesity, diabetes, and inflammatory bowel disease (IBD) [63]. Animal models have shown that the Western diet, which contains relatively little fiber, reduces the amount of *Bifidobacterium* and the diversity of the gut microbiota. [64]. It has been demonstrated that gut dysbiosis can be improved in rats fed a diet enriched with high-fat and sucrose by decreasing the *Firmicutes*-to-*Bacteroidetes* ratio and increasing the amount of *Lactobacillus* sp. [65].

3.2. Protein

Proteins are essential macronutrients that are necessary for several physiological functions, including immunological response, growth, and repair. The digestion and fermentation of dietary proteins by gut microbiota produce several byproducts, and they have the capacity to greatly impact microbial composition and gut health. Peptides and amino acids are absorbed into the bloodstream through the intestinal epithelium in the small intestine, where most protein digestion occurs. However, a fraction of dietary proteins and peptides escape digestion and reach the colon, where they undergo microbial fermentation [66]. In the colon, undigested proteins and peptides are broken down by gut bacteria, producing various byproducts that impact gut health and microbial composition [67]. Gut bacteria metabolize amino acids to generate a variety of compounds, such as SCFAs, branched-chain fatty acids (BCFAs), and gases like hydrogen and methane (Figure 5). Key amino acids involved in these processes include indoles, glutamine, lysine, and arginine [68]. Ammonia, amines (e.g., putrescine, cadaverine), phenolic compounds (e.g., p-cresol), and hydrogen sulfide can also be produced as a result of protein fermentation. These metabolites can have toxic effects on the gut epithelium and are associated with a higher risk of diseases such as colorectal cancer and IBD [66].

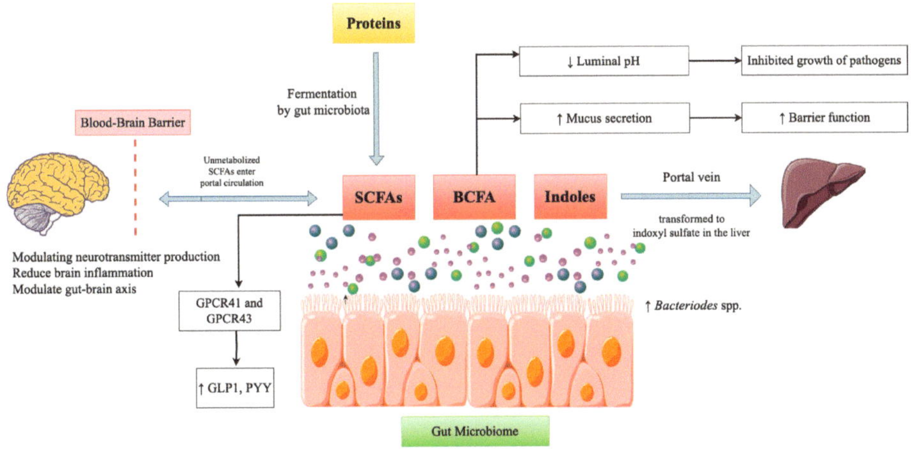

Figure 5. The role of proteins in the gut microbiome and brain function. ↑, increase; ↓, decrease.

The type and quantity of dietary protein ingested play an important role in altering the diversity and composition of gut microbiota. Diets that consist of more animal proteins tend to promote the growth of protein-fermenting bacteria such as *Bacteroides* and *Clostridia* species, among others, and suppress the growth of carbohydrate-fermenting bacteria such as *Bifidobacterium* and *Lactobacillus*. This particular shift may lead to the generation of more harmful by-products and fewer beneficial SCFAs [69]. Diets rich in plant proteins, especially those from pulses and cereals, have been linked to a healthier and more stable gut microbiome. In addition, plant proteins often contain dietary fiber, which helps in the establishment of healthy bacteria and the production of SCFAs, which are very important in ensuring a healthy gut [70]. Consuming a balanced mixture of both animal and vegetable proteins is suggested to support a diverse and healthy gut microbiome. The addition of vegetable proteins and dietary fiber can alleviate the adverse consequences of excess protein fermentation and support gut health [66].

Animal models have suggested that protein quality can modulate the composition of the gut microbiota. For instance, a preclinical trial conducted by Zhang et al. [71] showed that cheese whey proteins enhanced fecal *Lactobacilli* and *Bifidobacteria* counts as opposed to casein. Mung bean protein also increased the abundance of the *Ruminococcacea* family in a model of mice fed a high-fat diet. Based on this, the bile acid metabolism of *Ruminococcacea* family members was thought to be healthier in high-fat-diet mice [72].

3.3. Fats

The roles of dietary lipids extend even to the microscopic organisms thriving within the human stomach. There have been differences observed with regard to the health of the gut and the type of fats consumed by hosts in different organisms [73]. Positive effects on the gut microbiome can be expected with the use of unsaturated fatty acids. Unsaturated fats are, on the other hand, friendly to the intestinal microflora. For example, a link has been found between omega-3 fatty acids, increases in anti-inflammatory bacteria, and decreases in the level of pathogenic bacteria [74]. Wolters et al. [75] also found a possible association between the availability of MUFAs (palmitoleic, oleic, and eicosenoic acids) and genera of the family *Enterobacteriaceae*—*Prevotella*, *Turicibacter*, and *Parabacteroides*. Conversely, research has established that diets abundant in MUFAs, especially those in peanuts, sesame, pumpkin seeds, and rapeseed oil, as well as extra virgin olive oil, result in an improvement in the microbial composition of both healthy and unhealthy mice, including those susceptible to metabolic syndrome. MCFAs present in human breast milk, infant formulas, and virgin coconut oil have the potential to facilitate the supplementation of *Bifidobacterium* and *Lactobacillus* and improve cognitive and metabolic functions [76]. Polyunsaturated fatty acids (PUFAs) are called "essential fatty acids" since the human body cannot generate them and must obtain them through the diet. The main sources of PUFAs include fatty fish, nuts, seeds, and sunflower oil. Omega-3 PUFAs can be advantageous by increasing the number of *Lachnospiraceae* taxa that produce butyrate and re-establishing the optimal bacterial ratio [57]. As shown in Figure 6, certain bacterial species become more abundant as a result of PUFA metabolism. Many metabolites, including SCFAs such as butyrate, are consequently generated, leading to a decrease in endotoxin and interleukin (IL)-17 production, which may benefit human health by reducing inflammation. PUFA metabolism produces unmetabolized SCFAs that enter the systemic circulation and exert immunoregulatory effects, especially with the brain, influencing gut–brain functions.

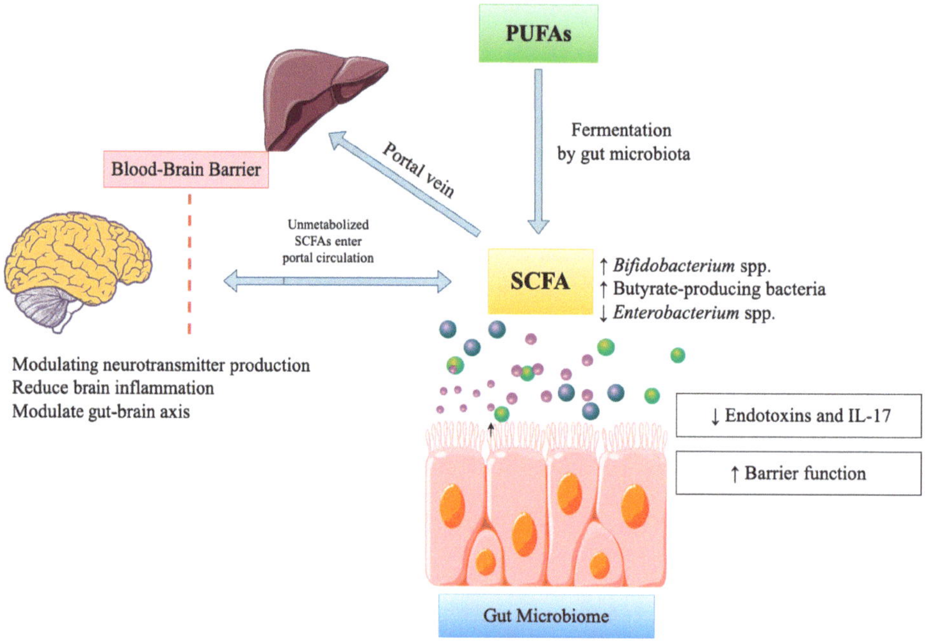

Figure 6. The role of PUFAs in relation to the gut microbiome and brain function. ↑, increase; ↓, decrease.

Diets high in saturated fats affect the gut microbiota in a negative way. These fats have the potential to decrease the number of good bacteria like *Bifidobacterium* and *Lactobacillus* while promoting the growth of hazardous ones. This imbalance, or dysbiosis, can lead to an increase in gut permeability, called "leaky gut", promoting systemic inflammation and contributing to metabolic disorders [77].

3.4. Micronutrients

Several micronutrients, including zinc and omega-3 fatty acids, may exert a significant impact on the functions of the brain and on development without affecting the gut microbiota [78]. Since single nutrients are never eaten in isolation in real life, nutritional research has shifted away from concentrating on them. Instead, it has been demonstrated that the most advantageous diet for both physical and mental health consists of a wide variety of various nutritional foods [79]. Dietary nutrients that support normal brain function include vitamins, minerals, and amino acids [78,80]. Many of these nutrients help with cell signaling, myelination, neurotransmitter production, and metabolic processes, among others, by acting as cofactors for enzymes [81]. Omega-3 fatty acids, folate, s-adenosyl methionine, inositol, and vitamins B3, B6, and C are among the nutrients with antidepressant effects that have been the subject of extensive research. Supplementation with these nutrients may be advantageous, but only as an addition to a diet that promotes gut health [78]. Because of the fact that eicosanoids produced by AA have inflammatory qualities and eicosapentaenoic acid has anti-inflammatory properties, a high ratio of omega-6 to omega-3 fatty acids may cause a pro-inflammatory condition. In addition, evidence suggests that a lack of omega-3 fatty acids may be responsible for some psychiatric disorders, such as depression [82]. For example, B vitamins are necessary for the production of neurotransmitters and for the preservation of the myelin sheath, while zinc and magnesium are crucial for the release of neurotransmitters and for synaptic plasticity. These micronutrient deficiencies can also result in gut dysbiosis, an overgrowth or imbalance of gut bacteria that usually presents with inflammation and the disrupted production of neurotransmitters, thereby worsening depressive symptoms [81]. Calcium, magnesium, potassium, and iron can affect the microbiome. These minerals have been proven to increase levels of healthy bacteria such as Bifidobacterium and Lactobacillus, which improve gut health [83]. Too much iron can lead to germ overgrowth, whereas too little iron can reduce the function and the variety of microbes present [84]. Zinc deficiency can result in dysbiosis and increased susceptibility to infections. Zinc supports the integrity of the gut barrier and promotes a balanced microbial environment [83].

To sum up, evidence suggests that a diet rich in fiber, polyphenols, and micronutrients positively alters gut microbial composition, reduces metabolic endotoxemia and neuroinflammation, and is associated with better cognitive health. There are numerous small-scale observational and interventional studies that have shown the advantages of fiber for brain function and health [85]. Eating patterns also contribute to the control and release of serotonin produced in the ENS, with the most important factor being the intake of complex carbohydrates with tryptophan. Some components, including polyunsaturated fatty acids, B vitamins, zinc, and folate, which are dietary micronutrients, may also promote the healthy growth and development of the brain and its functions, while a lack of such nutrients in the diet causes the opposite effect by promoting mental disorders and aggravating brain diseases.

3.5. Phytochemicals and Bioactive Compounds

The utilization of phytochemicals and bioactive agents such as polyphenols, flavonoids, and other plant-based materials has gained interest concerning their impacts on the gut microbiota and the brain, especially on depression. These substances are present in large quantities in fruits, vegetables, tea, and other herbs and represent antioxidants, anti-inflammatory agents, and prebiotics, thereby affecting the gut–brain axis considerably. For instance, it is known that polyphenols (and other compounds found in berries, red wine, and dark chocolate) affect gut microbiome composition through the stimulation of populations of probiotic bacteria such as *Lactobacillus* and *Bifidobacterium* and the suppression of pathogenic strains [86]. The enhancement of microbial populations correlated with health promotes systemic immunity so that inflammation is lessened, assists in the repair and healing of the gut lining, and increases the absorption of SCFA-like butyrate, which has neuroprotective functions and is associated with better mental state outcomes [87].

Flavonoids, another group of bioactive compounds found in foods like citrus fruits, onions, and green tea, are also necessary for supporting gut health and mitigating depressive symptoms. These compounds have been shown to improve the abundance and diversity of good gut bacteria, contributing to a healthier microbiome [88]. The fermentation of flavonoids by gut microorganisms produces numerous metabolites that have the ability to directly impact brain function by passing across the blood–brain barrier, modulate neurotransmitter levels, and exert anti-inflammatory actions. For example, the flavonoid quercetin has been observed to increase levels of serotonin and dopamine in the brain, which are crucial for mood regulation [89]. Additionally, flavonoids can inhibit the activation of microglia, the brain's resident immune cells, thereby reducing neuroinflammation, which is often elevated in individuals with depression [90].

Furthermore, the interplay between bioactive compounds and the gut microbiome can influence the gut–brain axis through the modulation of tryptophan metabolism. Tryptophan, an essential amino acid, is a precursor to serotonin, a neurotransmitter that plays a vital role in mood regulation [45]. Phytochemicals can impact the gut microbiota's ability to metabolize tryptophan, shifting its metabolic pathways toward the production of serotonin and other beneficial metabolites rather than kynurenine, which is associated with neurotoxicity and inflammation. This shift can help alleviate symptoms of depression by enhancing serotonin availability and reducing inflammatory markers in the brain [91].

In summary, phytochemicals and bioactive compounds exert multifaceted effects on the gut microbiota, promoting the growth of beneficial bacteria, enhancing the production of neuroprotective metabolites, and modulating inflammatory responses. These changes in the gut environment can positively influence the gut–brain axis, contributing to improved mental health and offering a promising avenue for the dietary management of depression. Continued research on the specific mechanisms and optimal dietary sources of these compounds is essential for developing effective nutritional strategies to support mental health through gut microbiota modulation.

3.6. The Influence of Dietary Patterns and Components on the Gut Microbiome in Modulating Depression

The composition and function of the gut microbiome can be greatly impacted by specific dietary patterns, leading to changes in mental health, including the severity and occurrence of depression [78]. The Western diet has been linked to detrimental effects because of the heavy consumption of processed foods, red meat, refined carbohydrates, and saturated fats, which has been associated with negative impacts on gut microbiota and mental health. Usually, this diet causes an increase in harmful bacteria and a decrease in microbial diversity, leading to gut dysbiosis. The resulting imbalance in the gut microbiome can enhance intestinal permeability, promote systemic inflammation, and adversely affect the gut–brain axis. These changes are often linked to increased levels of pro-inflammatory cytokines and neuroinflammation, which are associated with the development and exacerbation of depressive symptoms [92]. In contrast, the Mediterranean diet, rich in fruits, vegetables, whole grains, nuts, seeds, and olive oil, along with the moderate consumption of fish and poultry, has been shown to promote a healthy gut microbiome and support mental health. This diet increases the abundance of beneficial bacteria such as *Bifidobacterium* and *Lactobacillus* and enhances microbial diversity. The high content of fiber and polyphenols in the Mediterranean diet supports the production of SCFAs such as butyrate, which has anti-inflammatory properties and can improve the integrity of the gut barrier. These changes help to reduce systemic and neuroinflammation, contributing to a lower risk of depression [93]. The relationship between vegan and vegetarian diets and mental health outcomes, particularly depression, is complex and has yielded mixed findings in recent research. These diets are typically high in fiber, vitamins, minerals, and phytonutrients, which promote microbial diversity and the development of advantageous gut microorganisms. The high fiber content encourages the synthesis of SCFAs, which have anti-inflammatory and neuroprotective properties. Some studies suggest that individuals adhering to plant-based diets may experience lower incidences of depressive symptoms [94]. Conversely, other research findings indicate a potential association between plant-based diets and increased depressive symptoms [95]. With these conflicting results [96], the current body of research does not provide a definitive conclusion. Further studies are necessary to clarify these associations and to understand the underlying mechanisms involved. The ketogenic diet, characterized by high fat, moderate protein, and very low carbohydrate intake, has shown mixed effects on the gut microbiome and mental health. While some studies suggest that the ketogenic diet can reduce inflammation and support mental health by modulating neurotransmitter levels and reducing neuroinflammation, others indicate that it may reduce microbial diversity and negatively affect gut health due to low fiber intake. The impact of the ketogenic diet on depression remains an area of active research, with some evidence supporting its use in specific cases, such as treatment-resistant depression [97]. In summary, dietary patterns play a crucial role in shaping the gut microbiome and influencing mental health. Diets high in polyphenols, fiber, and healthy fats, like the Mediterranean, support a healthy gut microbiome and are associated with a reduced risk of depression. Conversely, diets high in processed foods and saturated fats, like the Western diet, can negatively impact gut health and exacerbate depressive symptoms.

A number of biological systems linked to depression may be directly impacted by the Western diet and other diets associated with metabolic endotoxemia. A nutritious diet full of fresh fruits and vegetables is strongly associated with subjective happiness and mental health, according to numerous studies [80,98]. One of the first interventional studies on this topic was a 12-week, parallel-group, single-blind randomized controlled trial (RCT) including thirty-three male and female participants with moderate to severe depression. The participants were assigned at random to either social assistance or dietary help treatments. In order to promote adherence to the suggested diet, the dietary intervention included tailored dietary advice and nutritional counseling, as well as goal setting, mindful eating, and motivational interviewing [99]. Compared to conventional therapy alone, the study's results showed that dietary intervention significantly reduced depression symptoms. This implies that changing the gut microbiota with an adjuvant dietary intervention may be a helpful way to treat depression.

In the MooDFOOD RCT, a diet fortified with calcium, vitamin D, selenium, and omega-3 fatty acids did not lessen major depressive episodes in people who were overweight or obese and had subsyndromal depressive symptoms [100], despite a recent RCT's confirmation of the benefits of a Mediterranean-style diet for depression [101]. But taking into consideration only those participants who adhered for a full year, a supplemental diet may be able to delay the onset of depression. The HELFIMED trial looked into how food interventions directly affected participants' self-reported depression. Patients who self-reported being depressed were put on a Mediterranean diet that included fish oil for a 6-month intervention period. To enable maximum dietary adherence, they also received nutritional coaching and cooking instruction. The control group received social support in order to take into consideration any possible non-dietary antidepressant benefits of nutritional counseling. The study's findings showed a substantial decline in depression, which was positively connected with greater adherence to the Mediterranean diet during both 3- and 6-month periods. The study also revealed an intriguing correlation between a lower ratio of omega-6 to omega-3 fatty acids and a reduction in depressive symptoms [100]. In conclusion, the above-mentioned advantages of a diet high in plants for depression likely include improved intestinal permeability, which lowers metabolic endotoxemia, and anti-inflammatory effects mediated by increased polyphenol and SCFA synthesis. Increasing the consumption of omega-3 fatty acids and trace minerals may also help to address deficiencies in certain important nutrients related to mood and overall brain health.

Numerous dietary components that are linked to either a higher or lower incidence of depression also change the gut flora (Table 1). It is possible that a dietary component's impact on gut microbiota may mediate its effect on mood, either fully or partially.

Table 1. Examples of dietary patterns linked to depression that have an immediate impact on the host while also interacting with the gut microbiome.

Dietary Component	Study Model	Effect	Reference
High-carbohydrate diet	18 SPF male C57BL/6J mice, 6–8 weeks old	Increased depression Increased levels of circulation and central nervous system inflammatory responses Abnormal expression of central 5-HT and its receptors Significantly decreased the diversity of intestinal flora	[102]
Dietary capsaicin	Male C57BL/6J mice, 4 weeks old	Reduced depression Relative abundances of *Ruminococcus* and *Prevotella* Increased levels of the monoamine neurotransmitter 5-HT Reduced levels of the inflammatory cytokine TNF-α	[103]
Herbal extracts	Male C57Bl6/J mice, 8 weeks old	Reduced depression Changes in the gut microbiome Increased SCFA production	[104]
Dietary fiber	Patients with hypertension from the Hospital of Soochow University and Jinchang Community, China	Reduced depression Protective effect on depression and anxiety	[105]
Dietary fiber	Male CD-1 mice, 3 months old	Reduced depression Maintained the integrity of the gut barrier Regulated the structure of the gut microbiota Production of related metabolites	[106]
Dietary fiber	Female C57BL/6J mice, 2 months old	Reduced depression Enhanced SCFA generation Restructured the gut microbiome	[107]
Apple polyphenol	Male C57BL/6 mice, 3–4 weeks old	Ameliorated depression Increased the richness and diversity of the gut microbiota Increased relative abundance of *Firmicutes* and *Bacteroidota* Decreased relative abundance of *Verrucomicrobiota* at the phylum level	[108]
Phenols	C57BL/6J mice, 5 weeks old	Reduced depression Regulated the structure of the gut microbiota Regulated microbial metabolism products such as 5-HTP	[109]
Oolong tea polyphenols	Male C57BL/6J mice, 6–8 weeks old	Reduced depression Reshaped intestinal microbiota dysbiosis Regulated cognition-related metabolites Strengthened mucosal integrity and intestinal barrier dysfunction by increasing tight junction protein expression	[110]
Orange flavonoids	Participants from Seoul and Gyeonggi-do, age 20–30 years	Reduced depression Changed the relative abundance of the gut microbiome, especially the butyrate-producing *Lachnospiraceae* family.	[111]

4. Potential Mechanisms Linking Diet, the Gut Microbiome, and Depression

4.1. Inflammatory Pathways

Progressive and persistent forms of depression are aggravated by inflammatory pathways, and more studies have provided evidence of the association between chronic inflammation and mood disorders. The complex triad of diet, the gut microbiome, and inflammation provides a hopeful opportunity to ameliorate neuroinflammation and depression. An altered diet seeking to modify the gut bacteria can have an extensive effect on inflammation as a result of the immune pathways targeted by bacterial metabolic products. These diet-induced bacterial metabolites, including, but not limited to, SCFAs, polyphenol metabolites, and secondary bile acids, have an effect on the production of pro-inflammatory and anti-inflammatory cytokines, which affect the CNS and also depression [2].

Increased levels of pro-inflammatory cytokines like IL-6, tumor necrosis factor-α (TNF-α), and IL-1β, which are implicated in starting inflammatory processes, are influenced by the CNS through several mechanisms that affect organs and, in turn, lead to depression [112]. The development of such a response is complicated further by the release of pro-inflammatory cytokines, which may also act on the HPA axis and increase the levels of cortisol. Studies have shown that the elevation of the stress hormone cortisol is linked with some psychiatric conditions, particularly mood disorders, cognitive deficits, and changes in the volume of mood-related brain areas such as the hippocampus and prefrontal cortex. Additionally, inflammatory cytokines have a direct effect on the bioavailability of critical neurotransmitters that are best known for their modulation of mood, mainly serotonin, dopamine, and glutamate [113]. Depressive symptoms are correlated with low levels of serotonin and an increase in neurotoxic metabolites like quinolinic acid, which leads to the overactivation of NMDA receptors, causing toxicity and consequent depressive symptoms [114]. Motivation and reward pathways, which are generally dysfunctional in depression, are also affected by chronic peripheral inflammation, which decreases dopamine and alters glutamate signaling [115]. The permanent inflammation of the peripheral system can lead to the activation of microglia, the immune system cells present in the brain, which leads to a state known as neuroinflammation. Activated microglia secrete inflammatory cytokines and generate ROS that are harmful to tissues in the brain, killing healthy neurons and inhibiting their regeneration [116]. The hippocampus, a functionally very important brain area in mood control and cognition, is most affected due to reduced neurogenesis, which has been associated with depressive conditions. Neuroinflammation also has implications for neuroplasticity; hence, it reduces the capacity of the brain to adjust and cope with stress and increases the severity of depressive states [117]. Inflammatory cytokines can reduce the levels of brain-derived neurotrophic factor (BDNF), a protein that promotes neurogenesis and supports the survival of neurons. BDNF is crucial for maintaining neuroplasticity and resilience to stress. Inflammatory markers like IL-1β inhibit BDNF production, particularly in the hippocampus, impairing neurogenesis and reducing the brain's capacity for stress adaptation. This reduced neuroplasticity makes the brain more susceptible to depressive states and less responsive to treatment. Inflammatory processes contribute to oxidative and nitrosative stress, which further impacts brain function and mood regulation [118]. Inflammation affects the gut–brain axis and increases gut permeability (leaky gut), allowing inflammatory endotoxins like LPS to enter the bloodstream. LPS triggers systemic inflammation, which can reach the brain and activate the HPA axis, as well as neuroinflammatory pathways, contributing to depressive symptoms. The gut–brain axis thus serves as an important route by which inflammation originating in the gut can influence brain function and mood [119].

A high-fiber diet is particularly effective in reshaping the microbiome to support anti-inflammatory functions by producing SCFAs. Butyrate plays a dual role: it acts directly on the gut epithelium to enhance tight junction integrity, reducing endotoxemia, and also interacts with immune cells to decrease the secretion of pro-inflammatory cytokines. Butyrate binds to G-protein-coupled receptors (GPCRs), specifically GPR41 and GPR43, which are expressed on colonic epithelial and immune cells, activating anti-inflammatory pathways through downstream signaling cascades such as the inhibition of the NF-κB pathway. This, in turn, suppresses microglial activation in the CNS, thereby modulating neuroinflammation associated with depressive symptoms [45]. Polyphenols, naturally present in a wide range of plant foods like berries, tea, and cocoa, undergo extensive metabolism by gut bacteria, yielding bioactive metabolites that cross the blood–brain barrier and exert neuroprotective effects. For example, gut-derived metabolites of polyphenols, such as urolithins (derived from ellagic acid) and phenyl-γ-valerolactones (from flavan-3-

ols), have been shown to inhibit pathways that generate pro-inflammatory mediators like IL-1β and TNF-α through the suppression of the JAK/STAT and MAPK pathways in immune cells. These metabolites also support antioxidant defenses by upregulating nuclear factor erythroid 2–related factor 2 (Nrf2) signaling, which reduces the oxidative stress that often coexists with neuroinflammatory conditions [120]. Importantly, polyphenol metabolites modulate the gut–brain axis by preventing the activation of peripheral immune cells that contribute to neuroinflammation, thereby potentially reducing depressive symptoms [121]. Fang et al. [122] also highlighted the role of alpha-linolenic acid, eicosapentaenoic acid, and docosahexaenoic acid in regulating gut microbiota to counter inflammatory processes. Along with food, omega-3 PUFAs, also commonly known as "healthy oils", are transformed into specialized pro-resolving mediators (SPMs), which include resolvins and protectins. These SPMs are features of distinct cells and target specific G proteins in immune cells. SPMs promote inflammation resolution and decrease the level of pro-inflammatory cytokine production [123].

Linking the availability of SCFAs, polyphenolic metabolites, and SPMs with the diet-modified microbiome provides an indication of precise mechanisms by which gut microbiome dietary modulation may alter inflammation in depression. Strategies aiming to modify the composition of the gut microbiome in an anti-inflammatory manner to enhance metabolite barriers and change the composition of the microbial community will have little net impact on the levels of pro-inflammatory cytokines such as IL-6 and TNF-α, which are known to aggravate depression. This approach to reducing inflammation provides a compelling foundation for dietary interventions aimed at treating or preventing depressive disorders, supporting the targeted modulation of gut-derived inflammatory pathways to improve mental health outcomes.

4.2. Neurotransmitter Synthesis

The gut microbiome is a major contributor to the production and modulation of various neuroactive substances, including serotonin, dopamine, and GABA, which have been proven important for mood and behavioral control [124]. The gut microbiome influences the generation of neurotransmitters using several mechanisms, including the production of the precursors of the neurotransmitters, the modulation of enzymes involved in neurotransmitter production, and the gut–brain axis. The microbiome can, therefore, modulate the peripheral and central nervous systems' neurotransmitter levels, mood and stress fluctuations, and psychiatric disorders (like depression and anxiety).

4.2.1. Serotonin Production

Approximately 90% of the serotonin in the human body is produced within the gastrointestinal tract, and certain gut bacteria are known to regulate its synthesis [33]. Various species, such as *Enterococcus* and *Streptococcus*, have been found to generate metabolites that promote the release of serotonin from enterochromaffin cells present in the gut lining [125]. This peripheral serotonin is not able to cross the blood–brain barrier, although it is known to assist with gut movement, immunity, and enteric nervous system function, among others. Tryptophan, a precursor of serotonin, is also influenced by certain microbial metabolites like SCFAs, which in turn affects the amount of serotonin produced. An optimal gut microbiota allows for more conversion of tryptophan to serotonin and less conversion to the neurotoxic products of the kynurenine pathway, which cause depression [126]. In addition, the stimulation of the microbial biosynthesis of SCFAs such as butyrate is effective in strengthening the gut lining, which in turn decreases peripheral inflammation and helps to maintain serotonin signaling within the central nervous system [125].

4.2.2. Dopamine Synthesis

Dopamine is known to play a key role in the regulation of reward, motivation, and mood states, and determining how the gut microbiota may also play a role is an area of active research. Some types of bacteria, such as *Bacillus* and *Escherichia coli*, have been shown to either synthesize dopamine or influence its synthesis from precursors such as tyrosine and phenylalanine [125]. Dopamine synthesized in the gut holds significance in mediating enteric nervous system signaling mechanisms, which in turn affect the motility and immune responses of the gut. The "gut–brain axis" also allows microbial metabolites, such as SCFAs, to reach the brain and modulate the dopaminergic system. For example, SCFAs are known to affect the function of the brain's reward center, where even a slight distortion leads to depressive symptoms, as the mesolimbic dopaminergic neurons are responsible for dopamine production. Also, microbial populations in the gut or their absence can alter the permeability of the blood–brain barrier, inhibiting or permitting the passage of dopamine precursors into the circulation and ultimately the brain, where they can increase the synthesis of dopamine, responsible for improving mood and increasing motivation [127].

4.2.3. GABA Production

As a neurotransmitter, GABA competes against the stimulative effects of natural brain activity. It is critical in decreasing levels of anxiety and facilitating relaxation, as well as moderating mood. Some gut microorganisms, for example, *Lactobacillus* and *Bifidobacterium*, synthesize GABA [125]. A part of the GABA synthesized in the gut is likely to be released to the brain via the vagus nerve, which connects the brain and the gut. Research has demonstrated that changes in the gut microbiome can lead to changes in GABA receptors in the brain associated with mood and stress changes. Specifically, it has been shown that the supplementation of *Lactobacillus rhamnosus* leads to increased GABA receptors in the amygdala and prefrontal cortex, which influences anxiety behavior. Additionally, this GABA modulation achieved via bacteria in the gut may also offset the "elevated" levels of the neurotransmitter "glutamate" associated with depression and anxiety [128].

4.2.4. Gut–Brain Communication and Neurotransmitter Modulation

The microbiome affects the levels of neurotransmitters that can be found in the brain mainly because of the gut–brain axis, which is a two-way communication channel that utilizes neural, hormonal, and immunological mechanisms to signal between the two organs. Such compounds as SCFAs, secondary bile acids, and derivatives of tryptophan have access to the ventral vagal nerve and then to the CNS, where they participate in the production and function of different neurotransmitters and their receptors. It has been observed that these interactions influence the level of BDNF, a neuroprotective factor that promotes adaptation in stress-provoking conditions and neuroplasticity. Changes in BDNF and neurotransmitter production are brought about by the gut microbiome, which, therefore, is capable of influencing the neuroplasticity of an individual, leading to changes in mood or behavior [91].

4.3. HPA Axis Modulation and the Gut Microbiome's Role in Stress Response

The HPA axis is at the core of how the body responds to stress, especially with the production of cortisol, which has a wide range of effects, including mood regulation, immune function, and protecting the body from stress and its effects. The gut microbiome is an important factor in determining the activity of the HPA axis in relation to stress, both in terms of its onset and maintenance. When stress is experienced by the body, the hypothalamus exerts a direct influence over the pituitary gland's activity, which induces the adrenal glands to secrete cortisol, the primary function of which is to help the body cope with stress. Cortisol enables the individual to cope with stress for a short duration; however, at times, it can cause an overdrive of the HPA axis, which is frequently triggered by prolonged stress or inflammation. HPA axis dysregulation then occurs, which is linked with mood disorders, especially anxiety and depression [88].

The HPA axis is also affected by the gut microbiota, in most cases by more than one mechanism. Some beneficial gut microflora, for example, species of *Lactobacillus* and *Bifidobacterium*, have been shown to exert protective effects against HPA activation caused by stress, at least in part by acting on the vagus nerve, which serves as a major part of the gut–brain axis. These microbes secrete SCFAs and other molecules that affect the immune and hormonal systems to prevent excessive inflammation of the HPA axis, which would lead to its excessive stimulation [129]. Furthermore, gut dysbiosis is associated with changes in gut function, such as increased gut permeability, which causes the translocation of endotoxins such as LPS into the circulation, evoking an immune response. The presence of increased levels of LPS then causes the activation of the HPA axis, which leads to the pathological production and accumulation of cortisol, which has been implicated in depression, inflammatory disorders of the gut, and issues linked to neurodegeneration [59].

Furthermore, LPS and peptidoglycan, which serve as microbe-associated molecular patterns (MAMPs) produced by microorganisms, along with the subsequently produced pro-inflammatory mediators, activate the HPA axis. The primary bacterial cell wall component, known as peptidoglycan, has an adverse effect on behaviors and brain development because it penetrates the blood–brain barrier and triggers the innate immune system's pattern recognition receptors. Similarly, LPS derived from gut microbes and peptidoglycan induces the activity of NOD1, a protein involved in the activation of the immune system [130].

The HPA axis and stress mechanisms involve interactions with different macro- and micronutrients. The consumption of diets rich in refined sugars and saturated fats has been found to correlate with elevated cortisol levels and altered HPA axis functioning [131]. Cortisol elevation and stress response failure caused by the deficiency of certain micronutrients like magnesium and vitamin B6 have been documented [132]. One of the beneficial effects of omega-3 fatty acids is the regulation of the HPA axis, which is manifested as reduced cortisol levels and an enhanced ability to cope with stress. Madison et al. [133] performed a meta-analysis and reported that cortisol levels were significantly decreased with omega-3 supplementation, with more effective stress responses overall. Such effects are thought to be connected with inflammation regulation via omega-3s, which shows an HPA axis-related effect.

4.4. Oxidative Stress and Neuroprotection

As defined, oxidative stress is an excess accumulation of ROS, which becomes detrimental since the body cannot detoxify them adequately and is a major contributor to the pathogenesis of mood disorders, especially neurovegetative disorders, e.g., depression and anxiety. Elevated oxidative stress factors can evoke neuronal injury and provoke inflammation and the disruption of neurotransmission, all of which are associated with mental health disturbances. The gut microbiome helps in enhancing oxidative stress and neuroprotection by producing metabolites, altering the immune system, and controlling the mechanisms of antioxidants [134].

Specific bacteria situated within the gut produce SCFAs such as acetate, propionate, and butyrate, which perform functions that involve being anti-inflammatory and being rich in antioxidants that inhibit oxidative stress. For example, butyrate aids in maintaining the structure of the blood–brain barrier, preventing the entry of ROS and peripheral inflammatory markers into the central nervous system [135]. Furthermore, complex fatty acids extend the activity of antioxidant enzymes such as superoxide dismutase (SOD), which decomposes ROS alongside glutathione peroxidase, thus shielding neurons against the harmful effects of oxidative stress. Moreover, some gut bacteria synthesize BDNF, a neuroactive compound that is important in coping with stress and promotes neuroplasticity [136].

On the other hand, dysbiosis of the gut refers to an alteration in the normal gut microflora; this is characterized by excess oxidative stress, both within the gut and outside it. Gut dysbiosis is therefore characterized by increased intestinal permeability, which paves the way for the entry of pro-inflammatory agents such as LPS into the circulation, resulting in inflammatory responses and inducing oxidative stress [45]. This excessive oxidative stress can also compromise neuroprotection and disturb the balance of the nervous system by causing neurodegeneration, which is one of the factors that contribute to the development of various mood disorders. This implies that maintaining a healthy balance of gut microbes helps the body manage oxidative stress with ease and provides neuroprotection, which highlights the relationship between gut health, oxidative stress, and mental health.

4.5. Depression-Related Changes in the Abundance of Gut Microbes

In recent years, there has been an increased interest in assessing the correlation between gut microbiome diversity and mental health, more specifically, depression. The existing research findings indicate that both the diversity and composition of gut microbiota contribute to the modulation of mood and mental well-being. Healthy gut microbiome diversity entails the composition and quantity of different microbial organisms present within the GI tract. It has been documented that higher diversity generally leads to healthier outcomes, such as a functional immune response and a regulated metabolic state. On the other hand, the decreased diversity of the microbiome has been shown to be related to several health challenges, such as gastrointestinal diseases, metabolic syndrome, and psychological disorders [64].

Research on germ-free animals, or animal models without sophisticated gut microbiota, has shed greater clarity on the relationship between gut microbiota and host neurobehavior. A reduction in *Parabacteroides* and *Bacteroides* was linked to an increase in immune-regulating genes and pro-inflammatory cytokines in a rat prenatal stress model. Anxiety developed as a result of these alterations, causing systemic inflammatory reactions [137]. Early life stress has been shown to increase the number of *Oscillibacter*, *Parasutterella*, *Treponema*, *Ruminiclostridium*, and *Helicobacter* while decreasing that of *Bacteroides*, *Rikenellaceae*, *E. ruminantium*, *Lactobacillus*, and *Parabacteroides*. Elevated corticosterone, adrenocorticotrophic hormone, and glucocorticoid receptor levels in the hippocampal region are linked to these alterations. These changes in the composition and number of gut microbes inhibit miR-124a and promote the production of miR-132. Furthermore, the increased expression of N-methyl-D-aspartate receptor (NR2A and NR2B) and glucocorticoid receptors was more frequent in rats treated with α-amino-3-hydroxy-5-methyl-4-isoxazolepropionic acid receptors (GluR1 and GluR2) [138]. Numerous studies have connected anxiety and depressive behaviors to the pathogenic influence of gut dysbiosis [124,139]. Patients with depression often express symptoms of GI disturbances, including bloating, nausea, vomiting, abdominal discomfort, and constipation [140]. Depression pathology is connected with the disproportionate abundance of *Enterobacteriaceae* and *Alistipes* species (increased), as well as *Faecalibacterium* species (reduced) [102]. The presence of pathogenic microbiota families, including *Ruminococcaceae*, *Shewanellaceae*, *Halomonadaceae*, and *Verrucomicrobiae*, was positively connected with patients' anxiety-like behavior, whereas the *Lachnospiraceae* and *Bacteroidaceae* families were associated with both anxiety and IBD [141]. The possible connection between gut dysbiosis and depressive-like behavior is clearly suggested by these clinical data (Table 2).

Table 2. Alterations in gut microbial abundance under depressive conditions.

Depression Model	Genus						Sample Site
	Lactobacillus	*Bacteroides*	*Clostridium*	*Bifidobacterium*	*Allobaculum*	*Turicibacter*	
CUMS	↑ [142]	↑ [143]	↑ [143]	↑ [144]	↑ [144]	-	Cecum
	↓ [145]	↓ [145]	↓ [146]	↓ [147]	-	↓ [146]	Fecal pellets
CRS	↓ [148]	↓ [148]	↓ [149]	↓ [149]	↓ [149]	↓ [148]	Fecal pellets
MS	↓ [138]	↓ [138]	↓ [138]	-	-	↓ [138]	Fecal pellets
CSDS	↓ [150]	-	-	↓ [150]	↓ [150]	↓ [150]	Fecal pellets

Abbreviations: CUMS, chronic unpredictable mild stress; CRS, chronic restraint stress; MS, maternal separation; CSDS, chronic social defeat stress; -, not investigated, ↑, increase; ↓, decrease.

5. Interventional Studies: Modifying the Diet to Influence Mental Health Through the Gut Microbiota

5.1. Dietary Interventions

To a greater degree than would have been the case some years ago, research now shows that dietary and eating patterns have an effect on mental well-being—especially depressive states. Adherence to dietary patterns such as the Mediterranean diet, characterized by the consumption of vegetables, fresh fruits, cereals and grains, nuts, and olive oil, is associated with a lower prevalence of depression, probably due to the high antioxidant properties of the above foods. Many of these foods seem to help lessen age-related oxidative stress, inflammation, and intestinal dysbiosis and promote favorable neurochemical processes. Following an average follow-up of 20.4 years among 49,261 Swedish women, those who strictly adhered to a Mediterranean diet pattern had a lower incidence of depression [151]. The Mediterranean diet is associated with improved gut microbial diversity, decreased GI inflammation, and intestinal barrier integrity. Two important biochemical connections between the Mediterranean diet and gut dysfunction are propionate and butyrate, two SCFAs produced by the microbiota [152]. On the flip side, the Western diet, which emphasizes processed food with high levels of added sugar, fats, oils, and other refined products, has been associated with an increased prevalence of depression. This dietary pattern may exacerbate the risks of inflammation, gut dysbiosis, and neurotoxicity, which have been found to be depressive. Also, it has been shown in several longitudinal studies and meta-analyses that adherence to a Western diet is associated with more mood disorders than Mediterranean and other health-promoting diets.

5.2. FMT (Fecal Microbiota Transplantation)

FMT is a viable intervention technique for chronic diseases linked with dysbiosis, since it is a quick way to reshape the patient's gut microbiota through the administration of fecal flora from healthy donors. It is well known from preclinical research that FMT can reduce depressive-like behavior. Mice with alcohol-induced depression-like behavior were cured when the microbiota from healthy donors was transplanted. Additionally, rats with stress-induced depression exhibited improved phenotypes due to neuroinflammation suppression, gut microbiota imbalance correction, and intestinal barrier restoration [153]. Bacteriophages may mediate the effectiveness of FMT [154]. Apart from the global microbiota transplantation performed during the FMT method, *Lactobacillus plantarum* mono-colonization shields *Drosophila* from depressive-like states [155]. The gut–brain axis may rely on the vagus nerve as a key signaling pathway to control the protective effects of FMT on depression. Regardless of the resolution of GI symptoms, FMT treatment progressively reduced depression symptoms in patients with diarrhea-predominant IBD [112]. An RCT further confirmed these beneficial treatment outcomes in patients with concurrent IBD, anxiety, and depression [69]. In a different RCT, the effectiveness of frozen oral FMT capsules was evaluated as a further treatment for MDD patients, and it was discovered that four weeks following transplantation, depressive symptoms considerably improved [156]. FMT therapy for depression also reduces GI symptoms and rebalances the gut environment, much like it does for autism. Despite negative effects and complications of FMT therapy being documented in some articles, this treatment approach is becoming more and more popular in clinical and scientific settings. This is especially the case for an alternative pill made from human feces that produces similar effects in a less invasive and more standardized manner [157].

5.3. Probiotics, Prebiotics, and Synbiotics

Recent clinical trials have shown that probiotics, prebiotics, and synbiotics may positively influence depression by modulating the gut–brain axis. When taken in sufficient quantities, living bacteria, known as probiotics, are beneficial to the health of the host. Probiotics containing beneficial bacteria, like *Lactobacillus* and *Bifidobacterium*, have shown reductions in depressive symptoms, possibly through anti-inflammatory and neurotransmitter-modulating effects [158]. As evidenced in Table 3, numerous studies have shown the effectiveness of different probiotic species (e.g., *L. helveticus*, *L. rhamnosus*, *B. longum*, and *B. breve* CCFM1025) in the treatment of clinical depression. It is crucial to remember that some research has produced unfavorable findings [159,160]. These contradictory results could be the result of heterogeneity in the cohort characteristics and/or the therapies employed, although this is unknown. Additionally, the synergistic effects of several species can boost the antidepressant effects of probiotics. According to Liu et al. [161], this implies that utilizing multi-species probiotics may be more advantageous than using single-species probiotics.

Prebiotics are substrates that the gut bacteria preferentially use to support host health. They may encourage the growth of certain advantageous microorganisms. Prebiotics, such as fructooligosaccharides (FOSs) and galactooligosaccharides (GOSs), promote beneficial bacterial growth and may improve mood by enhancing gut health [161]. Probiotic (*L. helveticus* and *B. longum*) and prebiotic (GOS) supplementation's effects on depression remission in patients with MDD were examined in a previous RCT. After eight weeks of treatment, it was found that probiotics, but not prebiotics, reduced the symptoms of depression [162]. In an RCT, administering patients with depression with 4G-β-D-galactosylsucrose for 24 weeks consistently increased their sense of self-efficacy without affecting their depressive symptoms [163]. By encouraging the growth of probiotics, prebiotics indirectly improve host health rather than having a direct impact on the body.

Synbiotics—combinations of probiotics and prebiotics—have shown promise as well, with some studies indicating enhanced effects on mood compared to those of probiotics alone. A synbiotic supplement (containing *L. acidophilus*, *L. casei*, and *B. bifidum* plus inulin) improved depressive symptoms in overweight and obese adults [164]. Probiotics and prebiotics have symbiotic connections. Thus, it is conceivable that synbiotics will lead to the next advances in the treatment of depression. It is thought that administering probiotics and prebiotics at the same time increases the activity of good gut bacteria. However, the careful selection of appropriate prebiotics and probiotic strains is necessary for the synthesis of successful synbiotics. Moreover, outcomes vary based on formulation, dosage, and individual gut microbiota composition, highlighting a need for further standardized research. Table 3 highlights the evidence from recent studies on prebiotic, probiotic, and synbiotic interventions.

Table 3. Recent prebiotic, probiotic, and synbiotic intervention studies.

Intervention	Dose	Study Design	Outcome	References
		Probiotic studies—preclinical studies		
Lactobacillus reuteri NK33 and *Bifidobacterium adolescentis* NK98	1×10^9 CFU for 5 days	Immobilization stress-induced depression in C57BL/6 mice	↓ Proteobacteria population and gut LPS production ↓ IL-6 and corticosterone levels ↓ Anxiety and depression phenotypes	[165]
Lactobacillus kefiranofaciens ZW3	1×10^7 CFU, 1×10^8 CFU, and 1×10^9 CFU for 2 weeks	Kunming male mice: CMS model of depression	↑ Stress-induced serum corticosterone levels ↑ Brain 5-HT levels ↑ Depression-like behavior	[166]
Lactobacillus helveticus MCC1848	1×10^{11} CFU ml^{-1} for 24 days	Male C57BL/6J (B6) mice: Subchronic and mild social defeat stress (sCSDS) model of depression	Restored normal sucrose consumption Restored nucleus accumbens dopamine and serotonin receptor gene expression	[167]
Bifidobacterium breve CCFM1025	1×10^9 CFU ml^{-1} for 5 weeks	Male adult C57BL/6 mice: CMS model of depression	↑ Metagenomic tryptophan biosynthesis and profile of SCFA producing bacteria Normalized the stress-induced expression of brain Nr3c1 glucocorticoid receptor ↑ BDNF and precursor levels	[168]
		Clinical studies		
Probiotic (*Lactobacillus helveticus* and *Bifidobacterium longum*) and prebiotic (galactooligosaccharide)	1×10^{10} CFU per 5 g sachet for 8 weeks	Patients with MDD: RCT for psychological outcomes	↓ Depression score ↓ Kynurenine/ tryptophan ratio	[162]
Bifidobacterium longum 1714	1×10^9 CFU ml^{-1} for 4 weeks	Randomized, double- blind, parallel-group design: Social stress-induced by "Cyberball" game	↑ Theta band power and beta-3 band power in cortex in resting stage Changes in the neural processing of social stress	[169]
		Prebiotic studies—preclinical studies		
Fructooligosaccharide from *Morinda officinalis*	50 mg kg^{-1} for 3 weeks	Male SD rats: 7-week CMS model of depression following 4 weeks of treatment	Recovery of sucrose intake ↑ Mobility and exploratory behavior ↓ Plasma corticosterone Repaired damage in the intestinal epithelium	[170]
		Synbiotic studies—clinical studies		
Synbiotic formulation 15 g of prebiotics and 5 g of probiotic containing *L. acidophilus* T16 and *B. bifidum* BIA-6,7,8	2.7×10^7 CFU g^{-1} each for 12 weeks	75 Hemodialysis patients: Serum BDNF was measured	↓ Clinical anxiety and depression scores ↑ Serum BDNF levels	[171]

Abbreviations: ↑, increase; ↓, decrease.

6. Limitations and Challenges

One of the major limitations in utilizing dietary interventions to manage depression is the significant variability in individual responses. This variability can arise from numerous factors, including genetic differences, baseline microbiome composition, lifestyle factors, and the presence of other health conditions. For instance, while some individuals may experience significant mood improvements and gut microbiome benefits from dietary changes, others may see little to no effect or even negative outcomes, highlighting the need for personalized dietary recommendations. Additionally, the long-term sustainability of dietary interventions poses another challenge. While short-term dietary changes can be beneficial, maintaining these changes in the long term can be difficult due to factors such as dietary preferences, socio-economic barriers, cultural influences, and the complexity of adhering to specific dietary regimens. This sustainability issue is compounded by the often restrictive nature of some therapeutic diets, which can lead to decreased adherence and potential nutrient deficiencies if not carefully managed. Thus, for dietary interventions to be effective in the long term, they must be feasible, culturally acceptable, and enjoyable, ensuring that individuals can maintain the recommended dietary patterns without feeling overly restricted or deprived.

7. Gaps in the Knowledge and Future Research Directions

7.1. The Need for Longitudinal and Large-Scale Studies

Despite the growing evidence linking diet, gut microbiome composition, and mental health, there is a pressing need for more longitudinal and large-scale studies to fully understand these complex relationships. Current research often relies on cross-sectional data or short-term interventions, which can provide only a snapshot of the interactions between diet, the microbiota, and mental health. Longitudinal studies are crucial for observing how changes in dietary patterns influence the gut microbiome and mental health outcomes over time, offering insights into causality rather than mere association. Additionally, there is a significant gap in the research regarding diverse populations. Many existing studies are conducted on homogeneous groups, limiting the generalizability of the findings across different ethnicities, ages, genders, and socio-economic backgrounds. Understanding the variations in dietary impacts across diverse populations can help to identify specific dietary recommendations tailored to different demographic groups, thereby enhancing the effectiveness and applicability of dietary interventions for mental health, particularly depression.

7.2. Mechanistic Studies

To advance our understanding of how diet influences mental health through the gut–brain axis, it is essential to delve deeper into the precise molecular mechanisms involved. Mechanistic studies are needed to elucidate the specific pathways by which dietary components affect the gut microbiota and how these microbial changes translate into alterations in brain function and mood regulation. For example, while we know that certain diets can increase the production of SCFAs and modulate tryptophan metabolism, the exact molecular interactions and regulatory pathways remain unclear. Understanding these mechanisms at a detailed level, including the role of specific microbial species, metabolites, and host factors, can help in the development of targeted therapies and interventions. Additionally, mechanistic studies can explore the interplay between diet-induced microbial changes and other physiological systems, such as the immune and endocrine systems, which are also implicated in the pathophysiology of depression. This comprehensive understanding is critical for identifying new therapeutic targets and optimizing dietary strategies for mental health improvement.

7.3. Personalized Nutrition

The idea of personalized nutrition represents a future perspective considering the feasibility of dietary management strategies based on an individual's microbiome and genetic constitution. Since the results of dietary interventions vary across individuals, personalized nutrition solutions can offer some dietary recommendations based on the gut microbiota composition, metabolic potential, and genetic factors of the individual. This is easy in theory because it involves mapping out how different foods and food patterns affect different individuals and what impact those individual effects have, if any, on mental health. In the present study, emphasis is placed on future projections concerning the emergence of effective and producible technologies providing a basis for predictive anthropometric strategies utilizing microbiome data. This can be seen as a large potential future step in the practical application of nutrition therapy, as it accepts that there are differences among individuals; therefore, it is not promoting a blanket approach. The last but most fundamental tool to be adopted for this will be advanced technologies such as the exploration of metagenomics and metabolomics, as well as the use of machine learning in understanding the interplay between food, microbiome composition, and mental health.

8. Conclusions

The interdependence between the diet, the gut microbiome, and depression emphasizes the importance of nutrition in mental health. Various dietary fibers, proteins, fats, and other nutrients are consumed by gut microbes and converted into different metabolites: SCFAs, indoles, and precursors of neurotransmitters, the effects of which can be seen on the gut–brain axis and, in turn, mood and cognitive abilities. Such metabolites produced by microbes have the ability to affect inflammation, the production of neurotransmitters, and the integrity of the gut barrier, all of which are important in understanding the mechanisms that cause depression. Especially interventions that include changes in fiber and omega-3s in the diet or the general use of a healthy balanced diet have been effective in positively impacting mental health through colonizing the gut with healthy bacteria and curtailing inflammation in the body. However, the differential effects on individuals and the compliance with these interventions for prolonged periods remain constraints. Future studies must be longitudinal or mechanistic and include functional strategies such as personalized nutrition in order to enable researchers to grasp the benefits of the gut microbiome for the treatment of mental disorders. The evidence calls for the incorporation of dietary modalities as a part of integrative mental healthcare beyond the use of medications, with the hope that nutrition and gut health will become the mainstay in curbing or treating depression in the future.

Author Contributions: Investigation, N.R.; resources, B.X.; writing—original draft preparation, N.R.; writing—review and editing, B.X.; supervision, B.X.; funding acquisition, B.X. All authors have read and agreed to the published version of the manuscript.

Funding: This project was jointly supported by two grants (project code: UICR0400015-24 and UICR0400016-24) from BNU-HKBU United International College, China.

Institutional Review Board Statement: Not applicable.

Informed Consent Statement: Not applicable.

Conflicts of Interest: The authors declare no conflicts of interest.

Abbreviations

The following abbreviations are used in this manuscript:

5-HT	5-hydroxytryptamine
BCFA	Branched-chain fatty acid
BDNF	Brain-derived neurotrophic factor
CNS	Central nervous system
CRS	Chronic restraint stress
CSDS	Chronic social defeat stress
CUMS	Chronic unpredictable mild stress
ENS	Enteric nervous system
FMT	Fecal microbiota transplantation
FOS	Fructooligosaccharides
GABA	Gamma-aminobutyric acid
GALT	Gut-associated lymphoid tissue
GI	Gastrointestinal
GLP	Glucagon-like peptide
GOS	Galactooligosaccharide
HPA	Hypothalamic–pituitary–adrenal
IBD	Inflammatory bowel disease
IDO	Indoleamine-2,3-dioxygenase
IL	Interleukin
LPS	Lipopolysaccharide

MDD	Major depressive disorder
MUFA	Monounsaturated fatty acid
MS	Maternal separation
PUFA	Polyunsaturated fatty acid
RCT	Randomized controlled trial
ROS	Reactive oxygen species
SCFA	Short-chain fatty acid
SPM	Specialized pro-resolving mediators
TNF-α	Tumor necrosis factor-alpha

References

1. Morais, L.H.; Schreiber, H.L., IV; Mazmanian, S.K. The gut microbiota–brain axis in behaviour and brain disorders. *Nat. Rev. Microbiol.* **2021**, *19*, 241–255. [CrossRef]
2. Vernocchi, P.; Del Chierico, F.; Putignani, L. Gut microbiota metabolism and interaction with food components. *Int. J. Mol. Sci.* **2020**, *21*, 3688. [CrossRef]
3. Beam, A.; Clinger, E.; Hao, L. Effect of diet and dietary components on the composition of the gut microbiota. *Nutrients* **2021**, *13*, 2795. [CrossRef] [PubMed]
4. Hrncir, T. Gut microbiota dysbiosis: Triggers, consequences, diagnostic and therapeutic options. *Microorganisms* **2022**, *10*, 578. [CrossRef] [PubMed]
5. Methiwala, H.N.; Vaidya, B.; Addanki, V.K.; Bishnoi, M.; Sharma, S.S.; Kondepudi, K.K. Gut microbiota in mental health and depression: Role of pre/pro/synbiotics in their modulation. *Food Funct.* **2021**, *12*, 4284–4314. [CrossRef] [PubMed]
6. Gomaa, E.Z. Human gut microbiota/microbiome in health and diseases: A review. *Antonie Van Leeuwenhoek* **2020**, *113*, 2019–2040. [CrossRef] [PubMed]
7. Parizadeh, M.; Arrieta, M.-C. The global human gut microbiome: Genes, lifestyles, and diet. *Trends Mol. Med.* **2023**, *29*, 789–801. [CrossRef]
8. Ali, Q.; Ma, S.; La, S.; Guo, Z.; Liu, B.; Gao, Z.; Farooq, U.; Wang, Z.; Zhu, X.; Cui, Y.; et al. Microbial short-chain fatty acids: A bridge between dietary fibers and poultry gut health—A review. *Anim. Biosci.* **2022**, *35*, 1461–1478. [CrossRef] [PubMed]
9. Tufail, M.A.; Schmitz, R.A. Exploring the Probiotic Potential of Bacteroides spp. Within One Health Paradigm. *Probiotics Antimicrob. Proteins* **2024**, 1–24. [CrossRef] [PubMed]
10. Afzaal, M.; Saeed, F.; Shah, Y.A.; Hussain, M.; Rabail, R.; Socol, C.T.; Hassoun, A.; Pateiro, M.; Lorenzo, J.M.; Rusu, A.V.; et al. Human gut microbiota in health and disease: Unveiling the relationship. *Front. Microbiol.* **2022**, *13*, 999001. [CrossRef] [PubMed]
11. Lin, H.; Guo, Q.; Wen, Z.; Tan, S.; Chen, J.; Lin, L.; Chen, P.; He, J.; Wen, J.; Chen, Y. The multiple effects of fecal microbiota transplantation on diarrhea-predominant irritable bowel syndrome (IBS-D) patients with anxiety and depression behaviors. *Microb. Cell Factories* **2021**, *20*, 233. [CrossRef] [PubMed]
12. Almeida, J.I.; Tenreiro, M.F.; Martinez-Santamaria, L.; Guerrero-Aspizua, S.; Gisbert, J.P.; Alves, P.M.; Serra, M.; Baptista, P.M. Hallmarks of the human intestinal microbiome on liver maturation and function. *J. Hepatol.* **2021**, *76*, 694–725. [CrossRef]
13. Guindo, C.O.; Drancourt, M.; Grine, G. Digestive tract methanodrome: Physiological roles of human microbiota-associated methanogens. *Microb. Pathog.* **2020**, *149*, 104425. [CrossRef]
14. Chidambaram, S.B.; Essa, M.M.; Rathipriya, A.; Bishir, M.; Ray, B.; Mahalakshmi, A.M.; Tousif, A.; Sakharkar, M.K.; Kashyap, R.S.; Friedland, R.P.; et al. Gut dysbiosis, defective autophagy and altered immune responses in neurodegenerative diseases: Tales of a vicious cycle. *Pharmacol. Ther.* **2022**, *231*, 107988. [CrossRef]
15. Lin, S.; Li, Q.; Jiang, S.; Xu, Z.; Jiang, Y.; Liu, L.; Jiang, J.; Tong, Y.; Wang, P. Crocetin ameliorates chronic restraint stress-induced depression-like behaviors in mice by regulating MEK/ERK pathways and gut microbiota. *J. Ethnopharmacol.* **2021**, *268*, 113608. [CrossRef]
16. Margoob, M.; Kouser, S.; Jan, N. Serotonin: The link between gut microbiome and brain. In *Serotonin-Neurotransmitter and Hormone of Brain, Bowels and Blood*; IntechOpen: London, UK, 2024. [CrossRef]
17. Abdolmaleky, H.M.; Zhou, J.-R. Gut Microbiota Dysbiosis, Oxidative Stress, Inflammation, and Epigenetic Alterations in Metabolic Diseases. *Antioxidants* **2024**, *13*, 985. [CrossRef] [PubMed]
18. Iannotti, F.A.; Di Marzo, V. The gut microbiome, endocannabinoids and metabolic disorders. *J. Endocrinol.* **2021**, *248*, R83–R97. [CrossRef] [PubMed]
19. Yang, H.-L.; Li, M.-M.; Zhou, M.-F.; Xu, H.-S.; Huan, F.; Liu, N.; Gao, R.; Wang, J.; Zhang, N.; Jiang, L. Links Between Gut Dysbiosis and Neurotransmitter Disturbance in Chronic Restraint Stress-Induced Depressive Behaviours: The Role of Inflammation. *Inflammation* **2021**, *44*, 2448–2462. [CrossRef] [PubMed]
20. Han, X.; Ma, Y.; Ding, S.; Fang, J.; Liu, G. Regulation of dietary fiber on intestinal microorganisms and its effects on animal health. *Anim. Nutr.* **2023**, *14*, 356–369. [CrossRef]

21. García-Montero, C.; Fraile-Martínez, O.; Gómez-Lahoz, A.M.; Pekarek, L.; Castellanos, A.J.; Noguerales-Fraguas, F.; Coca, S.; Guijarro, L.G.; García-Honduvilla, N.; Asúnsolo, A.; et al. Nutritional Components in Western Diet Versus Mediterranean Diet at the Gut Microbiota–Immune System Interplay. Implications for Health and Disease. *Nutrients* **2021**, *13*, 699. [CrossRef]
22. Shin, Y.-J.; Lee, D.-Y.; Kim, J.Y.; Heo, K.; Shim, J.-J.; Lee, J.-L.; Kim, D.-H. Effect of fermented red ginseng on gut microbiota dysbiosis- or immobilization stress-induced anxiety, depression, and colitis in mice. *J. Ginseng Res.* **2022**, *47*, 255–264. [CrossRef] [PubMed]
23. Mayer, E.A.; Nance, K.; Chen, S. The Gut–Brain Axis. *Annu. Rev. Med.* **2022**, *73*, 439–453. [CrossRef] [PubMed]
24. Margolis, K.G.; Cryan, J.F.; Mayer, E.A. The Microbiota-Gut-Brain Axis: From Motility to Mood. *Gastroenterology* **2021**, *160*, 1486–1501. [CrossRef]
25. Novelle, M.G. Decoding the Role of Gut-Microbiome in the Food Addiction Paradigm. *Int. J. Environ. Res. Public Health* **2021**, *18*, 6825. [CrossRef]
26. Han, Y.; Wang, B.; Gao, H.; He, C.; Hua, R.; Liang, C.; Zhang, S.; Wang, Y.; Xin, S.; Xu, J. Vagus Nerve and Underlying Impact on the Gut Microbiota-Brain Axis in Behavior and Neurodegenerative Diseases. *J. Inflamm. Res.* **2022**, *15*, 6213–6230. [CrossRef] [PubMed]
27. Martino, C.; Dilmore, A.H.; Burcham, Z.M.; Metcalf, J.L.; Jeste, D.; Knight, R. Microbiota succession throughout life from the cradle to the grave. *Nat. Rev. Microbiol.* **2022**, *20*, 707–720. [CrossRef] [PubMed]
28. Sun, L.-J.; Li, J.-N.; Nie, Y.-Z. Gut hormones in microbiota-gut-brain cross-talk. *Chin. Med. J.* **2020**, *133*, 826–833. [CrossRef] [PubMed]
29. Ghosh-Swaby, O.R.; Reichelt, A.C.; Sheppard, P.A.; Davies, J.; Bussey, T.J.; Saksida, L.M. Metabolic hormones mediate cognition. *Front. Neuroendocr.* **2022**, *66*, 101009. [CrossRef]
30. Frankiensztajn, L.M.; Elliott, E.; Koren, O. The microbiota and the hypothalamus-pituitary-adrenocortical (HPA) axis, implications for anxiety and stress disorders. *Curr. Opin. Neurobiol.* **2020**, *62*, 76–82. [CrossRef] [PubMed]
31. Topol, I.A.; Polyakova, I.S.; Elykova, A.V. Role of intestinal microbiota in regulation of immune reactions of gut-associated lymphoid tissue under stress and following the modulation of its composition by antibiotics and probiotics administration. *J. Microbiol. Epidemiol. Immunobiol.* **2023**, *99*, 722–733. [CrossRef]
32. Doney, E.; Cadoret, A.; Dion-Albert, L.; Lebel, M.; Menard, C. Inflammation-driven brain and gut barrier dysfunction in stress and mood disorders. *Eur. J. Neurosci.* **2021**, *55*, 2851–2894. [CrossRef]
33. Di Vincenzo, F.; Del Gaudio, A.; Petito, V.; Lopetuso, L.R.; Scaldaferri, F. Gut microbiota, intestinal permeability, and systemic inflammation: A narrative review. *Intern. Emerg. Med.* **2023**, *19*, 275–293. [CrossRef]
34. Dicks, L.M.T. Gut Bacteria and Neurotransmitters. *Microorganisms* **2022**, *10*, 1838. [CrossRef] [PubMed]
35. Guzel, T.; Mirowska-Guzel, D. The Role of Serotonin Neurotransmission in Gastrointestinal Tract and Pharmacotherapy. *Molecules* **2022**, *27*, 1680. [CrossRef] [PubMed]
36. Wiley, N.C.; Cryan, J.F.; Dinan, T.G.; Ross, R.P.; Stanton, C. Production of psychoactive metabolites by gut bacteria. *Mod. Trends Psychiatry* **2021**, *32*, 74–99. [CrossRef]
37. Yoo, J.Y.; Groer, M.; Dutra, S.V.O.; Sarkar, A.; McSkimming, D.I. Gut Microbiota and Immune System Interactions. *Microorganisms* **2020**, *8*, 1587. [CrossRef] [PubMed]
38. Mou, Y.; Du, Y.; Zhou, L.; Yue, J.; Hu, X.; Liu, Y.; Chen, S.; Lin, X.; Zhang, G.; Xiao, H.; et al. Gut Microbiota Interact With the Brain Through Systemic Chronic Inflammation: Implications on Neuroinflammation, Neurodegeneration, and Aging. *Front. Immunol.* **2022**, *13*, 796288. [CrossRef] [PubMed]
39. Cristofori, F.; Dargenio, V.N.; Dargenio, C.; Miniello, V.L.; Barone, M.; Francavilla, R. Anti-Inflammatory and Immunomodulatory Effects of Probiotics in Gut Inflammation: A Door to the Body. *Front. Immunol.* **2021**, *12*, 578386. [CrossRef]
40. Novellino, F.; Saccà, V.; Donato, A.; Zaffino, P.; Spadea, M.F.; Vismara, M.; Arcidiacono, B.; Malara, N.; Presta, I.; Donato, G. Innate Immunity: A Common Denominator between Neurodegenerative and Neuropsychiatric Diseases. *Int. J. Mol. Sci.* **2020**, *21*, 1115. [CrossRef] [PubMed]
41. Assan, D.; Mustapha, U.F.; Chen, H.; Li, Z.; Peng, Y.; Li, G. The Roles of Neuropeptide Y (*Npy*) and Peptide YY (*Pyy*) in Teleost Food Intake: A Mini Review. *Life* **2021**, *11*, 547. [CrossRef]
42. Kong, C.; Yang, M.; Yue, N.; Zhang, Y.; Tian, C.; Wei, D.; Shi, R.; Yao, J.; Wang, L.; Li, D. Restore Intestinal Barrier Integrity: An Approach for Inflammatory Bowel Disease Therapy. *J. Inflamm. Res.* **2024**, *17*, 5389–5413. [CrossRef]
43. Silva, Y.P.; Bernardi, A.; Frozza, R.L. The Role of Short-Chain Fatty Acids From Gut Microbiota in Gut-Brain Communication. *Front. Endocrinol.* **2020**, *11*, 25. [CrossRef]
44. Roth, W.; Zadeh, K.; Vekariya, R.; Ge, Y.; Mohamadzadeh, M. Tryptophan Metabolism and Gut-Brain Homeostasis. *Int. J. Mol. Sci.* **2021**, *22*, 2973. [CrossRef] [PubMed]
45. Bosi, A.; Banfi, D.; Bistoletti, M.; Giaroni, C.; Baj, A. Tryptophan Metabolites Along the Microbiota-Gut-Brain Axis: An Interkingdom Communication System Influencing the Gut in Health and Disease. *Int. J. Tryptophan Res.* **2020**, *13*, 1178646920928984. [CrossRef] [PubMed]

46. Sonali, S.; Ray, B.; Tousif, H.A.; Rathipriya, A.G.; Sunanda, T.; Mahalakshmi, A.M.; Rungratanawanich, W.; Essa, M.M.; Qoronfleh, M.W.; Chidambaram, S.B.; et al. Mechanistic Insights into the Link between Gut Dysbiosis and Major Depression: An Extensive Review. *Cells* **2022**, *11*, 1362. [CrossRef] [PubMed]
47. Więdłocha, M.; Marcinowicz, P.; Janoska-Jaździk, M.; Szulc, A. Gut microbiota, kynurenine pathway and mental disorders—Review. *Prog. Neuro-Psychopharmacol. Biol. Psychiatry* **2021**, *106*, 110145. [CrossRef] [PubMed]
48. Fila, M.; Chojnacki, J.; Pawlowska, E.; Szczepanska, J.; Chojnacki, C.; Blasiak, J. Kynurenine Pathway of Tryptophan Metabolism in Migraine and Functional Gastrointestinal Disorders. *Int. J. Mol. Sci.* **2021**, *22*, 10134. [CrossRef] [PubMed]
49. Lukić, I.; Ivković, S.; Mitić, M.; Adžić, M. Tryptophan metabolites in depression: Modulation by gut microbiota. *Front. Behav. Neurosci.* **2022**, *16*, 987697. [CrossRef]
50. Xiao, S.; Wang, Z.; Wang, B.; Hou, B.; Cheng, J.; Bai, T.; Zhang, Y.; Wang, W.; Yan, L.; Zhang, J. Expanding the application of tryptophan: Industrial biomanufacturing of tryptophan derivatives. *Front. Microbiol.* **2023**, *14*, 1099098. [CrossRef]
51. Kim, C.-S.; Jung, S.; Hwang, G.-S.; Shin, D.-M. Gut microbiota indole-3-propionic acid mediates neuroprotective effect of probiotic consumption in healthy elderly: A randomized, double-blind, placebo-controlled, multicenter trial and *in vitro* study. *Clin. Nutr.* **2023**, *42*, 1025–1033. [CrossRef]
52. Haleem, D.J. Nutritional importance of tryptophan for improving treatment in depression and diabetes. *Nutr. Rev.* **2022**, *81*, 133–149. [CrossRef]
53. Frings, C.; Domes, G.; Friehs, M.A.; Geißler, C.; Schneider, K. Food for Your Mind? The Effect of Tyrosine on Selective Attention. *J. Cogn. Enhanc.* **2019**, *4*, 285–295. [CrossRef]
54. He, W.; Wu, G. Metabolism of amino acids in the brain and their roles in regulating food intake. *Adv. Exp. Med. Biol.* **2020**, *1265*, 167–185. [CrossRef] [PubMed]
55. Hou, D.; Tang, J.; Feng, Q.; Niu, Z.; Shen, Q.; Wang, L.; Zhou, S. Gamma-aminobutyric acid (GABA): A comprehensive review of dietary sources, enrichment technologies, processing effects, health benefits, and its applications. *Crit. Rev. Food Sci. Nutr.* **2023**, *64*, 8852–8874. [CrossRef] [PubMed]
56. Rastall, R.; Diez-Municio, M.; Forssten, S.; Hamaker, B.; Meynier, A.; Moreno, F.J.; Respondek, F.; Stahl, B.; Venema, K.; Wiese, M. Structure and function of non-digestible carbohydrates in the gut microbiome. *Benef. Microbes* **2022**, *13*, 95–168. [CrossRef]
57. Rinninella, E.; Tohumcu, E.; Raoul, P.; Fiorani, M.; Cintoni, M.; Mele, M.C.; Cammarota, G.; Gasbarrini, A.; Ianiro, G. The role of diet in shaping human gut microbiota. *Best Pract. Res. Clin. Gastroenterol.* **2023**, *62–63*, 101828. [CrossRef] [PubMed]
58. Payling, L.; Fraser, K.; Loveday, S.; Sims, I.; Roy, N.; McNabb, W. The effects of carbohydrate structure on the composition and functionality of the human gut microbiota. *Trends Food Sci. Technol.* **2020**, *97*, 233–248. [CrossRef]
59. Ney, L.-M.; Wipplinger, M.; Grossmann, M.; Engert, N.; Wegner, V.D.; Mosig, A.S. Short chain fatty acids: Key regulators of the local and systemic immune response in inflammatory diseases and infections. *Open Biol.* **2023**, *13*, 230014. [CrossRef]
60. Majumdar, A.; Venkatesh, I.P.S.; Basu, A. Short-chain fatty acids in the microbiota–gut–brain axis: Role in neurodegenerative disorders and viral infections. *ACS Chem. Neurosci.* **2023**, *14*, 1045–1062. [CrossRef] [PubMed]
61. Peled, S.; Livney, Y.D. The role of dietary proteins and carbohydrates in gut microbiome composition and activity: A review. *Food Hydrocoll.* **2021**, *120*, 106911. [CrossRef]
62. Senés-Guerrero, C.; Gradilla-Hernández, M.S.; García-Gamboa, R.; García-Cayuela, T. Dietary fiber and gut microbiota. *Sci. Technol. Fibers Food Syst.* **2020**, 277–298. [CrossRef]
63. Dahl, W.J.; Rivero Mendoza, D.; Lambert, J.M. Diet, nutrients, and the microbiome. *Prog. Mol. Biol. Transl. Sci.* **2020**, *171*, 237–263. [CrossRef]
64. Schroeder, B.O.; Birchenough, G.M.H.; Ståhlman, M.; Arike, L.; Johansson, M.E.; Hansson, G.C.; Bäckhed, F. Bifidobacteria or Fiber Protects against Diet-Induced Microbiota-Mediated Colonic Mucus Deterioration. *Cell Host Microbe* **2018**, *23*, 27–40.e7. [CrossRef] [PubMed]
65. Jia, X.; Chen, Q.; Wu, H.; Liu, H.; Jing, C.; Gong, A.; Zhang, Y. Exploring a novel therapeutic strategy: The interplay between gut microbiota and high-fat diet in the pathogenesis of metabolic disorders. *Front. Nutr.* **2023**, *10*, 1291853. [CrossRef] [PubMed]
66. McQuilken, S.A. Digestion and absorption. *Anaesth. Intensiv. Care Med.* **2021**, *22*, 336–338. [CrossRef]
67. Wu, S.; Bhat, Z.F.; Gounder, R.S.; Ahmed, I.A.M.; Al-Juhaimi, F.Y.; Ding, Y.; Bekhit, A.E.-D.A. Effect of Dietary Protein and Processing on Gut Microbiota—A Systematic Review. *Nutrients* **2022**, *14*, 453. [CrossRef]
68. Wang, W.; Jiang, S.; Xu, C.; Tang, L.; Liang, Y.; Zhao, Y.; Zhu, G. Interactions between gut microbiota and Parkinson's disease: The role of microbiota-derived amino acid metabolism. *Front. Aging Neurosci.* **2022**, *14*, 976316. [CrossRef]
69. Chen, Y.; Yang, K.; Xu, M.; Zhang, Y.; Weng, X.; Luo, J.; Li, Y.; Mao, Y.-H. Dietary Patterns, Gut Microbiota and Sports Performance in Athletes: A Narrative Review. *Nutrients* **2024**, *16*, 1634. [CrossRef] [PubMed]
70. Martinez, T.M.; Meyer, R.K.; Duca, F.A. Therapeutic Potential of Various Plant-Based Fibers to Improve Energy Homeostasis via the Gut Microbiota. *Nutrients* **2021**, *13*, 3470. [CrossRef]
71. Zhang, C.; Zhang, Y.; Li, H.; Liu, X. The potential of proteins, hydrolysates and peptides as growth factors for *Lactobacillus* and *Bifidobacterium*: Current research and future perspectives. *Food Funct.* **2020**, *11*, 1946–1957. [CrossRef] [PubMed]

72. Nakatani, A.; Li, X.; Miyamoto, J.; Igarashi, M.; Watanabe, H.; Sutou, A.; Watanabe, K.; Motoyama, T.; Tachibana, N.; Kohno, M.; et al. Dietary mung bean protein reduces high-fat diet-induced weight gain by modulating host bile acid metabolism in a gut microbiota-dependent manner. *Biochem. Biophys. Res. Commun.* **2018**, *501*, 955–961. [CrossRef]
73. Ross, F.C.; Patangia, D.; Grimaud, G.; Lavelle, A.; Dempsey, E.M.; Ross, R.P.; Stanton, C. The interplay between diet and the gut microbiome: Implications for health and disease. *Nat. Rev. Microbiol.* **2024**, *22*, 671–686. [CrossRef] [PubMed]
74. Rousseau, G. Microbiota, a New Playground for the Omega-3 Polyunsaturated Fatty Acids in Cardiovascular Diseases. *Mar. Drugs* **2021**, *19*, 54. [CrossRef]
75. Wolters, M.; Ahrens, J.; Romaní-Pérez, M.; Watkins, C.; Sanz, Y.; Benítez-Páez, A.; Stanton, C.; Günther, K. Dietary fat, the gut microbiota, and metabolic health—A systematic review conducted within the MyNewGut project. *Clin. Nutr.* **2018**, *38*, 2504–2520. [CrossRef]
76. Roopashree, P.; Shetty, S.S.; Kumari, N.S. Effect of medium chain fatty acid in human health and disease. *J. Funct. Food* **2021**, *87*, 104724. [CrossRef]
77. Yang, Q.; Liang, Q.; Balakrishnan, B.; Belobrajdic, D.P.; Feng, Q.-J.; Zhang, W. Role of Dietary Nutrients in the Modulation of Gut Microbiota: A Narrative Review. *Nutrients* **2020**, *12*, 381. [CrossRef] [PubMed]
78. Horn, J.; Mayer, D.E.; Chen, S.; Mayer, E.A. Role of diet and its effects on the gut microbiome in the pathophysiology of mental disorders. *Transl. Psychiatry* **2022**, *12*, 164. [CrossRef]
79. Adan, R.A.; van der Beek, E.M.; Buitelaar, J.K.; Cryan, J.F.; Hebebrand, J.; Higgs, S.; Schellekens, H.; Dickson, S.L. Nutritional psychiatry: Towards improving mental health by what you eat. *Eur. Neuropsychopharmacol.* **2019**, *29*, 1321–1332. [CrossRef] [PubMed]
80. Emerson, S.D.; Carbert, N.S. An apple a day: Protective associations between nutrition and the mental health of immigrants in Canada. *Chest* **2018**, *54*, 567–578. [CrossRef]
81. Tardy, A.-L.; Pouteau, E.; Marquez, D.; Yilmaz, C.; Scholey, A. Vitamins and Minerals for Energy, Fatigue and Cognition: A Narrative Review of the Biochemical and Clinical Evidence. *Nutrients* **2020**, *12*, 228. [CrossRef] [PubMed]
82. Balić, A.; Vlašić, D.; Žužul, K.; Marinović, B.; Bukvić Mokos, Z. Omega-3 Versus Omega-6 Polyunsaturated Fatty Acids in the Prevention and Treatment of Inflammatory Skin Diseases. *Int. J. Mol. Sci.* **2020**, *21*, 741. [CrossRef]
83. Zhang, P. Influence of Foods and Nutrition on the Gut Microbiome and Implications for Intestinal Health. *Int. J. Mol. Sci.* **2022**, *23*, 9588. [CrossRef]
84. Seyoum, Y.; Baye, K.; Humblot, C. Iron homeostasis in host and gut bacteria—A complex interrelationship. *Gut Microbes* **2021**, *13*, 1874855. [CrossRef]
85. Berding, K.; Carbia, C.; Cryan, J.F. Going with the grain: Fiber, cognition, and the microbiota-gut-brain-axis. *Exp. Biol. Med.* **2021**, *246*, 796–811. [CrossRef] [PubMed]
86. Wang, X.; Qi, Y.; Zheng, H. Dietary Polyphenol, Gut Microbiota, and Health Benefits. *Antioxidants* **2022**, *11*, 1212. [CrossRef] [PubMed]
87. Zheng, S.; Zhang, H.; Liu, R.; Huang, C.-L.; Li, H.; Deng, Z.-Y.; Tsao, R. Do short chain fatty acids and phenolic metabolites of the gut have synergistic anti-inflammatory effects?—New insights from a TNF-α-induced Caco-2 cell model. *Food Res. Int.* **2020**, *139*, 109833. [CrossRef] [PubMed]
88. Ortega, M.A.; Fraile-Martínez, Ó.; García-Montero, C.; Alvarez-Mon, M.A.; Lahera, G.; Monserrat, J.; Llavero-Valero, M.; Gutiérrez-Rojas, L.; Molina, R.; Rodríguez-Jimenez, R.; et al. Biological Role of Nutrients, Food and Dietary Patterns in the Prevention and Clinical Management of Major Depressive Disorder. *Nutrients* **2022**, *14*, 3099. [CrossRef]
89. Wróbel-Biedrawa, D.; Grabowska, K.; Galanty, A.; Sobolewska, D.; Podolak, I. A Flavonoid on the Brain: Quercetin as a Potential Therapeutic Agent in Central Nervous System Disorders. *Life* **2022**, *12*, 591. [CrossRef] [PubMed]
90. Abu Tayab, M.; Islam, M.N.; Chowdhury, K.A.A.; Tasnim, F.M. Targeting neuroinflammation by polyphenols: A promising therapeutic approach against inflammation-associated depression. *Biomed. Pharmacother.* **2022**, *147*, 112668. [CrossRef]
91. Liaqat, H.; Parveen, A.; Kim, S.Y. Neuroprotective Natural Products' Regulatory Effects on Depression via Gut–Brain Axis Targeting Tryptophan. *Nutrients* **2022**, *14*, 3270. [CrossRef]
92. Clemente-Suárez, V.J.; Beltrán-Velasco, A.I.; Redondo-Flórez, L.; Martín-Rodríguez, A.; Tornero-Aguilera, J.F. Global Impacts of Western Diet and Its Effects on Metabolism and Health: A Narrative Review. *Nutrients* **2023**, *15*, 2749. [CrossRef] [PubMed]
93. Merra, G.; Noce, A.; Marrone, G.; Cintoni, M.; Tarsitano, M.G.; Capacci, A.; De Lorenzo, A. Influence of Mediterranean Diet on Human Gut Microbiota. *Nutrients* **2021**, *13*, 7. [CrossRef]
94. Shen, Y.C.; Chang, C.E.; Lin, M.N.; Lin, C.L. Vegetarian diet is associated with lower risk of depression in Taiwan. *Nutrients* **2021**, *13*, 1059. [CrossRef] [PubMed]
95. Ocklenburg, S.; Borawski, J. Vegetarian diet and depression scores: A meta-analysis. *J. Affect. Disord.* **2021**, *294*, 813–815. [CrossRef] [PubMed]
96. Jain, R.; Larsuphrom, P.; Degremont, A.; Latunde-Dada, G.O.; Philippou, E. Association between vegetarian and vegan diets and depression: A systematic review. *Nutr. Bull.* **2022**, *47*, 27–49. [CrossRef]

97. Shamshtein, D.; Liwinski, T. Ketogenic Therapy for Major Depressive Disorder: A Review of Neurobiological Evidence. *Recent Prog. Nutr.* **2021**, *2*, 003. [CrossRef]
98. Moreno-Agostino, D.; Caballero, F.F.; Martín-María, N.; Tyrovolas, S.; López-García, P.; Rodríguez-Artalejo, F.; Haro, J.M.; Ayuso-Mateos, J.L.; Miret, M. Mediterranean diet and wellbeing: Evidence from a nationwide survey. *Psychol. Health* **2019**, *34*, 321–335. [CrossRef] [PubMed]
99. Berk, M.; Jacka, F.N. Diet and Depression—From Confirmation to Implementation. *JAMA* **2019**, *321*, 842–843. [CrossRef]
100. Parletta, N.; Zarnowiecki, D.; Cho, J.; Wilson, A.; Bogomolova, S.; Villani, A.; Itsiopoulos, C.; Niyonsenga, T.; Blunden, S.; Meyer, B.; et al. A Mediterranean-style dietary intervention supplemented with fish oil improves diet quality and mental health in people with depression: A randomized controlled trial (HELFIMED). *Nutr. Neurosci.* **2019**, *22*, 474–487. [CrossRef]
101. 2021 Alzheimer's disease facts and figures. *Alzheimer's Dement.* **2021**, *17*, 327–406. [CrossRef]
102. Zou, L.; Tian, Y.; Wang, Y.; Chen, D.; Lu, X.; Zeng, Z.; Chen, Z.; Lin, C.; Liang, Y. High-cholesterol diet promotes depression- and anxiety-like behaviors in mice by impact gut microbe and neuroinflammation. *J. Affect. Disord.* **2023**, *327*, 425–438. [CrossRef] [PubMed]
103. Xia, J.; Gu, L.; Guo, Y.; Feng, H.; Chen, S.; Jurat, J.; Fu, W.; Zhang, D. Gut Microbiota Mediates the Preventive Effects of Dietary Capsaicin Against Depression-Like Behavior Induced by Lipopolysaccharide in Mice. *Front. Cell. Infect. Microbiol.* **2021**, *11*, 627608. [CrossRef] [PubMed]
104. Yang, Y.; Zhong, Z.; Wang, B.; Wang, Y. Xiaoyao San ameliorates high-fat diet-induced anxiety and depression via regulating gut microbiota in mice. *Biomed. Pharmacother.* **2022**, *156*, 113902. [CrossRef]
105. Liu, Y.; Ju, Y.; Cui, L.; Liu, T.; Hou, Y.; Wu, Q.; Ojo, O.; Du, X.; Wang, X. Association between Dietary Fiber Intake and Incidence of Depression and Anxiety in Patients with Essential Hypertension. *Nutrients* **2021**, *13*, 4159. [CrossRef]
106. Zou, H.; Gao, H.; Liu, Y.; Zhang, Z.; Zhao, J.; Wang, W.; Ren, B.; Tan, X. Dietary inulin alleviated constipation induced depression and anxiety-like behaviors: Involvement of gut microbiota and microbial metabolite short-chain fatty acid. *Int. J. Biol. Macromol.* **2024**, *259*, 129420. [CrossRef] [PubMed]
107. Liu, R.T.; Rowan-Nash, A.D.; Sheehan, A.E.; Walsh, R.F.; Sanzari, C.M.; Korry, B.J.; Belenky, P. Reductions in anti-inflammatory gut bacteria are associated with depression in a sample of young adults. *Brain Behav. Immun.* **2020**, *88*, 308–324. [CrossRef]
108. Xie, Y.; Wu, Z.; Qian, Q.; Yang, H.; Ma, J.; Luan, W.; Shang, S.; Li, X. Apple polyphenol extract ameliorates sugary-diet-induced depression-like behaviors in male C57BL/6 mice by inhibiting the inflammation of the gut–brain axis. *Food Funct.* **2024**, *15*, 2939–2959. [CrossRef]
109. Deng, L.; Zhou, X.; Tao, G.; Hao, W.; Wang, L.; Lan, Z.; Song, Y.; Wu, M.; Huang, J.-Q. Ferulic acid and feruloylated oligosaccharides alleviate anxiety and depression symptom via regulating gut microbiome and microbial metabolism. *Food Res. Int.* **2022**, *162*, 111887. [CrossRef] [PubMed]
110. Song, Z.; Zhang, X.; Hong, M.; Wu, Z.; Luo, S.; Cheng, K. Oolong tea polyphenols affect the inflammatory response to improve cognitive function by regulating gut microbiota. *J. Funct. Foods* **2023**, *105*, 105584. [CrossRef]
111. Park, M.; Choi, J.; Lee, H.-J. Flavonoid-Rich Orange Juice Intake and Altered Gut Microbiome in Young Adults with Depressive Symptom: A Randomized Controlled Study. *Nutrients* **2020**, *12*, 1815. [CrossRef]
112. Ávila, P.R.M.; Fiorot, M.; Michels, M.; Dominguini, D.; Abatti, M.; Vieira, A.; de Moura, A.B.; Behenck, J.P.; Borba, L.A.; Botelho, M.E.M.; et al. Effects of microbiota transplantation and the role of the vagus nerve in gut–brain axis in animals subjected to chronic mild stress. *J. Affect. Disord.* **2020**, *277*, 410–416. [CrossRef] [PubMed]
113. Dziurkowska, E.; Wesolowski, M. Cortisol as a Biomarker of Mental Disorder Severity. *J. Clin. Med.* **2021**, *10*, 5204. [CrossRef] [PubMed]
114. Hestad, K.; Alexander, J.; Rootwelt, H.; Aaseth, J.O. The Role of Tryptophan Dysmetabolism and Quinolinic Acid in Depressive and Neurodegenerative Diseases. *Biomolecules* **2022**, *12*, 998. [CrossRef]
115. Agus, A.; Planchais, J.; Sokol, H. Gut Microbiota Regulation of Tryptophan Metabolism in Health and Disease. *Cell Host Microbe* **2018**, *23*, 716–724. [CrossRef]
116. Sun, Y.; Koyama, Y.; Shimada, S. Inflammation From Peripheral Organs to the Brain: How Does Systemic Inflammation Cause Neuroinflammation? *Front. Aging Neurosci.* **2022**, *14*, 903455. [CrossRef] [PubMed]
117. Hassamal, S. Chronic stress, neuroinflammation, and depression: An overview of pathophysiological mechanisms and emerging anti-inflammatories. *Front. Psychiatry* **2023**, *14*, 1130989. [CrossRef] [PubMed]
118. Carniel, B.P.; da Rocha, N.S. Brain-derived neurotrophic factor (BDNF) and inflammatory markers: Perspectives for the management of depression. *Prog. Neuro-Psychopharmacol. Biol. Psychiatry* **2020**, *108*, 110151. [CrossRef] [PubMed]
119. Agirman, G.; Yu, K.B.; Hsiao, E.Y. Signaling inflammation across the gut-brain axis. *Science* **2021**, *374*, 1087–1092. [CrossRef]
120. Jaberi, K.R.; Alamdari-Palangi, V.; Savardashtaki, A.; Vatankhah, P.; Jamialahmadi, T.; Tajbakhsh, A.; Sahebkar, A. Modulatory Effects of Phytochemicals on Gut–Brain Axis: Therapeutic Implication. *Curr. Dev. Nutr.* **2024**, *8*, 103785. [CrossRef] [PubMed]
121. Xie, F.; Yang, W.; Xing, M.; Zhang, H.; Ai, L. Natural polyphenols-gut microbiota interactions and effects on glycolipid metabolism via polyphenols-gut-brain axis: A state-of-the-art review. *Trends Food Sci. Technol.* **2023**, *140*, 104171. [CrossRef]

122. Fang, J.; Zhang, Z.; Cheng, Y.; Yang, H.; Zhang, H.; Xue, Z.; Lu, S.; Dong, Y.; Song, C.; Zhang, X.; et al. EPA and DHA differentially coordinate the crosstalk between host and gut microbiota and block DSS-induced colitis in mice by a reinforced colonic mucus barrier. *Food Funct.* **2022**, *13*, 4399–4420. [CrossRef]
123. Harwood, J.L. Polyunsaturated Fatty Acids: Conversion to Lipid Mediators, Roles in Inflammatory Diseases and Dietary Sources. *Int. J. Mol. Sci.* **2023**, *24*, 8838. [CrossRef] [PubMed]
124. Liu, Y.; Xu, F.; Liu, S.; Liu, G.; Yang, X.; Gao, W.; Fan, K.; Zhao, H.; Ma, J. Significance of gastrointestinal tract in the therapeutic mechanisms of exercise in depression: Synchronism between brain and intestine through GBA. *Prog. Neuro-Psychopharmacol. Biol. Psychiatry* **2020**, *103*, 109971. [CrossRef] [PubMed]
125. Huang, F.; Wu, X. Brain Neurotransmitter Modulation by Gut Microbiota in Anxiety and Depression. *Front. Cell Dev. Biol.* **2021**, *9*, 649103. [CrossRef]
126. Legan, T.B.; Lavoie, B.; Mawe, G.M. Direct and indirect mechanisms by which the gut microbiota influence host serotonin systems. *Neurogastroenterol. Motil.* **2022**, *34*, e14346. [CrossRef] [PubMed]
127. Lucerne, K.E.; Osman, A.; Meckel, K.R.; Kiraly, D.D. Contributions of neuroimmune and gut-brain signaling to vulnerability of developing substance use disorders. *Neuropharmacology* **2021**, *192*, 108598. [CrossRef] [PubMed]
128. Dalziel, J.E.; Zobel, G.; Dewhurst, H.; Hurst, C.; Olson, T.; Rodriguez-Sanchez, R.; Mace, L.; Parkar, N.; Thum, C.; Hannaford, R.; et al. A Diet Enriched with *Lacticaseibacillus rhamnosus* HN001 and Milk Fat Globule Membrane Alters the Gut Microbiota and Decreases Amygdala GABA a Receptor Expression in Stress-Sensitive Rats. *Int. J. Mol. Sci.* **2023**, *24*, 10433. [CrossRef] [PubMed]
129. Zhang, H.; Wang, Z.; Wang, G.; Song, X.; Qian, Y.; Liao, Z.; Sui, L.; Ai, L.; Xia, Y. Understanding the Connection between Gut Homeostasis and Psychological Stress. *J. Nutr.* **2023**, *153*, 924–939. [CrossRef]
130. Fernández-García, V.; González-Ramos, S.; Martín-Sanz, P.; Portillo, F.G.-D.; Laparra, J.M.; Boscá, L. NOD1 in the interplay between microbiota and gastrointestinal immune adaptations. *Pharmacol. Res.* **2021**, *171*, 105775. [CrossRef] [PubMed]
131. Jamar, G.; Ribeiro, D.A.; Pisani, L.P. High-fat or high-sugar diets as trigger inflammation in the microbiota-gut-brain axis. *Crit. Rev. Food Sci. Nutr.* **2021**, *61*, 836–854. [CrossRef]
132. Pickering, G.; Mazur, A.; Trousselard, M.; Bienkowski, P.; Yaltsewa, N.; Amessou, M.; Noah, L.; Pouteau, E. Magnesium Status and Stress: The Vicious Circle Concept Revisited. *Nutrients* **2020**, *12*, 3672. [CrossRef] [PubMed]
133. Madison, A.A.; Belury, M.A.; Andridge, R.; Renna, M.E.; Shrout, M.R.; Malarkey, W.B.; Lin, J.; Epel, E.S.; Kiecolt-Glaser, J.K. Omega-3 supplementation and stress reactivity of cellular aging biomarkers: An ancillary substudy of a randomized, controlled trial in midlife adults. *Mol. Psychiatry* **2021**, *26*, 3034–3042. [CrossRef]
134. Shandilya, S.; Kumar, S.; Jha, N.K.; Kesari, K.K.; Ruokolainen, J. Interplay of gut microbiota and oxidative stress: Perspective on neurodegeneration and neuroprotection. *J. Adv. Res.* **2021**, *38*, 223–244. [CrossRef]
135. Dong, Y.; Cui, C. The role of short-chain fatty acids in central nervous system diseases. *Mol. Cell. Biochem.* **2022**, *477*, 2595–2607. [CrossRef] [PubMed]
136. Xu, Y.; Yu, Y.; Shen, Y.; Li, Q.; Lan, J.; Wu, Y.; Zhang, R.; Cao, G.; Yang, C. Effects of Bacillus subtilis and Bacillus licheniformis on growth performance, immunity, short chain fatty acid production, antioxidant capacity, and cecal microflora in broilers. *Poult. Sci.* **2021**, *100*, 101358. [CrossRef]
137. Gur, T.L.; Palkar, A.V.; Rajasekera, T.; Allen, J.; Niraula, A.; Godbout, J.; Bailey, M.T. Prenatal stress disrupts social behavior, cortical neurobiology and commensal microbes in adult male offspring. *Behav. Brain Res.* **2018**, *359*, 886–894. [CrossRef]
138. Karen, C.; Shyu, D.J.H.; Rajan, K.E. Lactobacillus paracasei Supplementation Prevents Early Life Stress-Induced Anxiety and Depressive-Like Behavior in Maternal Separation Model-Possible Involvement of Microbiota-Gut-Brain Axis in Differential Regulation of MicroRNA124a/132 and Glutamate Receptors. *Front. Neurosci.* **2021**, *15*, 719933. [CrossRef]
139. Kiecolt-Glaser, J.K.; Wilson, S.J.; Shrout, M.R.; Madison, A.A.; Andridge, R.; Peng, J.; Malarkey, W.B.; Bailey, M.T. The gut reaction to couples' relationship troubles: A route to gut dysbiosis through changes in depressive symptoms. *Psychoneuroendocrinology* **2021**, *125*, 105132. [CrossRef] [PubMed]
140. Sandes, S.; Figueiredo, N.; Pedroso, S.; Sant'Anna, F.; Acurcio, L.; Junior, M.A.; Barros, P.; Oliveira, F.; Cardoso, V.; Generoso, S.; et al. Weissella paramesenteroides WpK4 plays an immunobiotic role in gut-brain axis, reducing gut permeability, anxiety-like and depressive-like behaviors in murine models of colitis and chronic stress. *Food Res. Int.* **2020**, *137*, 109741. [CrossRef]
141. Breit, S.; Kupferberg, A.; Rogler, G.; Hasler, G. Vagus Nerve as Modulator of the Brain–Gut Axis in Psychiatric and Inflammatory Disorders. *Front. Psychiatry* **2018**, *9*, 44. [CrossRef] [PubMed]
142. Gao, K.; Farzi, A.; Ke, X.; Yu, Y.; Chen, C.; Chen, S.; Yu, T.; Wang, H.; Li, Y. Oral administration of *Lactococcus lactis* WHH2078 alleviates depressive and anxiety symptoms in mice with induced chronic stress. *Food Funct.* **2021**, *13*, 957–969. [CrossRef] [PubMed]
143. Zhang, Y.; Huang, J.; Xiong, Y.; Zhang, X.; Lin, Y.; Liu, Z. Jasmine Tea Attenuates Chronic Unpredictable Mild Stress-Induced Depressive-like Behavior in Rats via the Gut-Brain Axis. *Nutrients* **2021**, *14*, 99. [CrossRef] [PubMed]

144. Tian, P.; O'Riordan, K.J.; Lee, Y.-K.; Wang, G.; Zhao, J.; Zhang, H.; Cryan, J.F.; Chen, W. Towards a psychobiotic therapy for depression: Bifidobacterium breve CCFM1025 reverses chronic stress-induced depressive symptoms and gut microbial abnormalities in mice. *Neurobiol. Stress* **2020**, *12*, 100216. [CrossRef] [PubMed]
145. Song, X.; Wang, W.; Ding, S.; Liu, X.; Wang, Y.; Ma, H. Puerarin ameliorates depression-like behaviors of with chronic unpredictable mild stress mice by remodeling their gut microbiota. *J. Affect. Disord.* **2021**, *290*, 353–363. [CrossRef]
146. Chen, Y.-H.; Xue, F.; Yu, S.-F.; Li, X.-S.; Liu, L.; Jia, Y.-Y.; Yan, W.-J.; Tan, Q.-R.; Wang, H.-N.; Peng, Z.-W. Gut microbiota dysbiosis in depressed women: The association of symptom severity and microbiota function. *J. Affect. Disord.* **2020**, *282*, 391–400. [CrossRef] [PubMed]
147. Duan, J.; Huang, Y.; Tan, X.; Chai, T.; Wu, J.; Zhang, H.; Li, Y.; Hu, X.; Zheng, P.; Ji, P.; et al. Characterization of gut microbiome in mice model of depression with divergent response to escitalopram treatment. *Transl. Psychiatry* **2021**, *11*, 303. [CrossRef] [PubMed]
148. Xiao, Q.; Shu, R.; Wu, C.; Tong, Y.; Xiong, Z.; Zhou, J.; Yu, C.; Xie, X.; Fu, Z. Crocin-I alleviates the depression-like behaviors probably via modulating "microbiota-gut-brain" axis in mice exposed to chronic restraint stress. *J. Affect. Disord.* **2020**, *276*, 476–486. [CrossRef]
149. Wu, M.; Tian, T.; Mao, Q.; Zou, T.; Zhou, C.-J.; Xie, J.; Chen, J.-J. Associations between disordered gut microbiota and changes of neurotransmitters and short-chain fatty acids in depressed mice. *Transl. Psychiatry* **2020**, *10*, 350. [CrossRef]
150. Xie, R.; Jiang, P.; Lin, L.; Jiang, J.; Yu, B.; Rao, J.; Liu, H.; Wei, W.; Qiao, Y. Oral treatment with Lactobacillus reuteri attenuates depressive-like behaviors and serotonin metabolism alterations induced by chronic social defeat stress. *J. Psychiatr. Res.* **2019**, *122*, 70–78. [CrossRef] [PubMed]
151. Yin, W.; Löf, M.; Chen, R.; Hultman, C.M.; Fang, F.; Sandin, S. Mediterranean diet and depression: A population-based cohort study. *Int. J. Behav. Nutr. Phys. Act.* **2021**, *18*, 153. [CrossRef]
152. Seethaler, B.; Nguyen, N.K.; Basrai, M.; Kiechle, M.; Walter, J.; Delzenne, N.M.; Bischoff, S.C. Short-chain fatty acids are key mediators of the favorable effects of the Mediterranean diet on intestinal barrier integrity: Data from the randomized controlled LIBRE trial. *Am. J. Clin. Nutr.* **2022**, *116*, 928–942. [CrossRef] [PubMed]
153. Rao, J.; Xie, R.; Lin, L.; Jiang, J.; Du, L.; Zeng, X.; Li, G.; Wang, C.; Qiao, Y. Fecal microbiota transplantation ameliorates gut microbiota imbalance and intestinal barrier damage in rats with stress-induced depressive-like behavior. *Eur. J. Neurosci.* **2021**, *53*, 3598–3611. [CrossRef]
154. Lam, S.; Bai, X.; Shkoporov, A.N.; Park, H.; Wu, X.; Lan, P.; Zuo, T. Roles of the gut virome and mycobiome in faecal microbiota transplantation. *Lancet Gastroenterol. Hepatol.* **2022**, *7*, 472–484. [CrossRef]
155. Mayneris-Perxachs, J.; Castells-Nobau, A.; Arnoriaga-Rodríguez, M.; Martin, M.; de la Vega-Correa, L.; Zapata, C.; Burokas, A.; Blasco, G.; Coll, C.; Escrichs, A.; et al. Microbiota alterations in proline metabolism impact depression. *Cell Metab.* **2022**, *34*, 681–701.e10. [CrossRef] [PubMed]
156. Doll, J.P.K.; Vázquez-Castellanos, J.F.; Schaub, A.-C.; Schweinfurth, N.; Kettelhack, C.; Schneider, E.; Yamanbaeva, G.; Mählmann, L.; Brand, S.; Beglinger, C.; et al. Fecal Microbiota Transplantation (FMT) as an Adjunctive Therapy for Depression—Case Report. *Front. Psychiatry* **2022**, *13*, 815422. [CrossRef] [PubMed]
157. Servick, K. Alternatives to fecal transplants near approval. *Science* **2022**, *375*, 368–369. [CrossRef] [PubMed]
158. Liu, L.; Wang, H.; Zhang, H.; Chen, X.; Zhang, Y.; Wu, J.; Zhao, L.; Wang, D.; Pu, J.; Ji, P.; et al. Toward a Deeper Understanding of Gut Microbiome in Depression: The Promise of Clinical Applicability. *Adv. Sci.* **2022**, *9*, e2203707. [CrossRef]
159. Browne, P.D.; Bolte, A.C.; der Vaart, I.B.-V.; Claassen, E.; de Weerth, C. Probiotics as a treatment for prenatal maternal anxiety and depression: A double-blind randomized pilot trial. *Sci. Rep.* **2021**, *11*, 3051. [CrossRef]
160. Reininghaus, E.Z.; Platzer, M.; Kohlhammer-Dohr, A.; Hamm, C.; Mörkl, S.; Bengesser, S.A.; Fellendorf, F.T.; Lahousen-Luxenberger, T.; Leitner-Afschar, B.; Schöggl, H.; et al. PROVIT: Supplementary Probiotic Treatment and Vitamin B7 in Depression—A Randomized Controlled Trial. *Nutrients* **2020**, *12*, 3422. [CrossRef]
161. Liu, L.; Wang, H.; Chen, X.; Zhang, Y.; Zhang, H.; Xie, P. Gut microbiota and its metabolites in depression: From pathogenesis to treatment. *EBioMedicine* **2023**, *90*, 104527. [CrossRef] [PubMed]
162. Kazemi, A.; Noorbala, A.A.; Azam, K.; Eskandari, M.H.; Djafarian, K. Effect of probiotic and prebiotic vs placebo on psychological outcomes in patients with major depressive disorder: A randomized clinical trial. *Clin. Nutr.* **2019**, *38*, 522–528. [CrossRef] [PubMed]
163. Tarutani, S.; Omori, M.; Ido, Y.; Yano, M.; Komatsu, T.; Okamura, T. Effects of 4G-beta-D-Galactosylsucrose in patients with depression: A randomized, double-blinded, placebo-controlled, parallel-group comparative study. *J. Psychiatr. Res.* **2022**, *148*, 110–120. [CrossRef] [PubMed]
164. Hadi, A.; Sepandi, M.; Marx, W.; Moradi, S.; Parastouei, K. Clinical and psychological responses to synbiotic supplementation in obese or overweight adults: A randomized clinical trial. *Complement. Ther. Med.* **2019**, *47*, 102216. [CrossRef] [PubMed]
165. Jang, H.-M.; Lee, K.-E.; Kim, D.-H. The Preventive and Curative Effects of Lactobacillus reuteri NK33 and Bifidobacterium adolescentis NK98 on Immobilization Stress-Induced Anxiety/Depression and Colitis in Mice. *Nutrients* **2019**, *11*, 819. [CrossRef]

166. Sun, Y.; Geng, W.; Pan, Y.; Wang, J.; Xiao, P.; Wang, Y. Supplementation with *Lactobacillus kefiranofaciens* ZW3 from Tibetan Kefir improves depression-like behavior in stressed mice by modulating the gut microbiota. *Food Funct.* **2019**, *10*, 925–937. [CrossRef]
167. Maehata, H.; Kobayashi, Y.; Mitsuyama, E.; Kawase, T.; Kuhara, T.; Xiao, J.-Z.; Tsukahara, T.; Toyoda, A. Heat-killed *Lactobacillus helveticus* strain MCC1848 confers resilience to anxiety or depression-like symptoms caused by subchronic social defeat stress in mice. *Biosci. Biotechnol. Biochem.* **2019**, *83*, 1239–1247. [CrossRef]
168. Tian, T.; Xu, B.; Qin, Y.; Fan, L.; Chen, J.; Zheng, P.; Gong, X.; Wang, H.; Bai, M.; Pu, J.; et al. Clostridium butyricum miyairi 588 has preventive effects on chronic social defeat stress-induced depressive-like behaviour and modulates microglial activation in mice. *Biochem. Biophys. Res. Commun.* **2019**, *516*, 430–436. [CrossRef] [PubMed]
169. Wang, X.; Sun, G.; Feng, T.; Zhang, J.; Huang, X.; Wang, T.; Xie, Z.; Chu, X.; Yang, J.; Wang, H.; et al. Sodium oligomannate therapeutically remodels gut microbiota and suppresses gut bacterial amino acids-shaped neuroinflammation to inhibit Alzheimer's disease progression. *Cell Res.* **2019**, *29*, 787–803. [CrossRef]
170. Chi, L.; Khan, I.; Lin, Z.; Zhang, J.; Lee, M.S.; Leong, W.; Hsiao, W.W.; Zheng, Y. Fructo-oligosaccharides from Morinda officinalis remodeled gut microbiota and alleviated depression features in a stress rat model. *Phytomedicine* **2019**, *67*, 153157. [CrossRef] [PubMed]
171. Haghighat, N.; Rajabi, S.; Mohammadshahi, M. Effect of synbiotic and probiotic supplementation on serum brain-derived neurotrophic factor level, depression and anxiety symptoms in hemodialysis patients: A randomized, double-blinded, clinical trial. *Nutr. Neurosci.* **2019**, *24*, 490–499. [CrossRef]

International Journal of *Molecular Sciences*

MDPI

Review

A Comprehensive Review of the Triangular Relationship among Diet–Gut Microbiota–Inflammation

Nidesha Randeni [1,2], Matteo Bordiga [3] and Baojun Xu [1,*]

1 Food Science and Technology Program, Department of Life Sciences, BNU-HKBU United International College, Zhuhai 519087, China; nidesha.randeni96@gmail.com
2 Department of Agricultural and Plantation Engineering, Faculty of Engineering Technology, The Open University of Sri Lanka, Nawala, Nugegoda 10250, Sri Lanka
3 Department of Pharmaceutical Sciences, Università degli Studi del Piemonte Orientale "A. Avogadro", Largo Donegani 2, 28100 Novara, Italy; matteo.bordiga@uniupo.it
* Correspondence: baojunxu@uic.edu.cn; Tel.: +86-756-3620636

Abstract: The human gastrointestinal tract hosts a complex and dynamic community of microorganisms known as the gut microbiota, which play a pivotal role in numerous physiological processes, including digestion, metabolism, and immune function. Recent research has highlighted the significant impact of diet on the gut microbiota composition and functionality, and the consequential effects on host health. Concurrently, there is growing evidence linking the gut microbiota to inflammation, a key factor in many chronic diseases such as inflammatory bowel disease (IBD), obesity, diabetes, and cardiovascular diseases (CVDs). This review explores how dietary components influence the gut microbiota composition, how these microbial changes affect inflammatory pathways, and the therapeutic implications of modulating this axis for chronic inflammatory disease prevention and management. Beneficial dietary patterns, such as the Mediterranean diet (MD) and plant-based diets, promote a diverse and balanced gut microbiota composition, supporting anti-inflammatory pathways. Conversely, the Western diet (WD), high in saturated fats and refined sugars, is associated with dysbiosis and increased inflammation. With all the links between the three variables considered, this review attempts to offer a thorough examination of the triangle formed by inflammation, the gut microbiota, and food.

Keywords: gut microbiota; inflammation; diet; dysbiosis; metabolites

check for updates

Citation: Randeni, N.; Bordiga, M.; Xu, B. A Comprehensive Review of the Triangular Relationship among Diet–Gut Microbiota–Inflammation. *Int. J. Mol. Sci.* **2024**, *25*, 9366. https://doi.org/10.3390/ijms25179366

Academic Editor: Walter Wahli

Received: 17 July 2024
Revised: 21 August 2024
Accepted: 27 August 2024
Published: 29 August 2024

1. Introduction

The human gastrointestinal tract is home to a complex and dynamic community of microorganisms known as the gut microbiota. This intricate ecosystem plays a critical role in numerous physiological processes, including digestion, metabolism, immune function, and even behavior [1]. The composition and functionality of the gut microbiota are profoundly influenced by diet, with different dietary patterns inducing significant changes in microbial communities. For instance, diets rich in fiber, polyphenols, and healthy fats, such as the MD, promote the growth of beneficial bacteria and enhance microbial diversity. Conversely, the WD, characterized by a high intake of fats and refined sugars, is associated with decreased microbial diversity, an increase in pathogenic bacteria [2], and an increase in the number of mucin-degrading bacteria, which, in turn, can affect the intestinal mucus layer thickness [3].

In recent years, a growing body of evidence has emerged linking the gut microbiota to inflammation, which is a key factor in the pathogenesis of many chronic diseases, including IBD, obesity, diabetes, and CVDs [4]. The gut microbiota modulate the host's immune system and inflammatory responses through various mechanisms. Beneficial bacteria help to maintain gut barrier integrity, preventing the translocation of harmful pathogens and toxins into the bloodstream [5]. They also produce metabolites, such as short-chain fatty acids (SCFAs), which have anti-inflammatory effects and support immune homeostasis [6].

However, an imbalance in the gut microbiota, known as dysbiosis, can disrupt these protective mechanisms, leading to increased gut permeability, systemic inflammation [7], and impaired intestinal barrier structure and function [3]. Understanding the triangular relationship among diet, the gut microbiota, and inflammation is crucial for developing effective strategies to improve health outcomes. The interplay between these elements is mediated through several mechanistic pathways. Dietary components directly influence the composition and activity of the gut microbiota [8], which, in turn, produce metabolites that can modulate immune responses and inflammation. For instance, SCFAs produced from the fermentation of dietary fiber enhance the gut barrier function and regulate inflammatory pathways, while diets high in saturated fats can promote the growth of pro-inflammatory bacteria and increase endotoxin levels, leading to systemic inflammation [9].

This review aims to synthesize the current knowledge on the diet–gut microbiota–inflammation axis, providing insights into how dietary interventions can modulate the gut microbiota composition and function, and, consequently, influence inflammatory processes. By exploring the mechanistic pathways and therapeutic potential of this interconnected triad, we hope to contribute to the development of effective strategies for the prevention and management of chronic inflammatory diseases.

2. Gut Microbiome: General Concepts of Diet and Gut Microbiota

The microbial community of the gut microbiome is extremely complicated. While there are diverse microbial communities in more proximal parts of the gastrointestinal tract, the colon has the highest biomass [10]. The human gut contains 10 times the number of microbial cells of the human body as a whole; these bacteria weigh about 2 kg, represent up to 5000 different species, and total about 100 trillion [11,12]. The composition of the gut microbiota comprises fungus, bacteria, viruses, and parasites [1,11]. Moreover, *Prevotella*, *Ruminococcus*, Bacteroidetes, and Firmicutes are among the most prominent kinds of bacteria [13]. Firmicutes are the most prevalent in the average adult, followed by Bacteroidetes and Actinobacteria [14]. *Eubacterium*, *Ruminococcus*, and *Clostridium* are the three subgroups of Firmicutes. It has been demonstrated that the proportion of the bacterial species Firmicutes to Bacteroidetes (F/B) affects both health and disease [15]. There is mounting evidence that suggests that nonbacterial components may also be involved in health and disease, despite the majority of studies being on bacteria [16]. The microbiota inoculate humans for the first time during birth. It has been demonstrated that a variety of factors, including the method of delivery (vaginal versus caesarian section), food (breast versus formula feeding), and use of antibiotics, affect the composition of an infant's gut microbiota [17]. During the first few years of life, the microbiota quickly diversify [18]. They also play a significant role in the development of the immune system [19], providing defense against pathogenic microorganisms and bacterial overgrowth. In adults, the microbiota play a role in a number of processes, including bone density modulation [20], the modification and elimination of particular toxins and medications [21], and the optimization of intestinal barrier function [22]. The gut microbiome's bacteria are involved in producing neurotransmitters like serotonin, vitamins, and enzymes, as well as obtaining energy from meals and maintaining a balance between an opportunistic and helpful bacterial makeup. For example, the bacterially generated vitamin K is involved in immunological and metabolic processes. Therefore, disease may arise from an imbalance in bacterial species [23]. Numerous factors, including the host genetics and physiology, environmental exposures, age, and food, influence the makeup and function of the microbiota [24]. A shift in diet composition can have an impact on this ratio, because nutrition has been shown to be one of the factors that can change the bacterial composition the most [11]. The microbiota respond quickly (within 24 h) to changes in diet in both people and mice [25].

3. Diet and Gut Microbiota

3.1. Subsection Role of Diet in Shaping Gut Microbiota

Diet refers to the sum of foods and beverages consumed by an individual or group, playing a crucial role in maintaining health and well-being. It provides the essential nutrients required for various bodily functions, including energy production, growth, repair, and the maintenance of physiological processes. The composition of one's diet—encompassing macronutrients like carbohydrates, proteins, and fats, as well as micronutrients such as vitamins and minerals—significantly influences overall health outcomes [26].

Beyond providing essential nutrients, diet also has a profound impact on the gut microbiota, the diverse community of microorganisms living in the gastrointestinal tract. These microorganisms play key roles in digestion, immune function, and the production of essential metabolites. The interaction between diet and the gut microbiota is bidirectional: while diet shapes the composition and activity of the gut microbiota, the microbiota influence how nutrients and other dietary components are metabolized [27].

Different dietary patterns significantly influence health outcomes, particularly through their impact on the gut microbiota and inflammation [28]. The WD, characterized by high intakes of processed foods, red meats, sugars, and unhealthy fats, is often associated with adverse health effects, including a reduced microbial diversity and increased inflammation [29]. In contrast, the MD, rich in fruits, vegetables, whole grains, nuts, and olive oil, promotes a diverse and beneficial gut microbiota composition, contributing to anti-inflammatory effects and improved metabolic health [30]. Similarly, vegetarian and vegan diets, which emphasize plant-based foods, enhance gut health by increasing the abundance of fiber-fermenting bacteria and the production of anti-inflammatory metabolites [31]. Each of these dietary patterns offers unique insights into how dietary choices can shape the gut microbiota and influence inflammatory processes, thereby affecting overall health.

3.2. Effects of Dietary Patterns on Gut Microbiota

3.2.1. Mediterranean Diet

A varied diet rich in nuts, fruits, vegetables, and olive oils, with moderate fish, poultry, and wine intake and a minimal intake of processed and red meats, is the hallmark of the MD [32] (Figure 1). Dietary fiber content is high in multiple components. Because of its numerous food groups, ease of accessibility, and health advantages, it is becoming more and more popular among medical and non-medical personnel. The MD is the most recommended plant-based dietary pattern, with an abundance of evidence to support it. High levels of MD adherence have also been positively linked to changes in the microbiota and their metabolites [33], pointing to a possible role for the gut microbiota in the advantageous impacts of this dietary strategy.

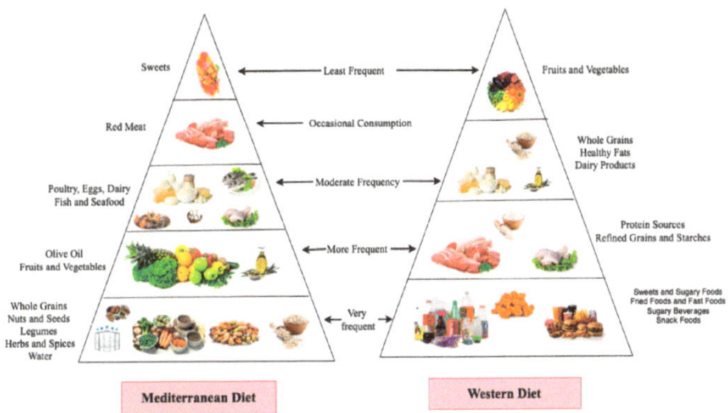

Figure 1. Mediterranean diet and Western diet.

Diverse effects have been observed in human research models investigating the influence of the MD on the gut microbiome. Numerous studies have reported increased *Bifidobacterium* abundance. *Prevotella, Bacteroides,* and *Enterococcus* were also elevated in one of these trials [34]. *Faecalibacterium prausnitzii, Roseburia,* and *Lachnospiraceae* were found to be more abundant in another study [35], while *Ruthenibacterium lactatiformans, Flavonifractor plautii, Parabacteroides merdae, Ruminococcus torques,* and *Ruminococcus gnavus* were found to be less abundant. Increases in Firmicutes and *Lactobacillus* [34] have been reported in other research. It has been discovered that some MD components individually change the composition of the gut microbiota. For instance, eating walnuts was linked to higher relative abundances of the *Eubacterium eligens* group, *Leuconostocaceae, Lachnospiraceae,* and *Roseburia* [36]. The study found an inverse correlation between the lipid profile, blood pressure, and the relative abundance of Lachnospiraceae. A higher abundance of the Clostridiales vadin BB60 group was linked to the monounsaturated oleic acid [36]. Significant increases in *Bifidobacterium* [37], *Lactobacillus,* and *Roseburia* [38] have been reported in response to polyunsaturated fatty acids (PUFAs). These findings are summarized in Table 1.

In a different study, researchers discovered that people following the MD had higher *Prevotella*-to-*Bacteroides* ratios, suggesting that a diet rich in natural fiber and resistant starch has a good effect on patients' bacterial composition [4]. A study that used a meal frequency questionnaire and a microbiota composition analysis was also carried out in a similar manner. Following completion, it was discovered that a greater F/B ratio was the result of poor diet adherence [2,39]. Individuals exhibited increased levels of SCFAs, *Bifidobacteria* counts, and Bacteroidetes when they adhered to the MD more closely [2].

Microbial metabolites provide some of the most compelling evidence for microbiota-mediated effects on human health. Tryptophan, an important amino acid included in many foods such as fish, chicken, dairy products, and grains [40], is one example of how it is metabolized by microbes in the MD. The microbiota have the ability to metabolize tryptophan into tiny compounds that can bind to the aryl hydrocarbon receptor (AhR) and cause enteroendocrine cells to secrete more glucagon-like peptide 1 (GLP-1) [40]. A decrease in AhR ligands and GLP-1 release, as well as compromised intestinal barrier function in animal models, have been linked to metabolic syndrome in both human and animal investigations [41]. The rise in microbiome gene representation linked to SCFA generation with the MD [35] is another example of the microbial-mediated influence of food; this is consistent with reported increases in fecal SCFAs [17].

A 12-month MD intervention was conducted in 612 non-frail or pre-frail subjects from five European countries (UK, France, The Netherlands, Italy, and Poland) as part of a study by Ghosh et al., which examined the effects of diet on inflammatory markers. There was a negative correlation found between the inflammatory markers of CPR, interleukin (IL)-17, and IL-2 and diet adherence. Positive levels of the anti-inflammatory cytokine IL-10 were also seen. Following the MD has been linked to several health benefits, such as an increased production of SCFAs and anti-inflammatory qualities that lower the risk of long-term inflammatory disorders like Type 2 diabetes mellitus (T2DM) [42].

3.2.2. Western Diet

Reduced amounts of fruits and vegetables and high amounts of animal proteins, refined carbohydrates, and sugars—all heavy in calories—are characteristics of the WD [43] (Figure 1). Dietary fiber has a major influence on the microbiota, yet a normal WD is poor in it [44]. The bacterial diversity in the gut microbiome is impacted by a WD that increases *Bacteroides* spp., *Alistipes* spp., and *Bilophila* spp. while decreasing *Lactobacillus* spp., *Roseburia* spp., and *E. rectale*, the beneficial bacteria [2]. In a recent study, mice fed with a high-fat, high-sugar diet or a low-fat, high-sugar diet were used. The results showed that the mice fed with the high-fat, high-sugar diet had more Firmicutes and Mollicutes and fewer Bacteroidetes [45]. Researchers carried out an experiment on mice in a different study. The mice were randomly allocated to receive either a high-fat diet (HFD) or a regular chow diet. The researchers observed comparable changes in the composition of the gut microbiota. This confirmed earlier findings that the high-fat diet group included significant concentrations of Firmicutes and Proteobacteria. Furthermore, the HFD group also had greater levels of *Enterobacteriaceae*, *Escherichia*, *Klebsiella*, and *Shigella* according to this study. Overall, these studies' findings indicate a relationship between alterations in the gut microbiota resulting from an HFD: a decrease in Bacteroidetes and an increase in Firmicutes [9]. A recent study by Suriyano et al. highlighted the significant impact of dietary macronutrients, specifically sugar and fat, on fecal bacterial counts and quantitative microbiome profiling in mice. The researchers found that fat, in particular, is a key factor driving changes in the gut microbiota composition. These alterations in gut bacteria are linked to the progression of obesity, diabetes, and local inflammation in various body tissues. The study emphasized that the type of macronutrients consumed can profoundly influence the gut microbiota, which play a critical role in the development of metabolic diseases and associated inflammatory conditions [46].

A WD is characterized by high quantities of additives produced from food processing, such as synthetic emulsifiers, and lower levels of natural foods containing dietary fiber. These emulsifiers have been demonstrated to modify the gut microbiota in mice, decrease the thickness of the mucus barrier, and encourage intestinal inflammation and metabolic disorders [47]. In a recent human randomized controlled clinical trial, microbial diversity, SCFA production, and free amino acid levels were observed to be reduced by dietary emulsifier supplementation during a brief dietary intervention in healthy participants. More specifically, there was an increase in *Lachnospiraceae* and *Roseburia* and a drop in *F. prausnitzii* and *Ruminococcus* levels. Lastly, some patients, but not all of them, experienced a decrease in their intestinal mucus thickness, even during a brief intervention [48].

High levels of animal-based proteins, which have the ability to alter the makeup and possible function of the gut microbiota, are another feature of a WD. Specifically, it has been observed that individuals following a WD had higher levels of trimethylamine N-oxide (TMAO) due to microbiota-dependent transformations converting l-carnitine and phosphatidylcholine, which are common in red meats [49]. Furthermore, high concentrations of ultra-processed foods and dangerous dietary additives, such as emulsifiers and artificial sweeteners, may contribute to the detrimental effects of the WD on the gut flora [50]. Junk foods that are characteristic of the WD are rich in artificial substances and preservatives, low in fiber, and can cause detrimental effects to the gut microbiota. These foods favor the proliferation of pathogenic bacteria and, acting at the same time, suppress beneficial bacteria, disrupting the microbiome's balance and weakening the integrity of the intestinal barrier. The environment that highly processed meals create in the gut appears to alter the microbiota in a way that supports a variety of inflammatory diseases, such as metabolic disorders, IBD, and obesity [51].

Also, a WD is very limited in terms of the consumption of whole grain, which provides part of the daily needed dietary fiber and nutrients. Whole grains are essential contributors to a healthy and diverse gut microbiome because they act as prebiotics, stimulating the production of SCFAs beneficial for gut and general health. A lack of whole grains in the diet considerably reduces the number of good microbes and the production of SCFAs, which leads to inflammation. A randomized controlled trial by Vanegas et al. examined the effects of whole grain consumption on the gut microbiota in humans. Participants who consumed a diet rich in whole grains, such as oats and barley, showed a significant increase in the abundance of beneficial gut bacteria, particularly *Bifidobacterium* and *Lactobacillus*. In contrast, those who consumed refined grains exhibited a decrease in microbial diversity and a reduction in the production of short-chain fatty acids (SCFAs), which are crucial for maintaining gut health and reducing inflammation [52].

Additionally, a low consumption of fruits and vegetables is another significant factor. These foods are vital sources of vitamins, minerals, fiber, and phytochemicals that support gut health. Fruits and vegetables contribute to a balanced gut microbiome by fostering microbial diversity and inhibiting the growth of pathogenic bacteria. Their antioxidants and anti-inflammatory properties further help to protect the gut lining and promote overall digestive health. A study by Lakshmanan et al. explored the impact of fruit and vegetable intake on the gut microbiota in a cohort of healthy adults. The researchers found that fruits and vegetables can provide long-term health benefits by controlling the relative abundance of bacteria such as L-*Ruminococcus* and unclassified bacteria from the Erysipelotrichaceae family, possibly through a reduction in the pro-inflammatory response [53]. These findings are highlighted in Table 1.

3.2.3. Vegetarian Diet

Vegetables, fruits, grains, legumes, and nuts make up the majority of a vegetarian diet, however, dairy and eggs are occasionally included [54]. Numerous food groups that are often included in a healthy vegetarian diet, such as those high in fiber, unsaturated fatty acids, and polyphenols, are also included in the MD. The saturated fatty acids obtained from animals are reduced as a result of not consuming red meat. A vegetarian diet has been linked to a decrease in the metabolites related to CVDs [49], such as acylcarnitine metabolites and l-carnitine [54]. These elements contribute to the explanation of some of the health advantages of a vegetarian diet, such as weight and cholesterol reductions [55]. A shift in the makeup of the gut microbiota, including a rise in alpha diversity, has also been linked to the vegetarian diet [56]. In particular, a vegetarian diet has been linked to higher relative abundances of the taxa that produce SCFAs, namely *Eubacterium biforme*, *F. prausnitzii*, *Eubacterium rectale* [56], and *Akkermansia* [54]. It has been shown that *Akkermansia* contributes to the preservation of the epithelial energy balance and integrity in the colons of animal models [57]. It is plausible that alterations in the microbiota linked to a vegetarian diet impact human health; however, more human data are required to substantiate a direct causal relationship.

According to the previous findings, high levels of *Prevotella* species have been associated with plant-based diets [58]. An animal study that used mice deficient in the miR-146a gene examined the effects of a plant-based diet rich in miR-146a on the microbial community. The study revealed that the microbiomes of the mice fed with the plant-based diet differed significantly from the microbiomes of the mice fed with the control diet, leading the researchers to conclude that a shift in microbial communities occurs when dietary fiber levels rise. For the mice consuming the plant-based diet, they discovered that *Bacteroides* and *Alloprevotella* significantly increased and Porphyromonadaceae and Erysipelotrichaceae decreased [59]. A human diet intervention study was carried out to investigate the bacterial composition based on what the participants indicated as their regular diet, in an effort to support the theory that a high-fiber diet alters the bacterial makeup. *Prevotella* enrichment was observed in ninety-eight vegetarian patients, but *Bacteroides* enrichment was observed in the microbiome environments of those who followed a conventional WD. The researchers also noticed that when 10 of the participants swapped diets, their microbiome compositions were altered within 24 h of the ingestion of the other diet [11]. Another related study produced comparable results. Additionally, they discovered that, in comparison to non-vegetarians, the vegetarian participants' microbiomes were enhanced with *Prevotella* [60]. These key findings are highlighted in Table 1.

Table 1. Impact of dietary patterns on gut microbiota in relation to depression.

Dietary Pattern	Key Components	Impact on Gut Microbiome	Health Outcome	References
MD	- ↑ fruits, vegetables, whole grains, legumes, nuts, and olive oil - moderate consumption of fish and poultry - ↓ red meat and processed foods - ↑ fiber, polyphenols, and omega-3 fatty acids	- ↑ microbial diversity - ↑ growth of beneficial bacteria, e.g., *Bifidobacterium*, *Lactobacillus* - ↑ levels of *Faecalibacterium prausnitzii*, linked to anti-inflammatory effects - ↑ production of SCFAs	**Reduced risk of depression:** - ↑ gut health - ↑ neurotransmitter-modulating bacteria - ↓ inflammation - ↓ inflammatory markers - ↓ obesity and diabetes - gut–brain axis modulation	[2,33,34,36,39]
WD	- ↑ red and processed meats, refined grains, high-fat dairy, and sugars - ↑ processed foods, artificial additives - ↓ fiber, fruits, vegetables, and whole grains	- ↓ microbial diversity - ↑ growth of pro-inflammatory bacteria, e.g., *Proteobacteria*, *Bacteroides* - ↓ levels of SCFA-producing bacteria, e.g., *Roseburia*, *Faecalibacterium*	**Increased risk of depression:** - ↑ pro-inflammatory cytokines - gut–brain axis disruption - ↑ metabolic syndrome, obesity, and inflammation - ↑ prevalence of gut dysbiosis and leaky gut syndrome	[2,9,45,47,49,51]
Vegetarian	- ↑ fruits, vegetables, whole grains, legumes, nuts, and seeds - no meat; some variations allow dairy and eggs - rich in fiber, phytochemicals, and antioxidants - ↓ saturated fats and cholesterol	- ↑ microbial diversity route - ↑ growth of beneficial bacteria, e.g., *Bifidobacterium*, *Prevotella* - ↑ levels of SCFAs, particularly butyrate, linked to gut health - ↓ levels of bile-tolerant bacteria, which are linked to inflammatory responses	**Reduced risk of depression:** - improved gut health; reduced inflammation - lower risk of cardiovascular diseases, obesity, and type 2 diabetes - enhanced protection against certain cancers - lower inflammation and improved insulin sensitivity	[11,49,55,57,58]

↑, increase; ↓, decrease.

3.3. Impact of Macronutrients on Gut Microbiota Composition

3.3.1. Carbohydrates

Food ingredients have a significant effect on the gut microbiota, affecting their composition in terms of richness and diversity. Consuming a lot of animal proteins, saturated fats, sugars, and salt may encourage the growth of pathogenic bacteria at the expense of beneficial bacteria, but eating complex polysaccharides and vegetable proteins may also increase the number of beneficial bacteria in the body [50]. Table 2 highlights the role of macronutrients in the gut flora.

Carbohydrates are primarily used by the body as a source of energy, and they significantly influence the composition and function of the gut microbiota. The gut microbiota, in turn, aid in the digestion of certain carbohydrates. The interaction between dietary carbohydrates and the gut microbiota plays a crucial role in shaping and altering the microbiota, affecting overall gut health and metabolic processes.

Dietary fiber is categorized as carbohydrates that are unable to be digested in the upper gastrointestinal system. They are not absorbed in the small intestine because they are not broken down by the host's digestive system. The gut microbiota's fermentation of carbohydrates is necessary for the breakdown of fibers. According to Rinninella et al., non-digestible oligosaccharides such as raffinose, stachyose, oligofructose, and inulin, resistant starches, and non-starch polysaccharides such as cellulose, hemicellulose, glucans, gums, and pectin make up fibers. Through saccharolytic fermentation, which is carried out by gut microbes in the colon, they produce monosaccharides, certain gases (methane and carbon dioxide), and SCFAs like butyrate, acetate, and propionate [61]. Dietary fiber consumption and the makeup of the gut bacteria influence the types and quantities of SCFAs that are present. Non-digestible complex carbohydrates have the ability to release particular microbial metabolites and encourage the growth of a wider range of microbial families and species with particular traits. Known also as microbiota-accessible carbohydrates

(MACs), the primary sources of these carbohydrates include cereals, some fruits and vegetables, and human milk (human milk oligosaccharides) when a person is breastfeeding. Bananas, chicory root, onions, garlic, artichokes, asparagus, and other foods include MACs such as oligofructose and inulin [50]. Furthermore, dietary fiber possesses various physicochemical properties that significantly influence the composition and functionality of the gut microbiota. One key property is its solubility, which allows soluble fibers to form viscous gels in the gut, facilitating the fermentation process by beneficial bacteria, thus increasing the production of SCFAs [62].

There is growing interest in utilizing these carbohydrates as prebiotics to support gut health. Prebiotics promote the growth of beneficial bacteria such as *Lactobacillus* and *Faecalibacterium prausnitzii*. These bacteria metabolize prebiotics into SCFAs, which have various positive effects on gut health, including enhancing gut barrier function, reducing inflammation, and providing energy to colon cells. Consequently, these impacts enhance the function of the intestinal barrier, elevate insulin sensitivity, and improve the lipid profile [63]. According to Singh et al., inulin dose-dependently changed the makeup of the fecal microbiota in male rats fed with an HFD by suppressing the growth of Firmicutes phyla (such as *Roseburia* and *Clostridium* clusters I, IV, and XIV) and boosting the number of *Bifidobacterium* species and Bacteroidetes [64].

It has been demonstrated in animal models that the WD, which has a comparatively lower fiber content, decreases the variety of the gut microbiota and the quantity of *Bifidobacterium* [65]. Reducing the F/B ratio and raising the quantity of *Lactobacillus* sp. in rats fed with a high-fat, sucrose-enriched diet have been demonstrated to ameliorate gut dysbiosis [66]. Furthermore, MAC intake may be able to slow down the spread of the opportunistic diarrheal bacterium *Clostridium difficile*, according to a new mouse model [67]. Remarkably, a high-fat-MAC diet ameliorated cognitive deficits in a mouse model by influencing the gut microbiota–brain axis, which is triggered by high-fat food intake [68].

In order to disrupt their gut flora, humanized mice were given a high-fiber diet in a preclinical experiment before being given a low-quality feed. However, providing the studied animals with a plant polysaccharide-rich diet with a 15% weight-neutral detergent fiber content did not restore their microbial diversity and composition. Furthermore, it was noted that this disruption persisted for a number of generations [13]. Studies conducted in clinical trials have repeatedly shown that a high-fiber diet intervention increases the abundance of various beneficial bacteria in the feces, including *Lactobacillus* sp., *Prevotella* sp., and *Bifidobacterium* sp. [69]. In a randomized, double-blind clinical trial, galacto-oligosaccharides have been shown to significantly increase *Bifidobacterium* sp. and *Lactobacillus* sp. after four weeks of treatment [70].

3.3.2. Fat

Microbial species modify the host's lipid profile and obesity by metabolizing ingested fatty acids into different fatty acids [57]. Increased insulin resistance, intestinal permeability, and the inflammation of adipose tissue are the outcomes of consuming an HFD, particularly one high in saturated fat acids, which causes dysbiosis with a drop in Bacteroidetes and a rise in Firmicutes and Proteobacteria [71].

According to Wolters et al. there may be a positive correlation between the Enterobacteriaceae family, *Prevotella*, *Turicibacter*, and *Parabacteroides* genera and monounsaturated fatty acids (MUFAs) (palmitoleic, oleic, and eicosenoic) [72]. Conversely, in both healthy and ill models, including people at risk of metabolic syndrome, a diet high in MUFAs—rich in sesame, pumpkin seeds, rapeseed, extra virgin olive oil, and peanuts—showed favorable health effects with an enhanced variety of gut flora. Virgin coconut oil, human milk, and baby formulas all include medium-chain fatty acids (MCFAs), which may promote the growth of *Lactobacillus* and *Bifidobacterium* to improve metabolic and cognitive processes [73]. Medium-chain triglycerides (MCTs) enhance the gut microbial balance and gut barrier integrity, which, in turn, promotes energy expenditure, weight reduction, and lipid catabolism. However, diets high in coconut oil may raise the ratio of F/B, *Allobaculum*,

Clostridium, *Lactobacillus*, and *Staphylococcus*, which may lead to metabolic problems and the inflammation of adipose tissue [61]. Since the human body is unable to produce PUFAs and must receive them from food, these fatty acids are referred to as "essential fatty acids". PUFAs are mostly found in sunflower oil, nuts, seeds, and fatty fish. By raising the number of butyrate-producing Lachnospiraceae taxa and restoring the F/B ratio, omega-3 PUFAs can have a beneficial effect [61].

Research has shown that the amount and type of dietary fat significantly impact the composition of the intestinal microbiota [13]. Additionally, the abundance of sulfate-reducing bacteria (SRB) can disrupt the intestinal barrier by breaking down the disulfide bonds in mucus, causing defects and increasing intestinal inflammation [50]. A recent review demonstrated that MUFAs do not affect richness or diversity indices, the distribution of phyla, or the ratio between Bacteroidetes and Firmicutes [8]. However, at the family and genus levels, diets high in monounsaturated fatty acids, such as those found in the MD, may positively correlate with the genera *Parabacteroides*, *Prevotella*, and *Turicibacter*, as well as the family Enterobacteriaceae. Conversely, these diets may negatively correlate with the genus *Bifidobacterium*.

It has been repeatedly demonstrated that a diet high in saturated and/or total fat negatively impacts the gut microbiota. Diets high in total and saturated fats have been found to have a detrimental impact on the richness and diversity of the gut microbiota, according to fifteen clinical reports (six of which were randomized controlled interventional trials and nine of which were observational studies) [72]. Carefully monitored feeding trials in mice confirmed these results, demonstrating that diets with fat ranging from 44% to 72% raised the gut microbiota's F/B ratio [74]. In a recent controlled-feeding clinical trial, 40% fat consumption by young adults in good health was found to be linked to unfavorable changes in the gut microbiota. Specifically, the intervention led to a decrease in the abundance of beneficial *Fecalibacterium* bacteria and an increase in harmful *Bacteroides* and Alistipes species, which are known to be prevalent in patients with T2DM. On the other hand, 20% fat consumption increased the amount of *Fecalibacterium* and *Blautia* spp. in the gut microbiota [75].

According to Yang et al., unsaturated fat enhances the abundance of *Akkermansia* and *Bifidobacterium* and decreases harmful bacteria like *Streptococcus* and *Escherichia* sp. Conversely, saturated fat consistently lowers health-beneficial microbes like *Bifidobacterium* and *Fecalibacterium*. Furthermore, depending on the fat quality, saturated and unsaturated fats might have different effects on human health. Saturated fat can raise the F/B ratio, whereas unsaturated fat can lower it [13].

3.3.3. Proteins

Proteins are linear chains of amino acids joined by peptide bonds. In the distal colon, the primary microbial phyla, Firmicutes, Bacteroidetes, and Proteobacteria, ferment amino acids. Proteolytic fermentation has the ability to produce branched-chain fatty acids like isobutyrate, 2-methyl butyrate, and isovalerate, as well as potentially toxic substrates like ammonia, nitrosamines, and TMAO in smaller amounts than saccharolytic fermentation [61]. The amount and quality of the dietary proteins that the gut microbiota digest, especially those with plant or animal origins, determine the balance of the gut microbiota populations and the generation of these metabolites. An increased amount of bile-tolerant anaerobic bacteria, such as *Bacteroides*, *Alistipes*, and *Bilophila*, may result from a diet high in animal proteins, such as red meat and dairy products. This could increase the amount of TMAO, a substance with proatherogenic potential that may contribute to CVD. More specifically, animal diets like red meat include large amounts of the amino acid L-carnitine, which may be a key factor in the elevated risk of CVD [76,77].

A traditional WD that includes a high amount of animal-based proteins may promote the growth of SRB, such as *Desulfovibrio* spp., which produces hydrogen sulfide from dietary inorganic sulfur and sulfated amino acids, such as taurine, methionine, and cysteine. This could lead to an increase in gut inflammation [78]. Conversely, consuming plant-based proteins like glycated pea proteins may lead to a rise in the number of good bacteria like *Lactobacillus* and *Bifidobacterium* and a drop in the number of *Bacteroides fragilis* and *Clostridium perfringens* [79]. Pulses, which mostly consist of lentils, beans, chickpeas, and peas, have garnered more interest in recent years due to their potential as a sustainable alternative to animal proteins. In fact, eating pulses has been linked to favorable changes in the gut flora in both humans and animals, including enhanced development of the butyrate- and acetate-producing species *Bifidobacterium, Faecalibacterium, Clostridium, Eubacterium,* and *Roseburia* [80]. To cut down on gut-inflammation-related proteins like cysteine and methionine, a diet high in plant-based proteins, such as pulses, is an excellent substitute for animal proteins. Moreover, they are rich in bioactive substances and resistant starches, both of which have a well-established beneficial effect on the gut microbial balance [81].

Data from animal models indicate that the makeup of the gut microbiota is influenced by the quality of the proteins. For instance, a preclinical study [82] revealed that cheese whey proteins, as opposed to casein, might function as growth factors for the fecal counts of *Lactobacilli* and *Bifidobacteria.* Mice with a HFD-induced F/B ratio were demonstrated to respond differently to mung bean protein. In a model of HFD mice, the mung bean protein also boosted the abundance of the Ruminococcacea family. The authors postulated that the Ruminococcacea family members' bile acid (BA) metabolism would have benefited the health of the HFD mice in light of this result [83].

3.4. *Impact of Micronutrients and Phytochemicals on Gut Microbiota Composition*

3.4.1. Vitamins

Vitamins are organic compounds that are essential in very small amounts for supporting normal physiological function. They frequently perform a number of functions in the body, the most significant of which is acting as cofactors for enzymes. Since our bodies cannot produce enough of certain vitamins to satisfy our daily needs, the diet is the main source of these nutrients. However, other vitamins, such as vitamin K and B group vitamins, are produced by the gut bacteria. They are necessary for immune system function, cell development and differentiation, and energy metabolism control. The gut microbiota are capable of producing thiamine, riboflavin, niacin, biotin, pantothenic acid, folate, and vitamin K, among other vitamins. Moreover, a number of studies have demonstrated that vitamin D may influence the microbiota's makeup, altering it and raising the number of potentially advantageous bacterial strains [61]. The gut microbiota may be influenced by antioxidants like carotenoids. According to recent research, lutein derived from black currants has been proven to both decrease Bacteroidetes and *Clostridium* spp. and increase bifidobacteria and lactobacilli. However, the gut microbiota mediate the anti-inflammatory benefits of beta-carotene [8].

3.4.2. Minerals

Essential micronutrients such as minerals and trace elements are involved in human metabolism and actively interact with the gut flora [84]. Human diseases can result from both excesses and shortages of certain micronutrients. For example, a lean phenotype has been linked to alterations in the gut microbiota that have been linked to high calcium consumption [85]. In a human trial, an eight-week course of 1000 mg of calcium per day increased the amount of *Clostridium* XVIII in men's fecal samples [86].

The effects of iron supplementation, which is frequently used to treat iron insufficiency, on the gut flora have been inconsistent. However, following iron consumption, several studies have consistently shown increases in harmful bacteria and decreases in good bacteria [87]. According to research on animals, excessive iron can lead to intestinal dysbiosis, which can decrease the number of some Lachnospiraceae family members and the genus *Allobaculum*, while increasing the number of bacteria from the Defluviitaleaceae, Ruminococcaceae, and Coprococcus families [88]. These results have been corroborated by in vitro investigations, which have demonstrated that elevated iron concentrations can raise harmful metabolites, reduce commensal bacteria, and boost pathogenic bacteria's pathogenicity [89].

Another important element that preserves the epithelial integrity is zinc, which may act by influencing the good gut microbiota [90]. According to Yang et al., grill chickens with a chronic zinc shortage experienced a considerable rise in Proteobacteria abundance and a decrease in Firmicutes. There are, however, insufficient clinical studies on the effects of dietary zinc on the human gut flora [13]. The impact of iodine supplementation on the gut flora seems to be influenced by the amount of fat in the diet. Iodine corrected the thyroid hormone status in an HFD animal model, but it also led to gut dysbiosis, which was characterized by a decrease in helpful bacteria like *Fecalibacterium prausnitzii* and an increase in dangerous bacteria [91]. On the other hand, the same iodine dose boosted good bacteria such as *Bifidobacterium*, *Lactobacillus*, *Fecalibacterium*, and *Allobaculum* in the setting of a low-fat diet (LFD) [91].

In conclusion, research on the effects of micronutrient shortage or supplementation is the main focus of current studies, even though there is little information on the precise processes by which minerals and trace elements affect the gut microbiota. It has been demonstrated that in experimental animals, calcium influences the gut microbiota, including *Ruminococcacea*, *Bifidobacterium*, and *Akkermansia*, as well as the ratio of Bacteroidetes to *Prevotella*. It is unknown how magnesium affects the gut microbiome, although some research has indicated that a sudden lack of the mineral may alter the balance of good gut bacteria. While some clinical trials have found no impact, iron supplementation generally increases harmful bacteria and lowers helpful bacteria, such as *Bifidobacterium*, in babies. Iron administration routes and chemical forms seem to have important effects. Table 2 highlights the roles of micronutrients in the gut flora. There is a clear need for more comprehensive preclinical and human intervention studies to fully understand the role of specific minerals and trace elements in modulating the gut microbiota [13].

Table 2. Impact of dietary nutrients on gut microbiota.

Nutrients	Dose and Treatment Duration/Test Model	Potentially Beneficial Microbiota	Potentially Detrimental Microbiota	Reference
Dietary carbohydrates	Model: almond-based low carbohydrate diet as reference/clinical trials (45 T2DM patients) Duration: 3 months Age: >18 years Location: Hospital of Soochow University, Suzhou, China	↑ *Roseburia* sp. ↑ *Ruminococcus* ↑ *Eubacterium*		[92]
	Model: oligofructose-enriched inulin (Synergy 1) (10 g/day), n = 18 and placebo (maltodextrin; 7 g/day), n = 16 as reference/clinical trials (34 pediatric celiac disease patients, 62% females, on a gluten-free diet) Duration: 3 months Age: mean age 10 years Location: University of Warmia and Mazury, Olsztyn, Poland	↑ *Bifidobacterium* sp.		[93]

Table 2. *Cont.*

Nutrients	Dose and Treatment Duration/Test Model	Potentially Beneficial Microbiota	Potentially Detrimental Microbiota	Reference
Dietary fat	Model: HFD (60% fat) and baseline as reference/mice model, n = 8 Duration: 14 weeks Age: 4 to 5 weeks Location: Jiangnan University, Wuxi, China	↓ *Bifidobacterium* sp. ↓ *Lactobacillus* sp. ↓ *Faecalibaculum*, sp. ↓ *Akkermansia.* sp.	↑ *Desulfovibrionaceae* sp. ↑ *Mucispirillum* sp.	[94]
	Model: fat diet (20%) and the baseline as reference/clinical trial (52% women) n = 217 Duration: 6 months Age: 18 to 35 years Location: Army General Hospital, north China and Zhejiang University, south China	↑ *Fecalibacterium* sp. ↑ *Parabacteroides* sp.		[75]
	Model: moderate-fat diet (30%) and the baseline as reference/clinical trial (52% women) n = 217 Duration: 6 months Age: 18 to 35 years Location: Army General Hospital, north China and Zhejiang University, south China	↓ F/B		[75]
	Model: HFD (40%) and the baseline as reference/clinical trial (52% women) n = 217 Duration: 6 months Age: 18 to 35 years Location: Army General Hospital, north China and Zhejiang University, south China	↓ F/B ↓ *Fecalibacterium* sp.	↑ *Alistipes* sp. ↑ *Bacteroides* sp.	[75]
Dietary protein	Model: high-animal-protein-based diet (514 g/kg) and baseline as reference/mice model Duration: 3 weeks Age: shortly after weaning Location: Czech academy of sciences, Prague		↑ *Escherichia* sp. ↑ *Staphylococcus* sp. ↑ *Enterococcus* sp.	[95]
	Model: plant-protein-based control diet (176 g/kg) and baseline as reference/mice model Duration: 3 weeks Age: shortly after weaning Location: Czech academy of sciences, Prague		↑ *Enterococcus* sp.	[95]
	Model: C57BL/6 DSS-treated mice with isocaloric diets with 53% protein and the diets with 30% protein as reference/DDS-treated mice model, n = 132 Duration: 3 days Age: 7 weeks Location: Université Paris-Saclay, France	↑ *Desulfovibrio* sp. ↑ *Bacteroides* sp.	↑ *Alloprevotella* sp. ↑ *Haemophilus* sp. ↑ *Klebsiella* sp.	[96]
Minerals	Model: supplementation of 1000 mg calcium + 1000 mg phosphorus/day and the supplementation of 1000 mg phosphorus/day as reference/clinical trials (men n = 30, women n = 32), Duration: 8 weeks Age: 29 ± 7 years Location: University Jena, Germany	↑ *Clostridium* sp.		[97]
	Model: high-iron-fortified formula (6.4 mg Fe/day) and iron drops (no-added-iron formula with liquid ferrous sulfate supplementation (5.7 mg Fe/day) as reference/clinical trials, n = 53 Duration: 45 days Age: 6 months Location: Sweden	↑ *Lactobacillus* sp.	↓ *Ruminococcus* sp.	[98]
	Model: 18 µg/day iodine and control group as reference/mice model, n = 24 Duration: 8 weeks Age: 3 weeks Location: Ningbo University, China	↑ *Bifidobacterium* sp. ↑ *Lactobacillus* sp. ↑ *Fecalibacterium* sp. ↑ *Allobaculum* sp. ↑ *Roseburia* sp.		[91]

Table 2. *Cont.*

Nutrients	Dose and Treatment Duration/Test Model	Potentially Beneficial Microbiota	Potentially Detrimental Microbiota	Reference
Vitamins	Model: one dose of 50,000 IU vitamin A and placebo as reference/clinical trial, Duration: 15 weeks Age: 48 h of birth for IU treatment, placebo–early (6–15 week) or late (2 year) infancy Location: Dhaka, Bangladesh	↑ *Bifidobacterium* sp. ↑ *Akkermansia* sp.		[99]
	Model: a dose of 40,000 IU vitamin D once weekly using two capsules of 20,000 IU (Plenachol, Encap) and baseline as reference/clinical trial (patients with vitamin D deficiency: 25[OH] vitamin D < 50 nmol/L) Duration: 1 month Age: NA Location: St Mark's Hospital, London, UK	↑ *Enterobacteria* sp.		[100]
Polyphenols	Model: HFD with blackberry anthocyanin rich extract (25 mg/kg body weight per day) and HFD as reference/rat model n = 6 Duration: 17 weeks Age: NA Location: NA	↑ *Akkermansia* sp.	↓ *Ruminococcus* sp.	[101]
	Model: tart cherry juice consumption (8 oz/day) and baseline as reference/clinical trials, n = 10 Duration: 5 days Age: 23–30 years Location: NA	↑ *Bifidobacterium* sp. ↑ *Prevotella* sp. ↑ *Bacteroides* sp.	↓ *Ruminococcus* sp.	[102]
	Model: different concentrations of grape phenolic compounds (2.5 and 5 mg/kg/day diluted in 0.1% Dimethyl Sulfoxide) and the control group (0.1% Dimethyl Sulfoxide alone) as reference/rat, n = 6 Duration: 2 months Age: NA Location: NA	↑ *Bifidobacterium* sp.		[103]

↑, increase; ↓, decrease; NA, not applicable.

3.5. Impact of Specific Dietary Components on Gut Microbiota Composition

3.5.1. Prebiotics

Prebiotics have been recognized as dietary components that host beneficial bacteria preferentially used to give a health advantage [104]. The majority of prebiotic substances are naturally occurring carbohydrates with a range of molecular structures found in food. Prebiotics include lactulose, inulin, galactooligosaccharides, and fructooligosaccharides. They are present in a variety of foods, including bananas, onions, garlic, and whole grains. In addition to carbohydrates, other compounds that may have prebiotic effects include polyphenols and PUFAs [8].

Prebiotics mainly function by giving beneficial microorganisms in the stomach a source of food. Since human enzymes cannot break them down, they pass through to the colon undigested, where the gut flora ferment them [105]. Some microorganisms, like those in the genus Bacteroides, can use high-molecular-weight polysaccharides, but others, like bifidobacteria, can metabolize low-molecular-weight carbohydrates very effectively, because they have a variety of cellular and extracellular glycosidases and specific transport systems [8]. During this fermentation process, SCFAs such as butyrate, propionate, and acetate are produced [106]. Prebiotics aid in maintaining a healthy gut microbiota composition by reducing the number of harmful bacteria and specifically promoting the development and activity of good gut bacteria like *Lactobacilli* and *Bifidobacteria*. Concurrently generated metabolites can boost gut barrier function, control the immune system, supply energy to the colon's lining cells, and even affect brain function via the gut–brain axis [107]. For instance, oligofructose increases the expression of many genes involved in mucus formation, glycosylation, and secretion, the expression of both transmembrane and secreted mucins, and the differentiation and quantity of goblet cells, preventing HFD-induced obesity in mice. These findings were linked to notable alterations in the composition of the gut microbiota, with oligofructose considerably raising the abundance of the bacterial taxa *Ruminococcaceae*, *Odoribacter*, *Akkermansia*, and two undiscovered *Muribaculaceae*. It is interesting to note that all of these bacterial species had a positive correlation with mucus layer indicators and a negative correlation with metabolic parameters [108].

3.5.2. Probiotics

Probiotics were defined as "live microorganisms which, when administered in adequate amounts, confer a health benefit on the host" by the Food and Agriculture Organization of the United Nations and the World Health Organization in 2001. This definition is currently the most widely accepted [105]. Probiotics are diverse organisms derived from many families of bacteria and yeasts. Along with the yeast *Saccharomyces cerevisiae*, some of the most well-known species of bacteria utilized as probiotics include *Lactobacillus*, *Bifidobacterium*, *Enterococcus*, and *Streptococcus*. There are several species in each genus, and there are numerous strains within each species [105]. Probiotics are typically thought to provide strain-specific health benefits. Probiotics work through a variety of intricate processes that vary according to the strain. Nevertheless, a few shared mechanisms have been identified. Probiotics can change the makeup of the gut microbiota, improve the operation of the intestinal barrier, modify the immune system, and compete with pathogens for nutrients and binding sites on the intestinal wall [8].

Additionally, they have the ability to produce metabolites and antimicrobial compounds that may have an impact on host health either directly or indirectly. By interacting through the gut–brain axis, they can affect the host's neurological system [109]. Probiotics have the previously mentioned capacity to alter the makeup and/or activity of the gut microbiota, which may aid in the prevention or treatment of a number of illnesses, including metabolic syndrome, irritable bowel syndrome (IBS), and IBD [105].

3.5.3. Polyphenols

Polyphenols include flavonoids, phenolic acids, stilbenes, and lignans from fruits, vegetables, cereals, tea, coffee, and wine [110]. Polyphenols are being increasingly recognized for their role in preventing diseases such as diabetes and obesity, attracting significant scientific interest. However, their absorption and bioavailability in humans are still subjects of debate. Researchers generally agree that the interactions between the intestinal microbiota and phenolic compounds significantly impact the bioavailability of these compounds [61]. Studies have shown that phenolic compounds can promote the growth of beneficial microorganisms; for instance, anthocyanins have been found to increase the populations of *Bifidobacterium* spp., *Lactobacillus*, and *Enterococcus* spp. Additionally, the microbiota are crucial in modulating the transformation of phenolic compounds into smaller metabolites, thereby influencing the bioavailability and beneficial properties of proanthocyanidins [111].

According to research conducted in vitro, polyphenols may influence the composition of the human gut microbiota by suppressing the growth of potentially harmful bacteria (like *Helicobacter pylori* and *Staphylococcus* sp.) and promoting the development of potentially helpful bacteria (like *Lactobacillus* and *Bifidobacteria*) [112]. According to Wang et al., polyphenols have been shown in animal studies to alter gut microorganisms, microbial diversity, and F/B ratio. This and other research has indicated that the primary mechanism underlying the health advantages of polyphenols in humans is their prebiotic-like properties [113].

3.5.4. Resistant Starch

Resistant starch is a type of carbohydrate that resists digestion in the small intestine and ferments in the large intestine, acting as a prebiotic to beneficial gut bacteria [114]. This fermentation process produces SCFAs, such as butyrate, which have significant health benefits, including anti-inflammatory properties and improved gut barrier function. Studies have shown that diets high in resistant starch can positively alter the gut microbiota composition by increasing the abundance of beneficial bacteria like *Bifidobacterium* and *Lactobacillus*. Additionally, resistant starch has been linked to improved metabolic health markers, such as a better insulin sensitivity and a reduced risk of colorectal cancer. These findings underscore the importance of incorporating resistant-starch-rich foods, such as legumes, unripe bananas, and cooked and cooled potatoes, into the diet to support gut health and reduce inflammation [106].

4. Gut Microbiota and Inflammation

4.1. Mechanisms of Gut Microbiota–Immune System Interaction

The diverse microbial community, referred to as the gut microbiota, inhabits the gastrointestinal tract and engages in a dynamic and reciprocal relationship with the host immune system. This crosstalk between the gut microbiota and the immune system is fundamental for the development and function of the immune system, as well as for the maintenance of gut homeostasis and overall health.

From birth, the gut microbiota are instrumental in shaping the host immune system. Early colonization by beneficial microbes influences the maturation of immune cells, such as T cells, and the development of gut-associated lymphoid tissue. These interactions help to train the immune system to distinguish between harmful pathogens and benign antigens, preventing overreactions that could lead to allergies or autoimmune diseases. Studies have shown that germ-free animals, which lack gut microbiota, have underdeveloped immune systems and are more susceptible to infections and immune-mediated disorders [115].

The gut microbiota influence the immune system's regulatory mechanisms, promoting anti-inflammatory responses and maintaining immune tolerance. Certain gut bacteria, such as *Bifidobacterium* and *Lactobacillus*, induce the production of anti-inflammatory cytokines like IL-10 and regulatory T cells (Tregs), which help to suppress excessive immune responses and maintain intestinal homeostasis [115]. Dysbiosis, or an imbalance in the gut microbiota, can disrupt these regulatory pathways, leading to chronic inflammation and contributing to various inflammatory diseases, including IBD, rheumatoid arthritis (RA), and metabolic syndrome [116].

4.2. Gut Barrier Function and Permeability

The intestinal barrier is a dynamic structure that may react to a variety of stimuli and interface with them. It is made up of several components. It plays a crucial role in maintaining homeostasis by allowing the absorption of nutrients and water while preventing the entry of harmful pathogens, toxins, and antigens [117]. The gut microbiota contribute to the integrity of the gut barrier, a crucial line of defense against pathogens [118]. Beneficial bacteria produce SCFAs like butyrate, which nourish colonocytes and strengthen the tight junctions between epithelial cells. This prevents the translocation of harmful bacteria and their toxins into the bloodstream, reducing the risk of systemic inflammation and infection. Additionally, the gut microbiota compete with pathogenic bacteria for nutrients and attachment sites, further protecting the host from infections [119].

The gut barrier comprises several layers that work together to maintain its integrity and function. The first line of defense is the mucus layer, which covers the epithelial cells lining the gut. This layer is composed of mucins and glycoproteins that are secreted by goblet cells [118]. The mucus traps pathogens and particles, preventing them from reaching the epithelial surface. It also provides a habitat for commensal bacteria, which play a role in barrier function and immune modulation [120]. Though mucus protects epithelial cells as the first line of defense and is composed of the same biological elements throughout the gastrointestinal tract, its characteristics fluctuate depending on the region. The mucus layers in the small and large intestines are not the same thickness. The small intestine's primary roles include the digestion of food and nutrient absorption; in addition, it receives far less exposure to the microbiota than the colon. Thus, it contains a single layer.

Conversely, in the large intestine, the quantity and kind of bacteria that reside there dictate the thickness of the mucus layer. The mucus is arranged into two layers: the loose outer layer and the stiff inner layer. Though there are notable morphological distinctions between these two levels, their peptide contents are nearly identical. In a steady state, the inner mucus layer is free of germs, because it is highly organized into a flat, lamellar structure and stays fixed to the epithelial cells. It also prevents bacteria from penetrating. The inner layer and the outer layer are separated by the relative demarcation line [118]. The middle epithelial cell layer is the primary physical barrier, consisting of a single layer of enterocytes, goblet cells, enteroendocrine cells, and paneth cells. These cells are interconnected by tight junctions, adherens junctions, and desmosomes, which regulate the permeability of the barrier. Tight junctions are dynamic structures composed of proteins such as claudins, occludins, and zonula occludens, which can open or close in response to various stimuli, thus controlling the paracellular transport of substances [7]. Underlying the epithelial layer is the lamina propria, which contains immune cells such as macrophages, dendritic cells, and lymphocytes, as well as structures like Peyer's patches. These immune cells constantly monitor and respond to microbial and antigenic stimuli, playing a crucial role in maintaining the integrity of the gut barrier and preventing systemic infections [121]. The gut microbiota, composed of trillions of microorganisms, form a complex and dynamic community that interacts with the gut barrier. Beneficial microbes help to maintain this barrier by producing metabolites such as SCFAs, which enhance the function of epithelial cells and tight junctions. Dysbiosis, or an imbalance in the gut microbiota, can lead to barrier dysfunction and increased permeability [120].

The integrity and function of the gut barrier are regulated by various factors, including the diet, microbial metabolites, immune responses, and genetic predispositions. Dietary components can significantly influence gut barrier function. Fiber-rich diets promote the production of SCFAs by the gut microbiota, which strengthens the epithelial barrier and modulates immune responses. Conversely, diets high in fat and sugar, typical of the WD, can disrupt the gut barrier by promoting inflammation and altering the composition of the microbiota [122]. SCFAs, such as butyrate, propionate, and acetate, produced by the fermentation of dietary fibers, play a crucial role in maintaining gut barrier integrity. Butyrate, in particular, is a primary energy source for colonocytes and enhances the expression of tight junction proteins, thereby reducing permeability. Other microbial metabolites, such as indole derivatives and secondary BAs, also contribute to barrier function [6]. Immune cells in the gut, including regulatory T cells (Tregs) and innate lymphoid cells, produce cytokines and other signaling molecules that modulate barrier function. Anti-inflammatory cytokines, such as IL-10, help to maintain barrier integrity, while pro-inflammatory cytokines, such as tumor necrosis factor-alpha (TNF-α), can disrupt tight junctions and increase permeability [123]. Genetic predispositions can affect the structure and function of the gut barrier. Mutations in genes encoding for tight junction proteins, mucins, and other components of the epithelial layer can lead to barrier dysfunction and increased susceptibility to diseases like IBD and celiac disease [7].

Dysfunction of the gut barrier, often referred to as "leaky gut" (Figure 2) is characterized by increased intestinal permeability, allowing for the translocation of pathogens, toxins, and antigens into the bloodstream. This can trigger systemic inflammation and contribute to the development of various diseases. Increased gut permeability is a hallmark of IBD, including Crohn's disease (CD) and ulcerative colitis (UC). Dysbiosis and inflammation disrupt the barrier, leading to chronic intestinal inflammation [124]. Gut barrier dysfunction is associated with metabolic disorders such as obesity, T2DM, and non-alcoholic fatty liver disease. Increased permeability allows for the translocation of endotoxins, such as lipopolysaccharides (LPS), which induce systemic inflammation and insulin resistance [125]. Leaky gut has been implicated in the pathogenesis of autoimmune diseases, including celiac disease, RA, and T1DM. Increased permeability allows for the passage of antigens that trigger autoimmune responses [126].

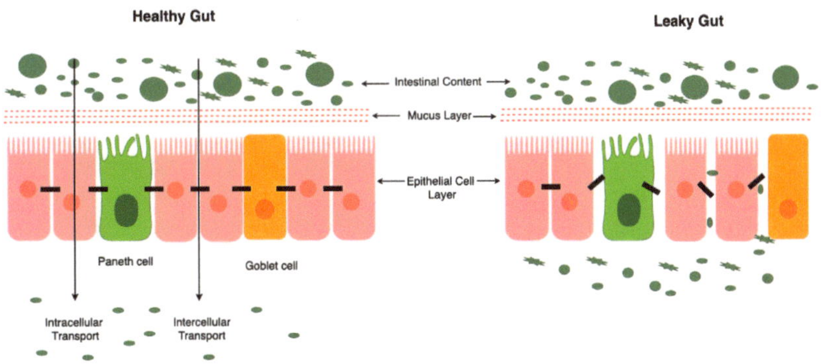

Figure 2. Gut barrier function and permeability (leaky gut situation). In a healthy gut, intestinal contents selectively pass through epithelial cells into the bloodstream, primarily via transcellular transport. This process is tightly regulated by tight junctions between epithelial cells, which restrict the passage of unwanted substances. In a leaky gut, the integrity of the intestinal barrier is compromised due to a reduced thickness of the mucus layer and the loosening of tight junctions. This compromised barrier allows pathogens and toxins to cross the intestinal lining and enter the bloodstream.

A significant contributing factor to leaky gut is a reduction in the thickness of the intestinal mucus layer. This mucus layer, secreted by goblet cells, acts as a protective barrier between the gut lumen and the epithelial cells. When the mucus layer becomes thinner, it provides less protection, allowing for pathogens and toxins to come into closer contact with the gut lining. Additionally, a decrease in the number of goblet cells further exacerbates this problem by reducing the production of mucus. This reduction compromises the gut's ability to maintain a robust barrier and can lead to increased intestinal permeability [127].

Furthermore, leaky gut is associated with changes in the microbial composition and function. A compromised mucus layer and fewer goblet cells disrupt the normal habitat for beneficial gut bacteria, leading to an imbalance in the microbiota. This imbalance can contribute to systemic inflammation and the development of various health issues, including inflammatory bowel disease, metabolic disorders, and even neuropsychiatric conditions. Therefore, maintaining the integrity of the intestinal mucus layer and ensuring adequate goblet cell function are essential for preserving the gut barrier function and overall health [127,128].

Emerging evidence suggests a link between gut barrier dysfunction and neurological disorders, such as autism spectrum disorder and depression. The gut–brain axis, a bidirectional communication system between the gut and the brain, is influenced by gut permeability and microbial metabolites [129]. The gut barrier is a complex and dynamic structure essential for maintaining intestinal and systemic health. Its integrity and function are regulated by various factors, including diet, microbial metabolites, immune responses, and genetic predispositions. Understanding the mechanisms underlying the gut barrier function and permeability is crucial for developing strategies to prevent and treat diseases associated with barrier dysfunction.

4.3. Production of Metabolites and Their Effects on Inflammation

The gut microbiota play a crucial role in the metabolism of dietary components, leading to the production of various metabolites that significantly influence host health. These metabolites include SCFAs, BAs, vitamins, amino acid derivatives, and other bioactive compounds. The production of these metabolites is a result of the fermentation and biotransformation processes carried out by the gut microbiota, and they can have profound effects on the host's physiology and immune responses.

4.3.1. Short-Chain Fatty Acids

In the large intestine, certain bacterial species digest carbohydrates to form short aliphatic tails of six carbons, which are called SCFAs. These SCFAs are produced under anaerobic conditions [130]. The main source of SCFAs, including acetate (C2), propionate (C3), and butyrate (C4) (Table 3), is dietary fibers, generally referred to as prebiotics. However, other nutrients, such as proteins and peptides, can also be converted into SCFAs. Bacteria digest oligofructose, arabinoxylan, inulin, and pectin to produce acetate, propionate, and butyrate in the proximal colon at considerable levels (70–140 mM) and at lesser amounts (20–70 mM) in the distal colon and distal ileum (20–40 mM) [131]. However, these proportions may differ based on variables such as the host genotype, fermentation location, microbiota makeup, and diet [132].

The portal vein delivers acetate and propionate to the liver, whereas the colonocytes mainly use butyrate. Propionate either remains in the liver or is discharged systemically into the peripheral venous system after being broken down by hepatocytes. As a result, often, only acetate is found in the peripheral blood [131]. Previous studies have shown that particular species with specific enzymes could be responsible for producing the different SCFAs. Microorganisms may also synthesize mono- and disaccharides from dietary fiber and other carbohydrates due to these enzymes. These saccharides are used by microbes to produce SCFAs [133].

The primary method by which enteric and acetogenic bacteria produce acetate is called reductive acetogenesis. As an acetogenesis process, the oxygen-sensitive Wood–Ljungdahl path is thought to be the most effective way for bacteria to generate acetate [131]. Bacteria metabolize carbohydrates in a few different ways to produce propionate. These routes consist of propanediol, acrylate, and succinate. Certain species of Firmicutes and Bacteroidetes favor the succinate route [134]. The synthesis of butyrate can proceed via two routes: butyrate synthesis from acetoacetyl-CoA, which is produced when two acetyl-CoA molecules combine. Butyryl-CoA: butyrate is produced by acetate CoA-transferase from butyryl-CoA. Butyryl-CoA: acetyl-CoA is extended to form butyrate by the enzyme acetate CoA-transferase, which is present in *Eubacterium*, *Roseburia*, *Anaerostipes*, and *Faecalibacterium prausnitzii* [135,136]. Butyrate kinase and phosphotransbutyrylase provide an additional pathway. For example, butyrate kinase is required by certain *Clostridium* species and several *Coprococcus* species in the Firmicutes family to produce butyrate [137] (Table 3).

In the organism, SCFAs are involved in a variety of physiological and immunological functions. SCFAs exert multiple anti-inflammatory effects through various mechanisms: 1. the regulation of immune cells, 2. the maintenance of gut barrier integrity, 3. epigenetic modulation, and 4. the modulation of metabolic pathways.

SCFAs influence the differentiation and function of immune cells, including T cells, macrophages, and dendritic cells. Butyrate and propionate enhance the differentiation of regulatory T cells (Tregs), which play a crucial role in maintaining immune tolerance and suppressing excessive inflammatory responses. SCFAs also inhibit the activation of pro-inflammatory macrophages and dendritic cells, reducing the production of pro-inflammatory cytokines such as IL-6 and TNF-α [138]. According to research by Souders et al., butyrate prevented TNF-α-induced reductions in Matrix metalloproteinase and mitochondrial-to-intracellular calcium ratios. This implies that butyrate may protect colonocytes from TNF-α-induced cytotoxicity by reducing the flow of calcium through the mitochondria [139].

Butyrate is particularly important for maintaining the integrity of the gut epithelial barrier. It provides energy to colonocytes and promotes the expression of tight junction proteins, which enhance the barrier function and prevent the translocation of pathogens and their toxins into the bloodstream. By maintaining the gut barrier integrity, SCFAs reduce systemic inflammation and protect against conditions like "leaky gut" syndrome [140]. SCFAs affect the host's enterocytes and digestive function locally. For example, at least 60–70% of the energy needed for colonic differentiation and proliferation is provided by butyrate, a key metabolic substrate for colonocytes [141].

In addition to supplying colonocytes' energy, SCFAs in the gut affect the colonic blood flow, colonic motility, and pH of the gastrointestinal environment, all of which have an impact on the absorption of nutrients and electrolytes. These effects may be mediated via the activation of G protein-coupled receptors, such as GPR41, GPR43, and GPR109A [142,143]. The activation of these receptors by SCFAs can lead to anti-inflammatory signaling pathways, further reducing inflammation and promoting metabolic health [144].

SCFAs, especially butyrate, act as histone deacetylase inhibitors, leading to the hyperacetylation of histones and changes in gene expression. This epigenetic modulation can suppress the expression of genes involved in inflammation and immune responses, contributing to the anti-inflammatory effects of SCFAs [145].

However, butyrate causes human monocytes to produce more IL-10 and less IL-12, which, in turn, prevents the synthesis of proinflammatory molecules, including nitric oxide, TNF-α, IL-1b, and nuclear factor kappa B (NF-kB) activity. Moreover, butyrate activates caspases 8 and 9, which, in turn, causes neutrophil death and inhibits the high mobility group box-1 [146]. In human goblet-like cells, line LS174T and T84 epithelial cells, butyrate, and propionate have been shown to induce the production and secretion of Mucin 2. It appears from this that SCFAs are essential bacterial compounds for maintaining gut integrity [147].

Table 3. SCFA production by microbes in the gut.

SCFA	Receptor	Pathway/Reaction	Producers	References
Acetate	GPR43	Via acetyl-CoA	*Akkermansia municuphila* *Bacteroides* spp. *Bifidobacterium* spp. *Prevotella* spp. *Ruminococcus* spp.	[148]
		Wood-Ljungdahl pathway	*Blautia hydrogenotrophica* *Clostridium* spp. *Streptococcus* spp.	[136]
Propionate	GPR43 GPR41	Succinate pathway	*Bacteroides* spp. *Phascolarctobacterium succinatutens* *Dialister* spp. *Veillonella* spp.	[148]
		Acrylate pathway	*Megasphaera elsdenii* *Coprococcus catus*	[136]
		Propanediol pathway	*Salmonella* spp. *Roseburia inulinivorans* *Ruminococcus obeum*	[136]
Butyrate	GPR41 GPR109A	Butyrate kinase route	*Coprococcus comes* *Coprococcus eutactus*	[136]
		Bytyryl-CoA	*Coprococcus catus* *Eubacterium rectale* *Roseburia* spp. *Faecalibacterium prausnitzii*	[135,136]

4.3.2. Bile Acids

BAs are steroid acids found predominantly in the bile of mammals and other vertebrates. When fat is present, the gut lumen secretes BAs. BAs come in two principal forms: cholic acid and chenodeoxycholic acid [149]. The gut microbiota transform primary BAs, which then interact with the Takeda-G-protein-receptor-5 (TGR-5) and the farnesoid X receptor (FXR) [150]. Through the FXR and TGR5, BAs exert impacts on metabolism [151]. According to Pedersen et al., the activation of the FXR and TGR5 enhances insulin sensitivity and glycogen production in the liver, raises pancreatic insulin output, and modulates brain satiety [150]. Taurocholic acid (TCA) is produced more often when animal fat is regularly consumed [151]. According to Agus et al., TCA prefers *Bilophila wadsworthia*, which is known to promote intestinal permeability and cause bacterial translocation. BA absorption may be hampered by this change in the microbiota. Consequently, there is a reduction in the expression of the FXR and FGF19, leading to an imbalance in BAs. Low-grade ongoing inflammation of the intestinal tract is related to this imbalance in BAs [151].

4.3.3. Tryptophan

Tryptamine and indole derivatives are essential for maintaining the intestinal epithelium's and immune cells' homeostasis. According to Hendrikx et al., these substances are produced by the gut microbiome's metabolism of tryptophan. These chemicals may encourage Th17 cells to reprogram into Treg cells, which would reduce inflammation. Nonetheless, a malfunction in the synthesis of the aryl hydrocarbon receptor ligand indole-3-propionic acid may result from a change in the host's gut bacterial makeup, most likely brought on by food [152]. Due to this ligand's malfunction, there is a reduction in GLP-1 and IL-22 release, breaking down intestinal permeability and causing LPS translocation and inflammation [151].

Microbial metabolites, such as SCFAs, secondary BAs, and tryptophan derivatives, play key roles in modulating immune responses. SCFAs, particularly butyrate, have been shown to enhance the production of Tregs and inhibit the activation of pro-inflammatory macrophages and dendritic cells [153]. Secondary BAs, produced by the microbial conversion of primary BA, can influence the differentiation and function of immune cells, further highlighting the intricate connection between microbial metabolism and immune regulation.

4.4. *Dysbiosis and Inflammatory Conditions*

Eubiosis, also known as "healthy microbiota", is the state in which the microbiota in the intestines are in balance and has positive effects on the entire body. Stable functional cores of the microbiome, a high taxonomic diversity, and high microbial gene richness are characteristics of healthy gut microbial communities. Large changes in the proportions of the five main phyla of the gut microbiota bacteria or the emergence of new bacterial taxa define dysbiosis, an imbalance that promotes illness. The two main characteristics of dysbiosis are the outgrowth of Gram-negative lipopolysaccharide-producing bacteria and a decrease in microbial richness and diversity [116,154].

The causes of dysbiosis are multifactorial and can be influenced by genetic, environmental, dietary, and lifestyle factors [154]. An increased intestinal permeability is typically a sign of dysbiosis. Under healthy settings, Th1 and Th17 cells eliminate the translocation of a small number of bacterial products, such as the polysaccharides of *Bacteroides* spp. or mucosa-adherent segmented filamentous bacteria (SFB). Conversely, a high concentration of invasive bacteria causes the Toll-like receptors to become overactivated. This, in turn, causes an overexpression of inflammatory cytokines, which, in turn, causes epithelial damage and chronic inflammation [155]. Surprisingly, mice with a diabetic genotype were protected from developing the disease by higher SFB levels, as seen in MyD88-deficient mice models (an adaptor for various innate immune receptors which recognize microbial stimuli) [156]. This suggests that the microbiota can have both promotion and inhibition effects. Although a dysbiotic gut community may be the distinguishing feature of a number of inflammatory illnesses, dysbiosis may also act as a catalyst for the disruption of intestinal homeostasis and the emergence of inflammation [155].

An expanding range of ailments, including autoimmune diseases, neurological disorders, obesity, metabolic disease, and IBD, have been linked to dysbiosis in terms of both onset and severity. Additionally, it may act as the catalyst for both *Clostridium difficile*-associated diarrhea, which primarily affects elderly people, and necrotizing enterocolitis, which is seen in infants [116].

4.5. Association between Gut Microbiota and Inflammatory Diseases

According to Feng et al., the gut microbiota and their metabolites may control the inflammatory conditions of the host. Numerous studies have connected inflammatory illnesses to the gut microbiome [21]. According to Forbes et al., the gut microbiota are altered by immune-mediated inflammatory illnesses such as CD, UC, multiple sclerosis, and RA [157]. Furthermore, an abundance of studies have demonstrated the pathophysiology of the gut microbiota in inflammatory illnesses, including obesity, T1DM and T2DM, and asthma [158].

4.5.1. Inflammatory Bowel Disease

IBD, encompassing CD and UC, is strongly associated with dysbiosis. Over 5 million people globally suffer from UC, the most prevalent kind of IBD [159,160]. Because of its mucous-layer-specific inflammation, the walls of the colon and rectum only sustain superficial damage [161]. Any portion of the gastrointestinal tract can be affected by CD, which is typified by erratic transmural inflammation that penetrates the intestinal wall into the serous layer and mostly affects the terminal ileum [124,161]. It has been shown in earlier research that abnormal intestinal permeability can predict the onset of IBD and is present in asymptomatic individuals years before the illness manifests [162]. Patients with IBD exhibit a reduced microbial diversity and altered composition of the gut microbiota, characterized by a decrease in beneficial bacteria such as *Faecalibacterium prausnitzii* and an increase in pathogenic bacteria like *Escherichia coli* [163].

These microbial shifts contribute to chronic intestinal inflammation through several mechanisms. Dysbiosis compromises the integrity of the gut barrier, leading to increased intestinal permeability, or "leaky gut". This allows bacteria and their metabolites to translocate into the bloodstream, triggering immune responses and inflammation [164]. An altered gut microbiota composition can lead to inappropriate activation of the mucosal immune system. For instance, dysbiosis in IBD is associated with increased levels of pro-inflammatory cytokines, such as TNF-α and IL-6, which exacerbate inflammation [162]. Dysbiosis affects the production of microbial metabolites like SCFAs, which have anti-inflammatory properties. Reduced levels of SCFAs can contribute to inflammation and disease progression [5].

4.5.2. Obesity and Metabolic Syndrome

Dysbiosis of the gut microbiota is also implicated in obesity and metabolic syndrome, conditions characterized by chronic low-grade inflammation. Obese individuals often have a less diverse gut microbiota and an altered ratio of F/B phyla, with a higher prevalence of Firmicutes [165]. This imbalance promotes metabolic inflammation through several pathways: dysbiosis can increase intestinal permeability, allowing LPS from Gram-negative bacteria to enter the circulation. This condition, known as metabolic endotoxemia, induces systemic inflammation and insulin resistance [166]. According to Ni et al., dysbiosis influences the inflammatory state of the adipose tissue. Certain gut bacteria produce metabolites that can modulate inflammation in the adipose tissue, contributing to insulin resistance and metabolic syndrome [163]. According to a study, obesity was linked to precancerous alterations in the transcriptome, while an elevated body mass index was linked to an increase in two proinflammatory colonic cytokines, TNF- α and IL6 [167].

4.5.3. Rheumatoid Arthritis

RA is an autoimmune disease characterized by chronic inflammation and the destruction of cartilage and bones. It has been documented that the gut microbiota play a role in the development and progression of RA. Patients with RA often show an altered gut microbiota composition, with an increase in pro-inflammatory bacteria such as *Prevotella copri* [168]. The altered gut microbiota in RA can affect systemic immune responses, promoting the differentiation of pro-inflammatory T cells and the production of autoantibodies that target joint tissues. The concept of the gut–joint axis suggests that microbial antigens and metabolites can influence joint inflammation. Dysbiosis may lead to the production of inflammatory mediators that travel from the gut to the joints, exacerbating RA symptoms [168].

According to a Chinese study, *Lactobacillus salivarius* was more prevalent in the saliva, teeth, and stomach of RA patients. On the other hand, *Haemophilus* species declined in these patients across all investigated areas [155]. A reduced intestinal microbial diversity was also seen in RA patients in a different study, and this finding was connected with both antibody production and the length of the disease. According to Maeda and Takeda, RA patients had higher relative abundances of *Collinsella aerofaciens* and *Eggerthella lenta* and a lower relative abundance of *Faecalibacterium* [169]. The genus *Collinsella* has been shown in vitro to increase intestinal permeability and induce the expression of IL-17A. This suggests that the growth of these microorganisms in the human gut can lead to an increase in proinflammatory conditions and make them a potential cause of arthritis [170].

4.5.4. Cardiovascular Diseases

CVDs, including atherosclerosis, have been associated with dysbiosis. Atherosclerosis is brought on by many risk factors, including metabolic syndrome, diabetes mellitus, tobacco use, and high blood pressure. Alongside these established risk factors for atherosclerosis, there is growing evidence to support the idea of a gut–systemic circulation axis in the atherosclerotic process. This axis is defined by the bloodstream passage of bacterially produced products such as LPS and TMAO [171]. The gut microbiota of patients with CVDs often show a reduced diversity and an increase in pro-inflammatory bacteria [172].

Certain gut bacteria metabolize dietary choline and carnitine into trimethylamine (TMA), which is further converted into TMAO in the liver. Elevated TMAO levels are associated with an increased risk of atherosclerosis and cardiovascular events. Dysbiosis can promote systemic inflammation and the activation of immune cells, contributing to the development and progression of atherosclerosis [172]. Numerous prospective studies have assessed the impact of low-grade endotoxemia on the risk of atherosclerosis, with the results showing that patients with high quantities of LPS are at a much higher risk [173]. It has also been noted that LPS contributes to the susceptibility of atherosclerotic plaques. Mice exposed to LPS displayed thrombus development and hemorrhaging. This result most likely stemmed from the creation of Leukotriene B4, a potent activator of leucocyte activation, and the activation of the arachidonic acid pathways, both of which are decreased in mice models of arterial inflammation [174].

The TMAO concentration has been suggested as a predictive and prognostic marker for CVDs due to the recent discovery of a positive correlation between it and acute coronary syndrome. Moreover, it has been documented that TMAO levels are correlated with the dimension of atherosclerotic plaques and the risk of cardiovascular events, including myocardial infarction, stroke, and mortality over three years. Through both inflammatory and metabolic processes, their levels have also been linked to unstable features of plaques, such as micro-vessels and a thinner fibrous cap [175].

4.5.5. Type 2 Diabetes Mellitus

The acquired condition known as T2DM is characterized by an increase in cardiovascular risk and mortality, as well as systemic inflammation. In obesity, several gene variants and environmental variables are mutually associated with the etiology of T2DM. The reported elevated risk of T2DM in individuals undergoing complete colectomy provides indirect evidence of the microbiota's role in T2DM development. As a result, resistance to diet-induced obesity has been observed in studies using germ-free mice; on the other hand, these mice showed an altered glucosidic tolerance and weight gain when exposed to bacteria specific to obesity, such as *Enterobacter cloacae*, or bacteria derived from obese donors [176].

AN increased intestinal permeability has been linked to T2DM, which may allow bacteria to pass through the gut barrier and cause low-grade systemic inflammation [177].

Remarkably, a long-term follow-up investigation revealed the prognostic significance of four bacterial species (*Clostridium citroniae*, *C. bolteae*, *Tyzzerella nexilis*, and *Ruminococcus gnavus*) in the development of T2DM. Numerous studies have shown altered gut microbiota in T2DM in the past, identifying a microbial signature for the illness and a connection between the gut microbiota and particular T2DM characteristics like insulin resistance [178].

A positive correlation has been observed in recent studies on diabetic patients between a few bacterial species and markers of systemic inflammation; specifically, the relative abundances of *Bifidobacterium adolescentis*, *Alistipes onderdonkii*, and *Eubacterium rectale* were positively correlated with IL-6, high-sensitivity C-reactive protein (CRP), and LPS-lipopolysaccharide binding protein; *Bacteroides thetaiotaomicron* was found to be positively correlated with the lipopolysaccharide-binding protein levels [179]. *Prevotella copri* and *Bacteroides vulgatus*, two bacteria that have been identified as being more prevalent in T2DM patients, have the ability to cause insulin resistance and enhance the availability of branched-chain amino acids in mice. *Ralstonia pickettii*-treated obese mice also demonstrated increased insulin resistance, pointing to a possible involvement of this bacterium in the development of T2DM [180].

4.5.6. Allergic Diseases

Allergic diseases, such as asthma and atopic dermatitis, are linked to dysbiosis, particularly during early life. A reduced microbial diversity and an altered gut microbiota composition in infancy are associated with an increased risk of developing allergic diseases [150]. The gut microbiota play a crucial role in the development of the immune system. Dysbiosis can impair the maturation of regulatory T cells (Tregs) and skew immune responses towards a pro-allergic phenotype. Dysbiosis can compromise the integrity of the gut barrier, allowing allergens to penetrate and trigger immune responses that lead to allergic inflammation [181].

5. Diet and Inflammation

Diet significantly influences the levels of biomarkers of inflammation in the body, which can serve as indicators of the overall inflammatory state and risk for various chronic diseases. Different dietary patterns impact key inflammatory biomarkers such as CRP, IL-6, TNF-α, and others [182].

CRP is produced by the liver in response to inflammation. Elevated CRP levels are associated with an increased risk of CVD, diabetes, and other inflammatory conditions [183]. IL-6 is a cytokine involved in inflammation and infection responses. High levels of IL-6 are linked to chronic inflammatory diseases and conditions like obesity [184] and metabolic syndrome [185]. TNF-α is a pro-inflammatory cytokine that plays a key role in systemic inflammation. Elevated levels are associated with autoimmune diseases, insulin resistance [184], and CVD [186].

Adherence to the MD has been consistently associated with lower levels of CRP, IL-6, and TNF-α. This diet's anti-inflammatory effects can be attributed to its high content of antioxidants, fiber, and healthy fats, particularly omega-3 fatty acids [42]. Rich in omega-3 polyunsaturated fats, the MD has anti-inflammatory and immunomodulatory effects by affecting the immune system function. These compounds decrease the production of pro-inflammatory factors like IL-1β, IL-6, TNF-α, VCAM-1, and MCP-1. They also lower ROS and nitrogen species levels while simultaneously increasing anti-inflammatory cytokines like IL-10 [187]. The MD pattern's high fiber intake is another characteristic. It also benefits the intestinal microbiota by modifying their composition and encouraging the release of metabolites like SCFAs, which control immune functions (e.g., acetate, propionate, and butyrate) [188]. For instance, butyrate reduces the synthesis of IL-1β, TNF-α, NF-kB, and IL-12 to have anti-inflammatory effects. By encouraging the colonization of Bacteroidetes and specific advantageous *Clostridium* groups and reducing Proteobacteria and Bacillaceae phyla, following the MD also helps to restore microbiota eubiosis [2]. The effects of the MD, rich in extra virgin olive oil (HQ-EVOO), on overweight/obese participants and normal-weight controls were investigated by Luisi et al. The authors investigated the potential effects of a three-month HQ-EVOO-enriched MD on the gut microbiota composition, oxidative stress metrics, and inflammation. It is interesting to note that this dietary approach reduced IL-6, TNF-α, and myeloperoxidase (a marker of inflammation and endothelial dysfunction) in both research groups [189].

Vegetarian and vegan diets are associated with lower levels of inflammatory biomarkers, including CRP and IL-6. Studies have demonstrated that plant-based diets can reduce CRP levels, potentially due to their high fiber content, low saturated fat, and abundance of anti-inflammatory phytochemicals [190]. Considering that CRP is a well-established biomarker of systemic low-grade inflammation connected to a number of disorders, Menzel et al. suggested that adopting a vegetarian or vegan diet may reduce the levels of circulating inflammatory biomarkers and improve inflammatory processes. For vegan or vegetarian populations, these anti-inflammatory qualities may lower the risk of chronic inflammatory disorders [191]. Remarkably, recent research has demonstrated the function of inflammasomes in controlling the gut microbiota and gut homeostasis [192]. These are a collection of protein complexes that stimulate the release of pro-inflammatory cytokines and are capable of identifying a wide range of stimuli that cause inflammation. These processes could impact immunological homeostasis associated with a corresponding decrease in the likelihood of developing metabolic disorders, such as atherosclerosis or metabolic syndrome. The role of inflammasomes in regulating the gut flora offers a novel and promising research topic that may help us to understand the mechanisms by which the diet affects the gut microbiota, inflammation, and health, even though more research is obviously needed [191].

The WD, characterized by a high intake of processed foods, red meats, and refined sugars, is linked to elevated levels of CRP, IL-6, and TNF-α. Research shows that individuals consuming a WD have higher levels of systemic inflammation markers, contributing to an increased risk of chronic diseases [71]. The main alteration in the microbiota composition caused by a HFD in both animal and human models is a rise in the F/B ratio. According to Velazquez et al., the most prevalent phyla in the gut microbiota of mice fed with an LFD and animals fed with an HFD were Bacteroidetes and Firmicutes, which accounted for 61% and 32% of the gut microbiota in LFD mice and 73% and 21% in old-HFD mice, respectively. However, compared to the LFD animals, the F/B ratio was greater in the HFD mice [193]. Increases in the abundances of Bacilli, Clostridiales, and Erysipelotrichales—all members of the Firmicutes phylum—are said to be the primary cause of the observed alterations in the F/B ratio [194]. Moreover, in a recent study, Yang et al. found that a HFD drives colorectal tumorigenesis by inducing gut microbial dysbiosis, metabolomic dysregulation with elevated lysophosphatidic acid, and gut barrier dysfunction in mice [195]. On the other hand, *A. muciniphila* lowers the pathogen load in diets heavy in fiber, demonstrating the context-dependent advantages of this mucin specialist. Increased mucus penetrability and altered behaviors of *A. muciniphila* and other community members are the causes of the increased vulnerability to pathogens, not changes in host immune systems or pathogen responses [196].

Apart from a HFD, diets high in added sugars are associated with an increased production of pro-inflammatory biomarkers like CRP and IL-6. A clinical study by Kim et al. indicated that a high sugar intake elevates CRP levels and other markers of inflammation, highlighting the inflammatory potential of excessive sugar consumption [197]. In a mouse model, high-sugar diet consumption significantly increased the abundance of *Escherichia coli* in fecal samples and promoted gut inflammation and a systemic immune response [198]. Moreover, in other mice studies, high-sugar diets induced colon inflammation compared with a standard diet by altering the composition of the gut microbiota, increasing the levels of *Akkermansia muciniphila*, known to produce enzymes degrading the mucus layer [199]. Short-term exposure to a high-sugar diet in mice increases their susceptibility to colitis by reducing the production of SCFAs and increasing the gut permeability [200].

6. The Interconnected Triangle: Diet, Gut Microbiota, and Inflammation

6.1. Interconnection of Diet and Gut Microbiota

Diet is one of the most significant factors shaping the composition and function of the gut microbiota. Different dietary patterns, such as the WD, MD, and plant-based diets, have distinct effects on gut microbial communities. For instance, a WD high in fats and sugars promotes an imbalance known as dysbiosis, characterized by a reduction in beneficial bacteria and an increase in pathogenic bacteria. In contrast, diets rich in fiber and polyphenols, like the MD, enhance the diversity and abundance of beneficial microbes such as bifidobacteria and lactobacilli (Figure 3). These beneficial bacteria produce SCFAs and other metabolites that support gut health and reduce inflammation [2,60].

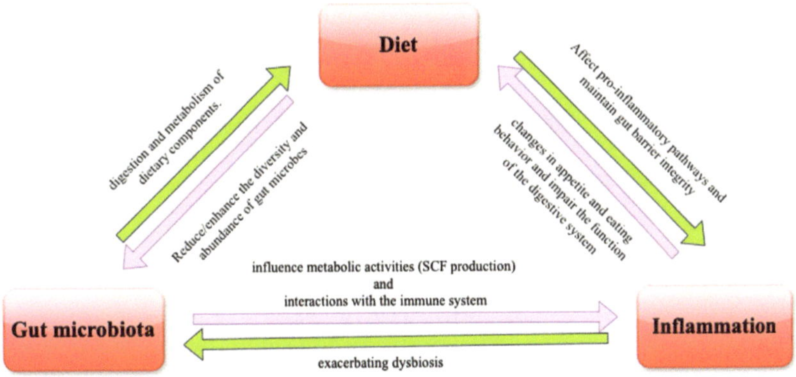

Figure 3. The interconnected triangle: diet, gut microbiota, and inflammation.

On the other hand, the gut microbiota play a crucial role in the digestion and metabolism of dietary components (Figure 3). Microbial enzymes break down complex carbohydrates, proteins, and fats that are indigestible by human enzymes alone. This breakdown results in the production of various metabolites, including SCFAs, which provide an energy source for colonocytes and influence the host metabolism [135]. Moreover, the gut bacteria synthesize essential vitamins such as vitamin K and B vitamins. The gut microbiota can also modulate appetite and eating behavior through the production of neurotransmitters and hormones, impacting dietary choices and nutrient intake [201].

6.2. Interconnection of Diet and Inflammation

Diet directly influences systemic inflammation through its impact on the gut microbiota and the immune system. Diets high in saturated fats and refined sugars promote pro-inflammatory pathways by increasing the production of endotoxins such as LPS [171] (Figure 3). Conversely, anti-inflammatory diets rich in omega-3 fatty acids, fiber, and polyphenols enhance the production of SCFAs, which have anti-inflammatory properties. These diets help to reduce the levels of pro-inflammatory cytokines and support the growth of beneficial bacteria that maintain the gut barrier integrity [5].

Chronic inflammation can alter dietary preferences and nutrient absorption. Inflammatory cytokines can affect the central nervous system, leading to changes in appetite and eating behavior, often resulting in a reduced food intake and altered taste preferences [77]. Inflammation can also impair the function of the digestive system, reducing the absorption of essential nutrients and leading to deficiencies that exacerbate inflammatory conditions (Figure 3). This cycle of inflammation and nutrient malabsorption can contribute to the progression of chronic diseases [164].

6.3. Interconnection between Gut Microbiota and Inflammation

Gut microbes significantly influence inflammatory processes through their metabolic activities and interactions with the immune system. Beneficial bacteria produce SCFAs, which enhance the gut barrier function, modulate immune responses, and inhibit the production of pro-inflammatory cytokines [144] (Figure 3). Dysbiosis, on the other hand, disrupts these processes, leading to an increased gut permeability and the translocation of bacterial endotoxins into the bloodstream. This endotoxemia triggers systemic inflammation and is associated with various inflammatory diseases such as IBD and RA [144,162].

On the other hand, chronic inflammation can negatively impact the gut microbiota, further exacerbating dysbiosis (Figure 3). Inflammatory conditions alter the gut environment, making it less hospitable for beneficial bacteria and promoting the growth of pathogenic microbes. This imbalance leads to a vicious cycle where inflammation begets dysbiosis and dysbiosis fuels further inflammation. Inflammatory diseases often show a reduced diversity of the gut microbiota, which is linked to worse clinical outcomes [157].

6.4. Role of the Host in the Triangular Relationship

In the triangular relationship between diet, the gut microbiome, and depression, the host plays a central role, with their genetic background, lifestyle factors, and environmental exposures significantly influencing this interaction. Variations in the abundance and composition of specific bacterial communities have been associated with specific genotypes of the host. Since genes affect the immune system function and prevalence of diseases, then certain alleles are associated with certain compositions. For instance, people with the rs651821 variant of the APOA5 gene tend to have more *Lactobacillus*, *Sutterella*, and *Methanobrevibacter*, which are associated with the risk of developing metabolic diseases [202]. According to the literature, specific variants of microbiome-associated genes are causally related to diseases including obesity, schizophrenia, Type 2 diabetes, amyotrophic lateral sclerosis, and inflammatory bowel disease [203]. Lifestyle factors, such as physical activity, stress levels, sleep patterns, and medication use (e.g., antibiotics and proton pump inhibitors), also impact the gut microbiome and its interaction with diet. Physical activity has been shown to promote gut microbiota diversity and SCFA production, while chronic stress can lead to dysbiosis by increasing the gut permeability and altering the microbial balance. Similarly, inadequate sleep and the use of medications can disrupt gut microbiota homeostasis, further complicating the relationship between diet and the microbiome [204]. Environmental exposures, including pollutants, toxins, and geographical factors, can modulate the gut microbiome and interact with both genetic predispositions and dietary factors. For instance, exposure to pollutants may exacerbate gut inflammation and disrupt microbial communities, which could amplify the effects of an unhealthy diet. Furthermore, factors such as urbanization, antibiotic use in food production, and water quality can have widespread effects on the microbiome compositions across populations, leading to variations in how individuals respond to dietary interventions [205]. Therefore, the host's genetic background, lifestyle choices, and environmental context collectively influence the diet–microbiome–depression link, making it a complex and highly individualized relationship.

6.5. Impact of Diet, Gut Microbiota, and Inflammation on the Brain and Behavior

The gut–brain axis is a complex and bidirectional communication network that links the gastrointestinal system with the central nervous system, playing a crucial role in maintaining homeostasis and influencing behavior and mental health. This axis involves multiple pathways, including neural, hormonal, immune, and metabolic signaling, through which the gut microbiota can impact brain function and mood [206]. Furthermore, the host's behavior is influenced by the relationship between the central nervous system and the enteric nervous system in the gut–brain axis. The enteric nervous system, sometimes called the "second brain", is made up of two nerve plexuses, the myenteric and submucosal plexuses, glial cells, and motor and sensory neurons [207]. Therefore, mediators of microbiota–gut–brain communication regulated by bacterial metabolism include short-chain fatty acids (SCFAs), such as butyrate, neurotransmitters, such as serotonin and gamma-aminobutyric acid, hormones, such as cortisol, and immunomodulators. These pathways may be modified by changes in the gut microbiota, which may lead to the onset or exacerbation of neuropsychiatric diseases and accompanying symptoms [208].

A number of pathophysiological changes, including an altered mood and impaired stress responses, are caused by functional abnormalities in the gut–brain axis [209]. The vagus nerve modulates the relationship between psychobiotics and their associated psychophysiological effects, based on the results of multiple animal investigations. This is mostly because psychobiotics, whether given after vagotomy or after the vagus nerve has been severed, do not cause a physiological reaction [210].

Diet plays a pivotal role in shaping the composition and functionality of the gut microbiota, which, in turn, can influence the gut–brain axis. For example, diets rich in fiber promote the growth of beneficial bacteria that produce short-chain fatty acids (SCFAs). These SCFAs can cross the blood–brain barrier and modulate brain activity, contributing to anti-inflammatory effects and the production of neurotransmitters like serotonin, which are vital for mood regulation [Bai]. Conversely, diets high in fat and sugar can lead to gut dysbiosis, a state of microbial imbalance, which is associated with an increased gut permeability ("leaky gut"), systemic inflammation, and altered brain function. Such changes can contribute to the development of mental health disorders, including depression and anxiety [211].

Behavioral and psychological factors also play a significant role in this relationship. Stress, for example, can disrupt the gut microbiota, leading to changes in gut barrier function and promoting inflammation, which can further influence mood and cognitive function. Chronic stress is known to alter the gut microbiota composition, reducing the abundance of beneficial bacteria and increasing the presence of harmful bacteria, thereby exacerbating the gut–brain axis dysfunction [212].

Additionally, psychological factors like anxiety and depression can alter eating behaviors, which may further impact gut health. Emotional eating or a preference for unhealthy, high-fat, and high-sugar foods can perpetuate a vicious cycle, where a poor diet exacerbates gut dysbiosis, leading to further mental health decline. Understanding the intricate interplay between diet, the gut microbiota, and the gut–brain axis highlights the importance of a balanced diet and stress management in promoting both gut health and mental well-being [212].

7. Therapeutic Implications

The growing understanding of the interconnected relationship between diet, gut microbiota, and inflammation opens up promising avenues for therapeutic interventions aimed at managing various health conditions. These therapeutic implications can be harnessed to prevent and treat chronic inflammatory diseases, optimize gut health, and enhance overall well-being.

Fecal microbiota transplantation (FMT) has the potential to restore the complete microbiota ecosystem. To replace lost activities in the gut microbiota, single helpful strains or groups of them (probiotics) might be added; in the meantime, undesirable or hazardous strains could be eliminated with the use of bacteriophages, antibiotics, or antifungals. Lastly, it may be possible to inhibit or stop the synthesis of toxic metabolites or increase the synthesis of advantageous metabolites by targeting microbial metabolic pathways. FMT, which involves the transfer of stool from a healthy donor to a recipient with dysbiosis, has shown promise in treating recurrent *Clostridioides difficile* infection and has cure rates exceeding 90% [213]. FMT is being explored for other conditions like IBD and metabolic syndrome [214]. FMT has also been used experimentally to treat other gastrointestinal disorders, such as UC, constipation, IBS, liver diseases such as cirrhosis with encephalopathy and alcoholic hepatitis, and neurological diseases such as multiple sclerosis and Parkinson's disease [201].

Probiotics are highly well-liked substances that affect the host health and gut microbiota. Probiotic microorganisms have a variety of actions that frequently cooperate. Their primary mechanisms include immune system modulation, enhanced intestinal barrier function, resistance to colonization, and metabolite synthesis that acts both locally (antimicrobials, enzymes, and organic acids) and remotely (hormones and neurochemicals). Strong evidence supports the safety and effectiveness of a number of probiotics, such as *Lactobacillus* species, *Bifidobacterium* species, and *Saccharomyces* species. Sanders et al. identified *Roseburia* spp. and *Faecalibacterium* spp. as further interesting options [215].

Promising therapeutic strategies also target the microbiome to either enhance the synthesis of protective metabolites or inhibit or reduce the production of harmful metabolites. One structural counterpart of choline that has been effectively used to prevent the microbial conversion of dietary choline to TMA is 3,3-dimethyl-1-butanol. Atherosclerosis and serious CVDs have been linked to TMAO, an oxidation product of TMA [201]. Remarkably, a new study indicates that, because of TMA's nephrotoxicity and cardiotoxicity, it may be the primary offender rather than TMAO [216].

With the potential for personalized nutrition and tailoring dietary recommendations based on an individual's unique gut microbiota composition, genetic profile, and specific health conditions, it is possible to achieve more effective and targeted outcomes. Personalized nutrition can help to manage conditions such as IBS, IBD, metabolic syndrome, and even mental health disorders like anxiety and depression. Advances in microbiome sequencing and bioinformatics tools facilitate the identification of specific microbial signatures associated with health and disease, allowing for the development of customized dietary interventions [217].

Probiotics can alleviate the symptoms of gastrointestinal disorders such as IBS and IBD, reducing bloating, abdominal pain, and diarrhea. Probiotics also play a role in enhancing immune responses by modulating the activity of immune cells and increasing the production of anti-inflammatory cytokines. Metabolic health can be improved through better glycemic control and lipid profiles, with certain probiotic strains aiding in weight management and reducing the risk of metabolic syndrome [8]. Moreover, emerging evidence suggests that probiotics can positively impact mental health by modulating the gut–brain axis, potentially alleviating symptoms of anxiety and depression [109]. Prebiotics support gut health by promoting beneficial bacterial growth, improving bowel regularity, and enhancing gut barrier function. They can reduce systemic inflammation and support immune homeostasis through the increased production of SCFAs [104].

However, the efficacy of probiotics and prebiotics is not without limitations. The benefits of probiotics are highly strain-specific, and not all strains provide the same health benefits, making it challenging to identify the most effective strains for specific conditions [218]. Additionally, probiotics must survive the harsh acidic environment of the stomach and bile salts to reach the intestines, where they must colonize effectively; otherwise, their benefits may be transient [219]. Individual variability, such as differences in existing gut microbiota, genetics, diet, and health status, also affects the efficacy of these treatments. Furthermore, the probiotic industry faces issues related to standardization, quality control, and regulatory approval, leading to variations in the potency, purity, and efficacy of probiotic products [219].

8. Research Gaps and Future Research Directions

Despite the growing body of evidence highlighting the importance of the diet–gut microbiota–inflammation axis, several research gaps remain. One significant gap is the lack of comprehensive, long-term studies that can conclusively establish causality rather than mere associations between diet, microbiota changes, and inflammation. Most existing studies are short-term and rely heavily on observational data, which can be influenced by numerous confounding factors. Furthermore, the individual variability in the gut microbiota composition and response to dietary interventions is not fully understood. This variability complicates the development of universal dietary guidelines and underscores the need for personalized nutrition strategies. Additionally, the mechanistic pathways through which specific dietary components influence microbial metabolites and subsequently modulate inflammatory responses are not completely elucidated. The interaction between different dietary components and their cumulative effects on the gut microbiota and inflammation also require further exploration.

To bridge these gaps, future research should focus on conducting long-term randomized controlled trials to establish causative relationships and understand the sustained impact of dietary interventions on the gut microbiota and inflammation. These studies should include diverse populations to account for genetic, environmental, and lifestyle factors that influence individual responses. Incorporating advanced technologies such as multi-omics (genomics, metabolomics, and proteomics) and machine learning can provide a more comprehensive understanding of the complex interactions within the diet–gut microbiota–inflammation axis. Detailed microbial and metabolic profiling before, during, and after dietary interventions can help to identify the key microbial players and metabolites involved in inflammatory processes.

Furthermore, studies should investigate the synergistic effects of combining various dietary components, such as fiber, polyphenols, and fermented foods, to determine the optimal dietary patterns for modulating the gut microbiota and reducing inflammation. Research into personalized nutrition should be intensified, utilizing individual microbiome data to tailor dietary recommendations and interventions. Exploring the role of lesser-studied bioactive compounds in foods, such as phytochemicals and micronutrients, and their specific effects on the gut microbiota and inflammation can also provide valuable insights.

Additionally, understanding the bidirectional relationship between the gut microbiota and diet, where the microbiota composition can influence dietary preferences and metabolic outcomes, is crucial. Investigating the impact of the gut microbiota on nutrient absorption and metabolism, and how this, in turn, affects inflammation, can provide a more holistic view of the triangular relationship. Finally, integrating behavioral and psychological aspects into research can help us to understand how diet and the microbiota influence mental health and vice versa, further elucidating the comprehensive impact of this interconnected triad on overall well-being.

As research advances in the fields of diet, the gut microbiome, and depression, there is a growing need for regulatory oversight to ensure the safety, efficacy, and ethical implementation of personalized interventions. Personalized nutrition and microbiome-targeted therapies hold great promise for improving mental health outcomes by tailoring dietary and therapeutic approaches to an individual's unique microbiome composition and genetic background. However, without proper regulation, there is a risk of unsubstantiated claims, inadequate safety evaluations, and unequal access to these innovations.

Regulatory agencies must establish clear guidelines for developing, testing, and marketing microbiome-based therapies, including probiotics, prebiotics, and other dietary supplements. This includes enforcing rigorous clinical trials to demonstrate their effectiveness and monitoring long-term safety, particularly as interventions targeting the microbiome are still in their early stages. In parallel, ensuring that healthcare providers receive adequate training on the scientific basis and potential risks of these interventions is crucial to avoid the misuse of and overreliance on emerging but unproven therapies.

Equitable access to personalized interventions is another critical issue. Personalized nutrition and microbiome-targeted treatments often require advanced diagnostic tools, such as genetic sequencing and microbiome profiling, which can be costly and may not be readily available to all individuals, particularly those from low-income or underserved populations. This could exacerbate existing health disparities by making these promising treatments accessible only to those who can afford them. Policymakers must consider strategies to make these interventions more widely available, such as subsidizing costs, integrating them into public health programs, and supporting research that focuses on diverse populations. Therefore, a combination of robust regulatory oversight and proactive measures to promote equitable access is essential for realizing the full potential of personalized interventions in managing depression and other health conditions linked to the gut microbiome.

9. Conclusions

The intricate relationship among diet, the gut microbiota, and inflammation underscores the importance of dietary choices in maintaining health and preventing chronic diseases. Evidence suggests that dietary patterns significantly influence the composition and functionality of the gut microbiota, which, in turn, play a crucial role in modulating inflammatory processes. Beneficial dietary components, such as fiber, polyphenols, and healthy fats, support a diverse and balanced gut microbiota, promoting anti-inflammatory pathways and overall health. Conversely, diets high in saturated fats and refined sugars can lead to dysbiosis and increased inflammation, contributing to the development and progression of chronic diseases. Therapeutic strategies, including probiotics, prebiotics, personalized nutrition, and fecal microbiota transplantation, hold promise in modulating the gut microbiota to reduce inflammation and improve health outcomes. Future research should continue to explore the mechanistic pathways of this triad, aiming to develop targeted interventions that harness the power of the gut microbiota for disease prevention and management. Understanding and leveraging the interconnected triangle of diet, the gut microbiota, and inflammation could pave the way for innovative approaches to enhance human health and treat chronic inflammatory conditions.

Author Contributions: Conceptualization, B.X.; methodology, B.X.; software, N.R. and B.X.; validation, B.X. and M.B.; investigation, N.R.; writing—original draft preparation, N.R.; writing—review and editing, B.X. and M.B.; supervision, B.X. and M.B. All authors have read and agreed to the published version of the manuscript.

Funding: This research was jointly funded by two grants (UICR0400015-24 and UICR0400016-24) from BNU-HKBU United International College, China.

Conflicts of Interest: The authors declare no conflicts of interest.

References

1. Vemuri, R.; Shankar, E.M.; Chieppa, M.; Eri, R.; Kavanagh, K. Beyond just bacteria: Functional biomes in the gut ecosystem including virome, mycobiome, archaeome and helminths. *Microorganisms* **2020**, *8*, 483. [CrossRef]
2. García-Montero, C.; Fraile-Martínez, O.; Gómez-Lahoz, A.M.; Pekarek, L.; Castellanos, A.J.; Noguerales-Fraguas, F.; Coca, S.; Guijarro, L.G.; García-Honduvilla, N.; Asúnsolo, A.; et al. Nutritional components in western diet versus Mediterranean diet at the gut microbiota–immune system interplay. Implications for health and disease. *Nutrients* **2021**, *13*, 699. [CrossRef] [PubMed]
3. Suriano, F.; Nyström, E.E.L.; Sergi, D.; Gustafsson, J.K. Diet, microbiota, and the mucus layer: The guardians of our health. *Front. Immunol.* **2022**, *13*, 953196. [CrossRef]
4. Potrykus, M.; Czaja-Stolc, S.; Stankiewicz, M.; Kaska, Ł.; Małgorzewicz, S. Intestinal microbiota as a contributor to chronic inflammation and its potential modifications. *Nutrients* **2021**, *13*, 3839. [CrossRef]
5. Aho, V.T.; Houser, M.C.; Pereira, P.A.; Chang, J.; Rudi, K.; Paulin, L.; Hertzberg, V.; Auvinen, P.; Tansey, M.G.; Scheperjans, F. Relationships of gut microbiota, short-chain fatty acids, inflammation, and the gut barrier in Parkinson's disease. *Mol. Neurodegener.* **2021**, *16*, 6. [CrossRef] [PubMed]
6. Gao, Y.; Davis, B.; Zhu, W.; Zheng, N.; Meng, D.; Walker, W.A. Short-chain fatty acid butyrate, a breast milk metabolite, enhances immature intestinal barrier function genes in response to inflammation in vitro and in vivo. *Am. J. Physiol. Gastrointest. Liver Physiol.* **2021**, *320*, G521–G530. [CrossRef]
7. Stolfi, C.; Maresca, C.; Monteleone, G.; Laudisi, F. Implication of intestinal barrier dysfunction in gut dysbiosis and diseases. *Biomedicines* **2022**, *10*, 289. [CrossRef]
8. Piccioni, A.; Covino, M.; Candelli, M.; Ojetti, V.; Capacci, A.; Gasbarrini, A.; Franceschi, F.; Merra, G. How do diet patterns, single foods, prebiotics and probiotics impact gut microbiota? *Microbiol. Res.* **2023**, *14*, 390–408. [CrossRef]
9. Bisanz, J.E.; Upadhyay, V.; Turnbaugh, J.A.; Ly, K.; Turnbaugh, P.J. Meta-analysis reveals reproducible gut microbiome alterations in response to a high-fat diet. *Cell Host Microbe* **2019**, *26*, 265–272. [CrossRef]
10. Kettle, H.; Louis, P.; Flint, H.J. Process-based modeling of microbial community dynamics in the human colon. *J. R. Soc. Interface* **2022**, *19*, 20220489. [CrossRef]
11. Beam, A.; Clinger, E.; Hao, L. Effect of diet and dietary components on the composition of the gut microbiota. *Nutrients* **2021**, *13*, 2795. [CrossRef]

12. Saleem, A. Role of Gut Microbiota in Type 2 Diabetic Patients. Master's Thesis, Quaid I Azam University Islamabad, Islamabad, Pakistan, 2023.
13. Yang, Q.; Liang, Q.; Balakrishnan, B.; Belobrajdic, D.P.; Feng, Q.J.; Zhang, W. Role of dietary nutrients in the modulation of gut microbiota: A narrative review. *Nutrients* **2020**, *12*, 381. [CrossRef]
14. Vaiserman, A.; Romanenko, M.; Piven, L.; Moseiko, V.; Lushchak, O.; Kryzhanovska, N.; Guryanov, V.; Koliada, A. Differences in the gut Firmicutes to Bacteroidetes ratio across age groups in healthy Ukrainian population. *BMC Microbiol.* **2020**, *20*, 221. [CrossRef]
15. Adak, A.; Khan, M.R. An insight into gut microbiota and its functionalities. *Cell. Mol. Life Sci.* **2019**, *76*, 473–493. [CrossRef]
16. Hou, K.; Wu, Z.X.; Chen, X.Y.; Wang, J.Q.; Zhang, D.; Xiao, C.; Zhu, D.; Koya, J.B.; Wei, L.; Li, J.; et al. Microbiota in health and diseases. *Signal Transduct. Target. Ther.* **2022**, *7*, 135. [CrossRef]
17. Perler, B.K.; Friedman, E.S.; Wu, G.D. The role of the gut microbiota in the relationship between diet and human health. *Annu. Rev. Physiol.* **2023**, *85*, 449–468. [CrossRef]
18. Roswall, J.; Olsson, L.M.; Kovatcheva-Datchary, P.; Nilsson, S.; Tremaroli, V.; Simon, M.C.; Kiilerich, P.; Akrami, R.; Krämer, M.; Uhlén, M.; et al. Developmental trajectory of the healthy human gut microbiota during the first 5 years of life. *Cell Host Microbe* **2021**, *29*, 765–776. [CrossRef] [PubMed]
19. Ansaldo, E.; Farley, T.K.; Belkaid, Y. Control of immunity by the microbiota. *Annu. Rev. Immunol.* **2021**, *39*, 449–479. [CrossRef] [PubMed]
20. de Sire, A.; de Sire, R.; Curci, C.; Castiglione, F.; Wahli, W. Role of dietary supplements and probiotics in modulating microbiota and bone health: The gut-bone axis. *Cells* **2022**, *11*, 743. [CrossRef]
21. Feng, W.; Liu, J.; Ao, H.; Yue, S.; Peng, C. Targeting gut microbiota for precision medicine: Focusing on the efficacy and toxicity of drugs. *Theranostics* **2020**, *10*, 11278. [CrossRef]
22. Su, M.; Tang, T.; Tang, W.; Long, Y.; Wang, L.; Liu, M. Astragalus improves intestinal barrier function and immunity by acting on intestinal microbiota to treat T2DM: A research review. *Front. Immunol.* **2023**, *14*, 1243834. [CrossRef] [PubMed]
23. Camelo-Castillo, A.; Rivera-Caravaca, J.M.; Orenes-Piñero, E.; Ramírez-Macías, I.; Roldán, V.; Lip, G.Y.; Marín, F. Gut microbiota and the quality of oral anticoagulation in vitamin k antagonists users: A review of potential implications. *J. Clin. Med.* **2021**, *10*, 715. [CrossRef] [PubMed]
24. Renson, A.; Herd, P.; Dowd, J.B. Sick individuals and sick (microbial) populations: Challenges in epidemiology and the microbiome. *Annu. Rev. Public Health* **2020**, *41*, 63–80. [CrossRef]
25. Oliver, A. *From a Remote Amazon Village to the Undergraduate Lab: How Diet and Lifestyle Shape the Human Microbiome*; University of California: Irvine, CA, USA, 2021.
26. Cena, H.; Calder, P.C. Defining a healthy diet: Evidence for the role of contemporary dietary patterns in health and disease. *Nutrients* **2020**, *12*, 334. [CrossRef] [PubMed]
27. Zhang, M.; Zhang, X.; Zhu, J.; Zhao, D.G.; Ma, Y.Y.; Li, D.; Ho, C.T.; Huang, Q. Bidirectional interaction of nobiletin and gut microbiota in mice fed with a high-fat diet. *Food Funct.* **2021**, *12*, 3516–3526. [CrossRef]
28. Bolte, L.A.; Vila, A.V.; Imhann, F.; Collij, V.; Gacesa, R.; Peters, V.; Wijmenga, C.; Kurilshikov, A.; Campmans-Kuijpers, M.J.E.; Fu, J.; et al. Long-term dietary patterns are associated with pro-inflammatory and anti-inflammatory features of the gut microbiome. *Gut* **2021**, *70*, 1287–1298. [CrossRef]
29. Clemente-Suárez, V.J.; Beltrán-Velasco, A.I.; Redondo-Flórez, L.; Martín-Rodríguez, A.; Tornero-Aguilera, J.F. Global impacts of western diet and its effects on metabolism and health: A narrative review. *Nutrients* **2023**, *15*, 2749. [CrossRef]
30. Barber, T.M.; Kabisch, S.; Pfeiffer, A.F.; Weickert, M.O. The effects of the Mediterranean diet on health and gut microbiota. *Nutrients* **2023**, *15*, 2150. [CrossRef]
31. Kang, J.W. *The Role of Targeted Prebiotics for Improving Gut Microbiome Composition and Function in Individuals Consuming Diets Low in Fiber, and Individuals at Risk for Chronic Disease*; University of California: Davis, CA, USA, 2022.
32. Ratajczak, A.E.; Festa, S.; Aratari, A.; Papi, C.; Dobrowolska, A.; Krela-Kaźmierczak, I. Should the Mediterranean diet be recommended for inflammatory bowel diseases patients? A narrative review. *Front. Nutr.* **2023**, *9*, 1088693. [CrossRef]
33. Boughanem, H.; Ruiz-Limón, P.; Pilo, J.; Lisbona-Montañez, J.M.; Tinahones, F.J.; Moreno Indias, I.; Macías-González, M. Linking serum vitamin D levels with gut microbiota after 1-year lifestyle intervention with Mediterranean diet in patients with obesity and metabolic syndrome: A nested cross-sectional and prospective study. *Gut Microbes* **2023**, *15*, 2249150. [CrossRef]
34. Merra, G.; Noce, A.; Marrone, G.; Cintoni, M.; Tarsitano, M.G.; Capacci, A.; De Lorenzo, A. Influence of Mediterranean diet on human gut microbiota. *Nutrients* **2020**, *13*, 7. [CrossRef]
35. Meslier, V.; Laiola, M.; De Filippis, F.; Roume, H.; Quinquis, B.; Giacco, R.; Mennella, I.; Ferracane, R.; Pons, N.; Pasolli, E.; et al. Mediterranean diet intervention in overweight and obese subjects lowers plasma cholesterol and causes changes in the gut microbiome and metabolome independently of energy intake. *Gut* **2020**, *69*, 1258–1268. [CrossRef] [PubMed]
36. Tindall, A.M.; McLimans, C.J.; Petersen, K.S.; Kris-Etherton, P.M.; Lamendella, R. Walnuts and vegetable oils containing oleic acid differentially affect the gut microbiota and associations with cardiovascular risk factors: Follow-up of a randomized, controlled, feeding trial in adults at risk for cardiovascular disease. *J. Nutr.* **2020**, *150*, 806–817. [CrossRef]
37. Vetrani, C.; Maukonen, J.; Bozzetto, L.; Della Pepa, G.; Vitale, M.; Costabile, G.; Riccardi, G.; Rivellese, A.A.; Saarela, M.; Annuzzi, G. Diets naturally rich in polyphenols and/or long-chain n-3 polyunsaturated fatty acids differently affect microbiota composition in high-cardiometabolic-risk individuals. *Acta Diabetol.* **2020**, *57*, 853–860. [CrossRef]

38. Guiducci, L.; Nicolini, G.; Forini, F. Dietary patterns, gut microbiota remodeling, and cardiometabolic disease. *Metabolites* **2023**, *13*, 760. [CrossRef] [PubMed]
39. Garcia-Mantrana, I.; Selma-Royo, M.; Alcantara, C.; Collado, M.C. Shifts on gut microbiota associated to Mediterranean diet adherence and specific dietary intakes on general adult population. *Front. Microbiol.* **2018**, *9*, 890. [CrossRef] [PubMed]
40. Ghiboub, M.; Verburgt, C.M.; Sovran, B.; Benninga, M.A.; de Jonge, W.J.; Van Limbergen, J.E. Nutritional therapy to modulate tryptophan metabolism and aryl hydrocarbon-receptor signaling activation in human diseases. *Nutrients* **2020**, *12*, 2846. [CrossRef]
41. Abdalqadir, N.; Adeli, K. GLP-1 and GLP-2 orchestrate intestine integrity, gut microbiota, and immune system crosstalk. *Microorganisms* **2022**, *10*, 2061. [CrossRef]
42. Ghosh, T.S.; Rampelli, S.; Jeffery, I.B.; Santoro, A.; Neto, M.; Capri, M.; Giampieri, E.; Jennings, A.; Candela, M.; Turroni, S.; et al. Mediterranean diet intervention alters the gut microbiome in older people reducing frailty and improving health status: The NU-AGE 1-year dietary intervention across five European countries. *Gut* **2020**, *69*, 1218–1228. [CrossRef]
43. Leeming, E.R.; Johnson, A.J.; Spector, T.D.; Le Roy, C.I. Effect of diet on the gut microbiota: Rethinking intervention duration. *Nutrients* **2019**, *11*, 2862. [CrossRef]
44. Cronin, P.; Joyce, S.A.; O'Toole, P.W.; O'Connor, E.M. Dietary fiber modulates the gut microbiota. *Nutrients* **2021**, *13*, 1655. [CrossRef]
45. Horne, R.G.; Yu, Y.; Zhang, R.; Abdalqadir, N.; Rossi, L.; Surette, M.; Sherman, P.M.; Adeli, K. High fat-high fructose diet-induced changes in the gut microbiota associated with dyslipidemia in Syrian hamsters. *Nutrients* **2020**, *12*, 3557. [CrossRef] [PubMed]
46. Suriano, F.; Vieira-Silva, S.; Falony, G.; de Wouters d'Oplinter, A.; Paone, P.; Delzenne, N.M.; Cani, P.D. Fat and not sugar as the determining factor for gut microbiota changes, obesity, and related metabolic disorders in mice. *Am. J. Physiol. Endocrinol. Metab.* **2023**, *324*, E85–E96. [CrossRef] [PubMed]
47. De Siena, M.; Raoul, P.; Costantini, L.; Scarpellini, E.; Cintoni, M.; Gasbarrini, A.; Rinninella, E.; Mele, M.C. Food emulsifiers and metabolic syndrome: The role of the gut microbiota. *Foods* **2022**, *11*, 2205. [CrossRef]
48. Chassaing, B.; Compher, C.; Bonhomme, B.; Liu, Q.; Tian, Y.; Walters, W.; Nessel, L.; Delaroque, C.; Hao, F.; Gershuni, V.; et al. Randomized controlled-feeding study of dietary emulsifier carboxymethylcellulose reveals detrimental impacts on the gut microbiota and metabolome. *Gastroenterology* **2022**, *162*, 743–756. [CrossRef] [PubMed]
49. Koeth, R.A.; Lam-Galvez, B.R.; Kirsop, J.; Wang, Z.; Levison, B.S.; Gu, X.; Copeland, M.F.; Bartlett, D.; Cody, D.B.; Dai, H.J.; et al. L-Carnitine in omnivorous diets induces an atherogenic gut microbial pathway in humans. *J. Clin. Investig.* **2019**, *129*, 373–387. [CrossRef]
50. Rinninella, E.; Cintoni, M.; Raoul, P.; Lopetuso, L.R.; Scaldaferri, F.; Pulcini, G.; Miggiano, G.A.D.; Gasbarrini, A.; Mele, M.C. Food components and dietary habits: Keys for a healthy gut microbiota composition. *Nutrients* **2019**, *11*, 2393. [CrossRef]
51. Narula, N.; Wong, E.C.; Dehghan, M.; Mente, A.; Rangarajan, S.; Lanas, F.; Lopez-Jaramillo, P.; Rohatgi, P.; Lakshmi, P.V.M.; Varma, R.P.; et al. Association of ultra-processed food intake with risk of inflammatory bowel disease: Prospective cohort study. *BMJ* **2021**, *374*, n1554. [CrossRef]
52. Vanegas, S.M.; Meydani, M.; Barnett, J.B.; Goldin, B.; Kane, A.; Rasmussen, H.; Brown, C.; Vangay, P.; Knights, D.; Jonnalagadda, S.; et al. Substituting whole grains for refined grains in a 6-wk randomized trial has a modest effect on gut microbiota and immune and inflammatory markers of healthy adults. *Am. J. Clin. Nutr.* **2017**, *105*, 635–650. [CrossRef]
53. Lakshmanan, A.P.; Mingione, A.; Pivari, F.; Dogliotti, E.; Brasacchio, C.; Murugesan, S.; Cusi, D.; Lazzaroni, M.; Soldati, L.; Terranegra, A. Modulation of gut microbiota: The effects of a fruits and vegetables supplement. *Front. Nutr.* **2022**, *9*, 930883. [CrossRef]
54. Djekic, D.; Shi, L.; Brolin, H.; Carlsson, F.; Särnqvist, C.; Savolainen, O.; Cao, Y.; Bäckhed, F.; Tremaroli, V.; Landberg, R.; et al. Effects of a vegetarian diet on cardiometabolic risk factors, gut microbiota, and plasma metabolome in subjects with ischemic heart disease: A randomized, crossover study. *J. Am. Heart Assoc.* **2020**, *9*, e016518. [CrossRef] [PubMed]
55. Crimarco, A.; Springfield, S.; Petlura, C.; Streaty, T.; Cunanan, K.; Lee, J.; Fielding-Singh, P.; Carter, M.M.; Topf, M.A.; Wastyk, H.C.; et al. A randomized crossover trial on the effect of plant-based compared with animal-based meat on trimethylamine-N-oxide and cardiovascular disease risk factors in generally healthy adults: Study with appetizing plantfood—Meat eating alternative trial. *Am. J. Clin. Nutr.* **2020**, *112*, 1188–1199. [CrossRef]
56. Van Soest, A.P.; Hermes, G.D.; Berendsen, A.A.; Van De Rest, O.; Zoetendal, E.G.; Fuentes, S.; Santoro, A.; Franceschi, C.; de Groot, L.C.P.G.M.; De Vos, W.M. Associations between pro-and anti-inflammatory gastrointestinal microbiota, diet, and cognitive functioning in Dutch healthy older adults: The NU-AGE study. *Nutrients* **2020**, *12*, 3471. [CrossRef]
57. Liu, M.J.; Yang, J.Y.; Yan, Z.H.; Hu, S.; Li, J.Q.; Xu, Z.X.; Jian, Y.P. Recent findings in Akkermansia muciniphila-regulated metabolism and its role in intestinal diseases. *Clin. Nutr.* **2022**, *41*, 2333–2344. [CrossRef]
58. Tomova, A.; Bukovsky, I.; Rembert, E.; Yonas, W.; Alwarith, J.; Barnard, N.D.; Kahleova, H. The effects of vegetarian and vegan diets on gut microbiota. *Front. Nutr.* **2019**, *6*, 447652. [CrossRef] [PubMed]
59. Spinler, J.K.; Oezguen, N.; Runge, J.K.; Luna, R.A.; Karri, V.; Yang, J.; Hirschi, K.D. Dietary impact of a plant-derived microRNA on the gut microbiome. *ExRNA* **2020**, *2*, 11. [CrossRef]
60. Trefflich, I.; Jabakhanji, A.; Menzel, J.; Blaut, M.; Michalsen, A.; Lampen, A.; Abraham, K.; Weikert, C. Is a vegan or a vegetarian diet associated with the microbiota composition in the gut? Results of a new cross-sectional study and systematic review. *Crit. Rev. Food Sci. Nutr.* **2020**, *60*, 2990–3004. [CrossRef] [PubMed]

61. Rinninella, E.; Tohumcu, E.; Raoul, P.; Fiorani, M.; Cintoni, M.; Mele, M.C.; Cammarota, G.; Gasbarrini, A.; Ianiro, G. The role of diet in shaping human gut microbiota. *Best Pract. Res. Clin. Gastroenterol.* **2023**, *62*, 101828. [CrossRef]
62. Guan, Z.W.; Yu, E.Z.; Feng, Q. Soluble dietary fiber, one of the most important nutrients for the gut microbiota. *Molecules* **2021**, *26*, 6802. [CrossRef]
63. Hughes, R.L.; Alvarado, D.A.; Swanson, K.S.; Holscher, H.D. The prebiotic potential of inulin-type fructans: A systematic review. *Adv. Nutr.* **2022**, *13*, 492–529. [CrossRef]
64. Singh, A.; Zapata, R.C.; Pezeshki, A.; Reidelberger, R.D.; Chelikani, P.K. Inulin fiber dose-dependently modulates energy balance, glucose tolerance, gut microbiota, hormones and diet preference in high-fat-fed male rats. *J. Nutr. Biochem.* **2018**, *59*, 142–152. [CrossRef] [PubMed]
65. Schroeder, B.O.; Birchenough, G.M.; Ståhlman, M.; Arike, L.; Johansson, M.E.; Hansson, G.C.; Bäckhed, F. *Bifidobacteria* or fiber protects against diet-induced microbiota-mediated colonic mucus deterioration. *Cell Host Microbe* **2018**, *23*, 27–40. [CrossRef] [PubMed]
66. Jia, X.; Chen, Q.; Wu, H.; Liu, H.; Jing, C.; Gong, A.; Zhang, Y. Exploring a novel therapeutic strategy: The interplay between gut microbiota and high-fat diet in the pathogenesis of metabolic disorders. *Front. Nutr.* **2023**, *10*, 1291853. [CrossRef] [PubMed]
67. Hryckowian, A.J.; Van Treuren, W.; Smits, S.A.; Davis, N.M.; Gardner, J.O.; Bouley, D.M.; Sonnenburg, J.L. Microbiota-accessible carbohydrates suppress *Clostridium difficile* infection in a murine model. *Nat. Microbiol.* **2018**, *3*, 662–669. [CrossRef]
68. Shi, H.; Wang, Q.; Zheng, M.; Hao, S.; Lum, J.S.; Chen, X.; Huang, X.F.; Yu, Y.; Zheng, K. Supplement of microbiota-accessible carbohydrates prevents neuroinflammation and cognitive decline by improving the gut microbiota-brain axis in diet-induced obese mice. *J. Neuroinflamm.* **2020**, *17*, 77. [CrossRef] [PubMed]
69. Rodriguez, C.I.; Isobe, K.; Martiny, J.B. Short-term dietary fiber interventions produce consistent gut microbiome responses across studies. *Res. Sq.* **2023**, *9*, e00133-24. [CrossRef]
70. Lee, J.H.; Kim, G.B.; Han, K.; Jung, E.J.; Suh, H.J.; Jo, K. Efficacy and safety of galacto-oligosaccharide in the treatment of functional constipation: Randomized clinical trial. *Food Funct.* **2024**, *15*, 6374–6382. [CrossRef]
71. Malesza, I.J.; Malesza, M.; Walkowiak, J.; Mussin, N.; Walkowiak, D.; Aringazina, R.; Bartkowiak-Wieczorek, J.; Mądry, E. High-fat, western-style diet, systemic inflammation, and gut microbiota: A narrative review. *Cells* **2021**, *10*, 3164. [CrossRef]
72. Wolters, M.; Ahrens, J.; Romaní-Pérez, M.; Watkins, C.; Sanz, Y.; Benítez-Páez, A.; Stanton, C.; Günther, K. Dietary fat, the gut microbiota, and metabolic health–A systematic review conducted within the MyNewGut project. *Clin. Nutr.* **2019**, *38*, 2504–2520. [CrossRef]
73. Roopashree, P.G.; Shetty, S.S.; Kumari, N.S. Effect of medium chain fatty acid in human health and disease. *J. Funct. Foods* **2021**, *87*, 104724. [CrossRef]
74. An, J.; Wang, Q.; Yi, S.; Liu, X.; Jin, H.; Xu, J.; Wen, J.; Zhu, J.; Tuo, B. The source of the fat significantly affects the results of high-fat diet intervention. *Sci. Rep.* **2022**, *12*, 4315. [CrossRef]
75. Wan, Y.; Wang, F.; Yuan, J.; Li, J.; Jiang, D.; Zhang, J.; Li, H.; Wang, R.; Tang, J.; Huang, T. Effects of dietary fat on gut microbiota and fecal metabolites, and their relationship with cardiometabolic risk factors: A 6-month randomized controlled-feeding trial. *Gut* **2019**, *68*, 1417–1429. [CrossRef]
76. Barrea, L.; Annunziata, G.; Muscogiuri, G.; Laudisio, D.; Di Somma, C.; Maisto, M.; Tenore, G.C.; Colao, A.; Savastano, S. Trimethylamine N-oxide, Mediterranean diet, and nutrition in healthy, normal-weight adults: Also a matter of sex? *Nutrition* **2019**, *62*, 7–17. [CrossRef]
77. Zhang, X.; Gérard, P. Diet-gut microbiota interactions on cardiovascular disease. *Comput. Struct. Biotechnol. J.* **2022**, *20*, 1528–1540. [CrossRef] [PubMed]
78. Dostal Webster, A.; Staley, C.; Hamilton, M.J.; Huang, M.; Fryxell, K.; Erickson, R.; Kabage, A.J.; Sadowsky, M.J.; Khoruts, A. Influence of short-term changes in dietary sulfur on the relative abundances of intestinal sulfate-reducing bacteria. *Gut Microbes* **2019**, *10*, 447–457. [CrossRef] [PubMed]
79. Boachie, R.T.; Capuano, E.; Oliviero, T.; Udenigwe, C.C.; Fogliano, V. Undigested glycated lentil proteins modulate the gut microbiota profile but not the metabolites in vitro. *J. Funct. Foods* **2023**, *107*, 105667. [CrossRef]
80. Graf, D.; Monk, J.M.; Lepp, D.; Wu, W.; McGillis, L.; Roberton, K.; Brummer, Y.; Tosh, S.M.; Power, K.A. Cooked red lentils dose-dependently modulate the colonic microenvironment in healthy C57Bl/6 male mice. *Nutrients* **2019**, *11*, 1853. [CrossRef]
81. Kadyan, S.; Sharma, A.; Arjmandi, B.H.; Singh, P.; Nagpal, R. Prebiotic potential of dietary beans and pulses and their resistant starch for aging-associated gut and metabolic health. *Nutrients* **2022**, *14*, 1726. [CrossRef]
82. Zhang, C.; Zhang, Y.; Li, H.; Liu, X. The potential of proteins, hydrolysates and peptides as growth factors for *Lactobacillus* and *Bifidobacterium*: Current research and future perspectives. *Food Funct.* **2020**, *11*, 1946–1957. [CrossRef]
83. Nakatani, A.; Li, X.; Miyamoto, J.; Igarashi, M.; Watanabe, H.; Sutou, A.; Watanabe, K.; Motoyama, T.; Tachibana, N.; Kohno, M.; et al. Dietary mung bean protein reduces high-fat diet-induced weight gain by modulating host bile acid metabolism in a gut microbiota-dependent manner. *Biochem. Biophys. Res. Commun.* **2018**, *501*, 955–961. [CrossRef]
84. Bielik, V.; Kolisek, M. Bioaccessibility and bioavailability of minerals in relation to a healthy gut microbiome. *Int. J. Mol. Sci.* **2021**, *22*, 6803. [CrossRef] [PubMed]
85. Zhang, F.; Ye, J.; Zhu, X.; Wang, L.; Gao, P.; Shu, G.; Jiang, Q.; Wang, S. Anti-obesity effects of dietary calcium: The evidence and possible mechanisms. *Int. J. Mol. Sci.* **2019**, *20*, 3072. [CrossRef] [PubMed]

86. Saadat, Y.R.; Niknafs, B.; Khatibi, S.M.H.; Ardalan, M.; Majdi, H.; Bahmanpoor, Z.; Abediazar, S.; Vahed, S.Z. Gut microbiota; An overlooked effect of phosphate binders. *Eur. J. Pharmacol.* **2020**, *868*, 172892. [CrossRef]
87. Sriwichaiin, S.; Thiennimitr, P.; Thonusin, C.; Sarichai, P.; Buddhasiri, S.; Kumfu, S.; Nawara, W.; Kittichotirat, W.; Fucharoen, S.; Chattipakorn, N.; et al. Deferiprone has less benefits on gut microbiota and metabolites in high iron-diet induced iron overload thalassemic mice than in iron overload wild-type mice: A preclinical study. *Life Sci.* **2022**, *307*, 120871. [CrossRef] [PubMed]
88. Fang, S.; Zhuo, Z.; Yu, X.; Wang, H.; Feng, J. Oral administration of liquid iron preparation containing excess iron induces intestine and liver injury, impairs intestinal barrier function, and alters the gut microbiota in rats. *J. Trace Elem. Med. Biol.* **2018**, *47*, 12–20. [CrossRef]
89. Murdoch, C.C.; Skaar, E.P. Nutritional immunity: The battle for nutrient metals at the host–pathogen interface. *Nat. Rev. Microbiol.* **2022**, *20*, 657–670. [CrossRef]
90. Ohashi, W.; Fukada, T. Contribution of zinc and zinc transporters in the pathogenesis of inflammatory bowel diseases. *J. Immunol. Res.* **2019**, *2019*, 8396878. [CrossRef] [PubMed]
91. Shen, H.; Han, J.; Li, Y.; Lu, C.; Zhou, J.; Li, Y.; Su, X. Different host-specific responses in thyroid function and gut microbiota modulation between diet-induced obese and normal mice given the same dose of iodine. *Appl. Microbiol. Biotechnol.* **2019**, *103*, 3537–3547. [CrossRef]
92. Ren, M.; Zhang, H.; Qi, J.; Hu, A.; Jiang, Q.; Hou, Y.; Feng, Q.; Ojo, O.; Wang, X. An almond-based low carbohydrate diet improves depression and glycometabolism in patients with type 2 diabetes through modulating gut microbiota and GLP-1: A randomized controlled trial. *Nutrients* **2020**, *12*, 3036. [CrossRef]
93. Drabińska, N.; Jarocka-Cyrta, E.; Markiewicz, L.H.; Krupa-Kozak, U. The effect of oligofructose-enriched inulin on fecal bacterial counts and microbiota-associated characteristics in celiac disease children following a gluten-free diet: Results of a randomized, placebo-controlled trial. *Nutrients* **2018**, *10*, 201. [CrossRef]
94. Wang, B.; Kong, Q.; Li, X.; Zhao, J.; Zhang, H.; Chen, W.; Wang, G. A high-fat diet increases gut microbiota biodiversity and energy expenditure due to nutrient difference. *Nutrients* **2020**, *12*, 3197. [CrossRef]
95. Kostovcikova, K.; Coufal, S.; Galanova, N.; Fajstova, A.; Hudcovic, T.; Kostovcik, M.; Prochazkova, P.; Jiraskova Zakostelska, Z.; Cermakova, M.; Sediva, B.; et al. Diet rich in animal protein promotes pro-inflammatory macrophage response and exacerbates colitis in mice. *Front. Immunol.* **2019**, *10*, 446008. [CrossRef] [PubMed]
96. Vidal-Lletjós, S.; Andriamihaja, M.; Blais, A.; Grauso, M.; Lepage, P.; Davila, A.M.; Viel, R.; Gaudichon, C.; Leclerc, M.; Blachier, F.; et al. Dietary protein intake level modulates mucosal healing and mucosa-adherent microbiota in mouse model of colitis. *Nutrients* **2019**, *11*, 514. [CrossRef] [PubMed]
97. Trautvetter, U.; Camarinha-Silva, A.; Jahreis, G.; Lorkowski, S.; Glei, M. High phosphorus intake, and gut-related parameters–results of a randomized placebo-controlled human intervention study. *Nutr. J.* **2018**, *17*, 23. [CrossRef]
98. Sjödin, K.S.; Domellöf, M.; Lagerqvist, C.; Hernell, O.; Lönnerdal, B.; Szymlek-Gay, E.A.; Sjödin, A.; West, C.E.; Lind, T. Administration of ferrous sulfate drops has significant effects on the gut microbiota of iron-sufficient infants: A randomised controlled study. *Gut* **2019**, *68*, 2095–2097. [CrossRef]
99. Huda, M.N.; Ahmad, S.M.; Kalanetra, K.M.; Taft, D.H.; Alam, M.J.; Khanam, A.; Raqib, R.; Underwood, M.A.; Mills, D.A.; Stephensen, C.B. Neonatal vitamin A supplementation and vitamin A status are associated with gut microbiome composition in Bangladeshi infants in early infancy and at 2 years of age. *J. Nutr.* **2019**, *149*, 1075–1088. [CrossRef]
100. Garg, M.; Hendy, P.; Ding, J.N.; Shaw, S.; Hold, G.; Hart, A. The effect of vitamin D on intestinal inflammation and fecal microbiota in patients with ulcerative colitis. *J. Crohn's Colitis* **2018**, *12*, 963–972. [CrossRef]
101. Marques, C.; Fernandes, I.; Meireles, M.; Faria, A.; Spencer, J.P.; Mateus, N.; Calhau, C. Gut microbiota modulation accounts for the neuroprotective properties of anthocyanins. *Sci. Rep.* **2018**, *8*, 11341. [CrossRef] [PubMed]
102. Mayta-Apaza, A.C.; Pottgen, E.; De Bodt, J.; Papp, N.; Marasini, D.; Howard, L.; Abranko, L.; Van de Wiele, T.; Lee, S.O.; Carbonero, F. Impact of tart cherries polyphenols on the human gut microbiota and phenolic metabolites in vitro and in vivo. *J. Nutr. Biochem.* **2018**, *59*, 160–172. [CrossRef]
103. Chacar, S.; Itani, T.; Hajal, J.; Saliba, Y.; Louka, N.; Faivre, J.F.; Maroun, R.; Fares, N. The impact of long-term intake of phenolic compounds-rich grape pomace on rat gut microbiota. *J. Food Sci.* **2018**, *83*, 246–251. [CrossRef]
104. Gibson, G.R.; Hutkins, R.; Sanders, M.E.; Prescott, S.L.; Reimer, R.A.; Salminen, S.J.; Scott, K.; Stanton, C.; Swanson, K.S.; Cani, P.D.; et al. Expert consensus document: The International Scientific Association for Probiotics and Prebiotics (ISAPP) consensus statement on the definition and scope of prebiotics. *Nat. Rev. Gastroenterol. Hepatol.* **2017**, *14*, 491–502. [CrossRef] [PubMed]
105. Ji, J.; Jin, W.; Liu, S.J.; Jiao, Z.; Li, X. Probiotics, prebiotics, and postbiotics in health and disease. *MedComm* **2023**, *4*, e420. [CrossRef] [PubMed]
106. Fusco, W.; Lorenzo, M.B.; Cintoni, M.; Porcari, S.; Rinninella, E.; Kaitsas, F.; Lener, E.; Mele, M.C.; Gasbarrini, A.; Collado, M.C.; et al. Short-chain fatty-acid-producing bacteria: Key components of the human gut microbiota. *Nutrients* **2023**, *15*, 2211. [CrossRef]
107. Barratt, M.J.; Nuzhat, S.; Ahsan, K.; Frese, S.A.; Arzamasov, A.A.; Sarker, S.A.; Slam, M.M.; Palit, P.; Islam, M.R.; Gordon, J.I.; et al. *Bifidobacterium infantis* treatment promotes weight gain in Bangladeshi infants with severe acute malnutrition. *Sci. Transl. Med.* **2022**, *14*, eabk1107. [CrossRef] [PubMed]

108. Paone, P.; Suriano, F.; Jian, C.; Korpela, K.; Delzenne, N.M.; Van Hul, M.; Salonen, A.; Cani, P.D. Prebiotic oligofructose protects against high-fat diet-induced obesity by changing the gut microbiota, intestinal mucus production, glycosylation and secretion. *Gut Microbes* **2022**, *14*, 2152307. [CrossRef] [PubMed]
109. Mörkl, S.; Butler, M.I.; Holl, A.; Cryan, J.F.; Dinan, T.G. Probiotics and the microbiota-gut-brain axis: Focus on psychiatry. *Curr. Nutr. Rep.* **2020**, *9*, 171–182. [CrossRef]
110. Prabhu, S.; Molath, A.; Choksi, H.; Kumar, S.; Mehra, R. Classifications of polyphenols and their potential application in human health and diseases. *Int. J. Phys. Educ.* **2021**, *6*, 293–301. [CrossRef]
111. Domínguez-Avila, J.A.; Villa-Rodriguez, J.A.; Montiel-Herrera, M.; Pacheco-Ordaz, R.; Roopchand, D.E.; Venema, K.; González-Aguilar, G.A. Phenolic compounds promote diversity of gut microbiota and maintain colonic health. *Dig. Dis. Sci.* **2021**, *66*, 3270–3289. [CrossRef]
112. Makarewicz, M.; Drożdż, I.; Tarko, T.; Duda-Chodak, A. The interactions between polyphenols and microorganisms, especially gut microbiota. *Antioxidants* **2021**, *10*, 188. [CrossRef]
113. Wang, L.; Zeng, B.; Liu, Z.; Liao, Z.; Zhong, Q.; Gu, L.; Wei, H.; Fang, X. Green tea polyphenols modulate colonic microbiota diversity and lipid metabolism in high-fat diet treated HFA mice. *J. Food Sci.* **2018**, *83*, 864–873. [CrossRef]
114. Addala, M.S.; Gudipati, M. Resistant starch and amylolytic enzymes from human gastrointestinal microflora-A review. *Transl. Cancer Res.* **2023**, *15*, 106.
115. Yoo, J.Y.; Groer, M.; Dutra, S.V.O.; Sarkar, A.; McSkimming, D.I. Gut microbiota and immune system interactions. *Microorganisms* **2020**, *8*, 1587. [CrossRef]
116. Chu, J.; Feng, S.; Guo, C.; Xue, B.; He, K.; Li, L. Immunological mechanisms of inflammatory diseases caused by gut microbiota dysbiosis: A review. *Biomed. Pharmacother.* **2023**, *164*, 114985. [CrossRef]
117. Schoultz, I.; Keita, Å.V. The intestinal barrier and current techniques for the assessment of gut permeability. *Cells* **2020**, *9*, 1909. [CrossRef]
118. Gieryńska, M.; Szulc-Dąbrowska, L.; Struzik, J.; Mielcarska, M.B.; Gregorczyk-Zboroch, K.P. Integrity of the intestinal barrier: The involvement of epithelial cells and microbiota—A mutual relationship. *Animals* **2022**, *12*, 145. [CrossRef]
119. Dong, Y.; Zhang, K.; Wei, J.; Ding, Y.; Wang, X.; Hou, H.; Wu, J.; Liu, T.; Wang, B.; Cao, H. Gut microbiota-derived short-chain fatty acids regulate gastrointestinal tumor immunity: A novel therapeutic strategy? *Front. Immunol.* **2023**, *14*, 1158200. [CrossRef] [PubMed]
120. Zhang, M.; Wu, C. The relationship between intestinal goblet cells and the immune response. *Biosci. Rep.* **2020**, *40*, BSR20201471. [CrossRef]
121. Brown, H.; Esterházy, D. Intestinal immune compartmentalization: Implications of tissue specific determinants in health and disease. *Mucosal Immunol.* **2021**, *14*, 1259–1270. [CrossRef]
122. Zhao, L.; Zhang, F.; Ding, X.; Wu, G.; Lam, Y.Y.; Wang, X.; Fu, H.; Xue, X.; Lu, C.; Ma, J.; et al. Gut bacteria selectively promoted by dietary fibers alleviate type 2 diabetes. *Sciences* **2018**, *359*, 1151–1156. [CrossRef]
123. Tajik, N.; Frech, M.; Schulz, O.; Schälter, F.; Lucas, S.; Azizov, V.; Dürholz, K.; Steffen, F.; Omata, Y.; Rings, A.; et al. Targeting zonulin and intestinal epithelial barrier function to prevent the onset of arthritis. *Nat. Commun.* **2020**, *11*, 1995. [CrossRef]
124. Jergens, A.E.; Parvinroo, S.; Kopper, J.; Wannemuehler, M.J. Rules of engagement: Epithelial-microbe interactions and inflammatory bowel disease. *Front. Med.* **2021**, *8*, 669913. [CrossRef] [PubMed]
125. Forlano, R.; Mullish, B.H.; Roberts, L.A.; Thursz, M.R.; Manousou, P. The intestinal barrier and its dysfunction in patients with metabolic diseases and non-alcoholic fatty liver disease. *Int. J. Mol. Sci.* **2022**, *23*, 662. [CrossRef] [PubMed]
126. Paray, B.A.; Albeshr, M.F.; Jan, A.T.; Rather, I.A. Leaky gut and autoimmunity: An intricate balance in individuals health and the diseased state. *Int. J. Mol. Sci.* **2020**, *21*, 9770. [CrossRef]
127. Camilleri, M. Leaky gut: Mechanisms, measurement and clinical implications in humans. *Gut* **2019**, *68*, 1516–1526. [CrossRef]
128. Gustafsson, J.K.; Johansson, M.E. The role of goblet cells and mucus in intestinal homeostasis. *Nat. Rev. Gastroenterol. Hepatol.* **2022**, *19*, 785–803. [CrossRef]
129. Shoubridge, A.P.; Choo, J.M.; Martin, A.M.; Keating, D.J.; Wong, M.L.; Licinio, J.; Rogers, G.B. The gut microbiome and mental health: Advances in research and emerging priorities. *Mol. Psychiatry* **2022**, *27*, 1908–1919. [CrossRef] [PubMed]
130. Mirzaei, R.; Zamani, F.; Hajibaba, M.; Rasouli-Saravani, A.; Noroozbeygi, M.; Gorgani, M.; Hosseini-Fard, S.R.; Jalalifar, S.; Ajdarkosh, H.; Abedi, S.H.; et al. The pathogenic, therapeutic and diagnostic role of exosomal microRNA in the autoimmune diseases. *J. Neuroimmunol.* **2021**, *358*, 577640. [CrossRef]
131. Rasouli-Saravani, A.; Jahankhani, K.; Moradi, S.; Gorgani, M.; Shafaghat, Z.; Mirsanei, Z.; Mehmandar, A.; Mirzaei, R. Role of microbiota short-chain fatty acids in the pathogenesis of autoimmune diseases. *Biomed. Pharmacother.* **2023**, *162*, 114620. [CrossRef]
132. Flynn, J.K.; Ortiz, A.M.; Herbert, R.; Brenchley, J.M. Host genetics and environment shape the composition of the gastrointestinal microbiome in nonhuman primates. *Microbiol. Spectr.* **2023**, *11*, e02139-22. [CrossRef]
133. Sun, Y.; Zhang, S.; Nie, Q.; He, H.; Tan, H.; Geng, F.; Ji, H.; Hu, J.; Nie, S. Gut firmicutes: Relationship with dietary fiber and role in host homeostasis. *Crit. Rev. Food Sci. Nutr.* **2023**, *63*, 12073–12088. [CrossRef]
134. Xu, Y.; Zhu, Y.; Li, X.; Sun, B. Dynamic balancing of intestinal short-chain fatty acids: The crucial role of bacterial metabolism. *Trends Food Sci. Technol.* **2020**, *100*, 118–130. [CrossRef]

135. Frolova, M.S.; Suvorova, I.A.; Iablokov, S.N.; Petrov, S.N.; Rodionov, D.A. Genomic reconstruction of short-chain fatty acid production by the human gut microbiota. *Front. Mol. Biosci.* **2022**, *9*, 949563. [CrossRef]
136. Liu, X.F.; Shao, J.H.; Liao, Y.T.; Wang, L.N.; Jia, Y.; Dong, P.J.; Liu, Z.Z.; He, D.D.; Li, C.; Zhang, X. Regulation of short-chain fatty acids in the immune system. *Front. Immunol.* **2023**, *14*, 1186892. [CrossRef]
137. Notting, F.; Pirovano, W.; Sybesma, W.; Kort, R. The butyrate-producing and spore-forming bacterial genus *Coprococcus* as a potential biomarker for neurological disorders. *Gut Microbiome* **2023**, *4*, e16. [CrossRef]
138. Kim, C.H. Complex regulatory effects of gut microbial short-chain fatty acids on immune tolerance and autoimmunity. *Cell. Mol. Immunol.* **2023**, *20*, 341–350. [CrossRef] [PubMed]
139. Souders, C.L.; Aristizabal-Henao, J.J.; Patuel, S.J.; Bowden, J.A.; Zubcevic, J.; Martyniuk, C.J. Interaction between butyrate and tumor necrosis factor α in primary rat colonocytes. *Biomolecules* **2023**, *13*, 258. [CrossRef]
140. Liu, H.; Wang, J.; He, T.; Becker, S.; Zhang, G.; Li, D.; Ma, X. Butyrate: A double-edged sword for health? *Adv. Nutr.* **2018**, *9*, 21–29. [CrossRef]
141. Salvi, P.S.; Cowles, R.A. Butyrate and the intestinal epithelium: Modulation of proliferation and inflammation in homeostasis and disease. *Cells* **2021**, *10*, 1775. [CrossRef]
142. Gasaly, N.; Hermoso, M.A.; Gotteland, M. Butyrate and the fine-tuning of colonic homeostasis: Implication for inflammatory bowel diseases. *Int. J. Mol. Sci.* **2021**, *22*, 3061. [CrossRef]
143. Martin-Gallausiaux, C.; Marinelli, L.; Blottière, H.M.; Larraufie, P.; Lapaque, N. SCFA: Mechanisms and functional importance in the gut. *Proc. Nutr. Soc.* **2021**, *80*, 37–49. [CrossRef]
144. Peng, K.; Xia, S.; Xiao, S.; Yu, Q. Short-chain fatty acids affect the development of inflammatory bowel disease through intestinal barrier, immunology, and microbiota: A promising therapy? *J. Gastroenterol. Hepatol.* **2022**, *37*, 1710–1718. [CrossRef] [PubMed]
145. Yao, Y.; Cai, X.; Fei, W.; Ye, Y.; Zhao, M.; Zheng, C. The role of short-chain fatty acids in immunity, inflammation and metabolism. *Crit. Rev. Food Sci. Nutr.* **2022**, *62*, 1–12. [CrossRef] [PubMed]
146. Chen, J.; Vitetta, L. The role of butyrate in attenuating pathobiont-induced hyperinflammation. *Immune Netw.* **2020**, *20*, E15. [CrossRef] [PubMed]
147. Maes, L. Characterisation of an In Vitro Intestinal Model of the Mucosal Interface in the Context of Crohn's Disease. Master's Thesis, Ghent University, Ghent, Belgium, 2020.
148. Wang, J.; Zhu, N.; Su, X.; Gao, Y.; Yang, R. Gut-microbiota-derived metabolites maintain gut and systemic immune homeostasis. *Cells* **2023**, *12*, 793. [CrossRef]
149. Perino, A.; Schoonjans, K. Metabolic messengers: Bile acids. *Nat. Metab.* **2022**, *4*, 416–423. [CrossRef]
150. Pedersen, K.J.; Haange, S.B.; Žížalová, K.; Viehof, A.; Clavel, T.; Leníček, M.; Engelmann, B.; Wick, L.Y.; Schaap, F.G.; Jehmlich, N.; et al. *Eggerthella lenta* DSM 2243 alleviates bile acid stress response in *Clostridium ramosum* and *Anaerostipes caccae* by transformation of bile acids. *Microorganisms* **2022**, *10*, 2025. [CrossRef]
151. Agus, A.; Clément, K.; Sokol, H. Gut microbiota-derived metabolites as central regulators in metabolic disorders. *Gut* **2021**, *70*, 1174–1182. [CrossRef]
152. Hendrikx, T.; Schnabl, B. Metabolites produced by intestinal bacteria capable of controlling liver disease manifestation. *J. Intern. Med.* **2019**, *286*, 32–40. [CrossRef]
153. Dupont, H.L.; Jiang, Z.D.; Dupont, A.W.; Utay, N.S. The intestinal microbiome in human health and disease. *Trans. Am. Clin. Climatol. Assoc.* **2020**, *131*, 178.
154. Perrotta, G. The intestinal microbiota: Towards a multifactorial integrative model. Eubiosis and dysbiosis in morbid physical and psychological conditions. *Arch. Clin. Gastroenterol.* **2021**, *7*, 024–035. [CrossRef]
155. Maciel-Fiuza, M.F.; Muller, G.C.; Campos, D.M.S.; do Socorro Silva Costa, P.; Peruzzo, J.; Bonamigo, R.R.; Veit, T.; Vianna, F.S.L. Role of gut microbiota in infectious and inflammatory diseases. *Front. Microbiol.* **2023**, *14*, 1098386. [CrossRef]
156. Solís González, V.E. Understanding the Role of Epithelial Myd88 Signals in the Early Development of Experimental Autoimmune Encephalomyelitis (EAE). Master's Thesis, Ludwig Maximilian University, Munich, Germany, 2022.
157. Forbes, J.D.; Chen, C.Y.; Knox, N.C.; Marrie, R.A.; El-Gabalawy, H.; de Kievit, T.; Alfa, M.; Bernstein, C.N.; Van Domselaar, G. A comparative study of the gut microbiota in immune-mediated inflammatory diseases—Does a common dysbiosis exist? *Microbiome* **2018**, *6*, 221. [CrossRef] [PubMed]
158. Meijnikman, A.S.; Gerdes, V.E.; Nieuwdorp, M.; Herrema, H. Evaluating causality of gut microbiota in obesity and diabetes in humans. *Endocr. Rev.* **2018**, *39*, 133–153. [CrossRef] [PubMed]
159. Alatab, S.; Sepanlou, S.G.; Ikuta, K.; Vahedi, H.; Bisignano, C.; Safiri, S.; Naghavi, M. The global, regional, and national burden of inflammatory bowel disease in 195 countries and territories, 1990–2017: A systematic analysis for the global burden of disease study 2017. *Lancet Gastroenterol. Hepatol.* **2020**, *5*, 17–30. [CrossRef]
160. Yang, C.; Long, D.; Sung, J.; Alghoul, Z.; Merlin, D. Orally administered natural lipid nanoparticle-loaded 6-shogaol shapes the anti-inflammatory microbiota and metabolome. *Pharmaceutics* **2021**, *13*, 1355. [CrossRef]
161. Arukha, A.P.; Freguia, C.F.; Mishra, M.; Jha, J.K.; Kariyawasam, S.; Fanger, N.A.; Zimmermann, E.M.; Fanger, G.R.; Sahay, B. *Lactococcus lactis* delivery of surface layer protein A protects mice from colitis by re-setting host immune repertoire. *Biomedicine* **2021**, *9*, 1098. [CrossRef]
162. Mehandru, S.; Colombel, J.F. The intestinal barrier, an arbitrator turned provocateur in IBD. *Nat. Rev. Gastroenterol. Hepatol.* **2021**, *18*, 83–84. [CrossRef]

163. Ni, Y.; Ni, L.; Zhuge, F.; Xu, L.; Fu, Z.; Ota, T. Adipose tissue macrophage phenotypes and characteristics: The key to insulin resistance in obesity and metabolic disorders. *Obesity* **2020**, *28*, 225–234. [CrossRef]
164. Fasano, A. All disease begins in the (leaky) gut: Role of zonulin-mediated gut permeability in the pathogenesis of some chronic inflammatory diseases. *F1000Research* **2020**, *9*, F1000 Faculty Rev-69. [CrossRef] [PubMed]
165. Magne, F.; Gotteland, M.; Gauthier, L.; Zazueta, A.; Pesoa, S.; Navarrete, P.; Balamurugan, R. The firmicutes/bacteroidetes ratio: A relevant marker of gut dysbiosis in obese patients? *Nutrients* **2020**, *12*, 1474. [CrossRef]
166. Mohammad, S.; Thiemermann, C. Role of metabolic endotoxemia in systemic inflammation and potential interventions. *Front. Immunol.* **2021**, *11*, 594150. [CrossRef] [PubMed]
167. Pfalzer, A.C.; Leung, K.; Crott, J.W.; Kim, S.J.; Tai, A.K.; Parnell, L.D.; Kamanu, F.K.; Liu, Z.; Rogers, G.; Shea, M.K.; et al. Incremental elevations in TNFα and IL6 in the human colon and precancerous changes in the mucosal transcriptome accompany adiposity. *Cancer Epidemiol. Biomarkers Prev.* **2018**, *27*, 1416–1423. [CrossRef]
168. Lin, L.; Zhang, K.; Xiong, Q.; Zhang, J.; Cai, B.; Huang, Z.; Yang, B.; Wei, B.; Chen, J.; Niu, Q. Gut microbiota in pre-clinical rheumatoid arthritis: From pathogenesis to preventing progression. *J. Autoimmun.* **2023**, *141*, 103001. [CrossRef] [PubMed]
169. Maeda, Y.; Takeda, K. Host–microbiota interactions in rheumatoid arthritis. *Exp. Mol. Med.* **2019**, *51*, 1–6. [CrossRef]
170. Jiao, Y.; Wu, L.; Huntington, N.D.; Zhang, X. Crosstalk between gut microbiota and innate immunity and its implication in autoimmune diseases. *Front. Immunol.* **2020**, *11*, 282. [CrossRef]
171. Violi, F.; Cammisotto, V.; Bartimoccia, S.; Pignatelli, P.; Carnevale, R.; Nocella, C. Gut-derived low-grade endotoxemia, atherothrombosis, and cardiovascular disease. *Nat. Rev. Cardiol.* **2023**, *20*, 24–37. [CrossRef]
172. Di Vincenzo, F.; Del Gaudio, A.; Petito, V.; Lopetuso, L.R.; Scaldaferri, F. Gut microbiota, intestinal permeability, and systemic inflammation: A narrative review. *Intern. Emerg. Med.* **2024**, *19*, 275–293. [CrossRef]
173. Joshi, C.; Bapat, R.; Anderson, W.; Dawson, D.; Hijazi, K.; Cherukara, G. Detection of periodontal microorganisms in coronary atheromatous plaque specimens of myocardial infarction patients: A systematic review and meta-analysis. *Trends Cardiovasc. Med.* **2021**, *31*, 69–82. [CrossRef]
174. Carnevale, R.; Nocella, C.; Petrozza, V.; Cammisotto, V.; Pacini, L.; Sorrentino, V.; Violi, F. Localization of lipopolysaccharide from *Escherichia Coli* into human atherosclerotic plaque. *Sci. Rep.* **2018**, *8*, 3598. [CrossRef]
175. Tang, W.H.W.; Li, X.S.; Wu, Y.; Wang, Z.; Khaw, K.T.; Wareham, N.J. Plasma trimethylamine N-oxide (TMAO) levels predict future risk of coronary artery disease in apparently healthy individuals in the EPIC-Norfolk prospective population study. *Am. Heart J.* **2021**, *236*, 80–86. [CrossRef]
176. Jensen, A.B.; Sørensen, T.I.; Pedersen, O.; Jess, T.; Brunak, S.; Allin, K.H. Increase in clinically recorded type 2 diabetes after colectomy. *eLife* **2018**, *30*, e37420. [CrossRef]
177. Fan, Y.; Pedersen, O. Gut microbiota in human metabolic health and disease. *Nat. Rev. Microbiol.* **2021**, *19*, 55–71. [CrossRef]
178. Ruuskanen, M.O.; Erawijantari, P.P.; Havulinna, A.S.; Liu, Y.; Méric, G.; Tuomilehto, J. Gut microbiome composition is predictive of incident type 2 diabetes in a population cohort of 5572. Finnish adults. *Diabetes Care* **2022**, *45*, 811–818. [CrossRef]
179. Sugawara, Y.; Kanazawa, A.; Aida, M.; Yoshida, Y.; Yamashiro, Y.; Watada, H. Association of gut microbiota and inflammatory markers in obese patients with type 2 diabetes mellitus: Post hoc analysis of a symbiotic interventional study. *Biosci. Microbiota Food Health* **2022**, *41*, 103–111. [CrossRef] [PubMed]
180. Vallianou, N.G.; Stratigou, T.; Tsagarakis, S. Metformin and gut microbiota: Their interactions and their impact on diabetes. *Hormones* **2019**, *18*, 141–144. [CrossRef] [PubMed]
181. Benamar, M.; Chen, Q.; Martinez-Blanco, M.; Chatila, T.A. Regulatory T cells in allergic inflammation. *Semin. Immunol.* **2023**, *70*, 101847. [CrossRef] [PubMed]
182. Aleksandrova, K.; Koelman, L.; Rodrigues, C.E. Dietary patterns and biomarkers of oxidative stress and inflammation: A systematic review of observational and intervention studies. *Redox Biol.* **2021**, *42*, 101869. [CrossRef]
183. Stanimirovic, J.; Radovanovic, J.; Banjac, K.; Obradovic, M.; Essack, M.; Zafirovic, S.; Gluvic, Z.; Gojobori, T.; Isenovic, E.R. Role of C-reactive protein in diabetic inflammation. *Mediators Inflamm.* **2022**, *2022*, 3706508. [CrossRef]
184. Wu, H.; Ballantyne, C.M. Metabolic inflammation and insulin resistance in obesity. *Circ. Res.* **2020**, *126*, 1549–1564. [CrossRef]
185. Khanna, D.; Khanna, S.; Khanna, P.; Kahar, P.; Patel, B.M. Obesity: A chronic low-grade inflammation and its markers. *Cureus* **2022**, *14*, e22711. [CrossRef]
186. DeMizio, D.J.; Geraldino-Pardilla, L.B. Autoimmunity and inflammation link to cardiovascular disease risk in rheumatoid arthritis. *Rheumatol. Ther.* **2020**, *7*, 19–33. [CrossRef] [PubMed]
187. Parikh, M.; Maddaford, T.G.; Austria, J.A.; Aliani, M.; Netticadan, T.; Pierce, G.N. Dietary flaxseed as a strategy for improving human health. *Nutrients* **2019**, *11*, 1171. [CrossRef]
188. Barrea, L.; Muscogiuri, G.; Frias-Toral, E.; Laudisio, D.; Pugliese, G.; Castellucci, B.; Garcia-Velasquez, E.; Savastano, S.; Colao, A. Nutrition and immune system: From the Mediterranean diet to dietary supplementary through the microbiota. *Crit. Rev. Food Sci. Nutr.* **2021**, *61*, 3066–3090. [CrossRef] [PubMed]
189. Luisi, M.L.E.; Lucarini, L.; Biffi, B.; Rafanelli, E.; Pietramellara, G.; Durante, M.; Vidali, S.; Provensi, G.; Madiai, S.; Gheri, C.F.; et al. Effect of Mediterranean diet enriched in high quality extra virgin olive oil on oxidative stress, inflammation and gut microbiota in obese and normal weight adult subjects. *Front. Pharmacol.* **2019**, *10*, 1366. [CrossRef] [PubMed]
190. Wang, Y.B.; Shivappa, N.; Hébert, J.R.; Page, A.J.; Gill, T.K.; Melaku, Y.A. Association between dietary inflammatory index, dietary patterns, plant-based dietary index and the risk of obesity. *Nutrients* **2021**, *13*, 1536. [CrossRef] [PubMed]

191. Menzel, J.; Jabakhanji, A.; Biemann, R.; Mai, K.; Abraham, K.; Weikert, C. Systematic review and meta-analysis of the associations of vegan and vegetarian diets with inflammatory biomarkers. *Sci. Rep.* **2020**, *10*, 21736. [CrossRef]
192. Watanabe, D.; Guo, Y.; Kamada, N. Interaction between the inflammasome and commensal microorganisms in gastrointestinal health and disease. *EMBO Mol. Med.* **2021**, *13*, e13452. [CrossRef]
193. Velázquez, K.T.; Enos, R.T.; Bader, J.E.; Sougiannis, A.T.; Carson, M.S.; Chatzistamou, I.; Carson, J.A.; Nagarkatti, P.S.; Nagarkatti, M.; Murphy, E.A. Prolonged high-fat-diet feeding promotes non-alcoholic fatty liver disease and alters gut microbiota in mice. *World J. Hepatol.* **2019**, *11*, 619. [CrossRef]
194. Velasquez, M.T. Altered gut microbiota: A link between diet and the metabolic syndrome. *Metab. Syndr. Relat. Disord.* **2018**, *16*, 321–328. [CrossRef]
195. Yang, J.; Wei, H.; Zhou, Y.; Szeto, C.H.; Li, C.; Lin, Y.; Coker, O.O.; Lau, H.C.H.; Chan, A.W.H.; Sung, J.J.Y.; et al. High-fat diet promotes colorectal tumorigenesis through modulating gut microbiota and metabolites. *Gastroenterology* **2022**, *162*, 135–149. [CrossRef]
196. Wolter, M.; Grant, E.T.; Boudaud, M.; Pudlo, N.A.; Pereira, G.V.; Eaton, K.A.; Desai, M.S. Diet-driven differential response of *Akkermansia muciniphila* modulates pathogen susceptibility. *Molecul. Syst. Biol.* **2024**, *20*, 596–625. [CrossRef] [PubMed]
197. Kim, H.; Hu, E.A.; Rebholz, C.M. Ultra-processed food intake and mortality in the USA: Results from the Third National Health and Nutrition Examination Survey (NHANES III, 1988–1994). *Public Health Nutr.* **2019**, *22*, 1777–1785. [CrossRef] [PubMed]
198. Fajstova, A.; Galanova, N.; Coufal, S.; Malkova, J.; Kostovcik, M.; Cermakova, M.; Pelantova, H.; Kuzma, M.; Sediva, B.; Hudcovic, T.; et al. Diet rich in simple sugars promotes a pro-inflammatory response via gut microbiota alteration and TLR4 signaling. *Cells* **2020**, *9*, 2701. [CrossRef]
199. Khan, S.; Waliullah, S.; Godfrey, V.; Khan, M.A.W.; Ramachandran, R.A.; Cantarel, B.L.; Behrendt, C.; Peng, L.; Hooper, L.V.; Zaki, H. Dietary simple sugars alter microbial ecology in the gut and promote colitis in mice. *Sci. Transl. Med.* **2020**, *12*, eaay6218. [CrossRef]
200. Laffin, M.; Fedorak, R.; Zalasky, A.; Park, H.; Gill, A.; Agrawal, A.; Keshteli, A.; Hotte, N.; Madsen, K.L. A high-sugar diet rapidly enhances susceptibility to colitis via depletion of luminal short-chain fatty acids in mice. *Sci. Rep.* **2019**, *9*, 12294. [CrossRef] [PubMed]
201. Hossain, K.S.; Amarasena, S.; Mayengbam, S. B vitamins and their roles in gut health. *Microorganisms* **2022**, *10*, 1168. [CrossRef]
202. Lim, M.Y.; You, H.J.; Yoon, H.S.; Kwon, B.; Lee, J.Y.; Lee, S.; Song, Y.M.; Lee, K.; Sung, J.; Ko, G. The effect of heritability and host genetics on the gut microbiota and metabolic syndrome. *Gut* **2017**, *66*, 1031–1038. [CrossRef]
203. Goodrich, J.K.; Davenport, E.R.; Clark, A.G.; Ley, R.E. The relationship between the human genome and microbiome comes into view. *Annu. Rev. Genetic.* **2017**, *51*, 413–433. [CrossRef]
204. Mundula, T.; Russo, E.; Curini, L.; Giudici, F.; Piccioni, A.; Franceschi, F.; Amedei, A. Chronic systemic low-grade inflammation and modern lifestyle: The dark role of gut microbiota on related diseases with a focus on COVID-19 pandemic. *Curr. Med. Chem.* **2022**, *29*, 5370–5396. [CrossRef]
205. Popli, S.; Badgujar, P.C.; Agarwal, T.; Bhushan, B.; Mishra, V. Persistent organic pollutants in foods, their interplay with gut microbiota and resultant toxicity. *Sci. Total Environ.* **2022**, *832*, 155084. [CrossRef]
206. Góralczyk-Bińkowska, A.; Szmajda-Krygier, D.; Kozłowska, E. The microbiota-gut-brain axis in psychiatric disorders. *Int. J. Molecul. Sci.* **2022**, *23*, 11245. [CrossRef] [PubMed]
207. Sharkey, K.A.; Mawe, G.M. The enteric nervous system. *Physiol. Rev.* **2023**, *103*, 1487–1564. [CrossRef]
208. Iannone, L.F.; Preda, A.; Blottière, H.M.; Clarke, G.; Albani, D.; Belcastro, V.; Carotenuto, M.; Cattaneo, A.; Citraro, R.; Ferraris, C.; et al. Microbiota-gut brain axis involvement in neuropsychiatric disorders. *Expert Rev. Neurother.* **2019**, *19*, 1037–1050. [CrossRef]
209. Kim, Y.K.; Shin, C. The microbiota-gut-brain axis in neuropsychiatric disorders: Pathophysiological mechanisms and novel treatments. *Curr. Neuropharmacol.* **2018**, *16*, 559–573. [CrossRef] [PubMed]
210. Bai, Y.; Zhou, X.; Zhao, J.; Wang, Z.; Ye, H.; Pi, Y.; Che, D.; Han, D.; Zhang, S.; Wang, J. Sources of dietary fiber affect the SCFA production and absorption in the hindgut of growing pigs. *Front. Nutri.* **2022**, *8*, 719935. [CrossRef] [PubMed]
211. Jamar, G.; Ribeiro, D.A.; Pisani, L.P. High-fat or high-sugar diets as trigger inflammation in the microbiota-gut-brain axis. *Crit. Rev. Food Sci. Nutri.* **2021**, *61*, 836–854. [CrossRef] [PubMed]
212. Doney, E.; Cadoret, A.; Dion-Albert, L.; Lebel, M.; Menard, C. Inflammation-driven brain and gut barrier dysfunction in stress and mood disorders. *Eur. J. Neurosci.* **2022**, *55*, 2851–2894. [CrossRef]
213. Hrncir, T. Gut microbiota dysbiosis: Triggers, consequences, diagnostic and therapeutic options. *Microorganisms* **2022**, *10*, 578. [CrossRef]
214. Malnick, S.D.; Fisher, D.; Somin, M.; Neuman, M.G. Treating the metabolic syndrome by fecal transplantation—Current status. *Biology* **2021**, *10*, 447. [CrossRef]
215. Sanders, M.E.; Merenstein, D.J.; Reid, G.; Gibson, G.R.; Rastall, R.A. Probiotics and prebiotics in intestinal health and disease: From biology to the clinic. *Nat. Rev. Gastroenterol. Hepatol.* **2019**, *16*, 605–616. [CrossRef]
216. Jaworska, K.; Hering, D.; Mosieniak, G.; Bielak-Zmijewska, A.; Pilz, M.; Konwerski, M.; Gasecka, A.; Kapłon-Cieślicka, A.; Filipiak, K.; Sikora, E.; et al. TMA, a forgotten uremic toxin, but not TMAO, is involved in cardiovascular pathology. *Toxins* **2019**, *11*, 490. [CrossRef] [PubMed]
217. Bush, C.L.; Blumberg, J.B.; El-Sohemy, A.; Minich, D.M.; Ordovás, J.M.; Reed, D.G.; Behm, V.A.Y. Toward the definition of personalized nutrition: A proposal by the American Nutrition Association. *J. Am. Coll. Nutr.* **2020**, *39*, 5–15. [CrossRef] [PubMed]

218. Pradhan, D.; Mallappa, R.H.; Grover, S. Comprehensive approaches for assessing the safety of probiotic bacteria. *Food Control* **2020**, *108*, 106872. [CrossRef]
219. Speranza, B.; Campaniello, D.; Petruzzi, L.; Altieri, C.; Sinigaglia, M.; Bevilacqua, A.; Rosaria Corbo, M. The inoculation of probiotics in vivo is a challenge: Strategies to improve their survival, to avoid unpleasant changes, or enhance their performances in beverages. *Beverages* **2020**, *6*, 20. [CrossRef]

International Journal of
Molecular Sciences

MDPI

Article

Bifidobacterium longum and *Chlorella sorokiniana* Combination Modulates IFN-γ, IL-10, and SOCS3 in Rotavirus-Infected Cells

Felizardo Velderrain-Armenta [1], Guadalupe González-Ochoa [1,*], Patricia Tamez-Guerra [2,*], Ricardo Romero-Arguelles [2], César I. Romo-Sáenz [2], Ricardo Gomez-Flores [2], Lilian Flores-Mendoza [1], Ramona Icedo-García [1] and José G. Soñanez-Organis [1]

[1] Department of Chemistry-Biology and Agriculture, Interdisciplinary Faculty of Biology Sciences and Health, University of Sonora, Navojoa C.P. 85880, Mexico; feli_fel@hotmail.com (F.V.-A.); lilian.flores@unison.mx (L.F.-M.); ramona.icedo@unison.mx (R.I.-G.); jose.organis@unison.mx (J.G.S.-O.)

[2] Laboratory of Immunology and Virology, Falculty of Biological Sciences, Autonomous University of Nuevo Leon, San Nicolás de los Garza C.P. 66455, Mexico; ricardoromeroarguelles@gmail.com (R.R.-A.); cirscirscirs1989@gmail.com (C.I.R.-S.); ricardo.gomezfl@uanl.edu.com (R.G.-F.)

* Correspondence: guadalupe.gonzalezochoa@unison.mx (G.G.-O.); patricia.tamezgr@uanl.edu.mx (P.T.-G.); Tel.: +52-64-2103-0734 (G.G.-O. & P.T.-G.)

Abstract: Rotavirus is the main cause of acute diarrhea in children up to five years of age. In this regard, probiotics are commonly used to treat or prevent gastroenteritis including viral infections. The anti-rotavirus effect of *Bifidobacterium longum* and *Chlorella sorokiniana*, by reducing viral infectivity and improving IFN-type I response, has been previously reported. The present study aimed to study the effect of *B. longum* and/or *C. sorokiniana* on modulating the antiviral cellular immune response mediated by IFN-γ, IL-10, SOCS3, STAT1, and STAT2 genes in rotavirus-infected cells. To determine the mRNA relative expression of these genes, HT-29 cells were treated with *B. longum* and *C. sorokiniana* alone or in combination, followed by rotavirus infection. In addition, infected cells were treated with *B. longum* and/or *C. sorokiniana*. Cellular RNA was purified, used for cDNA synthesis, and amplified by qPCR. Our results demonstrated that the combination of *B. longum* and *C. sorokiniana* stimulates the antiviral cellular immune response by upregulating IFN-γ and may block pro-inflammatory cytokines by upregulating IL-10 and SOCS3. The results of our study indicated that *B. longum*, *C. sorokiniana*, or their combination improve antiviral cellular immune response and might modulate pro-inflammatory responses.

Keywords: rotavirus; *Bifidobacterium*; *Chlorella*; gastroenteritis

Citation: Velderrain-Armenta, F.; González-Ochoa, G.; Tamez-Guerra, P.; Romero-Arguelles, R.; Romo-Sáenz, C.I.; Gomez-Flores, R.; Flores-Mendoza, L.; Icedo-García, R.; Soñanez-Organis, J.G. *Bifidobacterium longum* and *Chlorella sorokiniana* Combination Modulates IFN-γ, IL-10, and SOCS3 in Rotavirus-Infected Cells. *Int. J. Mol. Sci.* **2024**, *25*, 5514. https://doi.org/10.3390/ijms25105514

Academic Editors: Baojun Xu and Matteo Bordiga

Received: 1 April 2024
Revised: 14 May 2024
Accepted: 17 May 2024
Published: 18 May 2024

1. Introduction

Bifidobacterium species are probiotics commonly used as supplements to improve human health. They are particularly effective in treating or preventing gastrointestinal infections by shortening infectious diarrhea duration and severity [1,2]. The beneficial effect of probiotics is associated with the restoration of intestinal dysbiosis through the production of molecules such as short-chain fatty acids, bacteriocins, and others [1–4]. Bifidobacteria have also been reported to block infectivity or activate cellular immune responses against viral infections [2,4,5].

The health-beneficial effects of probiotics are limited by microbial load and viability [3]. They are usually combined with prebiotics or microalgae to improve probiotic viability [1,5]. In this regard, the use of the microalga *Chlorella sorokiniana* and *Chlorella vulgaris* has been shown to improve the viability and growth of probiotics, such as *Lactobacillus rhamnosus*, *Lactobacillus acidophilus*, and *Bifidobacterium lactis* [6–8]. In addition, *C. sorokiniana* has been shown to improve the viability and shelf life of probiotics, such as *Bifidobacteium longum* and *Lactobacillus plantarum* [5,9]. Furthermore, the microalga *Chlorella* is considered a functional food. It has been associated with antimicrobial activity and is increased in beneficial microorganisms' population in the intestines, which improves gut health [10].

On the other hand, viruses are a common cause of gastroenteritis worldwide [11]. In virus-infected cells, the first line of defense is the cellular immune response, mediated by interferons (IFNs) [12]. Type I IFNs comprise a large group of molecules within the IFN-α and IFN-β genes that directly induce antiviral response [12–14]. In addition, type II interferon gamma (IFN-γ) has antiviral activity and immunomodulatory functions. It has been related to the inhibition of viral infection by transmissible gastroenteritis virus and other viruses [15–18]. Although type I IFNs and IFN-γ directly activate the expression of interferon-stimulated genes (ISGs) and trigger antiviral activity, IFN-γ induces a different but partially overlapping set of genes [18]. Type I and II IFNs stimulate a significant expression and activation of STAT1, whereas IFN-α and IFN-β stimulate those of STAT2 [19]. STAT proteins mediate the signal transduction of cytokines involved in improving the antiviral response [19]. However, cytokine overstimulation mediated by JAK/STAT's signaling is associated with pro-inflammatory conditions, which may also cause tissue damage. In this regard, interleukin-10 (IL-10) is a regulatory cytokine that suppresses the immune response by blocking the action of pro-inflammatory cytokines. Suppression of cytokine signaling 3 (SOCS3) is also induced by IL-10 [20,21] and exerts negative feedback regulation on type I IFN and the JAK/STAT signaling pathway [20,22].

Within gastrointestinal pathogens, rotavirus causes acute gastroenteritis in children with symptoms such as watery diarrhea, vomiting, fever, and dehydration [23,24]. Rotavirus belongs to the *Reoviridae* family and possesses an icosahedral shape structure of 75 nm. Rotavirus has three concentric protein layers that surround a segmented genome of 11 fragments of double-stranded RNA (dsRNA). The viral genome encodes the structural proteins VP1-VP4, VP6, and VP7 and the non-structural proteins NSP1 to NSP6 [24]. The viral proteins VP7 and VP4 are part of the rotavirus external layer; VP6, the most abundant protein, is the intermediate layer, and VP2 is the inner layer of the viral particle. In the internal layer, VP1 is also present and functions as a viral polymerase, whereas VP3 is an RNA-capping enzyme [13]. The rotavirus pathogenesis is attributed to NSP1 and NSP4. In infected cells, NSP4 causes Ca^{2+}-dependent Cl^{-} secretion across the mammalian small intestinal mucosa, acting as a diarrhea-inducing enterotoxin. Furthermore, NSP1 participates in the evasion of the innate immune response by suppressing the type I IFN response and improving viral replication, hence increasing gastroenteritis severity [24–27].

Currently available vaccines against rotavirus include the human monovalent G1P [8] vaccine Rotarix® (GlaxoSmithKline Biologicals, Rixensart, Belgium) and the pentavalent human–bovine reassortant vaccine RotaTeq® (Merck & Co., West Point, PA, USA). Vaccines are effective against rotavirus severe diarrhea, but their disadvantages involve their limited distribution and effectiveness in developing countries. Rotavirus is still the main cause of gastroenteritis-associated morbidity and mortality in children worldwide [23]. Although these vaccines offer heterotypic protection against diverse genotypes (not-vaccinal genotypes), the rotavirus strain diversity would have an impact on the efficacy and effectiveness of a vaccine [28]. On the other hand, some reports have shown the effect of anti-rotavirus agents targeting host factors in different stages of viral pathogenesis, but those studies still require further investigation. At present, we lack antiviral agents approved against rotavirus [29].

On the other hand, *Lactobacillus* and *Bifidobacterium* are associated with restored antiviral signaling by up-regulating interferon levels [30–32]. Furthermore, microalgae *Chlorella* have demonstrated a beneficial immunomodulatory effect against enteric pathogens [6–8]. In previous reports, *Bifidobacterium longum* reduced rotavirus infectivity to 72%, and the combination of *B. longum* and *Chlorella sorokiniana* reduced the viral infectivity to 30%, thus indicating that the antiviral effect of *B. longum* was improved by *C. sorokiniana* [5]. Moreover, *B. longum* induces an in vitro antiviral response mediated by IFN-α in cells infected with rotavirus. The antiviral response induced by *B. longum* and *C. sorokiniana* was shown to be mediated by IFN-α and IFN-β [33]. Although a cellular response by type I IFNs in cells treated with *B. longum* and *C. sorokiniana* in rotavirus-infected cells was observed, the role of type II IFNs, STAT1/STAT2, IL-10, and SOCS3 in the immunomodulatory cellular effect of this probiotic and microalgae is still unknown. This work aimed to study the effect of *B. longum* and *C. sorokiniana* on modulating the antiviral cellular immune response mediated by IFN-γ, IL-10, SOCS3, STAT1, and STAT2 genes in rotavirus-infected cells.

2. Results

2.1. Rotavirus-Infected Cells

The mRNA levels of IFN-γ, IL-10, SOCS3, STAT1, and STAT2 genes were determined in HT-29 cells infected with rotavirus (without probiotic and microalga treatments). The results indicated that the relative expression levels of SOCS3, IFN-γ, STAT2, and IL-10 were lower in rotavirus-infected cells than in non-infected cells, whereas STAT1 was up-regulated compared to the control (non-infected cells) (Figure 1).

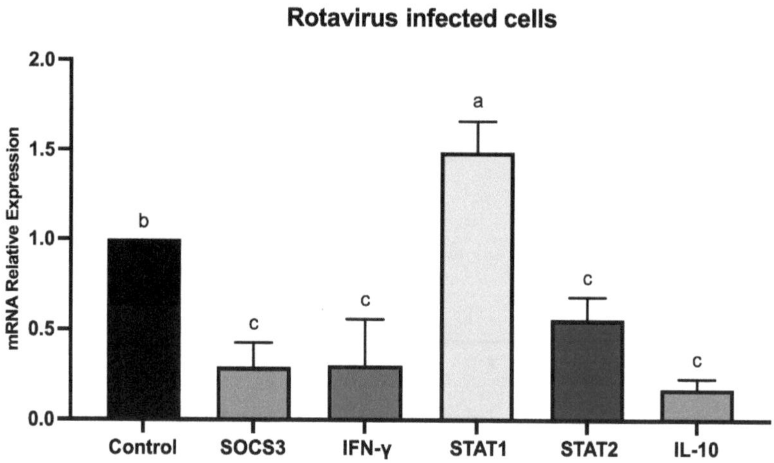

Figure 1. Rotavirus (RV)-infected cell assays. Relative expression level of SOCS3, IFN-γ, STAT1, STAT2, and IL-10 genes in HT-29 cells infected with rotavirus using uninfected HT-29 as the control. Data were analyzed by ANOVA with subsequent Tukey test using GraphPad Prism 10. Different letters indicate statistical significance between treatments. p values < 0.05 were considered statistically significant.

2.2. *B. longum* and Rotavirus Assays

To study the in vitro antiviral effect of *B. longum* in HT-29 cells (their viability was not affected by treatments) in the pre- and post-infection assays with rotavirus, the mRNA levels of IFN-γ, IL-10, SOCS3, STAT1, and STAT2 genes were measured. Our results showed a significant ($p < 0.05$) increase in mRNA of SOCS3, IFN-γ, IL-10, and STAT1 in cells incubated with the probiotic and infected with rotavirus. In addition, in cells infected and treated with the probiotic, we observed a significant ($p < 0.05$) increase in SOCS3, IL-10, and STAT2 relative expression (Figure 2).

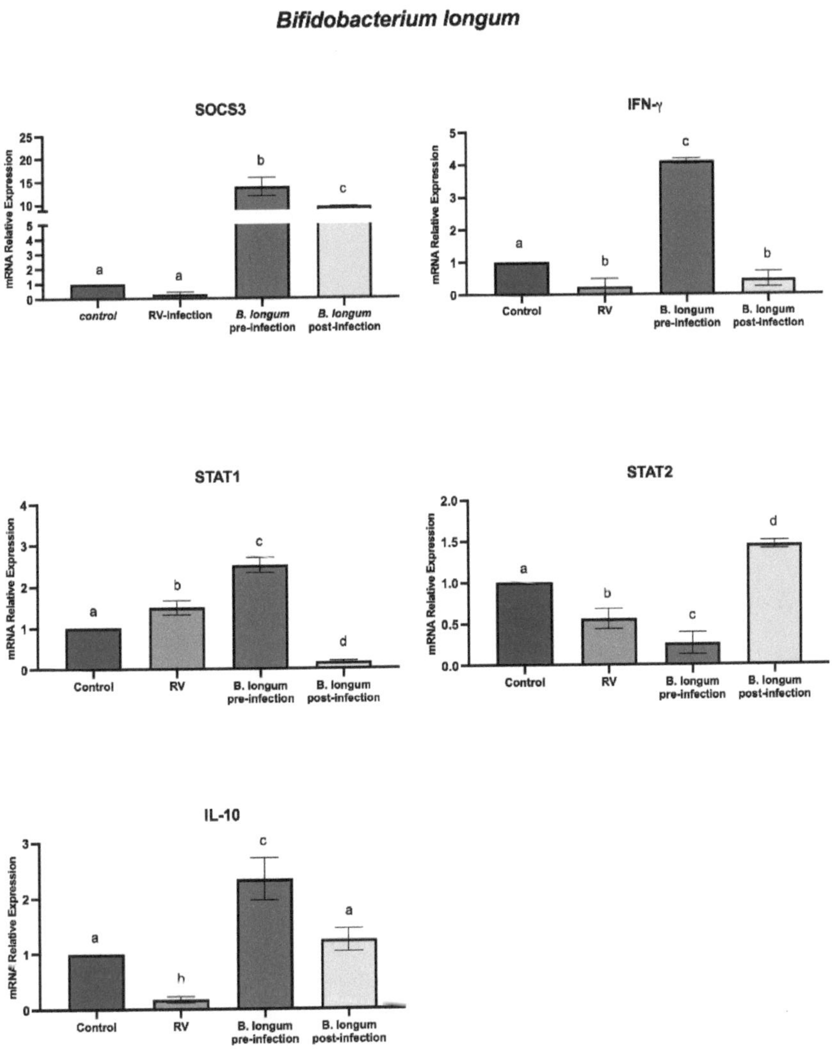

Figure 2. *B. longum* and rotavirus (RV) assays. The relative expression levels of SOCS3, IFN-γ, STAT1, STAT2, and IL-10 genes in HT-29 cells treated with *B. longum* before (pre-infection) and after (post-infection) rotavirus infection were analyzed by ANOVA with subsequent Tukey test using GraphPad Prism 10. Different letters indicate statistical significance between treatments. p values < 0.05 were considered statistically significant.

2.3. *C. sorokiniana and Rotavirus Assays*

To determine the effect of *C. sorokiniana* in HT-29 cells, we studied the mRNA levels of IFN-γ, IL-10, SOCS3, STAT1, and STAT2 genes in cells incubated with *C. sorokiniana* in the rotavirus pre- and post-infection assays. In cells with *C. sorokiniana* and infected with rotavirus, we observed a significant ($p < 0.05$) increase in mRNA levels of SOCS3 and IFN-γ. Furthermore, in the post-infection assays with rotavirus and *C. sorokiniana,* we observed a significant ($p < 0.05$) relative expression of SOCS3, STAT1, and STAT2. The mRNA level of IL-10 was downregulated in the pre-infection and post-infection assays (Figure 3).

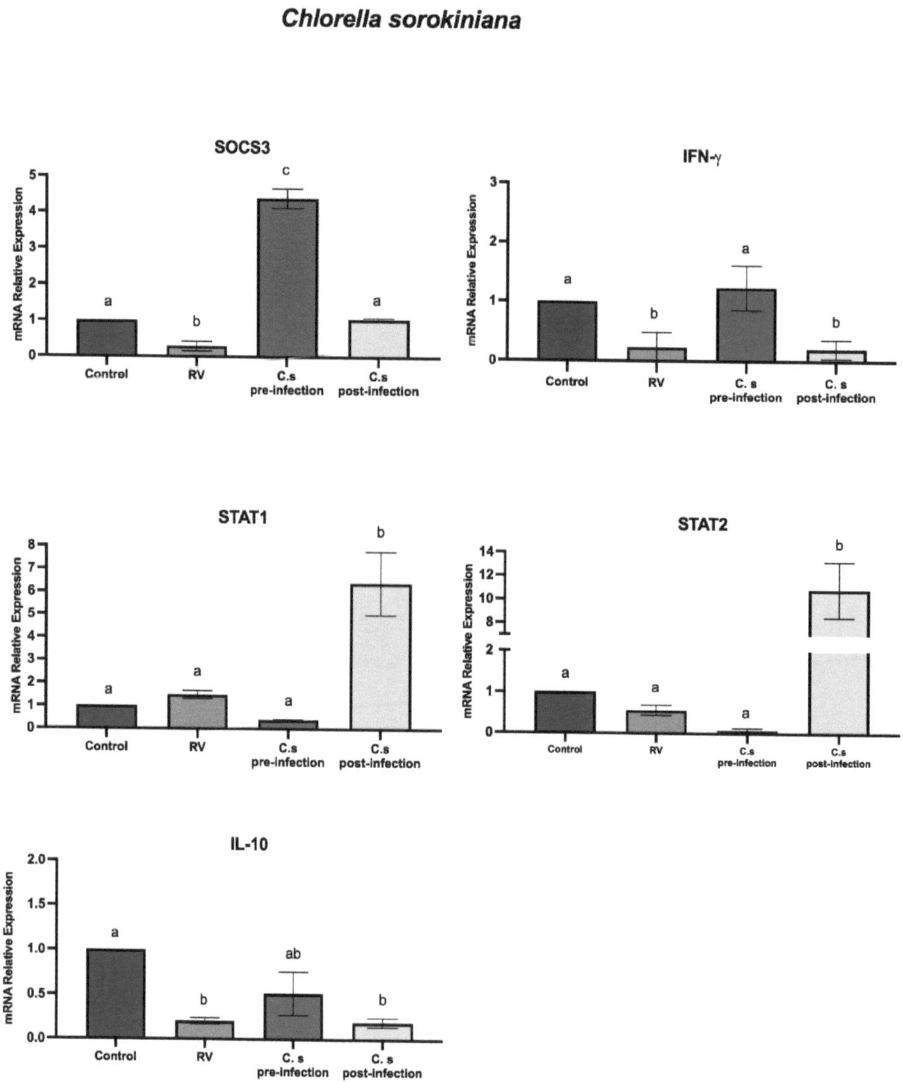

Figure 3. *C. sorokiniana* and rotavirus (RV) assays. The relative expression level of SOCS3, IFN-γ, STAT1, STAT2, and IL-10 genes in HT-29 cells treated with *B. longum* before (pre-infection) and after (post-infection) rotavirus infection. Data were analyzed by ANOVA with subsequent Tukey test using GraphPad Prism 10. Different letters indicate statistical significance between treatments. p values < 0.05 were considered statistically significant.

2.4. *B. longum* in Combination with *C. sorokiniana* and Rotavirus Assays

A combination of *B. longum* and *C. sorokiniana* was used to study its effect on the antiviral response in HT-29 by measuring the mRNA levels of IFN-γ, IL-10, SOCS3, STAT1, and STAT2 genes in cells with the combination of *B. longum*/*C. sorokiniana* in the pre- and post-infection assays with rotavirus. In cells incubated with both microorganisms and infected with rotavirus (pre-infection), we observed a significant ($p < 0.05$) upregulation of the mRNA levels of SOCS3, IFN-γ, STAT1, and IL-10. Moreover, in the post-infection assays (cells infected and further treated), the mRNA levels of SOCS3, IL-10, and IFN-γ were significantly ($p < 0.05$) upregulated, whereas STAT1 was downregulated (Figure 4). In Figure 5, a heat map summarizes the mRNA expression of SOCS3, IFN-γ, STAT-1, STAT-2, and IL-10 in cells with *B. longum* and/or *C. sorokiniana* and pre- or post-infected with rotavirus.

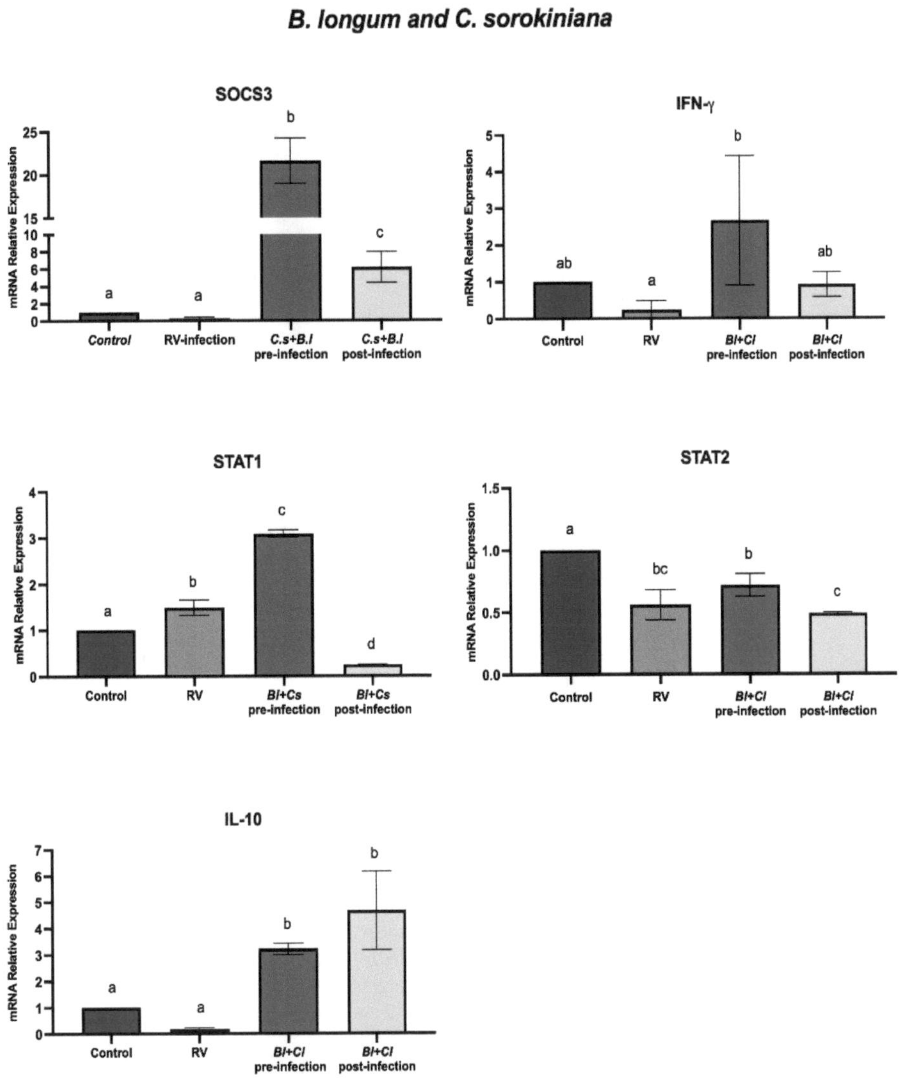

Figure 4. *B. longum* in combination with *C. sorokiniana* and rotavirus (RV) assays. The relative expression level of SOCS3, IFN-γ, STAT1, STAT2, and IL-10 genes in HT-29 cells treated with *B. longum* and *C. sorokiniana* before (pre-infection) and after (post-infection) rotavirus infection. Data were analyzed by ANOVA with subsequent Tukey test using GraphPad Prism 10. Different letters indicate statistical significance between treatments. p values < 0.05 were considered statistically significant.

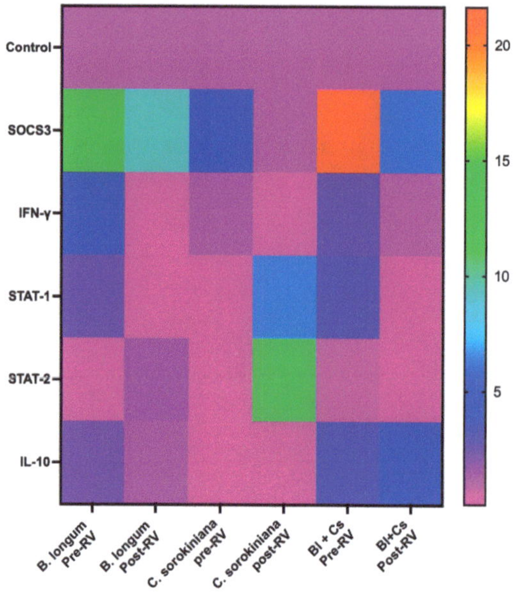

Figure 5. Heat map showing the mRNA expression of SOCS3, IFN-γ, STAT-1, STAT-2, and IL-10 in *B. longum* pre-infected with rotavirus (pre-RV): HT-29 cells infected with RV and treated with *B. longum* (Bl); *B. longum* post-infected with RV (post-RV): HT-29 cells treated with Bl and infected with RV; *C. sorokiniana* pre-RV: HT-29 cells infected with RV and treated with *C. sorokiniana* (Cs); *C. sorokiniana* post-RV: HT-29 cells treated with Cs and infected with RV; Bl + Cs pre-RV: HT-29 cells treated with *B. longum* in combination with *C. sorokiniana* and then infected with RV; Bl + Cs post-RV: HT-29 cells infected with RV and treated with *B. longum* in combination with *C. sorokiniana*. Data were analyzed by three-way ANOVA with subsequent Tukey test using GraphPad Prism 10.

3. Discussion

Rotavirus is the main cause of acute gastroenteritis in children worldwide [23]. In the present study, we showed that, in HT-29 cells infected with rotavirus, the mRNA relative expression of SOCS3, IFN-γ, and IL-10 was downregulated ($p < 0.05$), but a significant difference with STAT1 and STAT2, as compared with cells without rotavirus infection, was not observed (Figure 1). Previous studies reported that rotavirus block gene expression induced by type I and II IFNs. In addition, rotavirus was associated with STAT1/STAT2 inhibition, which decreased antiviral response in early infection, through the activity of NSP1 [34,35]. Rotavirus protein NSP1 inhibits IFN production by degrading the interferon precursors IRF3, IRF5, and IRF7. It also inhibits NFkB, which decreases IFN-β and cytokine production in infected cells, and IFN signaling by preventing the nuclear accumulation of STAT1/ STAT2 [35,36]. Some other viruses besides rotavirus express proteins associated with cellular immunity evasion [12]. On the other hand, probiotics might modulate antiviral immune responses. In this regard, *Lactobacillus mucosae* and *Bifidobacterium breve* have been associated with restoring antiviral signaling by a positive regulation of interferon levels [33].

In previous reports, we demonstrated that *B. longum* improved monolayer integrity, rotavirus load reduction, and in vitro anti-rotavirus response through IFN-α [33]. In this regard, the probiotics *Lactobacillus* spp. and *Bifidobacterium spp.* have been associated with reduced severity and shorter periods of diarrhea. In addition, in vitro assays with *B. longum* R0175 in porcine intestinal epitheliocyte cells before rotavirus infection showed a preventive effect against rotavirus infection [37]. In this study, we found a significant ($p < 0.05$) increase in the relative expression of SOCS3, IFN-γ, IL-10, and STAT1 in cells with *B. longum* and rotavirus, whereas a higher relative expression of SOCS3 and STAT2 was observed in cells infected with rotavirus and incubated with *B. longum*. In agreement with our analysis of SOCS3, studies in RAW264.7 cells exposed to *Bifidobacterium* species showed increased mRNA levels of SOCS1 or SOCS3. These results indicated that this probiotic may negatively modulate the levels of pro-inflammatory cytokines [20]. Furthermore, other studies have demonstrated the immunomodulatory effect of *Bifidobacterium* species through the up-regulation of IFN-γ and IL-10. Moreover, an anti-inflammatory effect of this probiotic on HT-29 cells has been described via modulation of JAK/STAT [38,39].

On the other hand, *Chlorella* genus is a widely studied microalga for its application in industries such as animal nutrition, pharmaceuticals, and health [6,40,41]. We have previously reported that cells treated with *C. sorokiniana* metabolites caused a significant reduction in rotavirus infectivity [5]. In this regard, *Chlorella* supplementation has shown an effect against hepatitis C by reducing viral load [33,42]. Furthermore, IFN-α relative gene expression was upregulated in cells with *C. sorokiniana* and rotavirus (post-infection assays). Nevertheless, there was a low relative expression of IFN-α in cells infected with rotavirus before the microalgae treatment. Our previous results indicated that *C. sorokiniana* blocked rotavirus infectivity and improved immune cellular response. In order to further study the effect of *C. sorokiniana* in rotavirus-infected cells, we investigated the relative expression of IFN-γ, IL-10, SOCS3, STAT1, and STAT2 genes. Our results indicated that, in HT-29 cells incubated with *C. sorokiniana* followed by rotavirus infection, the relative expression of SOCS3 and IFN-γ significantly increased as compared with that of infected cells without microalgae treatment, whereas in cells infected and treated with *C. sorokiniana*, the relative expression of SOCS3, STAT1, and STAT2 was significantly upregulated. This may indicate that the antiviral response induced by *C. sorokiniana* in rotavirus-infected cells was mediated by IFN-α and IFN-γ in the pre-infection assays and STAT1/STAT2 in the post-infection assays and regulated by SOCS3 in both assays. This result agrees with studies with *C. vulgaris* in mice infected with *Listeria monocytogenes*, which demonstrated an increased production of IFN-γ, thus enhancing intracellular resistance to pathogens [43]. In addition, in a randomized, double-blinded, placebo-controlled trial, a significant increase in IFN-γ was observed after eight weeks of *C. vulgaris* supplement intake (pills) [34].

The microalga *C. sorokiniana* has been related to probiotic viability enhancement and effectiveness against pathogens. *C. sorokiniana* and *C. vulgaris* are also associated with prebiotic activity by stimulating *L. rhamnosus* growth [6], and other species showed potential for improving *L. acidophilus* and *B. lactis* viability in yogurt [7,8]. We previously reported that *C. sorokiniana* improved the viability and anti-rotavirus effect of *B. longum* and *L. plantarum* [5]. In addition, we observed increased IFN-α levels in cells with the probiotic/microalgae after rotavirus infection, whereas IFN-β was elevated in cells infected and treated [33]. In the present study, we also evaluated the effect of *B. longum* and *C. sorokiniana*, including analyses of mRNA relative expression levels of IFN-γ, IL-10, SOCS3, STAT1, and STAT2 genes. Our results indicated that, in HT-29 cells with both microorganisms before rotavirus infection, the relative expression of SOCS3, IFN-γ, STAT1, and IL-10 was significantly ($p < 0.05$) upregulated. Furthermore, in the post-infection assays (cells infected and post-treated), the relative expression of SOCS3, IL-10, and IFN-γ was significantly upregulated. The effect of *B. longum* combined with *C. sorokiniana* against rotavirus in infected cells may be associated with blocking infectivity and inducing an antiviral cellular response mediated by type I and II IFNs.

The effect of the *B. longum*/*C. sorokiniana* combination in cells infected is mediated by the improvement of the cellular immune response. IFN-α, IFN-β, and IFN-γ significantly increase the expression and activation of STAT1, whereas IFN-α and IFN-β increase that of STAT2 [19]. In this regard, we observed an increased level of mRNA expression of IFN-γ and STAT1 in pre-infection with the probiotic alone or with the microalgae, which was associated with an improvement of antiviral cellular response, in comparison with rotavirus-infected cells alone. On the other hand, IL-10 is a regulatory cytokine that suppresses the immune response by blocking the action of pro-inflammatory cytokines. In addition, SOCS3 is induced by IL-10 [20,21]. SOCS3 exerts negative feedback regulation on IFN type I and the JAK/STAT signaling pathway [20,30]. In the present study, we observed an increase in the relative expression of IL-10 and SOCS3 in the pre- and post-infection assays with *B. longum* and *C. sorokiniana* as compared with rotavirus-infected cells without treatments, which may indicate that the protective effect of this probiotic and the microalga was also induced by IL-10 and SOCS3 by downregulating a pro-inflammatory response in rotavirus-infected cells. In this regard, the protective effect of *B. longum* in rotavirus-infected cells was improved with the combination of *C. sorokiniana*. Although some limitations of this study are that we only studied the in vitro assays and mRNA relative expression of SOCS3, IFN-γ, STAT-1, STAT-2, and IL-10, our results are consistent with the protective effect of *B. longum* and *C. sorokiniana* in rotavirus-infected cells and with previous reports of beneficial effects of probiotic and microalga against gastrointestinal pathogens [33,37–39,42].

4. Materials and Methods

4.1. Cells

HT-29 human colon tumor cells (ATCC HTB-38) were used in the assays with *B. longum* and *C. sorokiniana* in rotavirus-infected cells. MA104 rhesus monkey epithelial cells (ATCC CRL-2378) were used for rotavirus propagation and microtitration. HT-29 and MA104 cells were grown in RPMI-1640 and DMEM culture media (Gibco, Grand Island, NY, USA), respectively, and were supplemented with 5% fetal bovine serum (FBS; Mediatech Inc., Corning, NY, USA), 1% antibiotic and antimycotic solution (Caisson Laboratories, Smithfield, UT, USA), and 2 mM L-glutamine, and incubated at 37 °C and 5% CO_2.

4.2. Rotavirus Strain and Viral Titration

MA104 cells were infected with the human Rotavirus strain Wa for propagation and viral titration as follows. Rotavirus was activated with 10 µg/mL trypsin-EDTA solution 10X (Sigma-Aldrich, St. Louis, MO, USA) and incubated at 37 °C for 30 min, after which activated viruses were incubated with MA104 cells at 37 °C in 5% CO_2 for one hour. Next, the inoculum was replaced with DMEM, and the cells were incubated at 37 °C in 5% CO_2 for 24 h or until the monolayer lysis was observed. The viral lysates were stored at −20 °C. Rotavirus focus-forming units per mL (FFU/mL) were determined by an immunoperoxidase assay, as previously reported [44]. In brief, rotavirus lysates were used to infect MA104 cells and incubated at 37 °C in 5% CO_2 for 14 h. After incubation, the cellular monolayer was washed twice with PBS-Ca^{2+}, fixed with cold acetone-PBS (80% to 20%), and incubated for 45 min at room temperature. Next, the monolayer was washed twice with PBS-Ca^{2+} and incubated for one hour at 37 °C with the primary anti-rotavirus antibodies (Invitrogen, Carlsbad, CA, USA). After incubation, the cell monolayers were washed twice with PBS-Ca^{2+} and incubated for one hour at 37 °C with horseradish peroxidase (HRP)-anti-sheep IgG conjugate (Invitrogen), followed by incubation with the substrate (0.1 M sodium acetate buffer pH 5.0, 0.64 mg/mL aminoethyl carbazole (Sigma-Aldrich), and 0.36% hydrogen peroxide) for 15 min at 37 °C. After incubation, the reaction was stopped with three washes of water. FFU/mL was calculated as follows: FFU/mL = 20 (microscope objective) × 5.5 (well diameter) × average number of foci (duplicate determinations; 100 to 200 foci/well) × dilution (foci count). The number of viral particles per cell (MOI) was calculated with the number of viral particles used (FFU/mL) per well divided by the number of cells originally seeded in the well. Rotavirus MOI was 0.1 in each assay.

4.3. Probiotic

Bifidobacterium longum strain (ATCC® 15707) was grown on MPT-broth medium (5.0 g glucose, 10 g casein digest peptone, 0.50 g sodium chloride, 2.5 g yeast extract, 0.05 g ascorbic acid, 0.15 g potassium phosphate monobasic, 0.25 g potassium phosphate dibasic, 0.5 g L-cysteine hydrochloride, and 0.124 mg ferric ammonium citrate) [45]. Five milliliters was aliquoted in 13 × 100 mm borosilicate glass tubes and autoclaved for 15 min at 121 °C and 15 lbs pressure. The sterile MPT-broth was stored at −20 °C until use. *Bifidobacterium longum* was inoculated on MTP-broth medium and incubated at 37 °C. Colony-forming units (CFU) per milliliter were calculated by serial dilutions, and the bacterial inoculum for each assay was adjusted to 1 × 10^6 CFU/mL.

4.4. *C. sorokiniana*

The microalga *C. sorokiniana* was collected in the San Juan River in Cadereyta, Nuevo León, México [5,46]. It was grown in L-carnitine (LC) nutrient solution (5 mM KNO_3, 1 mM KH_2 PO_4, 2 mM $MgSO_4 \cdot 7H_2O$, 6.25 mM $Ca(NO_3)_2 \cdot 4H_2$ O, 46µM H_3 BO_3, 9.15 µM $MnCl_2 \cdot 4H_2O$, 765 nM $ZnSO_4 \cdot 7H_2$ O, 320 nM $CuSO_4 \cdot 5H_2O$, 15 nM $(NH_4)6Mo_7O_{24} \cdot 4H_2O$, 20 µM $FeSO_4 \cdot 7H_2O$, and 20 µM Na_2 EDTA) at 25 °C and 120 rpm under continuous light at 1400 lumens for 12 d [47].

4.5. *Cellular Viability Assay*

To demonstrate that treatments without rotavirus do not alter cellular viability, we determined that of HT-29 cells incubated with *B. longum* and/or *C. sorokiniana* by the colorimetric 3-(4, 5-dimethylthiazol-2-yl)-2, 5-diphenyltetrazolium bromide (MTT) reduction assay. For this, HT-29 cells were incubated with *C. sorokiniana* and/or *B. longum* in RPMI-1640 medium without FBS for 24 h at 37 °C and 5% CO_2 in 95% air, after which they were washed twice with PBS, and 20 µL of MTT (Sigma-Aldrich; 5 mg/mL final concentration) was added to the cells and incubated for three additional hours. MTT was then replaced by 10 µL of dimethyl sulfoxide (DMSO; Sigma-Aldrich) and incubated for three minutes under continuous shaking. Optical densities were then determined in a microplate reader (Multiskan GO, Thermo Fisher Scientific Inc., San Jose, CA, USA) at 570 nm [48].

4.6. *B. longum, C. sorokiniana, and Rotavirus Assays*

B. longum, C. sorokiniana, and their combination were used to treat rotavirus-infected cells. In the pre-infection assays, the cells were treated and post-infected with rotavirus as follows: HT-29 cells were treated with probiotics (1 × 10^6 CFU/mL), the microalgae biomass (1 × 10^6 cells/mL), or their combination for 24 h, after which they were infected with rotavirus (MOI 0.1) for one hour at 37 °C and 5% CO_2 in DMEM without FBS. Next, the inoculum was replaced with DMEM without FBS and incubated at 37 °C in 5% CO_2 for 24 h. Lysates were then stored at −20 °C until use. In the post-infection assays, the cells were first infected with rotavirus and post-treated with the probiotic and/or microalga. For this, HT-29 cells were infected with rotavirus (MOI 0.1) for one hour at 37 °C and 5% CO_2 in DMEM without FBS, followed by treatment with *B. longum* (1 × 10^6 CFU/mL), *Chlorella sorokiniana* (1 × 10^6 cells/mL), or their combination for 24 h. After this, the cells were stored at 20 °C until viral RNA purification and qPCR assays.

4.7. *qPCR Assay*

Total RNA extraction from rotavirus-infected cell lysates and/or treated with *C. sorokiniana* and/or *B. longum* was performed by the Trizol method (Life Technologies, Rockville, MD, USA). Total RNA was used as a template for cDNA synthesis (High-Capacity cDNA Reverse Transcription; Applied Biosystems, Foster City, CA, USA). Relative expression of IFN-γ, IL-10, SOCS3, STAT1, and STAT2 genes was determined by qPCR using PGK-1 as an endogenous gene (Table 1). Reactions were developed with the Sensi FAST SYBER Lo-ROX Kit (Bioline, London, UK), following the manufacturer's instructions. qPCR conditions were 95 °C for 5 min, 45 cycles at 58 °C for 5 s, and 60 °C for 10 s. Gene relative expression was calculated using the $2^{(-\Delta\Delta Ct)}$ method (Applied Biosystems).

Table 1. qPCR primer sequences.

Primer Name	Primer Sequences (5′ to 3′) Fwd	Rev	Product Length	Reference
SOCS3	5′-ACA ATC TGC CTC AAT CAC TCT G-3′	5′-TTG ACT TGG ATT GGG ATT TTG-3′	129	[49]
IFN-γ	5′-GGC ATT TTG AAG AAT TGG AAA G-3′	5′-TTT GGA TGC TCT GGT CAT CTT-3′	112	[49]
STAT1	5′-GAT CGC TTG CCC AAC TCT TG-3′	5′- ACT GTG ACA TCC TTG GGC TG-3′	198	[50]
STAT2	5′-GGC AGC GAA TCA CTC AAA GC-3′	5′-CACCAGAGTCAAGAAGCCGA-3′	159	[50]
IL-10	5′-TGG AGC AGG TGA AGA ATG-3′	5′-ATA GAA GCC TAC ATG ACA-3′	105	[49]
PGK1	5′-GAG ATG ATT ATT GGT GGT GGA A-3′	5′-AGT CAA CAG GCA AGG TAA TC-3′	160	[51]

4.8. Statistical Analysis

In each assay, three independent experiments were performed, and the results were reported as mean ± SD. Statistical analysis was calculated by the one-way ANOVA or three-way ANOVA and Tukey's multiple comparisons or Kruskal–Wallis and Dunn's multiple comparison tests using GraphPad Prism 10 (GraphPad Software Inc., San Diego, CA, USA). p values < 0.05 were considered statistically significant.

5. Conclusions

In this study, we observed that the probiotic *B. longum* increased the relative expression of SOCS3, IFN-γ, IL-10, and STAT1 in cells with *B. longum* and infected with rotavirus and SOCS3 and STAT2 in cells infected with rotavirus and incubated with *B. longum*. On the other hand, in the pre-infection assays with *C. sorokiniana*, the mRNA levels of SOCS3 and IFN-γ were increased, whereas in the post-infection assays, SOCS3, STAT1, and STAT2 were upregulated. Additionally, *B. longum* and *C. sorokiniana* improved the antiviral cellular immune response by upregulating SOCS3 and IFN-γ in the pre- and post-infection assays. This combination may block pro-inflammatory cytokines by upregulating IL-10 and SOCS3. Our results indicated that *B. longum*, *C. sorokiniana*, or their combination improve antiviral cellular immune response and might modulate pro-inflammatory responses. Although further research is needed, supplementation with probiotics such as *Bifidobacterium* and the microalgae *Chlorella* might be beneficial and become an alternative in the prevention or treatment of rotavirus gastroenteritis.

Author Contributions: Investigation, G.G.-O., R.I.-G. and J.G.S.-O.; methodology, F.V.-A. and R.R.-A.; project administration, P.T.-G. and G.G.-O.; validation, C.I.R.-S. and L.F.-M.; writing—original draft, G.G.-O.; writing—review and editing, P.T.-G., R.G.-F. and L.F.-M. All authors have read and agreed to the published version of the manuscript.

Funding: This research was supported by internal research support of the Universidad de Sonora projects USO513006130 and USO513007912 to G.G.-O.

Institutional Review Board Statement: This study did not require ethical approval.

Informed Consent Statement: Not applicable.

Data Availability Statement: The data presented in this study are available on request from the corresponding author.

Acknowledgments: We wish to thank Consejo Nacional de Ciencia y Tecnología (Conahcyt-Mexico) for scholarship 744803 to F.V.-A. and 893625 to R.R.-A; to Sistema Nacional de Investigadores e Investigadoras (SNII-Conahcyt-Mexico) for supporting through SNII-16614 to PT-G, SNII-9942 to R.G.-F., SNII-54907 to L.F.-M., SNII- 445572 to C.I.R.-S., SNII-165755 to J.G.S.-O., and SNII-62977 to G.G.-O.

Conflicts of Interest: The authors declare no conflicts of interest.

References

1. Quigley, E.M. Prebiotics and probiotics in digestive health. *Clin. Gastroenterol. Hepatol.* **2019**, *17*, 333–344. [CrossRef] [PubMed]
2. Ishizuka, T.; Kanmani, P.; Kobayashi, H.; Miyazaki, A.; Soma, J.; Suda, Y.; Aso, H.; Nochi, T.; Iwabuchi, N.; Xiao, J.-Z. Immunobiotic bifidobacteria strains modulate rotavirus immune response in porcine intestinal epitheliocytes via pattern recognition receptor signaling. *PLoS ONE* **2016**, *11*, e0152416. [CrossRef] [PubMed]
3. Abdi, M.; Ranjbar, R. A review on antiviral efficacy of Bifidobacterium species. *Rev. Res. Med. Microbiol.* **2022**, *33*, 74–81. [CrossRef]
4. Terpou, A.; Papadaki, A.; Lappa, I.K.; Kachrimanidou, V.; Bosnea, L.A.; Kopsahelis, N. Probiotics in food systems: Significance and emerging strategies towards improved viability and delivery of enhanced beneficial value. *Nutrients* **2019**, *11*, 1591. [CrossRef] [PubMed]
5. Cantú-Bernal, S.; Domínguez-Gámez, M.; Medina-Peraza, I.; Aros-Uzarraga, E.; Ontiveros, N.; Flores-Mendoza, L.; Gomez-Flores, R.; Tamez-Guerra, P.; González-Ochoa, G. Enhanced viability and anti-rotavirus effect of *Bifidobacterium longum* and Lactobacillus plantarum in combination with *Chlorella sorokiniana* in a dairy product. *Front. Microbiol.* **2020**, *11*, 875. [CrossRef] [PubMed]
6. Lai, Y.-C.; Chang, C.-H.; Chen, C.-Y.; Chang, J.-S.; Ng, I.-S. Towards protein production and application by using Chlorella species as circular economy. *Bioresour. Technol.* **2019**, *289*, 121625. [CrossRef] [PubMed]
7. Beheshtipour, H.; Mortazavian, A.M.; Haratian, P.; Darani, K.K. Effects of Chlorella vulgaris and Arthrospira platensis addition on viability of probiotic bacteria in yogurt and its biochemical properties. *Eur. Food Res. Technol.* **2012**, *235*, 719–728. [CrossRef]
8. Beheshtipour, H.; Mortazavian, A.M.; Mohammadi, R.; Sohrabvandi, S.; Khosravi-Darani, K. Supplementation of Spirulina platensis and Chlorella vulgaris algae into probiotic fermented milks. *Compr. Rev. Food Sci. Food Saf.* **2013**, *12*, 144–154. [CrossRef]
9. Yan, L.; Lim, S.; Kim, I. Effect of fermented chlorella supplementation on growth performance, nutrient digestibility, blood characteristics, fecal microbial and fecal noxious gas content in growing pigs. *Asian-Australas. J. Anim. Sci.* **2012**, *25*, 1742. [CrossRef]
10. Jin, J.B.; Cha, J.W.; Shin, I.S.; Jeon, J.Y.; Cha, K.H.; Pan, C.H. Supplementation with Chlorella vulgaris, Chlorella protothecoides, and Schizochytrium sp. increases propionate-producing bacteria in in vitro human gut fermentation. *J. Sci. Food Agric.* **2020**, *100*, 2938–2945. [CrossRef]
11. Meier, J.L. Viral acute gastroenteritis in special populations. *Gastroenterol. Clin.* **2021**, *50*, 305–322. [CrossRef]
12. Randall, R.E.; Goodbourn, S. Interferons and viruses: An interplay between induction, signalling, antiviral responses and virus countermeasures. *J. Gen. Virol.* **2008**, *89*, 1–47. [CrossRef]
13. Jefferies, C.A. Regulating IRFs in IFN driven disease. *Front. Immunol.* **2019**, *10*, 325. [CrossRef] [PubMed]
14. Chhabra, P.; Ranjan, P.; Cromeans, T.; Sambhara, S.; Vinjé, J. Critical role of RIG-I and MDA5 in early and late stages of Tulane virus infection. *J. Gen. Virol.* **2017**, *98*, 1016. [CrossRef] [PubMed]
15. Rhein, B.A.; Powers, L.S.; Rogers, K.; Anantpadma, M.; Singh, B.K.; Sakurai, Y.; Bair, T.; Miller-Hunt, C.; Sinn, P.; Davey, R.A. Interferon-γ inhibits Ebola virus infection. *PLoS Pathog.* **2015**, *11*, e1005263. [CrossRef]
16. Chaudhary, V.; Yuen, K.-S.; Chan, J.F.-W.; Chan, C.-P.; Wang, P.-H.; Cai, J.-P.; Zhang, S.; Liang, M.; Kok, K.-H.; Chan, C.-P. Selective activation of type II interferon signaling by Zika virus NS5 protein. *J. Virol.* **2017**, *91*, 10–1128. [CrossRef] [PubMed]
17. Sainz Jr, B.; Mossel, E.C.; Peters, C.; Garry, R.F. Interferon-beta and interferon-gamma synergistically inhibit the replication of severe acute respiratory syndrome-associated coronavirus (SARS-CoV). *Virology* **2004**, *329*, 11–17. [CrossRef]
18. Shan, L.; Fu, F.; Xue, M.; Li, L.; Liu, P. IFN-γ synergistically inhibit the replication of transmissible gastroenteritis virus through upregulating expression of type I interferon. *Chin. Vet. Sci./Zhongguo Shouyi Kexue* **2019**, *49*, 723–729.
19. Gao, Y.; Zhao, H.; Wang, P.; Wang, J.; Zou, L. The roles of SOCS 3 and STAT 3 in bacterial infection and inflammatory diseases. *Scand. J. Immunol.* **2018**, *88*, e12727. [CrossRef]
20. Okada, Y.; Tsuzuki, Y.; Hokari, R.; Komoto, S.; Kurihara, C.; Kawaguchi, A.; Nagao, S.; Miura, S. Anti inflammatory effects of the genus Bifidobacterium on macrophages by modification of phospho-IκB and SOCS gene expression. *Int. J. Exp. Pathol.* **2009**, *90*, 131–140. [CrossRef]
21. Ding, Y.; Chen, D.; Tarcsafalvi, A.; Su, R.; Qin, L.; Bromberg, J.S. Suppressor of cytokine signaling 1 inhibits IL-10-mediated immune responses. *J. Immunol.* **2003**, *170*, 1383–1391. [CrossRef] [PubMed]
22. Wang, X.; Jia, Y.; Ren, J.; Huo, N.; Liu, H.; Xiao, S.; Wang, X.; Yang, Z. Newcastle disease virus nonstructural V protein upregulates SOCS3 expression to facilitate viral replication depending on the MEK/ERK pathway. *Front. Cell. Infect. Microbiol.* **2019**, *9*, 317. [CrossRef] [PubMed]
23. Du, Y.; Chen, C.; Zhang, X.; Yan, D.; Jiang, D.; Liu, X.; Yang, M.; Ding, C.; Lan, L.; Hecht, R. Global burden and trends of rotavirus infection-associated deaths from 1990 to 2019: An observational trend study. *Virol. J.* **2022**, *19*, 166. [CrossRef] [PubMed]
24. Estes, M.; Greenberg, H. Rotaviruses. In *Fields Virology*, 6th ed.; Kinpe, D.M., Howley, P.M., Eds.; Lippincott: Philadelphia, PA, USA, 2013; pp. 1347–1401.
25. Omatola, C.A.; Olaniran, A.O. Rotaviruses: From pathogenesis to disease control—A critical review. *Viruses* **2022**, *14*, 875. [CrossRef] [PubMed]
26. Arnold, M.M.; Patton, J.T. Diversity of interferon antagonist activities mediated by NSP1 proteins of different rotavirus strains. *J. Virol.* **2011**, *85*, 1970–1979. [CrossRef] [PubMed]
27. Arnold, M.M.; Sen, A.; Greenberg, H.B.; Patton, J.T. The battle between rotavirus and its host for control of the interferon signaling pathway. *PLoS Pathog.* **2013**, *9*, e1003064. [CrossRef] [PubMed]

28. Varghese, T.; Kang, G.; Steele, A.D. Understanding rotavirus vaccine efficacy and effectiveness in countries with high child mortality. *Vaccines* **2022**, *10*, 346. [CrossRef] [PubMed]
29. Bhuinya, A.; Dass, D.; Banerjee, A.; Mukherjee, A. A tale of antiviral counterattacks in rotavirus infection. *Microbiol. Res.* **2022**, *260*, 127046. [CrossRef] [PubMed]
30. Wang, Q.; Fang, Z.; Li, L.; Wang, H.; Zhu, J.; Zhang, P.; Lee, Y.-k.; Zhao, J.; Zhang, H.; Lu, W. Lactobacillus mucosae exerted different antiviral effects on respiratory syncytial virus infection in mice. *Front. Microbiol.* **2022**, *13*, 1001313. [CrossRef]
31. Olímpio, F.; Andreata-Santos, R.; Rosa, P.C.; Santos, W.; Oliveira, C.; Aimbire, F. Lactobacillus rhamnosus Restores Antiviral Signaling and Attenuates Cytokines Secretion from Human Bronchial Epithelial Cells Exposed to Cigarette Smoke and Infected with SARS-CoV-2. *Probiotics and antimicrobial proteins* **2022**, *15*, 1513–1528. [CrossRef]
32. Sadeghpour Heravi, F.; Hu, H. Bifidobacterium: Host–Microbiome Interaction and Mechanism of Action in Preventing Common Gut-Microbiota-Associated Complications in Preterm Infants: A Narrative Review. *Nutrients* **2023**, *15*, 709. [CrossRef] [PubMed]
33. Romero-Arguelles, R.; Tamez-Guerra, P.; González-Ochoa, G.; Romo-Sáenz, C.I.; Gomez-Flores, R.; Flores-Mendoza, L.; Aros-Uzarraga, E. *Bifidobacterium longum* and *Chlorella sorokiniana* Improve the IFN Type I-Mediated Antiviral Response in Rotavirus-Infected Cells. *Microorganisms* **2023**, *11*, 1237. [CrossRef] [PubMed]
34. Kwak, J.H.; Baek, S.H.; Woo, Y.; Han, J.K.; Kim, B.G.; Kim, O.Y.; Lee, J.H. Beneficial immunostimulatory effect of short-term Chlorella supplementation: Enhancement of natural killer cell activity and early inflammatory response (randomized, double-blinded, placebo-controlled trial). *Nutr. J.* **2012**, *11*, 53. [CrossRef]
35. Holloway, G.; Dang, V.T.; Jans, D.A.; Coulson, B.S. Rotavirus inhibits IFN-induced STAT nuclear translocation by a mechanism that acts after STAT binding to importin-α. *J. Gen. Virol.* **2014**, *95*, 1723–1733. [CrossRef] [PubMed]
36. Holloway, G.; Coulson, B.S. Innate cellular responses to rotavirus infection. *J. Gen. Virol.* **2013**, *94*, 1151–1160. [CrossRef] [PubMed]
37. Leblanc, D.; Raymond, Y.; Lemay, M.-J.; Champagne, C.P.; Brassard, J. Effect of probiotic bacteria on porcine rotavirus OSU infection of porcine intestinal epithelial IPEC-J2 cells. *Arch. Virol.* **2022**, *167*, 1999–2010. [CrossRef]
38. Čitar, M.; Hacin, B.; Tompa, G.; Štempelj, M.; Rogelj, I.; Dolinšek, J.; Narat, M.; Matijašić, B.B. Human intestinal mucosa-associated Lactobacillus and Bifidobacterium strains with probiotic properties modulate IL-10, IL-6 and IL-12 gene expression in THP-1 cells. *Benef. Microbes* **2015**, *6*, 325–336. [CrossRef] [PubMed]
39. Aghamohammad, S.; Sepehr, A.; Miri, S.T.; Najafi, S.; Rohani, M.; Pourshafiea, M.R. The effects of the probiotic cocktail on modulation of the NF-kB and JAK/STAT signaling pathways involved in the inflammatory response in bowel disease model. *BMC Immunol.* **2022**, *23*, 8. [CrossRef] [PubMed]
40. Priyadarshani, I.; Rath, B. Commercial and industrial applications of micro algae—A review. *J. Algal Biomass Util.* **2012**, *3*, 89–100.
41. Wang, Z.; Yang, L.; Hu, F.; Yan, Y.; Yu, S.; Chen, T. Anti-inflammatory Effect of Acetone Extracts from Microalgae Chlorella sp. WZ13 on RAW264. 7 Cells and TPA-induced Ear edema in Mice. *Front. Mar. Sci.* **2022**, *9*, 920082.
42. Azocar, J.; Diaz, A. Efficacy and safety of Chlorella supplementation in adults with chronic hepatitis C virus infection. *World J. Gastroenterol. WJG* **2013**, *19*, 1085. [CrossRef] [PubMed]
43. Queiroz, M.L.; Bincoletto, C.; Valadares, M.C.; Dantas, D.C.; Santos, L.M. Effects of Chlorella vulgaris extract on cytokines production in Listeria monocytogenes infected mice. *Immunopharmacol. Immunotoxicol.* **2002**, *24*, 483–496. [CrossRef] [PubMed]
44. Chasey, D. Investigation of immunoperoxidase-labelled rotavirus in tissue culture by light and electron microscopy. *J. Gen. Virol.* **1980**, *50*, 195–200. [CrossRef] [PubMed]
45. González, M.B.; Quiñones-Gutiérrez, Y. Antibiosis of Cefotaxime/Clindamycin and Lactobacillus acidophiluson Related Bacteria to Diabetic Foot Ulcer. *Food Nutr. Sci.* **2018**, *9*, 277.
46. Reyna-Martinez, R.; Gomez-Flores, R.; López-Chuken, U.; Quintanilla-Licea, R.; Caballero-Hernandez, D.; Rodríguez-Padilla, C.; Beltrán-Rocha, J.C.; Tamez-Guerra, P. Antitumor activity of *Chlorella sorokiniana* and *Scenedesmus* sp. microalgae native of Nuevo León State, México. *PeerJ* **2018**, *6*, e4358. [CrossRef] [PubMed]
47. López-Chuken, U.J.; Young, S.D.; Guzmán-Mar, J.L. Evaluating a 'biotic ligand model'applied to chloride-enhanced Cd uptake by Brassica juncea from nutrient solution at constant Cd^{2+} activity. *Environ. Technol.* **2010**, *31*, 307–318. [CrossRef] [PubMed]
48. Mosmann, T. Rapid colorimetric assay for cellular growth and survival: Application to proliferation and cytotoxicity assays. *J. Immunol. Methods* **1983**, *65*, 55–63. [CrossRef] [PubMed]
49. Shen, Z.; He, H.; Wu, Y.; Li, J. Cyclosporin a inhibits rotavirus replication and restores interferon-beta signaling pathway in vitro and in vivo. *PLoS ONE* **2013**, *8*, e71815. [CrossRef] [PubMed]
50. Zhou, M.; Zhang, Y.; Chen, X.; Zhu, J.; Du, M.; Zhou, L.; Zhang, L.; Wang, W.; Sun, G. PTEN–Foxo1 signaling triggers HMGB1-mediated innate immune responses in acute lung injury. *Immunol. Res.* **2015**, *62*, 95–105. [CrossRef]
51. Jacobsen, A.V.; Yemaneab, B.T.; Jass, J.; Scherbak, N. Reference gene selection for qPCR is dependent on cell type rather than treatment in colonic and vaginal human epithelial cell lines. *PLoS ONE* **2014**, *9*, e115592. [CrossRef]

International Journal of
Molecular Sciences

MDPI

Article

Bergamot Polyphenolic Extract Combined with Albedo and Pulp Fibres Counteracts Changes in Gut Microbiota Associated with High-Fat Diet: Implications for Lipoprotein Size Re-Arrangement

Rocco Mollace [1,2,†], Roberta Macrì [1,*,†], Martina Nicita [1], Vincenzo Musolino [3], Micaela Gliozzi [1], Cristina Carresi [4], Irene Bava [1], Jessica Maiuolo [3], Annamaria Tavernese [1], Antonio Cardamone [1], Luigi Tucci [1], Giuseppe Trunfio [1], Elzbieta Janda [1], Ernesto Palma [4], Carolina Muscoli [1], Francesco Barillà [2], Massimo Federici [2], Federica Scarano [1] and Vincenzo Mollace [1,5,*]

1 Pharmacology Laboratory, Institute of Research for Food Safety and Health IRC-FSH, Department of Health Sciences, University Magna Graecia of Catanzaro, 88100 Catanzaro, Italy; rocco.mollace@gmail.com (R.M.); n.marty96@hotmail.com (M.N.); micaela.gliozzi@gmail.com (M.G.); irenebava@libero.it (I.B.); an.tavernese@gmail.com (A.T.); tony.c@outlook.it (A.C.); l.tucci@head-sa.com (L.T.); g.trunfio@head-sa.com (G.T.); janda@unicz.it (E.J.); muscoli@unicz.it (C.M.); federicascar87@gmail.com (F.S.)
2 Department of Systems Medicine, University of Rome Tor Vergata, 00133 Roma, Italy; francesco.barilla@uniroma2.it (F.B.); federicm@uniroma2.it (M.F.)
3 Pharmaceutical Biology Laboratory, Institute of Research for Food Safety and Health IRC-FSH, Department of Health Sciences, University Magna Graecia of Catanzaro, 88100 Catanzaro, Italy; v.musolino@unicz.it (V.M.); maiuolo@unicz.it (J.M.)
4 Veterinary Pharmacology Laboratory, Institute of Research for Food Safety and Health IRC-FSH, Department of Health Sciences, University Magna Graecia of Catanzaro, 88100 Catanzaro, Italy; carresi@unicz.it (C.C.); palma@unicz.it (E.P.)
5 Renato Dulbecco Institute, Lamezia Terme, 88046 Catanzaro, Italy
* Correspondence: roberta.macri@unicz.it or robertamacri85@gmail.com (R.M.); mollace@unicz.it or mollace@libero.it (V.M.)
† These authors contributed equally to this work.

Citation: Mollace, R.; Macrì, R.; Nicita, M.; Musolino, V.; Gliozzi, M.; Carresi, C.; Bava, I.; Maiuolo, J.; Tavernese, A.; Cardamone, A.; et al. Bergamot Polyphenolic Extract Combined with Albedo and Pulp Fibres Counteracts Changes in Gut Microbiota Associated with High-Fat Diet: Implications for Lipoprotein Size Re-Arrangement. *Int. J. Mol. Sci.* **2023**, *24*, 12967. https://doi.org/10.3390/ijms241612967

Academic Editors: Matteo Bordiga and Baojun Xu

Received: 21 July 2023
Revised: 12 August 2023
Accepted: 17 August 2023
Published: 19 August 2023

Abstract: Evidence exists that the gut microbiota contributes to the alterations of lipid metabolism associated with high-fat diet (HFD). Moreover, the gut microbiota has been found to modulate the metabolism and absorption of dietary lipids, thereby affecting the formation of lipoproteins occurring at the intestinal level as well as systemically, though the pathophysiological implication of altered microbiota composition in HFD and its role in the development of atherosclerotic vascular disease (ATVD) remain to be better clarified. Recently, evidence has been collected indicating that supplementation with natural polyphenols and fibres accounts for an improvement of HFD-associated intestinal dysbiosis, thereby leading to improved lipidaemic profile. This study aimed to investigate the protective effect of a bergamot polyphenolic extract (BPE) containing 48% polyphenols enriched with albedo and pulp-derived micronized fibres (BMF) in the gut microbiota of HFD-induced dyslipidaemia. In particular, rats that received an HFD over a period of four consecutive weeks showed a significant increase in plasma cholesterol, triglycerides and plasma glucose compared to a normal-fat diet (NFD) group. This effect was accompanied by body weight increase and alteration of lipoprotein size and concentration, followed by high levels of MDA, a biomarker of lipid peroxidation. Treatment with a combination of BPE plus BMF (50/50%) resulted in a significant reduction in alterations of the metabolic parameters found in HFD-fed rats, an effect associated with increased size of lipoproteins. Furthermore, the effect of BPE plus BMF treatment on metabolic balance and lipoprotein size re-arrangement was associated with reduced gut-derived lipopolysaccharide (LPS) levels, an effect subsequent to improved gut microbiota as expressed by modulation of the Gram-negative bacteria Proteobacteria, as well as Firmicutes and Bacteroidetes. This study suggests that nutraceutical supplementation of HFD-fed rats with BPE and BMP or with their combination product leads to restored gut microbiota, an effect associated with lipoprotein size re-arrangement and better lipidaemic and metabolic profiles.

Keywords: gut microbiota; atherosclerotic vascular disease; high-fat diet; lipoprotein assembly; lipid metabolism; bergamot extract; polyphenols; prebiotics

1. Introduction

Evidence has been accumulated showing that cardiovascular disease states, including myocardial infarction, hypertension, stroke and peripheral vascular diseases, may be affected by alterations of gut microbiota [1,2]. The pathophysiological basis of these correlations is still unknown. However, data have been provided that multiple bio-molecular pathways appear to be involved in changes occurring in gut microbiota which increase cardiometabolic risk, mostly related to infectious conditions, modifications of host bile acid and changes in lipidaemic profile [3–5]. These events are supposed to contribute to an enhanced passage of endotoxins or toxic metabolites from the intestine to the bloodstream, an effect potentially associated with pro-atherogenic effects, primarily through the acceleration of pre-existing atherosclerotic vascular lesions (e.g., inflammatory-related atherosclerotic plaque destabilization) [6,7]. Consistent data show that endotoxins such as *E. coli* lipopolysaccharide (LPS) deriving from gut Gram-negative bacteria represent potential enhancers for the development of atherosclerotic plaque, possibly via activation of smooth muscle cell proliferation [8–11]. In particular, LPS was found to activate cytokine release, an effect associated with the activation of Toll-like receptor (TLR) signalling, and promote LDL oxidation, thereby reducing constitutive NO release and promoting lipid peroxidation via harmful peroxynitrite generation [12]. Thus, the increase in circulating endotoxin levels as a consequence of gut microbiota disturbances could play a role in atherosclerosis development. On the other hand, bacterial metabolites, such as trimethylamine (TMA), are transformed in the liver into trimethylamine N-oxide (TMAO), which at high levels is associated with the development of non-alcoholic fatty liver disease (NAFLD) [13,14], an effect which is accompanied by altered metabolic balance, re-arrangement of lipoprotein composition and, finally, enhanced atherosclerosis development [15,16]. These data are partially confirmed by clinical observational studies in which modifications of gut microbiota seem to clearly correlate with increased cardiometabolic risk, though specific changes have not been identified yet and results are often not corroborated between studies [17,18].

Natural products claimed to produce potential benefits in modulating gut microbiota disorders have been correlated with an improvement of cardiometabolic profile in both animal models of cardiovascular disease and in patients [19–21]. Evidence has been shown that the use of prebiotics alone or in combination with natural antioxidants stimulates the growth of beneficial gut bacteria, leading to health benefit [22,23].

In particular, both pre-clinical and clinical studies have highlighted the possible beneficial responses to counteract the development of cardiovascular disease produced by several prebiotics, including oligosaccharides, inulin and pectins as the ones found in natural fibres deriving from plant derivatives [24]. These responses depend on an increased production of short-chain fatty acids (SCFAs) accounted for the growth of beneficial bacteria genera, such as *Lactobacillus* and *Bifidobacterium* [25]. In addition, SCFA increase leads to an antagonistic effect on the growth of pathogenic species [26,27] and to better intestinal epithelial cell layer protection [28]. Finally, SCFAs have been found to inhibit histone deacetylase, thereby contributing to resolving gut inflammatory conditions [29,30], leading also to better systemic metabolic balance via activating glucagon-like peptide-1 (GLP-1) production [31].

Besides these effects, the contribution of prebiotics to gut-microbiota-related disorders remains to be better clarified.

Alongside prebiotics, diet supplementation with natural plant-derived compounds, such as flavonoids, a family of active ingredients well represented in *Citrus* derivatives, berries, red wine, apples and olive oil, have widely been correlated with improvement of gut microbial dysbiosis [32,33]. The contribution of such natural antioxidants to prevent cardiometabolic disorders accompanying gut microbiota alterations may depend on various effects mostly related to their properties in scavenging the overproduction of free radical species (ROS) [34,35]. In particular, data exist that flavonoids such as naringin and hesperidin possess an inhibitory effect on TMA-lyase, thereby leading to reduced TMAO concentration [36,37]. Therefore, the combination of flavonoids and fibres with prebiotic properties may represent a valid solution for approaching cardiometabolic disorders thanks to their potential synergistic response under conditions such as high fat diet (HFD), in which an alteration of lipidaemic profile is associated with altered gut microbiota [38,39].

In recent years, evidence has been collected that *Citrus bergamia* Risso & Poiteau (bergamot), a *Citrus* species rich in flavonoids, may represent a potential source of natural antioxidants able to counteract HFD-induced alteration of lipidaemic profile both in animal models of hyperlipidaemia and in patients [40,41]. Moreover, a polyphenolic-rich extract derived from bergamot juice has been found to improve lipoprotein profiles in patients with liver steatosis alongside having a beneficial effect on liver function [42]. Finally, bergamot fibres alone or in combination with natural polyphenolic extracts were found to be able to antagonize fat accumulation and insulinaemic response both in animals and obese patients, suggesting that dietary supplementation with such products may have a beneficial response in metabolic regulation under HFD conditions [43,44].

Here, we studied the effect of a bergamot-derived polyphenolic-rich extract (BPE), combined with bergamot fibres (BMP) prebiotic-rich in pectin and oligosaccharides derived from bergamot albedo and pulp fibres, on the gut microbiota of rats fed a normal (NFD) and a high-fat diet (HFD). Data for this unique preparation on gut microbiota were correlated with lipidaemic profile, oxidative/inflammatory response and, finally, lipoprotein size and concentration.

2. Results

2.1. *Effect of BPE, BMF and BPE + BMF on Lipidaemic Profiles in Rats Fed a NFD and a HFD*

Supplementation of rats with an HFD was associated with metabolic changes and body weight alterations compared to rats fed a NFD. In particular, in rats fed a HFD over a period of four consecutive weeks, an increase in body weight, fasting plasma glucose, total cholesterol, LDL cholesterol, HDL cholesterol, triglycerides and MDA levels was found compared to rats receiving the standard diet (NFD) (Figure 1A,B). This effect was counteracted by supplementation of rats with BPE, BMF and BPE + BMF, as shown in Figure 1A,B. Indeed, in animals fed a HFD and supplemented with 20 mg/kg of BPE, BMF and BPE + BMF in a single daily administration via gastric gavage, a significant reduction in all parameters was found compared to the ones altered by HFD alone. No effect was found in rats receiving a standard diet (NFD) compared to groups supplemented with BPE, BMF and BPE + BMF at 20 mg/kg (Figure 1A,B). Moreover, all parameters at baseline were comparable in all eight groups of rats used throughout the study.

The effect of treatment with bergamot extracts was also found to be able to modify HFD-induced changes of lipoprotein profile compared to the NFD. Indeed, the HFD compared to the NFD produced a substantial re-arrangement of plasma levels of lipoprotein particles in rats as evaluated after 4 weeks compared to basal levels (Figure 2).

This effect was counteracted by treatment for 4 weeks with 20 mg/Kg of BPE, BMF or BPE + BMF (Figure 2). In particular, BPE + BMF was found to be able to decrease the mean concentration of IDL particles by 40.26%, to increase large LDLs by 27.6% and to decrease small LDLs by 24.4%. On the other hand, treatment with BPE + BMF led to a 40% increase in total HDL particles, mainly due to the increase in large HDL particles.

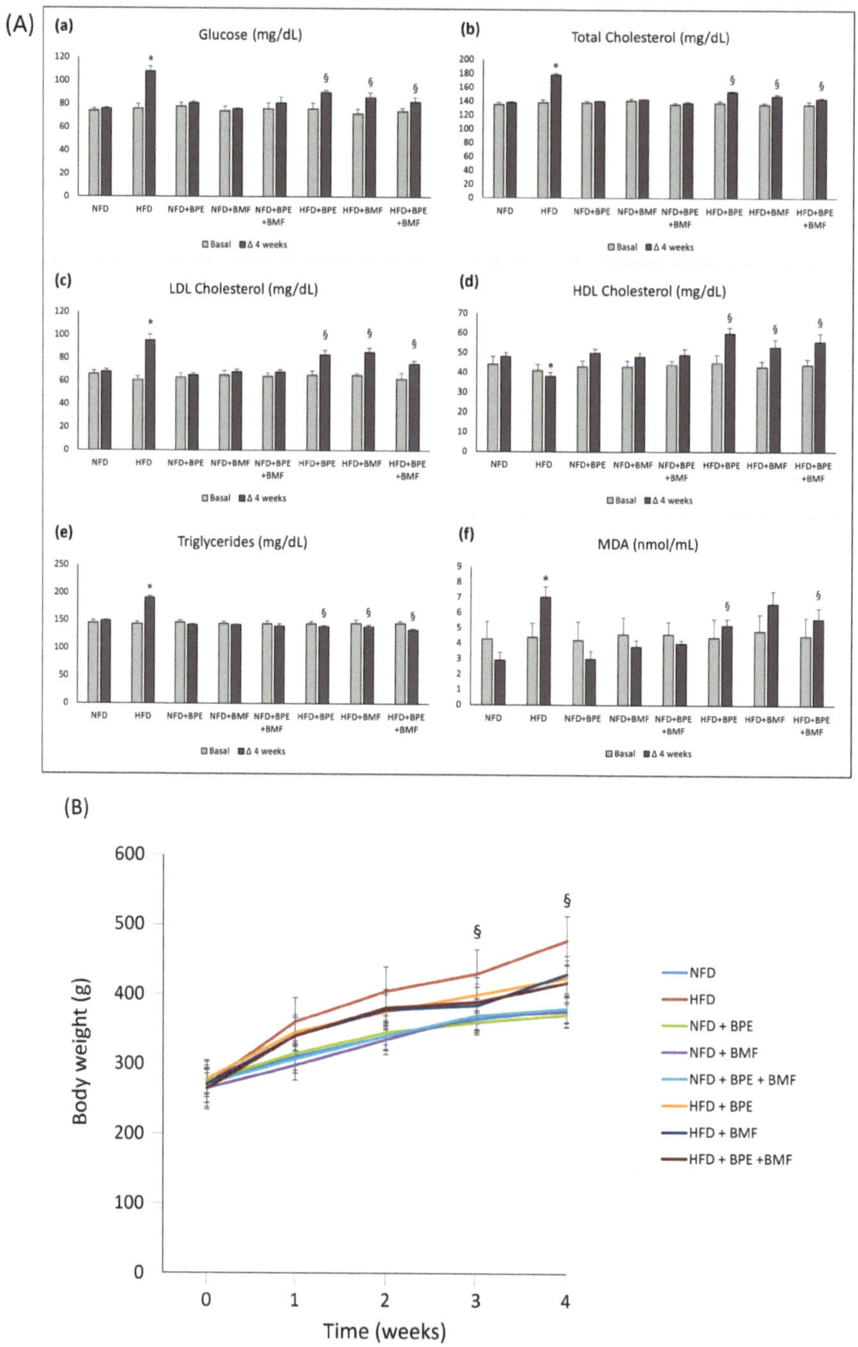

Figure 1. The effect of BPE, BMF or BPE + BMF (20 mg/Kg daily given orally over a period of 4 weeks on (**A**) plasma glucose (**a**), total cholesterol (**b**), LDL cholesterol (**c**), HDL cholesterol (**d**), triglycerides (**e**) and MDA (**f**) and on (**B**) body weight in NFD and HFD groups. Data are expressed as means ± SEs. *: $p < 0.05$ vs. NFD; §: $p < 0.05$ vs. HFD.

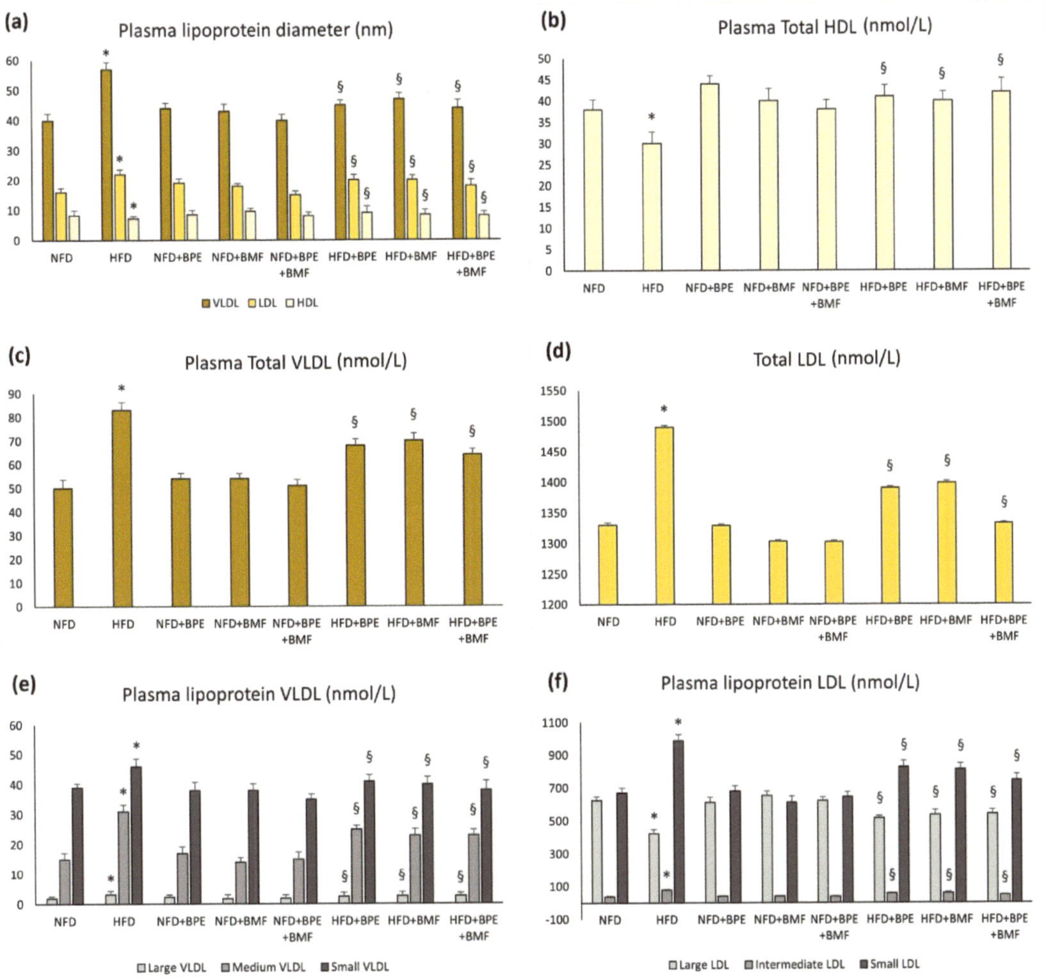

Figure 2. The effect of BPE, BMF or BPE + BMF (20 mg/Kg daily given orally over a period of 4 weeks) on lipoprotein diameter (**a**) and concentration (nmol/L) in NFD and HFD groups. Plasma total HDL (**b**); plasma total VLDL (**c**); plasma total LDL (**d**); plasma lipoprotein VLDL (large, medium and small) (**e**); plasma lipoprotein LDL (large, intermediate and small) (**f**). Data are expressed as means ± SEs. *: $p < 0.05$ vs. NFD; §: $p < 0.05$ vs. HFD.

2.2. Effect of BPE, BMF and BPE + BMF on Gut Microbiota in Rats NFD and a HFD

The studies on gut microbiota were carried out in order to verify the microbial composition and abundance at the phylum level in both NFD and HFD rats either untreated or treated with BPE, BMF and BPE + BMF (Figure 3).

In particular, under basal conditions, 11 phyla were found in eight groups, showing that Firmicutes and Bacteroidetes represented the largest proportions among detectable bacteria. However, in animals fed a HFD, compared to the NFD group, consistent changes were found in our study. Indeed, increased proportions of Firmicutes and Protobacteria were found as a consequence of HFD supplementation compared to the NFD and, in contrast, a significant decrease in Bacteroidetes was observed in the HFD group. This effect was counteracted by treating rats with BPE, BMF and BMF + BPE (Figure 4).

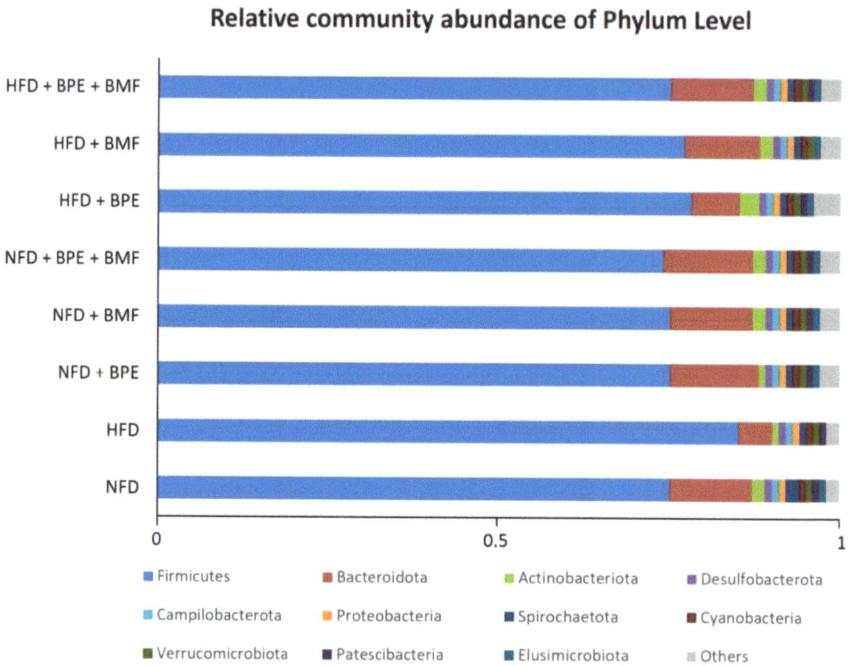

Figure 3. The modulatory effect of BPE, BMF or BPE + BMF on the gut microbiota composition (determined by sequencing the V3–V4 region of the 16S rRNA gene using the MiSeq Illumina system) expressed as relative community abundance at the phylum level. Among the identifiable bacterial phyla with ≥1% abundance in all samples, Firmicutes was the most abundant, followed by Bacteroidota and Actinobacteroidota.

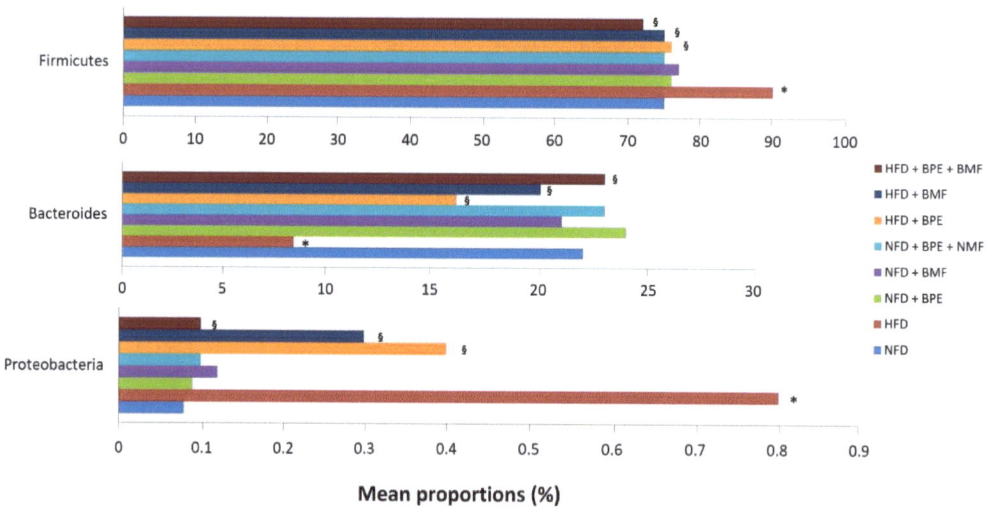

Figure 4. The effect of BPE, BMF or BPE + BMF on the relative abundances of Firmicutes, Bacteroidetes and Proteobacteria in NFD and HFD groups. Data are expressed as means ± SEs (n = 6). *: $p < 0.05$ vs. the NFD group; §: $p < 0.05$ vs. the HFD group.

Indeed, all bergamot derivatives almost completely attenuated changes produced by the HFD compared to the NFD, showing that the combination of polyphenols and bergamot fibres may contribute to restore the normal composition of gut microbiota.

2.3. Effect of BPE, BMF and BPE + BMF on Plasma LPS Levels in Rats Fed a NFD and a HFD

Plasma LPS levels detected after four weeks were significantly higher in the HFD group compared to the NFD group (Figure 4). Treatment of rats with BPE, BMF or BPE + BMF resulted in significant lowering of plasma LPS levels compared to the HFD group. No changes of plasma LPS concentration were found in NFD-fed rats either untreated or treated with BPE, BMF or BPE + BMF (Figure 5).

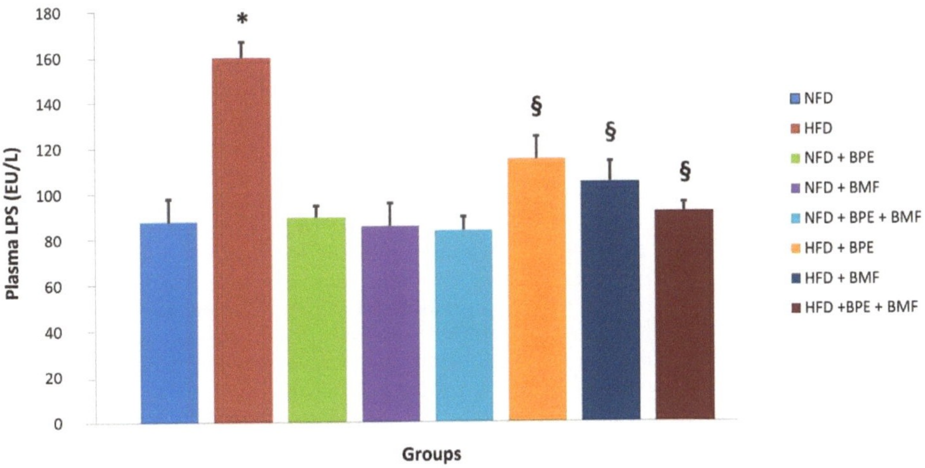

Figure 5. The effect of BPF, BMF or BPF + BMF (20 mg/Kg daily given orally over a period of 4 weeks) on plasma LPS (EU/L) concentration in NFD and HFD groups. Data are expressed as means ± SE. *: $p < 0.05$ HFD vs. NFD; §: $p < 0.05$ HFD.

3. Discussion

Our data show that a hyperlipidaemic diet (HFD) led to alterations of gut microbiota which resulted in a significant increase in the abundance of Firmicutes and Proteobacteria and a decreased abundance of Bacteroidetes compared to rats fed a NFD. This effect was associated with increased LPS levels in rats fed the HFD, thereby confirming that changes of gut microbiota subsequent to HFD lead to an inflammatory state combined with oxidative stress as detected by means of MDA measurements [45–47]. The changes of gut microbiota found after 4 weeks of HFD were associated with increased body weights and metabolic alterations represented by increased plasma glucose, cholesterol and triglycerides and changes of lipoprotein size and concentration compared with rats fed a normolipaemic diet (NFD). Moreover, oxidative stress biomarkers such as MDA were also found to be elevated in the blood of HFD rats, as previously shown by our and other groups [48–50].

These effects were counteracted by treating rats with BPE alone or in combination with BMF. In fact, in rats fed a HFD, where the diet was supplemented with BPE, BMP or a combination of both, the normal pattern of gut microbiota was restored, this effect being associated with reduction in body weight and in plasma levels of glucose, cholesterol and triglycerides. Interestingly, the improvement of lipidaemic profiles in rats receiving bergamot extracts alone or in combination with fibres was accompanied by an increased size of lipoproteins, mostly LDLs, which are known to play a key role in atherosclerosis development. On the other hand, the effect of natural antioxidants combined with BMF was associated with reduced LPS levels and with consistent reduction in MDA, thereby leading to an overall improvement of cardiometabolic risk profile in rats fed with HFD.

Our data are consistent with previous observations showing that the gut microbiota is a key player in modulating dietary lipid metabolism, affecting almost all the steps involved in the regulation of lipid digestion and absorption, being also involved in the generation of lipoproteins occurring at the intestinal level [15]. In particular, it has been shown that changes in gut microbiota composition, such as the one obtained by means of supplementation with Gram-negative bacteria, lead to obesity with increased LDL lipoproteins, an effect accompanied by elevation of plasma cholesterol and modulation of the lipid transfer protein system [51]. On the other hand, germ-free mice develop an obesity-resistant phenotype when fed a HFD, an effect which involves decreased fasting triglycerides and VLDL production when compared to conventionally reared mice [52].

Moreover, our data confirm previous evidence showing that HFD leads to modifications of gut microbiota [53]. On the other hand, our data confirm that restoring the equilibrium among several intestinal bacteria, as found when animals are fed a NFD, is accompanied by normalization of lipidaemic profile, lipoprotein re-arrangement and, finally, attenuated inflammation, which is associated with the altered lipidaemic profile.

The rationale of these responses is still to be better clarified. However, clear evidence exists that LPS affects the integrity of the intestinal mucosa by altering tight junctions and thereby impairing intestinal permeability [53–56]. In particular, evidence exists that alteration of gut microbiota induced by HFD leads to overproduction of LPS by Gram-negative bacteria, which is followed by impairment of the tight junction proteins, such as occludin, claudin-1 and ZO-1, which leads to LPS entering the portal circulation [57] and thereby producing at least two systemic responses: one is mediated by liver inflammation via TLR4 activation and cytokine release, which represent the key mechanisms of imbalanced packaging and release of lipoproteins from the liver [53,58,59]; the second one is represented by a condition of systemic inflammation and oxidative stress which leads to enhanced atherosclerosis and cardiometabolic risk [12,60,61].

These pathophysiological events are counteracted by bergamot extract and fibres. This is also consistent with previous data showing that BPE, a powerful antioxidant in vitro and in vivo, leads to significant protection of vasculature against oxidative damage subsequent to dyslipidaemia in both diet-induced metabolic syndrome and in patients undergoing increased cardiometabolic risk [48,62,63].

On the other hand, BPE has been found to produce relevant protection under conditions of liver inflammation, thereby preventing NASH and its deleterious effects in cardiometabolic risk, mostly due to its antioxidant properties [64].

Our data also show a synergistic response when prebiotic BMF is associated with BPE. Previous data have shown that BMF may produce a significant inhibition of post-prandial insulin response in patients and that this could account for a role of BMF in maintaining a normal metabolic balance in subjects suffering from metabolic syndrome [43].

On the other hand, the use of prebiotics is consistent with gut microbiota normalization able to reduce cardiometabolic risk [65–68].

Thus, it is likely that a combination of both antioxidant BPE and prebiotic BMF may better target inflammatory/oxidative damage and dysbiosis subsequent to HFD with a sequential response occurring via reduction in LPS release and subsequently by attenuating endotoxin-related systemic consequences.

In conclusion, our data confirm that HFD-related changes of gut microbiota are accompanied by increased body weight and alteration of lipoprotein size, an effect which is associated with modification of lipidaemic profile and imbalanced glucose levels. On the other hand, dysbiosis produced by alterations of gut microbiota and the subsequent alteration in lipidaemic profile and lipoprotein packaging contributes to the development of HFD-associated imbalance occurring in the mechanisms of lipid regulation and transport, occurring both at the intestinal and systemic level. This is also expressed by enhanced oxidative stress and increased LPS levels, thereby representing key mechanisms in systemic inflammation which may be found in animals fed HFD.

Our data also show, for the first time, that the alterations induced by HFD may be counteracted by supplementing rats with BPE, BMF or a combination of both, which restored gut microbiota and produced a re-arrangement of lipoprotein size, reduction in both LPS and MDA levels, and, finally, led to antagonism of diet-induced dyslipidaemia and metabolic imbalance. This suggests that combining antioxidants and prebiotics leads to sequential responses for better counteracting diet-induced alterations of gut microbiota and their deleterious effects on cardiometabolic risk profile.

4. Materials and Methods

4.1. Plant Material

Bergamot (botanical name: *Citrus bergamia* Risso & Poiteau, Variety Fantastico) is an endemic plant growing almost exclusively in the jonic cost of the Calabrian Region in the South of Italy. The bergamot polyphenolic-rich extract (BPE) and fibres (BMF) used throughout the study were obtained from fruits which were collected from December to January 2022 from plants located in a range of 90 Km from Bianco to Reggio Calabria, Italy (GPS coordinates: latitude: 38.0917; longitude: 16.15159).

In particular, bergamot juice (BJ) was obtained from peeled fruits by squeezing. The juice was oil-fraction-depleted through stripping and clarified through ultra-filtration, with subsequent loading on to a polystyrene resin column able to absorb polyphenol compounds of molecular weights between 300 and 600 Da (Mitsubishi Chemical, Cleansui Co., Ltd., Tokyo, Japan). Polyphenol fractions were eluted by a 1 mM KOH solution. The basic eluate was incubated on a rocking platform to reduce the furocoumarin amount. The phytocomplex was neutralized through filtration on cationic resin at acidic pH, vacuum dried and minced to the desired particle size to obtain BPE powder. The toxicological analyses performed, including heavy metal, pesticide, phthalate and synephrine content analyses, revealed the absence of toxic compounds at significant levels [40]. Standard microbiological tests showed that the final BPE and BMF were free of mycotoxins and contaminating bacteria. Moreover, studies for detecting acute and chronic toxicity in rodents for both products were conducted under (Good Laboratory Practice) GLP conditions. In particular, the study N.152-0041 was carried out in compliance with the EU Directive 2004/9/EC and Directive 2004/9/EC for GLP Guidelines and with OECD Guidelines for Repeated Dose 30-day Oral Toxicity Study in Rodents [69]. In addition, the study N.152-0043 was conducted with OECD Guidelines for Repeated Dose 90-day Oral Toxicity Study in Rodents [70]. Finally, micronuclei test, aberration test and reversed mutation measurements confirmed that both BPE and BMF were unable to produce toxicological or mutagenic effects in rodents according to current regulatory guidelines.

The antioxidant analysis of bergamot extract was performed prior to the administration of products in animals by means of EPR analysis (Supplementary S1, Figures S1 and S2).

Fibres obtained by bergamot albedo were micronized, and HPLC and nutritional analysis of the powder was performed. All materials were provided by H&AD srl (Herbal and Antioxidants Derivatives srl, Bianco, Italy). Test products were dissolved in water. A specification sheet of both extracts is shown in the Supplementary Materials (Supplementary S3 and S4).

4.2. High-Pressure Liquid Chromatography (HPLC)

HPLC analysis was performed by a Fast 1200 HPLC system (Agilent Technologies, 5301 Stevens Creek Blvd, Santa Clara, CA, USA) equipped with a DAD detector and a ZORBAX Eclipse XDB-C18 column—50 mm. A quantity of 2 μL of sample was injected and eluted with a two-solvent gradient of water and acetonitrile. Different gradients were used to determine flavonoid, furocoumarin and other polyphenol contents. The flow rate was 3 mL/min, and the column was maintained at 35 °C. The detector was monitored at 280 nm. Flavonoid and furocoumarin pure standards were purchased from Sigma-Aldrich (Burlington, MA, USA). Brutieridin and melitidin were identified according to Di Donna [40]. In the Supplementary Methods section, it is possible to view the chromatograms of HPLC analyses related to BPE and BMF powders. Method validation parameters, such as linearity, limits of detection, precision and accuracy, are available in Supplementary S2.

The main flavonoids identified were standardized at 48% in BPE powder (Supplementary S3). The estimated concentrations of the five main flavonoids were: neoeriocitrin (131,942 ppm), naringin (138,945 ppm), neohesperidin (134,617 ppm), melitidin (27,950 ppm) and brutieridin (56,496 ppm) (Supplementary S3).

The concentrations of the main flavonoids in BMF powder were standardized at 17%: neoeriocitrin (4398 ppm), naringin (5184 ppm), neohesperidin (3966 ppm), melitidin (1233 ppm) and brutieridin (2286 ppm) (Supplementary S4).

4.3. Animals

Male Sprague-Dawley 3-month-old rats (Charles River, Milan, Italy) weighing 270–290 g were used throughout the study. All animals were housed and cared for in accordance with the Italian National Health Ministry Guidelines on Laboratory Animal Welfare following the Italian regulations for the protection of animals used for experimental and other scientific purposes (D.L. 26/2014) as well as with the European Community guidelines [71]. The study was conducted according to the guidelines of the Declaration of Helsinki and approved by the local Ethics Committee (Calabrian Region Prot. 324: 12 October 2021) (Supplementary S5).

Rats were housed and maintained under controlled conditions of temperature (21 °C) and humidity (60–65%) with a 12 h light/12 h dark cycle and allowed food ad libitum.

The high-fat diet TD.88137 Total Fat (21% by weight; 42% kcal from fat) was purchased from Harlan Laboratories, Rossdorf, Germany; the BPE and BMF, both micronized and, when required, co-grinded, were kindly provided by H&AD srl (Herbal and Antioxidants Derivatives srl, Bianco, Italy).

4.4. Study Design

After adaptation, the rats were randomly allocated in eight experimental groups:

Control group, fed a normal-fat diet (NFD) for 4 consecutive weeks (n = 6);
HFD group, fed a high-fat diet (HFD) for 4 consecutive weeks (n = 6);
NFD receiving 20 mg/Kg of BPE for 4 consecutive weeks (n = 6);
NFD receiving 20 mg/Kg of BMF for 4 consecutive weeks (n = 6);
NFD receiving 20 mg/Kg of BPE + BMF for 4 consecutive weeks (n = 6);
HFD receiving 20 mg/Kg of BPE for 4 consecutive weeks (n = 6);
HFD receiving 20 mg/Kg of BMF for 4 consecutive weeks (n = 6);
HFD receiving 20 mg/Kg of BPE + BMF for 4 consecutive weeks (n = 6);

Quantities of 20 mg/Kg of the BPE powder were administered alone or in combination with 20 mg/Kg mg of power derived from dried albedo and pulp fibres (BMF).

All treatments were given via gastric gavage once daily over a period of four weeks. The doses of BPE and BMF used throughout the study and the treatment time were chosen according to previous data from our and other research groups [40,72,73].

Body weight, plasma glucose, total cholesterol, triglycerides, HDL, LDL, MDA, plasma lipoprotein size and concentrations were measured as previously shown [40,74] before starting treatment and at the end of the feeding period. Blood samples were collected by means of venipuncture of the tail vein and were stored as previously described [40].

At the same experimental times, rat faeces were collected from the anus of each rat using sterile EP tubes and immediately preserved in liquid nitrogen at basal and after 4 weeks of treatment in animals fed the NFD and HFD according to the treatment schedule.

4.5. Laboratory Measurements

At the baseline and after 4 weeks of treatment, a 12 h fasting morning blood sample was collected. All plasma marker concentrations or activities were measured using classical methods and commercial assay kits, according to the manufacturers' instructions. Assay kits for total cholesterol, LDL-C, HDL-C, triglycerides, malondialdehyde (MDA), paraoxonase and glutathione peroxidase were purchased from Novamedical Srl (Reggio Calabria, Italy). All the laboratory tests were performed in a blinded manner in respect to the assigned treatment.

At the baseline and after 4 weeks of treatment with BPE, BMP and BPE + BMP, plasma samples were collected in EDTA tubes after a 12 h overnight fast: fasting plasma glucose (mg/dL), total cholesterol (mg/dL), low-density lipoprotein cholesterol (LDL cholesterol) (mg/dL), high-density lipoprotein cholesterol (HDL cholesterol) (mg/dL) and triglycerides (TGs) (mg/dL).

Plasma lipoprotein particles were detected as previously described [42] by means of the proton NMR spectroscopy technique with the simultaneous concentration measure of lipoprotein subclasses of different sizes. NMR provides calculated values for mean very-low-density lipoprotein (VLDL), LDL and HDL particle sizes plus estimates of total and VLDL, TGs and HDL cholesterol.

The oxidative stress index was assessed by plasma lipid peroxidation product malondialdehyde (MDA) through a lipid peroxidase assay kit (Sigma-Aldrich, Saint Louis, MO, USA) according to the manufacturer's protocol. Briefly, the lipid peroxidation is determined through the reaction of MDA with thiobarbituric acid (TBA) to obtain a colorimetric (532 nm)/fluorometric (λex = 532/λem = 553 nm) product proportional to the MDA concentration in the sample [41].

LPS plasma levels at basal and after 4 weeks of treatment with BPE, BMF and BPE + BMF were detected through a competitive inhibition enzyme immunoassay. A Low Sample Volume Lipopolysaccharides (LPS) ELISA Kit (abbexa, abx355419, Leiden, NL) was used for the in vitro LPS quantification.

4.6. 16S rRNA Gene Sequencing and Analysis

4.6.1. DNA Extraction and PCR Amplification

Rat faeces samples were processed using the E.Z.N.A.® Stool DNA Kit (Omega Biotek, Norcross, GA, USA) to extract total microbial genomic DNA. Before being used, the DNA quality and concentration were assessed using 1.0% agarose gel electrophoresis analysis and a NanoDrop® ND-2000 spectrophotometer (Thermo Scientific Inc., Waltham, MA, USA). An ABI GeneAmp® 9700 PCR thermocycler (Thermo Scientific Inc., Waltham, MA, USA) was used to amplify the hypervariable region V3-V4 of the bacterial 16S rRNA gene using primer pairs 338F (5′-ACTCCTACGGGAGGCAGCAG-3′) and 806R (5′-GGACTACHVGGGTWTCTAAT-3′). Each sample was amplified in triplicate.

The PCR products were run on a 2% agarose gel and then purified through the GeneJET DNA Gel Extraction Kit (Thermo Fisher Scientific, Waltham, MA, USA). The quantification was performed using an AxyPrep DNA Gel Extraction Kit (Axygen Biosciences, Union City, CA, USA) according to the manufacturer's instructions.

4.6.2. Illumina MiSeq Sequencing

Once the amplicons were purified, they were combined in equimolar proportions and paired-end sequenced on the Illumina MiSeq PE300 platform (Illumina, San Diego, CA, USA) in accordance with standard protocols (Majorbio Bio-Pharm Technology Co. Ltd., Shanghai, China). The raw sequences were pre-analysed according to the BIPES protocol [75], and QIIME (1.8.0) was used to perform the following analyses. The representative sequences were aligned through PyNAST algorithms [76]. Phylogenetic relationships were assessed by FastTree sequence alignment [77], and the taxonomic assignments were combined to construct the BIOM file [78].

4.7. Statistical Analysis

Data analyses were performed using GraphPad PRISM 9.3.1 (GraphPad Software, Inc., La Jolla, CA, USA). The differences among treatments were evaluated using GraphPad PRISM 9.3.1 (GraphPad Software, Inc., La Jolla, CA, USA). The Shapiro–Wilk test was used to test normality. Normally distributed data were analyzed by one-way ANOVA followed by Tukey's test; data not normally distributed were analyzed through the Kruskal–Wallis test followed by Dunn's tests. The results are shown as means ± SEs. A p-value < 0.05 was considered statistically significant.

Supplementary Materials: The following supporting information can be downloaded at: https://www.mdpi.com/article/10.3390/ijms241612967/s1.

Author Contributions: R.M. (Rocco Mollace), R.M. (Roberta Macrì), V.M. (Vincenzo Mollace) and F.S.: Conceptualization. R.M. (Rocco Mollace), R.M. (Roberta Macrì), F.S. and V.M. (Vincenzo Mollace): Writing—Original Draft. M.N., A.T., A.C., V.M. (Vincenzo Musolino), M.G., C.C., I.B., J.M., L.T. and G.T.: Investigation, Methodology and Formal Analysis. E.J. and E.P.: Writing—Review and Editing. V.M. (Vincenzo Mollace), C.M., F.B. and M.F.: Supervision. V.M. (Vincenzo Mollace): Funding. All authors have read and agreed to the published version of the manuscript.

Funding: The work was supported by the public resources from the Italian Ministry of Research: PON-MIUR 03PE000_78_1 and PONMIUR 03PE000_78_2. POR Calabria FESR FSE 2014–2020 Asse 12-Azioni 10.5.6 e 10.5.12.

Institutional Review Board Statement: The study was conducted according to the guidelines of the Declaration of Helsinki and approved by local Ethics Committee (Calabrian Region Prot. 324: 12 October 2021).

Data Availability Statement: The data presented in this study are available upon request from the corresponding author.

Conflicts of Interest: The authors declare no conflict of interest.

References

1. Rahman, M.M.; Islam, F.; Or-Rashid, M.H.; Mamun, A.A.; Rahaman, M.S.; Islam, M.M.; Meem, A.F.K.; Sutradhar, P.R.; Mitra, S.; Mimi, A.A.; et al. The gut microbiota (microbiome) in cardiovascular disease and its therapeutic regulation. *Front. Cell. Infect. Microbiol.* **2022**, *12*, 903570. [CrossRef]
2. Katsimichas, T.; Theofilis, P.; Tsioufis, K.; Tousoulis, D. Gut microbiota and coronary artery disease: Current therapeutic perspectives. *Metabolites* **2023**, *13*, 256. [CrossRef]
3. Krishnan, S.; Alden, N.; Lee, K. Pathways and functions of gut microbiota metabolism impacting host physiology. *Curr. Opin. Biotechnol.* **2015**, *36*, 137–145. [CrossRef] [PubMed]
4. Maciel-Fiuza, M.F.; Muller, G.C.; Campos, D.M.S.; do Socorro Silva Costa, P.; Peruzzo, J.; Bonamigo, R.R.; Veit, T.; Vianna, F.S.L. Role of gut microbiota in infectious and inflammatory diseases. *Front. Microbiol.* **2023**, *14*, 1098386. [CrossRef]
5. Yntema, T.; Koonen, D.P.Y.; Kuipers, F. Emerging roles of gut microbial modulation of bile acid composition in the etiology of cardiovascular diseases. *Nutrients* **2023**, *15*, 1850. [CrossRef] [PubMed]
6. Shen, X.; Li, L.; Sun, Z.; Zang, G.; Zhang, L.; Shao, C.; Wang, Z. Gut microbiota and atherosclerosis-focusing on the plaque stability. *Front. Cardiovasc. Med.* **2021**, *8*, 668532. [CrossRef] [PubMed]
7. Nakajima, A.; Mitomo, S.; Yuki, H.; Araki, M.; Seegers, L.M.; McNulty, I.; Lee, H.; Kuter, D.; Ishibashi, M.; Kobayashi, K.; et al. Gut microbiota and coronary plaque characteristics. *JAHA* **2022**, *11*, e026036. [CrossRef]
8. Yin, Q.; Jiang, D.; Li, L.; Yang, Y.; Wu, P.; Luo, Y.; Yang, R.; Li, D. LPS promotes vascular smooth muscle cells proliferation through the TLR4/Rac1/Akt signalling pathway. *Cell. Physiol. Biochem.* **2017**, *44*, 2189–2200. [CrossRef] [PubMed]
9. Salvemini, D.; Mollace, V.; Pistelli, A.; Anggard, E.; Vane, J. Metabolism of glyceryl trinitrate to nitric oxide by endothelial cells and smooth muscle cells and its induction by *Escherichia coli* lipopolysaccharide. *Proc. Natl. Acad. Sci. USA* **1992**, *89*, 982–986. [CrossRef] [PubMed]
10. Curcio, A.; Torella, D.; Indolfi, C. Mechanisms of smooth muscle cell proliferation and endothelial regeneration after vascular injury and stenting: Approach to therapy. *Circ. J.* **2011**, *75*, 1287–1296. [CrossRef] [PubMed]
11. Jiang, D.; Yang, Y.; Li, D. Lipopolysaccharide induced vascular smooth muscle cells proliferation: A new potential therapeutic target for proliferative vascular diseases. *Cell Prolif.* **2017**, *50*, e12332. [CrossRef] [PubMed]
12. Gliozzi, M.; Scicchitano, M.; Bosco, F.; Musolino, V.; Carresi, C.; Scarano, F.; Maiuolo, J.; Nucera, S.; Maretta, A.; Paone, S.; et al. Modulation of nitric oxide synthases by oxidized LDLs: Role in vascular inflammation and atherosclerosis development. *Int. J. Mol. Sci.* **2019**, *20*, 3294. [CrossRef] [PubMed]
13. Flores-Guerrero, J.L.; Post, A.; van Dijk, P.R.; Connelly, M.A.; Garcia, E.; Navis, G.; Bakker, S.J.L.; Dullaart, R.P.F. Circulating trimethylamine-N-oxide is associated with all-cause mortality in subjects with nonalcoholic fatty liver disease. *Liver Int.* **2021**, *41*, 2371–2382. [CrossRef] [PubMed]
14. Theofilis, P.; Vordoni, A.; Kalaitzidis, R.G. Trimethylamine N-Oxide levels in non-alcoholic fatty liver disease: A systematic review and meta-analysis. *Metabolites* **2022**, *12*, 1243. [CrossRef] [PubMed]
15. Yu, Y.; Raka, F.; Adeli, K. The Role of the gut microbiota in lipid and lipoprotein metabolism. *J. Clin. Med.* **2019**, *8*, 2227. [CrossRef]

16. Toh, J.Z.K.; Pan, X.H.; Tay, P.W.L.; Ng, C.H.; Yong, J.N.; Xiao, J.; Koh, J.H.; Tan, E.Y.; Tan, E.X.X.; Dan, Y.Y.; et al. A Meta-Analysis on the global prevalence, risk factors and screening of coronary heart disease in nonalcoholic fatty liver disease. *Clin. Gastroenterol. Hepatol.* **2022**, *20*, 2462–2473.e10. [CrossRef]
17. Tousoulis, D.; Guzik, T.; Padro, T.; Duncker, D.J.; De Luca, G.; Eringa, E.; Vavlukis, M.; Antonopoulos, A.S.; Katsimichas, T.; Cenko, E.; et al. Mechanisms, therapeutic implications, and methodological challenges of gut microbiota and cardiovascular diseases: A position paper by the ESC working group on coronary pathophysiology and microcirculation. *Cardiovasc. Res.* **2022**, *118*, 3171–3182. [CrossRef]
18. Yin, J.; Liao, S.X.; He, Y.; Wang, S.; Xia, G.H.; Liu, F.T.; Zhu, J.J.; You, C.; Chen, Q.; Zhou, L.; et al. Dysbiosis of gut microbiota with reduced Trimethylamine-N-Oxide level in patients with large-artery atherosclerotic stroke or transient ischemic attack. *JAHA* **2015**, *4*, e002699. [CrossRef] [PubMed]
19. Zhao, X.; Oduro, P.K.; Tong, W.; Wang, Y.; Gao, X.; Wang, Q. Therapeutic potential of natural products against atherosclerosis: Targeting on gut microbiota. *Pharmacol. Res.* **2021**, *163*, 105362. [CrossRef]
20. Zhu, L.; Zhang, D.; Zhu, H.; Zhu, J.; Weng, S.; Dong, L.; Liu, T.; Hu, Y.; Shen, X. Berberine treatment increases Akkermansia in the gut and improves high-fat diet-induced atherosclerosis in Apoe$^{-/-}$ mice. *Atherosclerosis* **2018**, *268*, 117–126. [CrossRef]
21. McGrail, L.; Garelnabi, M. Polyphenolic compounds and gut microbiome in cardiovascular diseases. *Curr. Pharm. Biotechnol.* **2020**, *21*, 578–586. [CrossRef]
22. Vacca, M.; Celano, G.; Lenucci, M.S.; Fontana, S.; Forgia, F.M.; Minervini, F.; Scarano, A.; Santino, A.; Dalfino, G.; Gesualdo, L.; et al. In Vitro selection of probiotics, prebiotics, and antioxidants to develop an innovative synbiotic (NatuREN G) and testing its effect in reducing uremic toxins in fecal batches from CKD patients. *Microorganisms* **2021**, *9*, 1316. [CrossRef] [PubMed]
23. Everard, A.; Lazarevic, V.; Gaïa, N.; Johansson, M.; Ståhlman, M.; Backhed, F.; Delzenne, N.M.; Schrenzel, J.; François, P.; Cani, P.D. Microbiome of prebiotic-treated mice reveals novel targets involved in host response during obesity. *ISME J.* **2014**, *8*, 2116–2130. [CrossRef]
24. Kaur, A.P.; Bhardwaj, S.; Dhanjal, D.S.; Nepovimova, E.; Cruz-Martins, N.; Kuča, K.; Chopra, C.; Singh, R.; Kumar, H.; Şen, F.; et al. Plant prebiotics and their role in the amelioration of diseases. *Biomolecules* **2021**, *11*, 440. [CrossRef] [PubMed]
25. Liu, L.; Li, Q.; Yang, Y.; Guo, A. Biological function of short-chain fatty acids and its regulation on intestinal health of poultry. *Front. Vet. Sci.* **2021**, *8*, 736739. [CrossRef] [PubMed]
26. Jacobson, A.; Lam, L.; Rajendram, M.; Tamburini, F.; Honeycutt, J.; Pham, T.; Van Treuren, W.; Pruss, K.; Stabler, S.R.; Lugo, K.; et al. A gut commensal-produced metabolite mediates colonization resistance to *Salmonella* infection. *Cell Host Microbe* **2018**, *24*, 296–307.e7. [CrossRef]
27. Lamas, A.; Regal, P.; Vázquez, B.; Cepeda, A.; Franco, C.M. Short chain fatty acids commonly produced by gut microbiota influence *Salmonella enterica* motility, biofilm formation, and gene expression. *Antibiotics* **2019**, *8*, 265. [CrossRef] [PubMed]
28. Pérez-Reytor, D.; Puebla, C.; Karahanian, E.; García, K. Use of short-chain fatty acids for the recovery of the intestinal epithelial barrier affected by bacterial toxins. *Front. Physiol.* **2021**, *12*, 650313. [CrossRef]
29. Kim, C.H. Immune regulation by microbiome metabolites. *Immunology* **2018**, *154*, 220–229. [CrossRef]
30. Segain, J.P.; Raingeard de la Blétière, D.; Bourreille, A.; Leray, V.; Gervois, N.; Rosales, C.; Ferrier, L.; Bonnet, C.; Blottière, H.M.; Galmiche, J.P. Butyrate inhibits inflammatory responses through NFkappaB inhibition: Implications for Crohn's disease. *Gut* **2000**, *47*, 397–403. [CrossRef] [PubMed]
31. Tolhurst, G.; Heffron, H.; Lam, Y.S.; Parker, H.E.; Habib, A.M.; Diakogiannaki, E.; Cameron, J.; Grosse, J.; Reimann, F.; Gribble, F.M. Short-chain fatty acids stimulate glucagon-like peptide-1 secretion via the G-protein-coupled receptor FFAR2. *Diabetes* **2012**, *61*, 364–371. [CrossRef] [PubMed]
32. Wang, X.; Qi, Y.; Zheng, H. Dietary polyphenol, gut microbiota, and health benefits. *Antioxidants* **2022**, *11*, 1212. [CrossRef] [PubMed]
33. Wang, M.; Zhao, H.; Wen, X.; Ho, C.T.; Li, S. *Citrus* flavonoids and the intestinal barrier: Interactions and effects. *Compr. Rev. Food Sci. Food Saf.* **2021**, *20*, 225–251. [CrossRef] [PubMed]
34. Bianchi, F.; Cappella, A.; Gagliano, N.; Sfondrini, L.; Stacchiotti, A. Polyphenols-gut-heart: An impactful relationship to improve cardiovascular diseases. *Antioxidants* **2022**, *11*, 1700. [CrossRef]
35. Jones, R.M.; Mercante, J.W.; Neish, A.S. Reactive oxygen production induced by the gut microbiota: Pharmacotherapeutic implications. *Curr. Med. Chem.* **2012**, *19*, 1519–1529. [CrossRef] [PubMed]
36. Micek, A.; Godos, J.; Del Rio, D.; Galvano, F.; Grosso, G. Dietary flavonoids and cardiovascular disease: A comprehensive dose-response meta-analysis. *Mol. Nutr. Food Res.* **2021**, *65*, e2001019. [CrossRef]
37. Zhou, P.; Zhao, X.N.; Ma, Y.Y.; Tang, T.J.; Wang, S.S.; Wang, L.; Huang, J.L. Virtual screening analysis of natural flavonoids as trimethylamine (TMA)-lyase inhibitors for coronary heart disease. *J. Food Biochem.* **2022**, *46*, e14376. [CrossRef]
38. Hernandez-Baixauli, J.; Puigbò, P.; Torrell, H.; Palacios-Jordan, H.; Ripoll, V.J.R.; Caimari, A.; Del Bas, J.M.; Baselga-Escudero, L.; Mulero, M. A pilot study for metabolic profiling of obesity-associated microbial gut dysbiosis in male wistar rats. *Biomolecules* **2021**, *11*, 303. [CrossRef] [PubMed]
39. Flaig, B.; Garza, R.; Singh, B.; Hamamah, S.; Covasa, M. Treatment of dyslipidemia through targeted therapy of gut microbiota. *Nutrients* **2023**, *15*, 228. [CrossRef]

40. Mollace, V.; Sacco, I.; Janda, E.; Malara, C.; Ventrice, D.; Colica, C.; Visalli, V.; Muscoli, S.; Ragusa, S.; Muscoli, C.; et al. Hypolipemic and hypoglycaemic activity of bergamot polyphenols: From animal models to human studies. *Fitoterapia* **2011**, *82*, 309–316. [CrossRef]
41. Musolino, V.; Gliozzi, M.; Scarano, F.; Bosco, F.; Scicchitano, M.; Nucera, S.; Carresi, C.; Ruga, S.; Zito, M.C.; Maiuolo, J.; et al. Bergamot polyphenols improve dyslipidemia and pathophysiological features in a mouse model of non-alcoholic fatty liver disease. *Sci. Rep.* **2020**, *10*, 2565. [CrossRef]
42. Gliozzi, M.; Carresi, C.; Musolino, V.; Palma, E.; Muscoli, C.; Vitale, C.; Gratteri, S.; Muscianisi, G.; Janda, E.; Muscoli, S.; et al. The effect of bergamot-derived polyphenolic fraction on LDL small dense particles and nonalcoholic fatty liver disease in patients with metabolic syndrome. *Adv. Biol. Chem.* **2014**, *4*, 129–137. [CrossRef]
43. Mare, R.; Mazza, E.; Ferro, Y.; Gliozzi, M.; Nucera, S.; Paone, S.; Aversa, I.; Pujia, R.; Marafioti, G.; Musolino, V.; et al. A new breakfast brioche containing bergamot fiber prevents insulin and glucose increase in healthy volunteers: A pilot study. *Minerva Endocrinol.* **2021**, *46*, 214–225. [CrossRef] [PubMed]
44. Musolino, V.; Gliozzi, M.; Bombardelli, E.; Nucera, S.; Carresi, C.; Maiuolo, J.; Mollace, R.; Paone, S.; Bosco, F.; Scarano, F.; et al. The synergistic effect of *Citrus bergamia* and *Cynara cardunculus* extracts on vascular inflammation and oxidative stress in non-alcoholic fatty liver disease. *J. Tradit. Complement. Med.* **2020**, *10*, 268–274. [CrossRef] [PubMed]
45. Ley, R.E.; Turnbaugh, P.J.; Klein, S.; Gordon, J.I. Microbial ecology: Human gut microbes associated with obesity. *Nature* **2006**, *444*, 1022–1023. [CrossRef] [PubMed]
46. Cani, P.D.; Bibiloni, R.; Knauf, C.; Waget, A.; Neyrinck, A.M.; Delzenne, N.M.; Burcelin, R. Changes in gut microbiota control metabolic endotoxemia-induced inflammation in high-fat diet-induced obesity and diabetes in mice. *Diabetes* **2008**, *57*, 1470–1481. [CrossRef] [PubMed]
47. Kunst, C.; Schmid, S.; Michalski, M.; Tümen, D.; Buttenschön, J.; Müller, M.; Gülow, K. The influence of gut microbiota on oxidative stress and the immune system. *Biomedicines* **2023**, *11*, 1388. [CrossRef]
48. Gliozzi, M.; Walker, R.; Muscoli, S.; Vitale, C.; Gratteri, S.; Carresi, C.; Musolino, V.; Russo, V.; Janda, E.; Ragusa, S.; et al. Bergamot polyphenolic fraction enhances rosuvastatin-induced effect on LDL-cholesterol, LOX-1 expression and protein kinase B phosphorylation in patients with hyperlipidemia. *Int. J. Cardiol.* **2013**, *170*, 140–145. [CrossRef]
49. Oppedisano, F.; Muscoli, C.; Musolino, V.; Carresi, C.; Macrì, R.; Giancotta, C.; Bosco, F.; Maiuolo, J.; Scarano, F.; Paone, S.; et al. The protective effect of *Cynara Cardunculus* extract in diet-induced NAFLD: Involvement of OCTN1 and OCTN2 transporter subfamily. *Nutrients* **2020**, *12*, 1435. [CrossRef] [PubMed]
50. Yang, R.L.; Shi, Y.H.; Hao, G.; Li, W.; Le, G.W. Increasing oxidative stress with progressive hyperlipidemia in human: Relation between malondialdehyde and atherogenic index. *J. Clin. Biochem. Nutr.* **2008**, *43*, 154–158. [CrossRef]
51. Martinez, I.; Wallace, G.; Zhang, C.; Legge, R.; Benson, A.K.; Carr, T.P.; Moriyama, E.N.; Walter, J. Diet-induced metabolic improvements in a hamster model of hypercholesterolemia are strongly linked to alterations of the gut microbiota. *Appl. Environ. Microbiol.* **2009**, *75*, 4175–4184. [CrossRef] [PubMed]
52. Leone, V.; Gibbons, S.M.; Martinez, K.; Hutchison, A.L.; Huang, E.Y.; Cham, C.M.; Pierre, J.F.; Heneghan, A.F.; Nadimpalli, A.; Hubert, N.; et al. Effects of diurnal variation of gut microbes and high-fat feeding on host circadian clock function and metabolism. *Cell Host Microbe* **2015**, *17*, 681–689. [CrossRef] [PubMed]
53. Tang, W.; Pan, L.; Cheng, J.; Wang, X.; Zheng, L.; Wang, S.; Zhou, Y.; Wang, H. High-fat-diet-induced gut microbiome changes in mice. *Stress Brain* **2022**, *2*, 17–30. [CrossRef]
54. Binienda, A.; Twardowska, A.; Makaro, A.; Salaga, M. Dietary carbohydrates and lipids in the pathogenesis of leaky gut syndrome: An overview. *Int. J. Mol. Sci.* **2020**, *21*, 8368. [CrossRef]
55. Rohr, M.W.; Narasimhulu, C.A.; Rudeski-Rohr, T.A.; Parthasarathy, S. Negative effects of a high-fat diet on intestinal permeability: A review. *Adv. Nutr.* **2020**, *11*, 77–91. [CrossRef] [PubMed]
56. Stephens, M.; von der Weid, P.Y. Lipopolysaccharides modulate intestinal epithelial permeability and inflammation in a species-specific manner. *Gut Microbes* **2020**, *11*, 421–432. [CrossRef]
57. Allam-Ndoul, B.; Castonguay-Paradis, S.; Veilleux, A. Gut microbiota and intestinal trans-epithelial permeability. *Int. J. Mol. Sci.* **2020**, *21*, 6402. [CrossRef]
58. Ghosh, S.S.; Wang, J.; Yannie, P.J.; Ghosh, S. Intestinal barrier dysfunction, LPS translocation, and disease development. *J. Endocr. Soc.* **2020**, *4*, bvz039. [CrossRef] [PubMed]
59. Hoyles, L.; Fernandez-Real, J.M.; Federici, M.; Serino, M.; Abbott, J.; Charpentier, J.; Heymes, C.; Luque, J.L.; Anthony, E.; Barton, R.H.; et al. Molecular phenomics and metagenomics of hepatic steatosis in non-diabetic obese women. *Nat. Med.* **2018**, *24*, 1070–1080. [CrossRef]
60. Petrucci, G.; Rizzi, A.; Hatem, D.; Tosti, G.; Rocca, B.; Pitocco, D. Role of oxidative stress in the pathogenesis of atherothrombotic diseases. *Antioxidants* **2022**, *11*, 1408. [CrossRef] [PubMed]
61. Higashi, Y. Roles of oxidative stress and inflammation in vascular endothelial dysfunction-related disease. *Antioxidants* **2022**, *11*, 1958. [CrossRef]
62. Carresi, C.; Gliozzi, M.; Musolino, V.; Scicchitano, M.; Scarano, F.; Bosco, F.; Nucera, S.; Maiuolo, J.; Macrì, R.; Ruga, S.; et al. The effect of natural antioxidants in the development of metabolic syndrome: Focus on bergamot polyphenolic fraction. *Nutrients* **2020**, *12*, 1504. [CrossRef]

63. Mollace, V.; Scicchitano, M.; Paone, S.; Casale, F.; Calandruccio, C.; Gliozzi, M.; Musolino, V.; Carresi, C.; Maiuolo, J.; Nucera, S.; et al. Hypoglycemic and hypolipemic effects of a new lecithin formulation of bergamot polyphenolic fraction: A double blind, randomized, placebo-controlled study. *Endocr. Metab. Immune Disord. Drug Targets* **2019**, *19*, 136–143. [CrossRef]
64. Parafati, M.; Lascala, A.; La Russa, D.; Mignogna, C.; Trimboli, F.; Morittu, V.M.; Riillo, C.; Macirella, R.; Mollace, V.; Brunelli, E.; et al. Bergamot polyphenols boost therapeutic effects of the diet on non-alcoholic steatohepatitis (NASH) induced by "junk food": Evidence for anti-inflammatory activity. *Nutrients* **2018**, *10*, 1604. [CrossRef]
65. de Paulo Farias, D.; de Araújo, F.F.; Neri-Numa, I.A.; Pastore, G.M. Prebiotics: Trends in food, health and technological applications. *Trends Food Sci.* **2019**, *93*, 23–35. [CrossRef]
66. Gerdes, V.; Gueimonde, M.; Pajunen, L.; Nieuwdorp, M.; Laitinen, K. How strong is the evidence that gut microbiota composition can be influenced by lifestyle interventions in a cardio-protective way? *Atherosclerosis* **2020**, *311*, 124–142. [CrossRef]
67. Nicolucci, A.C.; Hume, M.P.; Martínez, I.; Mayengbam, S.; Walter, J.; Reimer, R.A. Prebiotics reduce body fat and alter intestinal microbiota in children who are overweight or with obesity. *Gastroenterology* **2017**, *153*, 711–722. [CrossRef]
68. Oniszczuk, A.; Oniszczuk, T.; Gancarz, M.; Szymańska, J. Role of gut microbiota, probiotics and prebiotics in the cardiovascular diseases. *Molecules* **2021**, *26*, 1172. [CrossRef]
69. OECD. *Test No. 407: Repeated Dose 28-Day Oral Toxicity Study in Rodents*; OECD guidelines for the testing of chemicals, section 4; OECD Publishing: Paris, France, 2008. [CrossRef]
70. OECD. *Test No. 408: Repeated Dose 90-Day Oral Toxicity Study in Rodents*; OECD guidelines for the testing of chemicals, section 4; OECD Publishing: Paris, France, 2018. [CrossRef]
71. Council Directive 86/609/EEC of 24 November 1986 on the approximation of laws, regulations and administrative provisions of the Member States regarding the protection of animals used for experimental and other scientific purposes. *Off. J. Eur. Commun.* **1986**, *L358*, 1–28.
72. Musolino, V.; Gliozzi, M.; Nucera, S.; Carresi, C.; Maiuolo, J.; Mollace, R.; Paone, S.; Bosco, F.; Scarano, F.; Scicchitano, M.; et al. The effect of bergamot polyphenolic fraction on lipid transfer protein system and vascular oxidative stress in a rat model of hyperlipemia. *Lipids Health Dis.* **2019**, *18*, 115. [CrossRef]
73. Perna, S.; Spadaccini, D.; Botteri, L.; Girometta, C.; Riva, A.; Allegrini, P.; Petrangolini, G.; Infantino, V.; Rondanelli, M. Efficacy of bergamot: From anti-inflammatory and anti-oxidative mechanisms to clinical applications as preventive agent for cardiovascular morbidity, skin diseases, and mood alterations. *Food Sci. Nutr.* **2019**, *7*, 369–384. [CrossRef]
74. Mollace, R.; Macrì, R.; Tavernese, A.; Gliozzi, M.; Musolino, V.; Carresi, C.; Maiuolo, J.; Fini, M.; Volterrani, M.; Mollace, V. Comparative effect of bergamot polyphenolic fraction and red yeast rice extract in rats fed a hyperlipidemic diet: Role of antioxidant properties and PCSK9 expression. *Nutrients* **2022**, *14*, 477. [CrossRef]
75. Caporaso, J.G.; Kuczynski, J.; Stombaugh, J.; Bittinger, K.; Bushman, F.D.; Costello, E.K.; Fierer, N.; Peña, A.G.; Goodrich, J.K.; Gordon, J.I.; et al. QIIME allows analysis of high-throughput community sequencing data. *Nat. Methods* **2010**, *7*, 335–336. [CrossRef]
76. Caporaso, J.G.; Bittinger, K.; Bushman, F.D.; DeSantis, T.Z.; Andersen, G.L.; Knight, R. PyNAST: A flexible tool for aligning sequences to a template alignment. *Bioinformatics* **2010**, *26*, 266–267. [CrossRef]
77. Price, M.N.; Dehal, P.S.; Arkin, A.P. FastTree: Computing large minimum evolution trees with profiles instead of a distance matrix. *Mol. Biol. Evol.* **2009**, *26*, 1641–1650. [CrossRef]
78. Bolyen, E.; Rideout, J.R.; Dillon, M.R.; Bokulich, N.A.; Abnet, C.C.; Al-Ghalith, G.A.; Alexander, H.; Alm, E.J.; Arumugam, M.; Asnicar, F.; et al. Reproducible, interactive, scalable and extensible microbiome data science using QIIME 2. *Nat. Biotechnol.* **2019**, *37*, 852–857. [CrossRef]

International Journal of *Molecular Sciences*

MDPI

Review

Polyphenols as Drivers of a Homeostatic Gut Microecology and Immuno-Metabolic Traits of *Akkermansia muciniphila*: From Mouse to Man

María Carolina Rodríguez-Daza [1] and Willem M. de Vos [1,2,*]

1 Laboratory of Microbiology, Wageningen University & Research, Stippeneng 4, 6708 WE Wageningen, The Netherlands
2 Human Microbiome Research Program, Faculty of Medicine, University of Helsinki, 00014 Helsinki, Finland
* Correspondence: willem.devos@wur.nl

Abstract: *Akkermansia muciniphila* is a mucosal symbiont considered a gut microbial marker in healthy individuals, as its relative abundance is significantly reduced in subjects with gut inflammation and metabolic disturbances. Dietary polyphenols can distinctly stimulate the relative abundance of *A. muciniphila*, contributing to the attenuation of several diseases, including obesity, type 2 diabetes, inflammatory bowel diseases, and liver damage. However, mechanistic insight into how polyphenols stimulate *A. muciniphila* or its activity is limited. This review focuses on dietary interventions in rodents and humans and in vitro studies using different phenolic classes. We provide critical insights with respect to potential mechanisms explaining the effects of polyphenols affecting *A. muciniphila*. Anthocyanins, flavan-3-ols, flavonols, flavanones, stilbenes, and phenolic acids are shown to increase relative *A. muciniphila* levels in vivo, whereas lignans exert the opposite effect. Clinical trials show consistent findings, and high intervariability relying on the gut microbiota composition at the baseline and the presence of multiple polyphenol degraders appear to be cardinal determinants in inducing *A. muciniphila* and associated benefits by polyphenol intake. Polyphenols signal to the AhR receptor and impact the relative abundance of *A. muciniphila* in a direct and indirect fashion, resulting in the restoration of intestinal epithelial integrity and homeostatic crosstalk with the gut microbiota by affecting IL-22 production. Moreover, recent evidence suggests that *A. muciniphila* participates in the initial hydrolysis of some polyphenols but does not participate in their complete metabolism. In conclusion, the consumption of polyphenol-rich foods targeting *A. muciniphila* as a pivotal intermediary represents a promising precision nutritional therapy to prevent and attenuate metabolic and inflammatory diseases.

Keywords: *Akkermansia muciniphila*; gut microbiota; polyphenols; ecological niche; AhR; IL-22; polyphenol degraders; phenolic metabolites

check for updates

Citation: Rodríguez-Daza, M.C.; de Vos, W.M. Polyphenols as Drivers of a Homeostatic Gut Microecology and Immuno-Metabolic Traits of *Akkermansia muciniphila*: From Mouse to Man. *Int. J. Mol. Sci.* **2023**, *24*, 45. https://doi.org/10.3390/ijms24010045

Academic Editors: Baojun Xu and Matteo Bordiga

Received: 1 November 2022
Revised: 12 November 2022
Accepted: 14 December 2022
Published: 20 December 2022

1. Introduction

The human gut houses trillions of microbes, including viruses, fungi, archaea, and bacteria, the latter being the major representative [1,2]. The gut microbiota plays an essential role in human health. Deviations in the composition and functions of the gut microbiota have been associated with many diseases, including gastrointestinal, cardiovascular [3–6], hepatic [7], immunological, and neurological disorders [8,9].

A. *muciniphila*, a member of the phylum Verrucomicrobia, is an anaerobic and mucosa-associated colonic bacterium with mucin-degrading capabilities that has been identified as a microbial marker of a homeostatic intestine [10]. It has been linked to potential benefits by counteracting gut barrier disruption, obesity, and related metabolic disturbances [10–13]. Recently in humans [11] and rodents [14–16], treatment with *A. muciniphila* was shown to enhance obesity and related disorders, including glucose tolerance, insulin resistance, fat mass gain, total plasma cholesterol, liver dysfunction, and low-grade inflammation

(Figure 1). A specific protein present on the outer membrane of this bacterium, termed Amuc_1100, has been demonstrated to be implicated in the interaction with the host mucosa via toll-like receptor-2 (TLR2) signaling, restoring the intestinal barrier function and decreasing metabolic endotoxemia in mice [10,14,17].

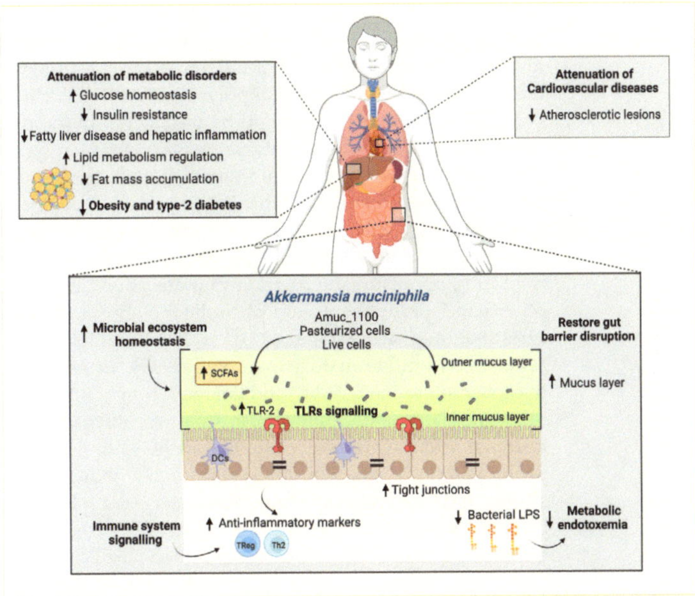

Figure 1. Beneficial impact of *Akkermansia muciniphila* on host health. *A. muciniphila* protects against cardiovascular diseases, metabolic disorders, and intestinal microecological disruption. An increased relative proportion of *A. muciniphila* in the gut microbiota has been inversely associated with obesity and type 2 diabetes. The administration of active *A. muciniphila*; its membrane protein, Amuc_1100; and pasteurized cells can reverse metabolic disorders associated with a high-fat diet, obesity, and type 2 diabetes. Clinical and preclinical studies highlight the role of *A. muciniphila* in attenuating fat mass accumulation, adipose tissue inflammation, insulin resistance, and glucose homeostasis. *A. muciniphila* alleviates chronic intestinal inflammation and metabolic endotoxemia (reduces the level of lipopolysaccharide (LPS)) and induces immune cell and Toll-like receptor (TLR)-2 signaling. Owing to its glycan-degrading capabilities and proximity to the host colonic epithelium, this bacterium contributes to the maintenance of the mucus layer barrier function and intestinal homeostasis. The graph was created with BioRender.

Obesity, type 2 diabetes (T2D), and chronic consumption of a high-fat diet (HFD) have detrimental effects on the gut microbiota composition, especially on relative levels of *A. muciniphila,* altering the integrity of the intestinal epithelium and symbiotic host–microbe crosstalk [11,18–21]. For instance, a long-term HFD leads to mucin (Muc2) protein misfolding and endoplasmic reticulum (ER) stress and reduces Muc2 production and O-glycosylation by goblet cells (GCs) [22–24]. Hence, the accumulation of non-glycosylated Muc2 precursors weakens the protective mucus layer and exacerbates the adhesion and mucus degradation by opportunistic pathogens, in addition to reducing mucosa-associated symbionts, such as *A. muciniphila* [24,25]. These gut ecological disturbances were reported to notably reduce the *A. muciniphila* population but were shown to be prevented and restored by the consumption of polyphenols [26,27].

Polyphenols are chemically diverse molecules synthesized by plants as secondary metabolites. They exert pharmacological properties, such as antioxidant, anticarcinogenic, antimicrobial, and anti-inflammatory activities [28–32]. Most of the bioactivity of polyphenols is thought to be mediated by their interaction with the host gut microbiota. Indeed, several reports have demonstrated the poor bioavailability of polyphenols by using cellular cultures mimicking the epithelium of the small and large intestines or by measuring the metabolites from gut microbial degradation of these compounds in vivo [33–36]. Accumulating studies suggest that dietary polyphenols may stimulate the growth of *A. muciniphila* in vivo and show this bacterium to be a critical player boosting the bioactivities of polyphenols in different health conditions, including health-steady status, cardiovascular diseases, liver damage, and gut inflammation [37–44]. However, the basic mechanisms of the potential prebiotic effects of polyphenols on *A. muciniphila* are unclear.

Distinct mechanisms might be implicated in the promoting effects of polyphenols on *A. muciniphila*. Polyphenols can restore gut microbiota disbalance and mucus barrier depletion in obesity and inflammatory bowel diseases (IBD) [44,45]. In the colon, these molecules interact with more than 1000 species of bacteria that might have the catalytic and hydrolytic ability to metabolize polyphenols to the corresponding aglycone and derived metabolites [36,46]. On the one hand, phenolics and derivatives can act as antimicrobials, inhibiting opportunistic bacteria that are abundant in metabolic diseases [28]. On the other hand, polyphenols and phenolic metabolites can simultaneously act locally in the intestinal mucosa, be absorbed through the epithelium into the systemic circulation, and regulate anti-inflammatory markers [27–30]. In line with this, the activation of the aryl hydrocarbon receptor (*AhR*) and the induction of interleukin (IL)-22 appear to be involved in the mechanisms of action, leading to a protective O-glycan-rich mucus layer and reduced mucosal inflammation by polyphenols [47–49]. However, in this context, it should be mentioned that *A. muciniphila* can also directly induce IL-22, as shown recently [50]. Recent studies indicate that polyphenols induce goblet cell (GC) differentiation, increase mucus layer thickness, and reduce colitis. However, it remains unclear whether this is a direct effect or whether *A. muciniphila* is involved [26,27,51].

The strong association between the consumption of polyphenols and the relative abundance of *A. muciniphila* in humans and animals also suggests the trophic utilization of phenolics by this bacterium. Indeed, individuals with a polyphenol-degrading metabotype that yields the production of phenolic metabolites, such as urolithin A, present with an increased relative abundance of *A. muciniphila* and a better metabolic outcome and health status than those lacking a specific polyphenol-degrading species [43,52,53]. Several enzymes have been identified to participate in the polyphenol metabolism by gut bacteria [44]. However, none has been described and tested to date in *A. muciniphila*. A few in vitro studies on the growth of *A. muciniphila* in polyphenol-enriched media suggest the nutritional utilization of phenolics [43,54]. However, diverging findings on the produced metabolites and the reduced growth of *A. muciniphila* in the presence of polyphenols as a unique nutritional source or combined with mucin and glucose prevent the confirmation of such a hypothesis.

The aim of the present review is to provide an overview of polyphenols that shape the composition of the gut microbiota, particularly promoting the relative abundance of *A. muciniphila*, with a protective role against metabolic and inflammatory disorders. We identified compelling trials in rodents and humans in *A. muciniphila* was shown to be selectively induced by different classes of dietary polyphenols. We focused on the impact of polyphenols and derived metabolites on the mucosal immunomodulation and re-establishment of the gut microenvironment, improving the ecological niche of *A. muciniphila* as a potential non-prebiotic mechanism. Additionally, hints with respect to the metabolism of polyphenols by *A. muciniphila* and its possible molecular adaptation to these compounds are discussed. Preclinical and clinical dietary interventions demonstrate that *A. muciniphila* potentiates many polyphenol-induced processes by regulating host metabolism and inflammation, contributing to the maintenance of gut homeostasis.

2. *Akkermansia muciniphila*, the Gut Microenvironment, and Immune Regulation

The intestinal epithelial barrier is enhanced by the presence of a mucus layer and immune factors produced by the host [55]. This mucus layer secreted by GCs consists mainly of a mesh-like network of gel-forming Muc2 mucin, with tandem repeated sequences rich in threonine and serine residues acting as the attachment sites of O-glycans [56]. These O-glycans primarily consist of N-acetylgalactosamine, N-acetylglucosamine, fucose, galactose, sialic acid (N-acetylneuraminic acid), mannose, and sulfate. The post-translational O-glycosylation provides resistance to proteolysis by traditional proteases, furnishes selective commensal microbes, and prevents pathogen colonization [25]. *A. muciniphila* is a prominent O-glycan utilizer species capable of hydrolyzing glycosidic linkages and degrading the colonic mucin. Its genome is equipped with specialized glycosyl hydrolases, including α-fucosidases, α-sialidases, and endo-α-N-acetylgalactosaminidases [57,58]. All these features give *A. muciniphila* the selective trophic advantage to prevail, signal immunological response in the intestinal mucosa, and contribute to regular mucus turnover. Its relative abundance positively correlates with mucus layer thickness, intestinal integrity, reduced endotoxemia and intestinal inflammation, and improved metabolic phenotype in humans and animals (Figure 1) [10,11,59]. Mechanistic studies conducted in vitro and in rodent models demonstrate that *A. muciniphila* increases transepithelial electrical resistance; stimulates GCs and mucin secretion; alleviates chronic intestinal inflammation; regulates glucose and lipid metabolism; and induces immune responses, including T-cell receptors, B-cell receptors, NF-κB activation, and TLR2 signaling [14,17,59,60].

The structure and glycosylation pattern of mucin are critical modulators of the gut microbiome, particularly *A. muciniphila*. O-glycosylation defects drive a disrupted niche in the intestinal mucosa and relate to the onset of gut inflammation and mucin binding and degradation by microbial pathogens. For instance, minimal sulfate modification and a reduced abundance of *A. muciniphila* have been reported in subjects with ulcerative colitis [61–63]. A reduced proportion of fucosylated glycans and an increased proportion of hyposialylated T-antigen glycans have been associated with exacerbated inflammation [61,64]. Interestingly, such abnormalities are attenuated by the exogenous administration of mucin O-glycans, modulating gut microbiota alterations, notably by inhibiting pathogens and favoring *A. muciniphila* [65].

The regulation of gut immunity and host glycans is fundamental to maintain a protective gut mucosal barrier. Signals regulating the innate immune response, such as induction of the aryl hydrocarbon receptor (*AhR*), have been shown to bridge homeostasis between the gut microbiota and the intestinal epithelium. The ligand-activated *AhR* leads to the transcription of several genes, such as *Rort* and *Foxp3,* and modulates interleukin (IL)-22 secretion, with a pivotal role in mucosal defense [66]. For instance, *IL-22* can induce fucosylation of intestinal epithelial cells and prevent the overcolonization of opportunistic pathogens [67]. The role of these markers in signaling colonic innate response makes them essential regulators of the gut mucus barrier and the colonization of *A. muciniphila*. The *AhR* senses dietary compounds and microbial metabolites, including those of tryptophan metabolism. The modulation of the gut microbiota, especially the stimulation of tryptophan-metabolizing species, has been associated with the enrichment of *AhR* ligands and improved mucosal microenvironment [68,69]. Recently, *A. muciniphila* was also shown to metabolize tryptophan to indole derivatives and *AhR* agonists [70], in addition to being found to induce IL22 in a MyD88-dependent fashion in mice [50]. Similarly, the live and pasteurized cells of *A. muciniphila* and its membrane protein, Amuc_1100, were shown to induce *AhR*-targeted genes, including *CYP1A1*, *IL-10*, and *IL-22*, protecting against ulcerative colitis in mice [71]. Additionally, integrative analysis of fecal metagenomes and untargeted serum metabolomes from patients and mice with ulcerative colitis highlighted a significant correlation between *A. muciniphila* and the restoration of tryptophan metabolism [71]. A synergic effect might exist between *AhR* modulators and the abundance of *A. muciniphila* in triggering gut immune regulation by shapes on *AhR* signaling, thereby attenuating colonic inflammation and metabolic alterations.

The increased abundance of *A. muciniphila* might be sustained by the polyphenol-driven restoration of mucosal integrity. However, it is not possible to exclude the crosstalk between polyphenols and the gut microbiota from the mechanisms underpinning the stimulation of *A. muciniphila*. The sections below tackle the role of different polyphenol classes as modulators of gut microbiota composition, especially *A. muciniphila*. Furthermore, mechanistic insights into the effects of polyphenols in regulating immune markers linked to the attenuation of gut inflammation and improved intestinal barrier function are discussed.

3. Interactions between Polyphenols and the Gut Microbiota and Their Impact on the Attenuation of Chronic Diseases

Polyphenols are categorized into two main structural classes—flavonoids and non-flavonoids—which vary in the number of phenolic rings and their type of linkages, as well as the main dietary sources (Figure 2). The flavonoid group is subcategorized into anthocyanins, flavonols, flavanones, isoflavones, and flavan-3-ols. The non-flavonoid group includes phenolic acids, stilbenes, and lignans [44]. Vegetables and fruits, such as pomegranates, berries, grapes, tea, cocoa, and coffee, are rich sources of polyphenols.

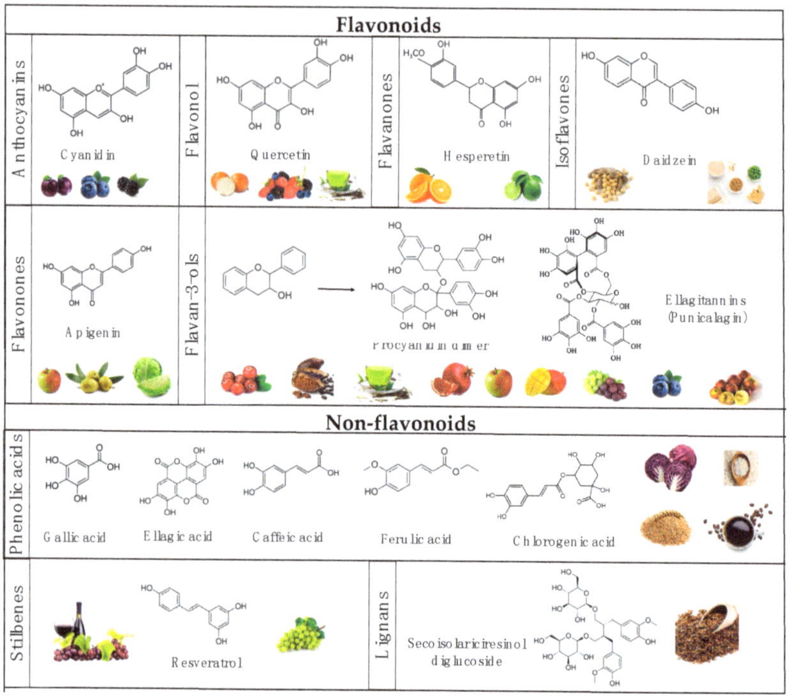

Figure 2. Chemical structures of polyphenol classes of flavonoids and non-flavonoids and their main food sources.

Owing to the poor biological availability of dietary polyphenols [72], growing evidence highlights the gut microbiota as a critical intermediary involved in priming host cell signaling in the intestinal mucosa and distal organs; these interactions are associated with diverse biological roles in preventing and attenuating the chronicity of several diseases [31,73–79]. Approximately 90–95% of polyphenol intake accumulates in the intestinal lumen up to the millimolar range, where, together with bile conjugates, it is subjected to the enzymatic activities by the gut microbiota, sequentially producing metabolites with different physiological significance [35]. Ottaviani et al. (2016) [36] evaluated the pharmacokinetic profile of polyphenol metabolism in humans, outlining the distinguished absorption of different phenolic compounds in the proximal gastrointestinal tract and the colon. The authors assessed the absorption, distribution, metabolism, and excretion of (−)-epicatechin (EC), flavanol monomers, and procyanidins, with a degree of polymerization (DP) of 2–10. Based on the timing of the appearance of the various radiolabeled phenolic compounds in plasma and urine, it was demonstrated that most EC and procyanidin (EC polymer) absorption occurred in the colon. For instance, EC-derived metabolites were absent from plasma after 6–8 h, indicating that the ingested EC and polymers continued down the gastrointestinal tract and reached the colon. In the colon, gut microbiota–polyphenol interactions, which start with the C-ring opening, yielded the formation of O-glucuronidated and sulfated derivatives of phenyl- γ-valerolactones (γVL). Here, more than 70% of the ingested, radiolabeled (−)epicatechin flavanols were shown to be absorbed into the circulatory system via the colon, compared to 20% absorbed from the small intestine. The phenolic metabolites were further submitted to phase II metabolism by enzymes in the colon mucosa and the liver, resulting in increased levels of a mixture of γVLs and phenolic acids in the blood circulation [36]. This provides insights into the role of the gut microbiome in generating phenolic bioactive metabolites in the human body. Many of these metabolic outputs are involved in various physiological processes, as reviewed by Man and colleagues (2020) [80].

For instance, the biological roles of parent phenolic compounds and the de novo generated bioactive molecules by the gut microbiota have been demonstrated in the regulation of inflammation [74], oxidative stress, [75], thermogenesis [31], apoptosis; proliferation of cancer cells [76]; blood pressure regulation; mitochondrial functions; and glucose and lipid metabolism [77–79].

Polyphenols exert beneficial effects on the gut microbiota by acting as prebiotics. In 2017, the ISAPP (International Scientific Association for Probiotics and Prebiotics) updated the prebiotic concept by considering phenolics/phytochemicals between the potential prebiotic candidates acting similarly as prebiotics. To fit in this category, phenolics must be selectively utilized by host microbes, conferring a health benefit [81]. The stimulation of beneficial microbes is related to their enzymatic capacity and adaptation to degrade and tolerate phenolics with different structural features. Increased attention is given to the stimulation of beneficial species by polyphenol intake in humans and animals, especially *A. muciniphila,* as a promising dietary strategy to counteract obesity and diabetes (table). *A. muciniphila* is relatively abundant in healthy individuals, accounting for approximately 4% of the gut microbiota. However, its abundance declines significantly in certain metabolic diseases [11,14,60].

A. muciniphila appears to be essential in boosting polyphenol-mediated effects to cope with obesity. Specifically, this species has been correlated with higher levels of a phenolic metabolite, urolithin A, and polyphenol-degrader species, such as *Gordonibacter* spp. [53]. Although a gap in the enzymatic ability of *A. muciniphila* to degrade polyphenols is yet to be filled, this species is regularly implicated in immune–metabolic processes promoted by dietary polyphenols in preclinical, in vitro, and clinical trials. Here, we compiled major findings on the effects of polyphenol-rich foods on the modulation of *A. muciniphila* abundance in the context of several chronic diseases.

4. Modulation of *A. muciniphila* by Polyphenol-Rich Foods In Vitro and in Animal Models

4.1. Anthocyanin-Rich Foods

Anthocyanins are glycosides of anthocyanidins, one of the major flavonoids widely found in berries. Plum, beans, and pomegranate are also good sources of anthocyanins. The identified anthocyanins are cyanidin, malvidin, petunidin, pelargonidin, peonidin, and delphinidin [82]. These compounds are mainly described as poorly available in the small intestine and have been extensively studied due to their potent antioxidant properties and benefits on human health [83] (Figure 3). The conversion of anthocyanins by gut bacteria is related to their β-glucosidase enzymatic capability [84]. Bacterial species such as *Lantiplantibacillus plantarum, Lactobacillus acidophilus, Streptococcus thermophiles, Lactobacillus delbrueckii* subsp. *bulgaricus, Bacteroides thetaiotaomicron, Blautia producta, Erysipelatoclostridium ramosum, Bifidobacterium animalis,* and *Bifidobacterium* spp. are involved in degrading anthocyanins [84,85].

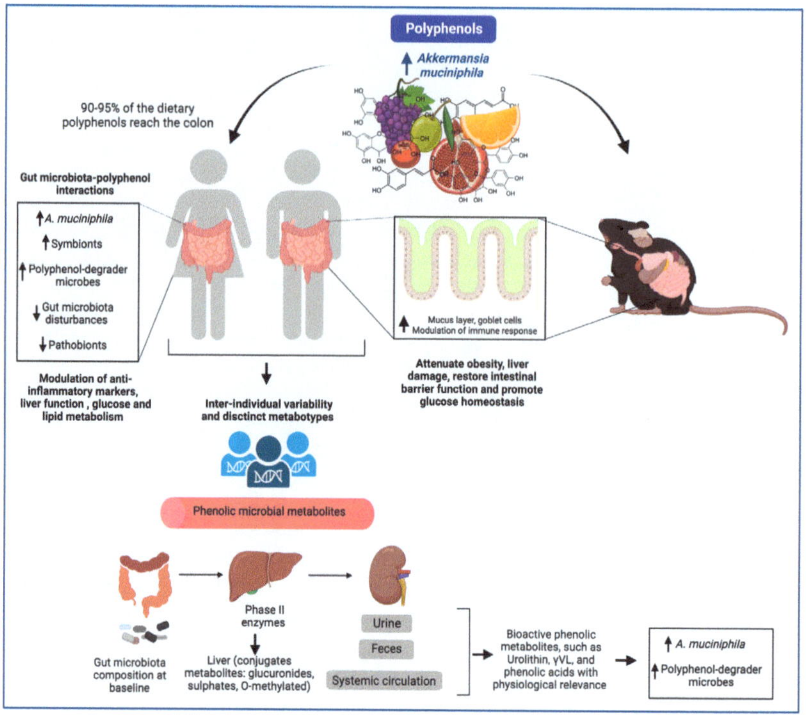

Figure 3. Schematic illustration of the modulatory effects of polyphenols on the gut microbiota, especially on *A. muciniphila* in humans and animals. A vast proportion of dietary polyphenols reach the colon, interacting with the gut microbiota. Polyphenols reduce gut microbiota disturbances and promote the relative abundance of *A. muciniphila* in humans and rodents. These changes are associated with improved metabolic phenotypes in the host. Interindividual variability in humans influences the effects of polyphenols on the gut microbiota and the production of bioactive metabolites, such as phenyl-γ-valerolactones (γVL), urolithins, and phenolic acids. Mechanistic insights into animal models sustain the role of polyphenols in signaling anti-inflammatory immune markers, increasing mucin-secreting goblet cells, and restoring intestinal barrier function. The graph was created with BioRender.

The intake of anthocyanin-rich fruits has been shown to stimulate *A. muciniphila* and attenuate host gut inflammation, liver injury, and obesity-associated metabolic alterations (Table 1). Recently, Song and colleagues (2020) investigated whether anthocyanins play a crucial role in the antidiabetic effect of açai fruits (*Euterpe oleracea* Mart.) in high-fat diet (HFD)-induced obesity in mice [86]. The anthocyanin-rich açai extract significantly shifted the overall structure and composition of the gut microbiota, in addition to ameliorating HFD-induced glucose intolerance and insulin resistance. The relative abundance of *A. muciniphila* was mainly induced by the açai extract and highlighted as a key player driving the anthocyanin-induced antiobesity effects. Increased *A.muciniphila* abundance was significantly associated with lower serum triglyceride (TG) levels, glucose, and insulin regulation. Likewise, it was negatively related to the expression of genes involved in lipid metabolisms, such as *Scd1*, *Srebf1*, and *Ppara*.

Table 1. Modulation of the gut microbiota by anthocyanin-rich foods in animal models.

	Polyphenol-Rich Food	Experimental Design	Main Findings in the Gut Microbiota	*A. muciniphila* Modulation	Impact on Host Health	Ref.
Anthocyanins	Açai extract	SPF C57BL/6J mice fed a HFD for 14 weeks with daily gavage of 150 mg/kg of anthocyanin-rich açai extract	↑ *Parabacteroides distasonis* and *B. acidifaciens.*	↑ *A. muciniphila* was significantly and negatively associated with serum TG, glucose, and insulin	Enhanced liver damage, glucose intolerance, and insulin resistance	[86]
	Davidson's plum	HCHF-induced obesity in Wistar rats supplemented with 8 mg of anthocyanins equivalent/kg/day for 8 weeks	↓ Clostridiaceae, ↑ *Turicibacter* spp.	↑ *A. muciniphila*	Reduction in visceral fat accumulation, inflammation, and plasma TG	[87]
	Bilberry	Western diet (WD)-induced NAFLD in male C57BL/6N mice for 18 weeks supplemented with 2% bilberry anthocyanins	↓ F/B, *Prevotella* spp., Lactobacillales, and Clostridiales; ↑*B. acidifaciens*; ↑*Parabacteroides* spp.	↑ *A. muciniphila* was negatively correlated with liver injury	Attenuated liver injury and dyslipidemia	[37]
	Black raspberries (BRBs)	5% whole BRB powder, 0.2% BRB anthocyanins or 2.25% of the residue fraction supplemented in F-344 rats under a standard diet for 6 weeks	5% whole BRB powder; ↑ *Anaerostipes*; ↑ *Ruminococcus*; ↑ *Coprobacillus*; ↓ *Acetivibrio*; ↓ *Anaerotruncus* spp.	↑ *A. muciniphila* by the whole BRB and the residue	Enhanced inflammatory biomarkers	[88]

The results relate to the relative abundance of taxa in fecal or cecal samples. BW, body weight; HFD, high-fat diet; HCHF, high-carbohydrate, high-fat diet; F/B, Firmicutes-to-Bacteroidota ratio; TG, triglycerides; non-alcoholic fatty liver disease (NAFLD).

Nakano and colleagues (2020) demonstrated the role of bilberry anthocyanins (BAs) in preventing the development of non-alcoholic fatty liver disease (NAFLD) induced by a 3-month intake of a Western diet (WD) in mice [37]. The study targeted hepatic markers associated with lipid accumulation, inflammation, and oxidative status, as well as their link with changes in the gut microbiota. Bilberry anthocyanins lowered the serum levels of liver damage indices, including aspartate aminotransferase (AST) and alanine aminotransferase (ALT) in both the normal diet (ND) and WD-fed mice. Interestingly, observations of host phenotypes correlated with an increased relative proportion of *A. muciniphila*. The WD-induced disrupted gut microbiome was attenuated when supplemented with anthocyanin-rich açai, increasing the relative abundance of *A. muciniphila*, *Bacteroides acidifaciens*, and *Parabacteroides*. Furthermore, John et al. (2019) [87] demonstrated the potential of Davidson's plum, which is rich in anthocyanins, including cyanidin sambubioside, cyanidin glucoside, peonidin sambubioside, and peonidin glucoside, to ameliorate the symptoms of metabolic syndrome in a high-carbohydrate, high-fat diet (HFHD)-fed rat model. Supplementation of Davidson's plum (8 mg of anthocyanin equivalent/kg/day) for eight weeks reduced plasma TG, visceral and abdominal fat mass, and inflammatory markers. Davidson's plum supplementation increased the abundance of *A. muciniphila* in rats fed the obesogenic and corn starch diets but was significantly higher in the latter group. Whereas in this study, we did not analyze the mucus layer, increased mucus secretion can be induced by the additive action of fibers and anthocyanins, providing a better ecological niche for *A. muciniphila*. Many anthocyanin compounds have been demonstrated to improve mucosa condition and injury and to hamper inflammatory opportunistic microbes [89,90].

In another study, Pan et al. reported an increase in relative levels of *Akkermansia* spp., along with *Anaerostipes* and *Ruminococcus* spp. and a decrease in *Acetivibrio* spp. abundance in F-344 rats in response to the dietary administration of anthocyanin-rich black raspberry (BRB) powder for six weeks [88]. This work sustained the modulation of inflammatory biomarkers by the whole BRBs, the BRB-derived anthocyanins, or the BRB-derived residue fraction. The metabolism of phenolics by the gut microbiota was not analyzed in that study; however, the authors indicated the potential of berry anthocyanins and fiber-bound phenolics to reduce gut inflammation, which is a possible indirect mechanism concomitantly sustaining a potential beneficial gut microbiota composition. It is worth noting that species of *Anaerostipes* spp., which were also increased by BRB, have the potential to degrade phenolic acids and are known producers of beneficial short-chain fatty acids such as butyrate and propionate. Notably, this genus harbors a functional gallate decarboxylase capability required to degrade gallic acid, as detected in silico and evaluated in vitro [91]. Interestingly, syntrophic interactions have been demonstrated between butyrate-producing species, such as *Anaerostipes caccae* and *A. muciniphila*, in a mucin-enriched medium [92]. In this sense, the butyrate producer benefits from the mucolytic activity of *A. muciniphila*, and in turn, the produced butyrate fuels the colonocytes and the goblet cells to support mucus production. Comparable complex cross-feeding chains might occur between polyphenol-degrader genera and *A. muciniphila*, although this hypothesis warrants further investigation. Subsequent in vitro mechanistic studies on the degradation of anthocyanins by *A. muciniphila* and identified polyphenol degraders are needed to translate the actual impact of the consumption of anthocyanin-rich foods on the stimulation of *A. muciniphila* and the enhancement of human metabolic diseases.

4.2. Flavan-3-Nol-Rich Foods

Flavanols are broadly characterized by the absence of a double bond between C-2 and C-3 and the absence of a carbonyl group on the C ring (C-4); instead, they have a hydroxyl group(s) on C-3 or C-4, where flavan-3-ols are the most reported [93] (Figure 2). Flavan-3-ol can contain a galloyllated and gallated group, including flavan-3-ol monomers, (+)-catechin, and (−)-epicatechin (EC), as well as oligomers, such as procyanidins, (−)-epigallocatechin (EGC)], (−)-epigallocatechin-3-gallate (EGCG), and (−)-epicatechin-3-

gallate]. These compounds are widely distributed in foods such as cacao, teas, beans, red wine, apples, pomegranates, and berries [94].

The modulation of the gut microbiome by flavan-3-ols and the derived metabolites beneficially influences the metabolic host's health. Dietary flavan-3-ols are widely associated with a cardioprotective and anti-inflammatory roles in coping with chronic diseases, including T2D, obesity, NAFLD, inflammatory bowel diseases (IBDs), and colon cancer [38]. Non-gallated flavan-3-ols, such as proanthocyanidins (PACs), also known as condensed tannins, are catabolized by the gut microbiota to phenylvaleric acid and valerolactones as primary metabolites [32,95]. Galloyllated flavan-3-ols, such as gallotannins and ellagitannins (ET), well-known as hydrolyzable tannins, are metabolized into ellagic acid and urolithin metabolites. Among the gut bacterial species known to metabolize flavan-3-ols are Lactobacilli, *Gordonibacter* spp., *Eggerthella lenta* (Eggerthellaceaea), and *Adlercreutzia equolifaciens* (Coriobacteriaceae) [44]. The derived flavan-3-ols metabolites are endowed with numerous biological properties, including antioxidant, antiatherosclerotic, antimicrobial, and anti-inflammatory activities [29–32].

The antiobesity effect of peach, a fruit rich in flavan-3-ols, was evaluated in female ICR mice fed an HFD [96]. The oral administration of peach peel extract (PPE), which is mainly composed of epicatechin 3-O-glucoside (28%) and proanthocyanins (12.7%), was shown to enhance intestinal and metabolic alterations. Unsurprisingly, the obesogenic diet resulted in a significant reduction in the relative proportion of *Akkermansia* spp., together with species of *Bacteroides, Bifidobacterium*, Lachnospiraceae_NK4A136_group, Prevotellaceae_UCG-001, *Alloprevotella*, and *Roseburia*. Moreover, it increased *Blautia, Bilophila, Odoribacter*, and *Ruminiclostridium* spp., among others. An intake of 300 mg/kg body weight (BW) of PPE and a lower concentration of 150 mg/kg BW significantly reversed the HFD-induced gut microbial changes after 12 weeks of intervention. Interestingly, the relative levels of *A. muciniphila* were recovered and even rose to levels higher than in the non-obese counterparts, doubling in abundance with the highest concentration of PPE of 300 mg/kg BW. These changes prevented obesity by reducing BW gain compared to the untreated HFD-fed mice, which is consistent with the reduced adipose tissue accumulation, decreased oxidative damage, and improved serum and liver lipid levels.

Accumulating studies point to proanthocyanins and ellagitannins as promising phenolic compounds to efficiently target the relative abundance of *A. muciniphila* [44] (Tables 2 and 3). This might be related to the structural features of tannins, such as the degree of polymerization, the high proportion of hydroxyl and galloyllated groups, and their high affinity with membrane proteins [97–99]. Owing to these characteristics, tannins exhibit significant antimicrobial action against opportunistic pathogens found flourishing in metabolic and inflammatory diseases [98]. In previous work, we demonstrated that berry polyphenolic fractions mainly composed of anthocyanidins, as well as phenolic acids, oligomeric PACs, and polymers of PACs, distinctly modulated the gut microbiota of mice fed an obesogenic diet for eight weeks [100]. In this study, blueberry polyphenolic fractions were administered in equivalent proportions to those found in the whole extract. Remarkably, *A. muciniphila* was mainly stimulated by blueberry PAC-rich fractions but not by anthocyanins and phenolic acids in HFD-fed mice.

Table 2. Impact of flavan-3-ol consumption on the relative colonic abundance of *A. muciniphila* and other bacteria.

	Polyphenol-Rich Foods	Experimental Design	Main Findings in the Gut Microbiota	*A. muciniphila* Modulation	Impact on Host Health	Ref.
Flavan-3-ols	Wild blueberry and PACs fractions	HFHS-induced obesity in male C57BL/6N mice gavaged daily with 200 mg/kg BW of WBE or equivalent PACs for 8 weeks	↑2-fold *Adlercreutzia equolifaciens** (Coriobacteriaceae) by the WBE and PACs	↑2.5-fold *A. muciniphila* (24.8%)	Enhanced glucose tolerance, ↑GC, restored colon mucus layer	[100]
	Cranberry juice (57% PACs)	Germ-free C57BL/6 with a simplified human microbiome gavaged daily with 200 mg/kg BW	↑*Bacteroides ovatus*, ↑*Clostridium hiranonis*	↑*A. muciniphila*	↑Intestinal mucus accumulation	[101]
	Cranberry (CP) and blueberry (BP) extracts (PACs) and their fibrous residues (CF and BF)	HFHS-induced obesity in male C57BL/6N mice fed 200 mg/kg berry powder or the equivalent fibrous fractions for 8 weeks	↓F/B, ↑*Dubosiella newyorkensis*, ↑*Angelakisella* spp., ↑Coriobacteriaceae* ↑Eggerthellaceae,*. ↓Lachnospiraceae, ↓Ruminococcaceae, ↓Peptostreptococcaceae	↑*A. muciniphila* by CP and BP correlated with lower BW	↓Fat mass depots, ↓BW, ↑mucus layer thickness	[102]
	Peach peel extract (PPE) (28% epicatechin 3-O-glucoside and 12.7% PACs)	HFD-induced obesity in female ICR mice fed 300 mg/kg (HPP) or 150 mg/kg (LPP) BW for 12 weeks	↓F/B. ↑*Lactobacillus* spp.,* ↑*Bifidobacterium* spp.,* ↑*Roseburia* spp., ↑*Bacteroides* spp.,* ↑Lachnospiraceae,* ↑Prevotellaceae, and ↑*Alloprevotella* spp.	↑*A. muciniphila* by HPP and LPP	↓BW, ↓oxidative stress, ↓hepatic lipid accumulation, ↑butyrate	[96]
	Cranberry extract (CP) (PACs) and agave inulin (AG)	HFHS-induced obesity in male C57BL/6N mice gavaged with either 200 mg/kg BW CP, 1 g/kg BW AG, or both for 9 weeks	↑*Muribaculum* spp., ↑*Faecalibaculum rodentium*, ↑*Roseburia* spp., ↑*Alistipes* spp. ↑Bacteroidaceae*, ↓*Ruminiclostridium* spp., ↓Lachnospiraceae, ↓Peptococcaceae	↑5.0-fold *A. muciniphila* only by CP	↑GC number, ↑Nlrp6, improved glucose tolerance, ↑TLR2, ↑AhR, ↓colon inflammation	[103]

* Some species belonging to this genus or family exhibit polyphenol-degrading abilities. BW, body weight; HFD, high-fat diet; HFHS, high-fat, high-sucrose diet; IP, intraperitoneal; F/B, Firmicutes-to-Bacteroidota ratio; NA, not applicable; TJPs, tight-junction proteins; GC, goblet cells; LBP, lipid and lipopolysaccharide-binding protein.

In agreement with the selective effect of PACs on *Akkermansia*, Neto et al. (2021) [101] showed that a cranberry juice extract enriched with 57% PACs increased relative levels of *A. muciniphila* and intestinal mucus accumulation in mice with a simplified human microbiome (consisting of 25 predefined species) after 10 days of treatment. Interestingly, this expansion coincided with an increase in the relative abundance of *Bacteroides ovatus* and butyrate-producing taxa such as *Clostridium hiranonis*. In fact, cranberry juice supplementation did not increase the relative levels of *A. muciniphila* when it was inoculated alone in a gnotobiotic mouse model; in contrast, it increased the levels of *A. muciniphila* when it was in a microbial community. This observation suggests that *A. muciniphila* could proliferate at the expense of other microbes and PAC-induced mucin secretion. Various studies using blueberry and cranberry polyphenols, as well as fibers, were performed in obese mice, confirming that berry polyphenols but not fibers selectively promote the relative abundance of *A. muciniphila*, bearing health benefits [102]. This outcome was reproduced by oral supplementation of cranberry flavan-3-ols-rich extract administered separately but not with agave inulin in HFD-fed mice for eight weeks [103]. Indeed, the presence of inulin in the diet abolished the stimulating effects of polyphenols on the *A. muciniphila* levels, although this diet promoted other beneficial microbes. Interestingly, only the diet enriched with polyphenols but not inulin significantly increased the mucin-secreting goblet cells in obese mice [103].

Grape PACs have also been shown to increase the relative proportions of *A. muciniphila*, *Bacteroides* spp. and *S24-7* and to affect those of *Clostridium* spp., Lachnospiraceae, and Ruminococcaceae [104,105]. In addition, the increase in the relative *A. muciniphila* proportion by PACs might depend on the relative abundance of this bacterium at the baseline of the dietary treatment [106]. Masumoto et al. (2016) [107] showed a similar shift in relative levels of *A. muciniphila* by administering apple polymeric PACs (0.5%) to mice fed an HFHS diet for 20 weeks. The PAC-rich diet increased the relative proportions of *Akkermansia* spp., as well as *Adlercreutzia*, *Roseburia*, *S24-7*, and *Bacteroides* spp. taxa. In contrast, the proportion of *Clostridium*, *Lachnospiraceae*, and *Bifidobacterium* spp. was reduced. Here, the Firmicutes/Bacteroidota ratio was reduced to the level observed in the mice fed a standard diet. The PAC polymers significantly improved lipid metabolism and decreased the gene expression of hepatic lipopolysaccharide (LPS) receptors (*Tlr4* and *Cd14*), concurrent with reduced inflammation, enhanced intestinal permeability, and the inhibition of HFD-induced opportunistic microbes in the gut microbiota.

Camu-camu (*Myrciaria dubia*) is an Amazonian fruit rich in ET, galloylated PACs, and ellagic acid. The dietary supplementation of camu-camu extract (200 mg kg^{-1}) in male HFHS-fed C57Bl/6J mice prevented obesity and improved glucose homeostasis. Gut microbiota analysis showed that flavan-3-ol-rich camu-camu decreased the abundance of *Lactobacillus* spp. and promoted *A. muciniphila*, *Barnesiella* spp. and *Turicibacter* spp. [108]. In agreement with this study, Abot and colleagues (2022) confirmed the promoting effects of camu-camu polyphenolic extract on *A. muciniphila*, even when administrated at a lower dose of 62.5 mg/kg to mice with HFD-induced obesity [109].

The stimulation of *A. muciniphila* by flavan-3-ols has also been shown to play a role in the attenuation of liver-related diseases. For instance, Xia and colleagues (2021) [110] recently evaluated the potential of catechin-rich Zhenjiang aromatic vinegar extract to modify gut microbiota composition and liver injury induced by long-term alcohol consumption in ICR male mice. Notably, ethanol administration increased the Firmicutes/Bacteroidota ratio, which was decreased by administering the vinegar catechin-rich extract. At the genus level, the relative abundances of *Akkermansia*, Lachnospiraceae_NK4A136_group, and *Bacteroides* spp. were significantly increased. These results demonstrate the role of polyphenols in stimulating *A. muciniphila* and reversing ethanol-induced gut disruption.

Table 3. Impact of flavan-3-ols and phenolic metabolites on the relative colonic abundance of *A. muciniphila* and other bacteria.

	Polyphenol-Rich Foods	Experimental Design	Main Findings in the Gut Microbiota	*A. muciniphila* Modulation	Impact on Host Health	Ref.
Flavan-3-ols	Apple polymeric proanthocyanins (PACs)	C57BL/6J mice fed an HFHS diet for 20 weeks supplemented with 0.5% polymeric PACs	↓F/B., ↑*Adlercreutzia* spp.,* ↑*Roseburia* spp., ↑S24-7, ↑*Bacteroides* spp., ↓*Clostridium*, ↓Lachnospiraceae, ↓*Bifidobacterium* spp.*	↑8.0-fold *A. muciniphila*	↓Dyslipidemia, ↓liver damage, ↓insulin resistance; ↓inflammation, ↓intestinal permeability	[107]
	Grape polyphenols (GP) (catechins)	HFD-induced obesity in male C57BL/6J mice fed 1% GP for 13 weeks	↓F/B, ↑*Alistipes* spp., ↑*Raouterella* spp. ↓*Lactobacillus* spp., ↓*Turicibacter* spp., ↓Lachnospiraceae, ↓Clostridiales	↑*A. muciniphila* 49% and 54.8% in cecum and feces, respectively	↓BW gain, ↓adiposity, endotoxemia, and improved glucose intolerance	[111]
	Camu-camu(CC) (galloylated PACs and ellagitannins)	HFHS-induced obesity in male C57Bl/6J fed 200 mg/kg CC for 8 weeks	↓*Lactobacillus* spp., ↑*Barnesiella* spp., ↑*Bifidobacterium* spp., ↑*Turicibacter* spp.	↑*A. muciniphila* and correlated with plasma bile acids	↓Hepatosteatosis, ↓metabolic endotoxemia, ↓glucose intolerance	[108]
	Camu-camu polyphenolic extract (CCE)	HFD-induced obesity in male C57BL/6J mice fed 200 mg/kg or 62.5 mg/kg CCE for 5 weeks	NA	↑*A. muciniphila* by 62.5 mg/kg CCE	↓Dyslipidemia, ↓BW, ↓hepatosteatosis	[109]
	Catechin-rich Zhenjiang aromatic vinegar extract (ZAVE)	Long-term alcohol consumption in ICR male mice gavaged daily with 200 or 800 mg/kg BW of ZAVE for 30 days	↓F/B, ↑Lachnospiraceae_NK4A136_group, ↑*Bacteroides* spp., ↓*Bilophila* and ↓*Butyricimonas* spp.	↑*A. muciniphila*, correlated with ↑antimicrobial peptides, ↓oxidative stress, and ↓inflammation	↓Gut inflammation, ↑IL-22, ↑Reg3g, ↓liver damage	[110]
	Green tea (GTE) (48% EGCG)	HFD-induced obesity in male C57BL/6J mice fed 2% GTE for 8 weeks	↓F/B, ↑microbial diversity, ↑Actinobacteria, ↑Coriobacteriales, ↑Turicibacterales, ↓Clostridiales, ↑*B. pseudolongum**, ↑*B. adolescentis**	↑*A. muciniphila* by all the tea infusions	↓Adipose inflammation, ↓metabolic endotoxemia, ↑TJPs in ileum and colon	[112]
	Green, oolong, and black teas (flavan-3-ols)	Male C57BL/6J mice fed an HFD supplemented with 45% energy from fat as food and tea infusions as drinking water for 13 weeks	↑Microbial diversity, ↑*Alistipes*, ↑Lachnospiraceae, ↑*Rikenella microfusus*, ↓*Allobaculum* spp., ↓*B. acidifaciens*, ↓*Clostridium leptum*, ↓*Parabacteroides goldsteinii*	↑*A. muciniphila* and negatively correlated with serum LBP levels	↓BW, ↓fat tissue accumulation, ↓metabolic endotoxemia, ↑lipid and glucose metabolism	[113]
Phenolic Metabolites	Urolithin A (UA) and urolithin B (UB)	Male Wistar rats fed a normal diet receiving an IP injection of UA or UB (2.5 mg/kg each) for 4 weeks	↓F/B, ↑Flavobacteriales by UA, ↓Lactobacillales and, ↓Clostridiales by UA and UB	↑*A. muciniphilia* by UA and UB	↓Serum AST	[114]

* Some species belonging to this genus or family exhibit polyphenol-degrading abilities. BW, body weight; HFD, high-fat diet; HFHS, high-fat, high-sucrose diet; IP, intraperitoneal; F/B, Firmicutes-to-Bacteroidota ratio; NA, not applicable; TJPs, tight-junction proteins; GC, goblet cells; LBP, lipid and lipopolysaccharide-binding protein.

Polyphenol-rich teas are consumed worldwide and associated with health benefits. Teas mainly contain gallated flavan-3-ols, such as EGCG and ECG, as well as non-gallated flavan-3-ols, such as EC and EGC [115]. These compounds effectively prevent cardiovascular diseases. For instance, supplementing 2% green tea extract (GTE) containing EGCG (48%) was shown to impact the gut microbiota significantly, mainly by promoting *Akkermansia* and *Bifidobacterium* spp. in male C57BL/6J mice fed an HFD for eight weeks. At the phylum level, the GTE increased the relative abundance of Actinobacteria and Verrucomicrobia but reduced that of the Firmicutes. Species-level analysis indicated that *A. muciniphila, Bifidobacterium pseudolongum*, and *Bifidobacterium adolescentis* were significantly stimulated by the catechin-rich green extract compared to the HFD-fed untreated mice. It is worth noting that *Bifidobacterium* spp. are reported to be involved in flavan-3-ol metabolism and the release of valerolactones metabolites [116,117]. The derived metabolites exert additive protective effects against intestinal inflammation, improving the mucosa microenvironment status. GTE enhanced adipose inflammation and TJPs in the ileum and colon of HFD-induced obese mice, preventing metabolic disorders associated with obesity [112].

The modulation of the gut microbiota by different teas, including green, oolong, and black tea, was evaluated by Liu and colleagues in a mouse model of HFD-induced obesity for 13 weeks (Table 3) [113]. Although the specific proportion of phenolic classes was not detailed in the tea infusions, flavan-3-ols and flavonols were the major groups among the identified compounds. The green tea contained 3332.35 $\pm$ 70.91 mg L^{-1} total polyphenols, oolong tea contained 2911.52 $\pm$ 51.51 mg L^{-1}, and black tea contained 2732.11 $\pm$ 23.64 mg L^{-1}. All tea infusions recovered the microbial diversity and increased relative levels of *Akkermansia* spp., together with that of *Alistipes* spp., Lachnospiraceae, and *Rikenella microfusus*, and decreased those of *Allobaculum* spp., *Bacteroides acidifaciens, Clostridium leptum*, and *Parabacteroides goldsteinii* compared to HFD. The tea infusions prevented BW gain and fat tissue accumulation and decreased serum glucose, lipid, and lipopolysaccharide-binding protein (LBP) to levels comparable to the group fed a low-fat diet [113]. In the study, the relative increase in *Akkermansia* associated with tea-based drinks was significantly and negatively correlated with sserum LBP levels.

In concordance with the above findings, the daily administration of 0.32% EGCG to male HFD-fed C57BL/6N mice for eight weeks increased the abundance of the phyla Verrucomicrobia and Actinobacteria and the genera *Adlercreutzia* and *Akkermansia* but decreased the abundance of phyla Deferribacteres, Proteobacteria, and Firmicutes, as well as an unclassified genus of Desulfovibrionaceae, compared to the untreated HFD group [118]. The EGCG treatment inhibited BW gain, hepatic lesions size, and TG content in the liver of HFD-induced obese mice. Accordingly, a recent study evaluated whether the relative increase in *Akkermansia* was attributed to the EGCG compound from a green tea extract compared to the whole flavan-3-ol-rich extract. Again, the increased *Akkermansia* proportion was reproduced by an equivalent concentration of EGCG in obese and lean mice [119].

Most of the above studies show, as a common denominator, the stimulation of the relative proportion of *A. muciniphila* concomitantly with that of polyphenol-degrader taxa, such as *Adlercreutzia* and *Bifidobacterium* (Tables 2 and 3) [120–122]. This observation suggests that *A. muciniphila* might benefit from flavan-3-ols degradation byproducts through the action of other microbes and the impact these compounds exert on its mucosal ecological niche. In another mechanistic study, it was demonstrated that polyphenols directly interact with intestinal mucin and reinforce the mucus layer. Feng et al. (2022) [123] evaluated the effects of EC, EGCG, and tannic acid (TA) on the intestinal mucin and barrier function through isothermal titration calorimetry and multiple-particle tracking using ex vivo mucus and Caco-2/HT29-MTX cocultures. Strikingly, it was demonstrated that gallated polyphenols and TA strongly bind to intestinal mucin, reinforce the mucus layer, and counteract inflammation [123].

Because polyphenols can significantly increase the viscoelasticity properties of the mucin-rich mucus layer, the colonization of opportunistic pathogens and LPS penetration might be substantially limited, thus contributing to the amelioration of the gut microbiota and intestinal disruption. These results can explain the stimulation of *A. muciniphila* by ellagitannins, in addition to providing a selective landscape for this bacterium to thrive in vivo.

Recently, Xia et al. (2021) [54] assessed the in vitro growth of *A. muciniphila* in a culture medium enriched with EGCG alone as a unique carbon source and in a medium containing EGCG and either mucin or glucose. Whereas *A. muciniphila* could not grow in the EGCG-enriched medium, it was reported to grow better when both EGCG and mucin were present in the medium than with only mucin. The growth improvement of *A. muciniphila* was dose-dependent, showing a gradual increase proportional to increasing concentrations of 50, 350, and 500 mg L^{-1} of EGCG. However, it should be noted that in this study, the *A. muciniphila* growth curve was measured by optical density (OD600), which is not a suitable method when high concentrations of polyphenols are added into the media because the presence of increased polyphenol interactions with mucin and proteins in the growing media provoke the formation of clumps, resulting in increased OD when the concentration of polyphenols increases. In this work, phenolic products were also analyzed in the media with and without EGCG. The presence of catechin gallate, gallocatechin, epigallocatechin, gallic acid, esculetin, hydroxyhydroquinone, and 3,4-dihydroxybenzaldehyde was reported only in *Akkermansia*-inoculated EGCG-enriched media, suggesting that *A. muciniphila* partly degrades EGCG in a yet unknown pathway.

Moreover, it is unclear whether the phenolic products described in the EGCG fermentation by *A. muciniphila* with mucin or glucose might also result from the spontaneous hydrolysis [124] of this compound in the non-inoculated media, as the metabolic output of the control medium is not detailed in this study. Although this is an interesting approach to examining the prebiotic effect of polyphenols on *A. muciniphila*, the microbial enzymes responsible for ellagitannin trophic utilization, such as tannases and gallate decarboxylase, remain to be identified and tested in *A. muciniphila*.

In 2017, Henning and colleagues [43] reported the participation of *A. muciniphila* only in the hydrolysis of pomegranate ellagitannins (ETs) (punicalagin and ellagic acid) but not in their complete catabolism pathways to further produce the final metabolite, urolithin A. It has been demonstrated that in the human intestine, ETs and ellagic acid follow a lactone ring cleavage, decarboxylation, and dihydroxylation to form various metabolites, including urolithins. The notion of ET utilization by *A. muciniphila* arose from evidence for the stimulatory effects of ET (1000 mg of pomegranate extract daily) on *A. muciniphila* in humans associated with a healthier metabolic status and better clinical outcomes. A compelling finding from this study was that the relative levels of *Akkermansia* spp. were significantly higher in stool samples of urolithin-A producers (so-called metabotype A) compared to non-producing subjects [52,53]. Consequently, Henning and colleagues aimed to confirm the in vitro degradation of these phenolics in monoculture with *A. muciniphila* [43]. The authors demonstrated that *A. muciniphila* growth was significantly inhibited by the pomegranate ET (0.18 mg/mL and 0.28 mg/mL) and slightly reduced when grown only with ellagic acid (50 μM). Notwithstanding, ellagic acid increased when *A. muciniphila* was grown in a medium enriched with pomegranate ET, suggesting ellagitannin hydrolysis by *A. muciniphila*. Similarly, ellagic acid levels decreased when *A. muciniphila* was grown with only ellagic acid. However, phenolic metabolites other than urolithin A could be produced, as this was not detected in the media. Collectively, these findings the premise that *A. muciniphila* might prime phenolic degradation by other microbes by initiating the hydrolysis of parent compounds. In turn, it takes advantage of the effects of the phenolic metabolites in the mucosal microenvironment and microbial counterparts.

4.3. Flavonol-Rich Foods

Flavonols have a plane 3-hydroxyflavone base distinguished by hydroxyl groups in benzene rings [125]. Quercetin, myricetin, and kaempferol are good representatives of this phenolic group. Quercetin is widely found in apples, teas, berries, and onions. The gut bacterial species of *Eubacterium ramulus, Escherichia coli, Streptococcus lutetiensis,*

L. acidophilus, Weissella confusa, Enterococcus gilvus, Clostridium perfringens, Bacteroides fragilis, and *Bacillus subtilis* have been reported to utilize quercetin [126,127]. The quercetinases, including a family of cupin-type dioxygenases, flavonol 2,4-dioxygenase, and quercetin 2,3-dioxygenases, have been reported to be involved in this activity [128–130].

Quercetin is a polyphenol with anti-inflammatory and cardioprotective effects. For instance, the role of quercetin in improving hypercholesteremia was demonstrated in LDLR$^{-/-}$ (LDL receptor-deficient) C57BL/6 mice fed an HFD [131]. Quercetin supplementation (100 µg per day) reduced BW gain, atherosclerotic lesions, inflammation, and oxidative stress. The gut microbiota composition was significantly changed, with an increase in the relative abundances of Actinobacteria and Bacteroidota and a reduction in Firmicutes. At the genus level, quercetin supplementation increased the relative levels of the *Akkermansia, Bacteroides, Parabacteroides,* and *Ruminococcus* genera [132]. Similarly, Etxeberria et al. (2015) [133] investigated whether quercetin administration could prevent gut microbiota disruption induced by an HFHS diet for 6 weeks. Quercetin impacted the gut microbiota at different taxonomic levels. On the one hand, the Firmicutes/Bacteroidota ratio and the growth of bacterial species associated with diet-induced obesity were decreased, including Erysipelotrichaceae, *Bacillus* spp. and *Eubacterium cylindroides*. In contrast, relative levels of *A. muciniphila* were shown to be increased by quercetin compared to the relative abundance detected in diet-induced overweight control rats.

The coadministration of quercetin with other phenolics has also been demonstrated to reverse obesity and gut microbial dislocation, in addition to stimulating *A. muciniphila* (Table 4). For instance, 30 mg kg^{-1} BW per day of quercetin jointly administrated with 15 mg kg^{-1} BW per day of resveratrol in HFD-fed Wistar rats for ten weeks improved obesity by modulating the gut microbiota. The combination of the two phenolics significantly increased the microbial diversity and the relative abundance of *A. muciniphila*, Ruminococcaceae_UCG-014, Bacteroidales_S24-7_group_norank, Ruminococcaceae_UCG-005, and the Christensenellaceae family in HFD-fed rats. Additionally, a reduction in *Lachnoclostridium* and *Bilophila* spp. in the quercetin and resveratrol group was observed compared to the HFD group. The rat obesity phenotype was enhanced by a reduction in BW, in addition to the attenuation of serum lipids and inflammatory markers [133]. Dietary supplementation with a *Smilax china* extract rich in quercetin and neochlorogenic acid to C57BL/6J mice for 12 weeks improved glucose tolerance and reduced BW, inflammation, and lipid metabolism. Within the gut microbiota, the relative abundance of Akkermansiaceae increased, and that of Desulfovibrionaceae, Lachnospiraceae, and Streptococcaceae decreased compared to the HFHS-fed group [134].

Quercetin was found to promote *A. muciniphila* either administered alone or combined with other phenolic compounds in rodents (Table 4). In previous work, we screened the presence of quercetinases in the *A. muciniphila* ATCC BAA-835 genome, finding two proteins with a conserved domain of quercetinases, notably a cupin domain-containing protein encoded by Amuc_0801 and γ-carboxymuconolactone decarboxylase encoded by Amuc_1806 [44]. However, the enzymatic activity involved in quercetin degradation remains to be established.

Recently, it was demonstrated that a synergic effect of *A. muciniphila* (2×10^8 colony-forming units (CFU)/200 µL) conjointly administrated with quercetin (37.5 mg kg^{-1} day^{-1}) could counteract obesity and NAFLD in an HFD-induced obese Wistar rat model [39]. After six weeks of HFD-induced obesity, the three-week dietary treatment with *A. muciniphila* in Wistar rats was associated with less body fat; however, coadministration of *A. muciniphila* with quercetin boosted steatosis remission and improved hepatic lipogenesis modulation. Significantly, *A. muciniphila* intestinal colonization was increased when coadministered with quercetin, suggesting that the gut microenvironment created by quercetin intake determined the ability of *A. muciniphila* to proliferate within the gut microbiota. In particular, *A. muciniphila* correlated with liver lipid and bile acid metabolisms, supporting the role of this bacterium in many comorbidities, including metabolic syndrome and hepatic damage.

Flavonols also occur in glycosylated form, bound to glucose, rhamnose, or rutinose. The most common glycosidic form of quercetin is rutin (a glycoside combining quercetin and the disaccharide rutinose). The α-L-rhamnosidases intervene in rutin utilization via gut microbial species; *Bifidobacterium catenulatum* and *B. pseudocatenulatum* were shown to degrade glycosylated flavanols [135]. Recently, Riva and colleagues [136] identified which gut microbes were metabolically active after the amendment of rutin using a fluorescence-based single-cell activity measure (biorthogonal non-canonical ammino-acid-tagging (BONCAT)) combined with fluorescence-activated cell sorting (FACS). Lachnospiraceae (*Lachnoclostridium* and *Eisenbergiella*), Enterobacteriaceae, Tannerellaceae, and Erysipelotrichaceae species were included in the rutin-responsive fraction of the gut microbiota. Specifically, Enterobacteriaceae was associated with the conversion of rutin to quercetin-3-glucoside (Q-glc), and Lachnospiraceae was associated with quercetin (Q) production [136]. Although Verrucomicrobiaceae was not included within the rutin-responsive group, this study uncovered the contribution of other bacterial taxa in degrading flavonols and how they might support *A. muciniphila* colonization in the gut.

Flavonol glycosides have also been shown to protect against intestinal barrier disruption, inflammatory responses, and colitis by regulating gut microbiota composition, especially by increasing *A. muciniphila.* Bu et al. (2021) [137] evaluated the effect of flavonol-rich extract of *Abelmoschus manihot* flowers containing quercetin-3-O-robinobioside, gossypetin-3-O-glucoside, quercetin-3′-O-glucoside, isoquercetin, hyperoside, myricetin, gossypetin, and quercetin in DSS-induced colitis in C57BL/6J mice. The flavonol glycoside-rich extract remarkably increased the relative proportion of *A. muciniphila*, along with *Gordonibacter*, a well-known polyphenol-degrader genus (Table 4). These changes were reflected in the attenuation of the colonic inflammatory response and intestinal epithelial barrier dysfunction. The role of *Akkermansia* in protecting against colitis was further confirmed by daily administration of *A. muciniphila* cells (3×10^8) to DSS-induced mice, demonstrating that this bacterium can reduce inflammatory markers, restore intestinal barrier function, and relieve colitis. Overall, these studies highlight *A. muciniphila* as an intermediary player in the potentialization of the protective role of flavonols in treating intestinal disruption and gut inflammation.

Table 4. Modulation of the gut microbiota by flavonols and flavonol glycosides in animal models.

	Polyphenol-Rich Foods	Experimental Design	Main Findings in the Gut Microbiota	*A. muciniphila* Modulation	Impact on Host Health	Ref.
Flavonols	Quercetin	HFD-induced obesity in LDLR−/− (LDL receptor-deficient) C57BL/6 mice fed 100 μg of quercetin daily for 12 weeks	↑Microbial diversity, ↑Actinobacteria, ↑Bacteroidota ↓Firmicutes, ↓*Lactobacillus*,* ↑*Bacteroides*,* ↑Parabacteroides, ↑*Ruminococcus*	↑2.0-fold *A. muciniphila*	↓BW, ↓Intestinal cholesterol, ↓oxidative stress, ↓inflammation	[131]
		HFHS-induced obesity in Wistar rats fed 30 mg/kg BW quercetin for 6 weeks	↓F/B, ↓Erysipelotrichaceae, ↓*Bacillus*, ↓*Eubacterium cylindroides*, ↑*Barnesiella*, ↑*Bacteroides dorei*, ↑*Bacteroides chinchillae*, ↑*Prevotella*	↑1-8-fold *A. muciniphila* (13.84%)	↓Insulin resistance, ↑TJPs	[132]
	Quercetin + resveratrol	HFD-induced obesity in Wistar rats fed 30 mg/kg BW quercetin, 15 mg/kg BW of resveratrol, or both for 10 weeks	↑Microbial diversity, ↑Bacteroidales_S24-7_group, ↑Ruminococca-ceae, ↑Christensenellaceae, ↓*Lachnoclostridium*, ↓*Bilophila*	↑ *A. muciniphila*	↓BW, ↓serum lipids, ↓inflammatory markers	[133]
	Quercetin (Q) + *A. muciniphila* cells	HFD-induced obesity in Wistar rats fed with 2 × 10^8 CFU/200 μL and 37.5 mg/kg Q for 3 weeks	↑Cyanobacteria, ↑*Oscillospira* spp., ↓Actinobacteria, ↓*Lactococcus* spp., ↓*Lactobacillus* spp.,* ↓*Blautia* spp., ↓*Rothia* spp., ↓*Roseburia* spp.	↑*A. muciniphila* only when co-administered with Q. It correlated with ↓BW, lipid, and bile acid metabolism	↓BW, ↓fat mass depot, ↓NAFLD	[39]
	Abelmoschus manihot flowers (TFA) (Flavonol glycosides)	DSS-induced colitis in C57BL/6J mice gavaged with 125 mg/kg or 62.5 mg/kg of TFA for 7 days	↑*Gordonibacter* spp.,* ↑*Erysipelatoclostridium* spp., ↓Tenericutes, ↓Proteobacteria	↑*A. muciniphila* (only by 125 mg/kg TFA), correlated with ↓gut inflammation, ↑*Muc2*, ↑barrier function	↓ Colonic inflammatory, ↓intestinal epithelial barrier dysfunction	[137]

* Some species belonging to his genus or family exhibit polyphenol-degrading abilities. Modulation relates to relative colonic or cecal abundance. BW, body weight; HFD, high-fat diet; HFHS, high-fat, high-sucrose diet; NAFLD, non-alcoholic fatty liver disease; DSS, dextran sulfate sodium; F/B, Firmicutes-to-Bacteroidota ratio; TJPs, tight-junction proteins.

4.4. Flavanone and Flavonone-Rich Foods

Flavanones, such as hesperidin, eriodictyol, and naringin, are the most studied compounds in lemons, oranges, and berries. Naringin and hesperidin reach the colon intact [19], where, through the action of gut microbial α-rhamnosidase and β-glucosidase, the aglycones hesperetin and naringenin are generated. Hesperetin and naringenin can be subsequently metabolized into various phenolics, including dihydrocaffeic acid, isoferulic acid, 4-hydroxyphenylacetic acid, dihydroferulic acid, ferulic acid, resorcinol, phloroglucinol, phloretic acid, phloroglucinol acid, hydrocinnamic acid, 3-(3′-hydroxyphenyl), propionic acid, protocatechuic acid, and hippuric acid [138]. Among gut microbial species, *B. catenulatum* and *B. pseudocatenultum* have been reported to have the ability to hydrolyze hesperidin [135]. These compounds have the therapeutic potential to modulate several cardiovascular disease (CVD) risk factors. For instance, glucose-lowering, obesity-preventing, lipid-regulating, antioxidant, and anti-inflammatory effects have been reported in diabetic and obese models. More information about the health benefits of flavanones and derived metabolites can also be found in a review by Mas-Capdevila (2020) [139]. Liu et al. (2020) [140] evaluated the effects of hesperidin (100 or 200 mg/kg BW) in HFD-fed male C57BL/6 mice for 10 weeks. *A. muciniphila* was reduced in HFD-fed mice, but unexpectedly, it failed to be changed by hesperidin; instead, the treatment enriched *Lactobacillus salivarius* and *Desulfovibrio* and decreased *Helicobacter ganmani, Helicobacter hepaticus, B. pseudolongum*, and *Mucispirillum schaedleri* in HFD-fed mice. Whereas the relative abundance of *A. muciniphila* was not significantly modulated by hesperidin, gut microbiota modulation was correlated with reduced BW and obesity-related disturbances.

Examples of flavonones include luteolin and apigenin; apples, olives, and artichokes are good sources of these compounds. Within the gut microbiota, *Eubacterium ramulus* has been shown to metabolize apigenin, as well as quercetin, naringenin, daidzein, and genistein. This is attributed to the action of a phloretin-hydrolase to break the phloretin C-C bond [130,141,142]. Apigenin can enhance the production of butyrate and propionate, which are associated with mucin production [143], protecting the gut lining and facilitating the growth of mucin-dependent gut bacteria such as *A. muciniphila*. The stimulation of *A. muciniphila* by apigenin supplementation in animal models supports the above notion. It was demonstrated that apigenin administration to male C57BL/6 J mice for 16 weeks could alleviate metabolic endotoxemia by improving intestinal disruption and restoring gut barrier function induced by an HFD [144]. Augmentation of *Akkermansia* at the genus level potentially mediated the protective effects of apigenin on metabolic syndrome. Indeed, the metabolic phenotype, including BW loss and changes in gut microbiota composition, could be transferred from apigenin-fed mice to HFD-feeding mice via fecal microbiota transplantation [144].

4.5. Phenolic Acid-Rich Foods

Phenolic acids are predominantly found bound to the fiber fractions of vegetables, fruits, and cereal. Berries, red cabbage, rice, and wheat bran are good sources of phenolic acids [145]. Free phenolic acids are also found in coffee and cacao. Chlorogenic acid and caffeic acid, the predominant phenolic acid in coffee, can modify the gut microbiota structure, reducing intestinal and systemic inflammation in colitis. Notably, in the colon, chlorogenic acid is hydrolyzed into caffeic acid metabolite by mucosal and microbial esterase in the intestinal tract [146,147]. Both chlorogenic and caffeic acids significantly increased the relative abundance of *A. muciniphila* and the microbial richness, in addition to reducing the ratio of Firmicutes to Bacteroidota in mice with DSS-induced colitis [148,149] (Table 5). For instance, the effects of dietary caffeic acid (1 mM, CaA) administered for 15 days on murine experimental colitis were shown to contribute to coping with DSS-induced intestinal inflammation and gut microbiota dislocations. CaA significantly suppressed inflammatory markers, such as the secretion of IL-6, TNFα, and IFNγ and the colonic infiltration of CD3+ T cells, CD177+ neutrophils, and F4/80+ macrophages. Furthermore, caffeic acid has been shown to hamper the abundance of inflammatory Ruminococcaceae species [149,150].

Released fiber-bound phenolic acids, after in vitro digestion and colonic fermentation, also showed a significant prebiotic effect on *A. muciniphila*, *Faecalibacterium prausnitzii*, *Bifidobacterium* spp. and *Lactobacillus* spp. in in vitro studies [151]. The bound phenolics, including p-coumaric acid, hydroxybenzoic acid, and ferulic acid, exhibited excellent antioxidant and hypoglycemic activities. Interestingly, the dietary fibers removed by polyphenols did not show the same promoting effects on the gut microbiota [151]. This observation highlights the phenolic compounds as an important factor supporting trophic interactions with other gut microbes, leading to *A. muciniphila* proliferation. The enzymes involved in phenolic acid metabolism, such as feruloyl esterases (also called hydroxycinnamoyl esterases) and phenolic acid reductases, have been reported in *Bifidobacterium* and *Lactobacillus* genera. In contrast, owing to its polysaccharide-degrading activity, *F. prausnitzii* intervenes in releasing fiber-bound phenolic acids. The ability to utilize phenolic acids has been reported in *B. longum*, *Lactobacillus helveticus*, *Lactobacillus johnsonni*, *Limosilactobacillus reuteri*, *L. acidophilus*, *Limosilactobacillus fermentum*, *Latilactobacillus curvatus*, *Furfurilactobacillus rossiae*, and *L. plantarum* [152–154]. However, phenolic acid reductase and feruloyl esterase activities remain to be assessed in *A. muciniphila*.

Table 5. Modulation of the gut microbiota and *A. muciniphila* by flavones and phenolic acids in animals.

	Polyphenol-Rich Foods	Experimental Design	Main Findings in the Gut Microbiota	A. *muciniphila* Modulation	Impact on Host Health	Ref.
Flavanones	Hesperidin	HFD-induced obesity in male C57BL/6 mice gavaged with 100 or 200 mg/kg BW hesperidin for 10 weeks	↑*Lactobacillus salivarius*, ↑Desulfovibrio _C21_c20, ↓*Helicobacter* spp., ↓*B. pseudolongum* and ↓*Mucispirillum schaedleri*	Failed to change *A. muciniphila*	↓BW, ↓inflammation, ↓plasma LBP, ↑intestinal integrity	[140]
Flavonones	Apigenin	HFD-induced obese male C57BL/6J mice gavaged with 50 mg/kg BW apigenin for 16 weeks	↓F/B, ↑Bacteroidaceae, ↓Erysipelotrichaceae	↑*A. muciniphila* (Akkermansiaceae)	↓Metabolic endotoxemia, ↓inflammation, ↓liver injury, ↓hepatosteatosis, ↑intestinal integrity	[144]
Phenolic acids	Caffeic acid (CaA)	DSS-induced colitis in female C57BL/6 mice gavaged with 1 mM CaA for 15 days	↑Microbial diversity, ↓F/B, ↑Tenericutes	↑*A. muciniphila* (25%)	↓Gut and serum inflammatory markers, ↓NF-κB signaling pathways	[149]
	Chlorogenic acid (ChA)	DSS-induced colitis in female C57BL/6 mice gavaged with 1 mM ChA for 15 days	↓F/B, ↑microbial diversity	↑*A. muciniphila* (38%)	↓Diarrhea and rectal bleeding, ↓mucin depletion, ↓gut inflammation	[148]
	Rice bran fiber-bound phenolic acids (RBDF) (p-coumaric acid, hydroxybenzoic acid, and ferulic acid)	In vitro colonic fermentation of GI-digested RBDF	↑*F. prausnitzii*, ↑*Bifidobacterium*,* ↑*Lactobacillus** spp.	↑*A. muciniphila* only by the fiber-bound phenolics but not by the phenolic-free fibers	↑Antioxidant and hypoglycemic activities	[151]

* Some species belonging to this genus or family exhibit polyphenol-degrading abilities. Modulation relates to relative colonic or cecal abundance. BW, body weight; HFD, high-fat diet; HFHS, high-fat, high-sucrose diet; DSS, dextran sulfate sodium; F/B, Firmicutes-to-Bacteroidota ratio; TJPs, tight-junction proteins; GC, goblet cells; LBP, lipid- and lipopolysaccharide-binding protein.

4.6. Stilbene-Rich Foods

Stilbenes are found in grapes, cranberries, and red fruits. Resveratrol is a stilbene compound with several therapeutic effects on health, including protective properties against atherosclerosis and metabolic syndromes [155]. The microbial degradation of resveratrol yields 3,4′-dihydroxy-trans-stilbene and 3,4′-dihydroxybibenzyl (lunularin) metabolites.

Recently, lunularin was reported to be further dehydroxylated by the gut microbiota, although only at the 3 position, to yield 4-hydroxydibenzyl, a novel metabolite found in human urine after resveratrol intake in healthy individuals [156]. The microbial species *Slackia equolifaciens*, *Adlercreutzia equolifaciens*, *Eggerthella lenta* (ATCC 43055), and *Bacteroides uniformis* (ATCC 8492) have been identified as dihydroresveratrol producers [156,157]. Regarding the impact of resveratrol and its relationship with the microbiota, there is evidence of it stimulatory effects on *A. muciniphila* [51] (Table 6). Resveratrol attenuated trimethylamine-N-oxide (TMAO)-induced atherosclerosis in ApoE$^{-}/^{-}$mice, decreasing TMAO concentration by modulation of the gut microbiota. Primarily, resveratrol increased the relative abundance of *Akkermansia*, *Bacteroides*, *Lactobacillus*, and *Bifidobacterium* spp.; instead, the relative levels of *Prevotella*, Ruminococcaceae, *Anaerotruncus*, *Alistipes*, and Peptococcaceae were decreased [158].

Resveratrol attenuated gut microbiota and intestinal disruption and reduced metabolic endotoxemia and colon inflammation in HFD-fed rats for six weeks [159]. This coincided with an increase in the relative abundance of *A. muciniphila* and reduced bacterial invasion in the distal colon. Likewise, increased tight-junction proteins (TJPs) and decreased FAK, MyD88, and IRAK4 were observed. Furthermore, resveratrol inhibited HFD-induced cannabinoid receptor type 1 (CB1) mRNA and suppressed CB2 mRNA levels in the colon. These data indicate that the endocannabinoid system plays a crucial role in resveratrol-induced protection against non-alcoholic steatohepatitis by maintaining gut barrier integrity and inhibiting gut inflammation. Resveratrol was also shown to attenuate colonic inflammation and clinical symptoms in a murine model of 2,4,6-trinitrobenzenesulfonic acid (TNBS)-induced colitis [160]. It restored the gut microbiota to homeostatic levels and increased butyrate production. At the species level, the relative levels of *A. muciniphila* were significantly increased in the TNBS + resveratrol group compared to the TNBS group [160].

Supplementation with resveratrol and pterostilbene in combination with a strictly calorie-restricted diet drives a beneficial microbial profile characterized by a relative increase in the level of the *Akkermansia* [161,162]. The activity of pterostilbene, a stilbene and a dimethoxy resveratrol derivative, may differ from that of resveratrol, but both molecules have been reported to exhibit similar cellular effects and mimic calorie restriction at the molecular level [162,163].

4.7. Lignan-Rich Foods

Lignans, such as secoisolariciresinol diglucoside (SDG), are present in berries, legumes, cereals, and especially in flaxseed. Most of the biological activities of lignans depend upon colonic bacterial transformations. The removal of glucose moiety from SDG, which the first step of the change, is performed by strains producing β-glucosidases. Lignan-hydrolyzing taxa include *Lactobacillus* spp., *Bifidobacterium* spp., Bacteroidales spp. and Clostridiales spp. Other species, such as *E. faecalis*, *E. lenta*, *B. producta*, *Eubacterium limosum*, *Clostridium scindens*, and *Lactonifactor longoviformis*, have been shown to utilize lignans [164,165]. Aglycones are further transformed into the enterolignans enterodiol (ED) and enterolactone (EL), exerting improved bioavailability and biological activity. This is the case of phytoestrogens, which were shown to protect against a chemically induced breast cancer model [166]. Contrary to other phenolic categories, lignans have been shown to decrease relative levels of *Akkermansia* spp. (Table 6). For instance, flaxseed lignans decreased the relative levels of *A. muciniphila* by 30-fold, increasing those of *Prevotella* and *Roseburia* spp. by 20-fold and 10-fold, respectively, in mice fed a basal diet for three weeks [167]. Another recent study confirmed the reduction in *Akkermansia* spp. and the increase in *Lactobacillus* and *Bifidobacterium* spp. in association with feeding with a high dose (50 mg/kg BW) of lignan syringaresinol in middle-aged mice [168]. The mechanisms underlying this coexclusion relationship between lignan derivatives and *Akkermansia* require further investigation.

Table 6. Modulation of the gut microbiota and *A. muciniphila* by stilbenes and lignans in animals.

	Polyphenol-Rich Foods	Experimental Design	Main Findings in the Gut Microbiota	A. *Muciniphila* Modulation	Impact on Host Health	Ref.
Stilbenes	Resveratrol (RSV)	TMAO-induced atherosclerosis in ApoE−/− female C57BL/6 mice fed a chow diet with 0.4%RSV for 30 days	↑*Bacteroides*,* ↑*Lactobacillus* spp.,* ↑*Bifidobacterium* spp.,* ↓*Prevotella* spp., ↓Ruminococcaceae, ↓*Anaerotruncus* spp., ↓*Alistipes* spp., ↓Peptococcaceae	↑*A. muciniphila*	Protected against atherosclerosis, ↓gut microbial TMA production	[158]
		HFD-induced obesity in male Sprague–Dawley rats fed 100 mg/kg RSV for 6 weeks	↑Ruminococcaceae, ↑Lachnospiraceae, ↓*Desulfovibrio* spp.	↑*A. muciniphila*, correlated with endocannabinoid system modulation.	↑TJPs, ↓CB1, ↓CB2, ↓steatohepatitis, ↓gut inflammation, ↓metabolic endotoxemia	[159]
		TNBS-induced colitis in female BALB/c mice fed 100 mg/kg RSV for 5 days	↓*B. acidifaciens*, ↑*Ruminococcus gnavus*	↑4.5-fold *A. muciniphila*	Attenuated colitis, ↓gut inflammation, ↑butyrate	[160]
	Pterostilbene (Pst)	Zucker (fa/fa) rats fed a standard diet and gavaged with 15 mg/kg BW Pst for 6 weeks	↓F/B, ↑*Odoribacter splanchnicus*	↑*A. muciniphila*, correlated with ↓BW	↓BW, ↓fat mass, ↓serum insulin, ↓glucose intolerance	[162]
Lignans	Flaxseed (FS) (secoisolariciresinol diglucoside)	C57BL/6 mice fed a standard diet with 10% FS for 3 weeks	↑20-fold *Prevotella* spp., ↑10-fold *Roseburia* spp.	↓30-fold *A. muciniphila*	↑GC, ↑Muc2, ↑RegIIIγ in the colon	[167]
	Syringaresinol (SYR)	Middle-aged male C57BL/6 mice fed 10 or 50 mg/kg BW of SYR for 10 weeks	↓F/B, ↑*Lactobacillus* spp.,* ↑*B. pseudolongum*,* ↓Bacteroidaceae*	↓*A. muciniphila* by high dose of SYR	↓LBP, ↑Foxp3+ regulatory T cells	[168]

* Some species belonging to this genus or family exhibit polyphenol-degrading abilities. Modulation relates to relative colonic or cecal abundance. BW, body weight; HFD, high-fat diet; HFHS, high-fat, high-sucrose diet; DSS, dextran sulfate sodium; F/B, Firmicutes-to-Bacteroidota ratio; TJPs, tight-junction proteins; GC, goblet cells; LBP, lipid- and lipopolysaccharide-binding protein; TMAO, trimethylamine-N-oxide; GI, gastrointestinal; TNBS, 2,4,6-trinitrobenzenesulfonic acid; TMA, trimethylamine; CB1- CB2, cannabinoid receptor type 1 and 2.

5. Modulation of *A. muciniphila* by Polyphenol-Rich Foods in Human Intervention Trials

A substantial body of literature has shown the modulatory effects of polyphenols on the gut microbiome in animal models. Despite the limited number of clinical trials, consistent data on the gut microbiota composition and the stimulation of *A. muciniphila* have been observed. Accumulating studies show evidence of the effect of polyphenols on improving metabolic alterations in the context of cardiovascular diseases [169,170]. Again, most studies focusing on the bioavailability of phenolics and the metabolome of individuals consuming polyphenol-rich foods point to the gut microbiota composition as an essential dynamic to extract polyphenol benefits to counteract metabolic diseases. However, factors such as interindividual variations on the baseline gut microbiota composition, genotypes, race, sex, and uncontrolled dietary habits in humans make it challenging to elucidate the selective mechanisms of polyphenols in boosting *A. muciniphila* health benefits in the host.

The duration of the dietary intervention, the health conditions, the presence of inhabitant microbes with the ability to break down phenolics (so-called metabotypes), the dose, and the phenolic class are essential variables to consider when translating the outcomes from experimental animal studies to human clinical trials. For instance, different profiles of the metabolic fate of labelled polyphenols (i.e., glucuronidation, sulfation, and methylation) have been reported across species. EC metabolites differ between humans and rats and, to a somewhat lesser degree, between humans and mice, as demonstrated by the kinetics of the metabolism of the flavanol (−)-epicatechin [36]. Despite the concerns about the species-dependent metabolic equivalence between humans and rodents, the role of the gut microbiome in polyphenol catabolism remains the most relevant factor in both models. In this section, we review compelling human studies that have assessed the effects of polyphenols on host health, in addition to the gut microbiota composition and the identification of *A. muciniphila* (Table 7).

Table 7. Modulation of relative colonic levels of *A. muciniphila* and other bacteria by polyphenol-rich foods in human intervention trials.

	Polyphenol-Rich Foods	Experimental Design	Main Findings in the Gut Microbiota	*A. muciniphila* Modulation	Impact on Host Health	Ref.
Phenolic acids	Oats rich in (β-glucans and polyphenol s)	Intake of 80 g of oat comprising by β-glucans (3.0 g) and polyphenols (56.8 mg) in mildly hypercholesterolemic subjects for 45 days	↑*Roseburia* spp., ↑*Prevotella* spp., ↑*Paraprevotella* spp., ↑*Dialister succinatiphilus*, ↑*Roseburia hominis*, ↑*Butyrivibrio crossotus*, ↑*B. pseudocatenulatum*,* ↑*Clostridium symbiosum*, ↓*Megamonas hypermegale*, ↓*Clostridium nexile*, ↓*Roseburia inulinivorans*	↑*A. muciniphila*, correlated with ↓HDL-C	↓Dyslipidemia, ↑propionate, and ↑acetate	[171]
Flavan-3-ols and stilbenes	Epigallocatechin-3-gallate (EGCG) + resveratrol (RSV)	Intake of EGCG (282 mg/day) and RES (80 mg/day) in overweight and obese men and women for 12 weeks	↓Bacteroidota in men	No changes were detected in *A. muciniphila*	Improved fat oxidation in men	[172]
	Resveratrol (RSV)	Intake of 1g of RSV twice daily in obese men with MetS for 30 days	↓*Alistipes*, ↓*Collinsiella*, ↓*Christensenella*, ↓*Holdemania*, and ↓*Turicibacter* spp. in Caucasian men	↑*A. muciniphila* only in Caucasian men	Improved glucose homeostasis only in Caucasian men	[161]
	Magnolia berry Schisandra Chinensis (SCF)	Intake of 100 mL of juice twice a day containing 12 mg total phenolics and 3.34 mg total flavonoids in obese woman for 12 weeks	↑*Roseburia*, ↑*Prevotella*, ↑*Bifidobacterium*,* and ↑*Bacteroides** spp.	↑*A. muciniphila*	↓Fat mass, ↓fasting blood glucose, ↓TG, ↓AST, and ↓ALT	[173]
Flavanones	Orange juice (hesperidin)	Intake of orange juice (300 mL d^{-1}) for 60 days in healthy women	↑*Lactobacillus* spp.,* ↑*Ruminococcus* spp.	↑*A. muciniphila*, correlated with improved glycemia and dyslipidemia	↓Glucose intolerance, ↓insulin resistance, ↓TG	[40]
Flavan-3-ols	Mango (quercetin and kaempferol glycosides, gallic acid and gallotannins)	Intake of 300–400 g of mango pulp in patients with IBD for 8 weeks	↑ *L. plantarum*,* ↑*L. lactis*, ↑ *L. reuteri**	No changes were detected in *A. muciniphila*	↓Plasma levels of proinflammatory cytokines, ↑butyrate	[174]
	Pomegranate juice (PJ) (Ellagitannins) and urolithin A (UA)	Intake of PJ (240 mL) for 3 weeks and UA (500 mg) for 48 h in healthy subjects	↓F/B and ↑microbial diversity, ↑Clostridiales, and ↑Ruminococcaceae in UA producers; ↑F/B and ↓in non-UA producers	↑*A. muciniphilia* in high UA producers	6.0-fold ↑plasma UA glucuronide after UA intake compared to PJ	[42]
	Pomegranate extract	Intake of POM (1 g/day) in healthy volunteers for 4 weeks categorized as urolithin A (UA) producers and non-producers	↑Actinobacteria, ↓Firmicutes, ↑*Butyrivibrio* spp., ↑*Enterobacter* spp., ↑*Escherichia* spp., ↑*Lactobacillus* spp.,* ↑*Prevotella* spp., ↓*Collinsella* spp. in UA producers	47-fold ↑*A. muciniphilia* in UA producers	↑Plasma UA glucuronide	[43,52]

* Some species belonging to this genus or family exhibit polyphenol-degrading abilities. F/B, Firmicutes-to-Bacteroidota ratio; IBD, inflammatory bowel disease; TG, triglycerides; AST, aspartate aminotransferase; ALT, alanine aminotransferase.

The effects of various polyphenols, either pure or in combination with other food compounds, on gut microbiota and metabolic phenotypes were studied in humans with cardiovascular diseases. For instance, the consumption of 80 g of oats rich in β-glucans (3.0 g) and polyphenol (56.8 mg) in mildly hypercholesterolemic subjects for 45 days significantly increased the relative abundance of *A. muciniphila* and *Roseburia* spp. [171]. Oat intake significantly reduced total cholesterol and LDL-C. In particular, *A. muciniphila, Roseburia, Bifidobacterium*, and *F. prausnitzii* correlated with oat-induced changes in plasma lipids and short-chain fatty acids (SCFAs). In another study, intake of a combination of EGCG and resveratrol (RES) (282 and 80 mg day^{-1}, respectively) or placebo was evaluated in overweight and obese men and women for 12 weeks. EGCG capsules contained 94% EGCG (141 mg per capsule), and RES capsules contained 20% trans-resveratrol (40 mg per capsule). Analyses of taxon abundance by real-time qPCR assays showed that *A. muciniphila* was not significantly affected by the administration of EGCG+RES in men or women, alongside Firmicutes, Actinobacteria, Gammaproteobacteria, and sulfate-reducing and acetogenic bacteria. In this study, the relative abundance of Bacteroidota was higher in men than women; EGCG+RES decreased its abundance in the same group, which correlated with improved fat oxidation [172]. These findings highlight that the effect of polyphenols relies on the interindividual differences linked not only to sex but also to the baseline of the microbiota composition. However, because qPCR assays target the changes of only specific taxa, a complete analysis of the whole gut microbiota is required to better understand the effects of polyphenol consumption on the abundance and function of other microbes.

In another study, Walker and colleagues [161] conducted a double-blind, placebo-controlled clinical trial and further assessed whether the impact of polyphenols on the gut microbiota was race-dependent (Table 7). The modulatory effects of polyphenols in the gut microbiota, especially the stimulation of relative levels of *A. muciniphila* and its impact on glucose metabolism and inflammation, were studied in obese men with metabolic syndrome [161]. The authors showed that oral supplementation of 1000 mg of resveratrol twice daily for 30 days improved glucose homeostasis in Caucasians but not in non-Caucasian obese men. Only in Caucasians, significant increases in relative levels of *A. muciniphila* were observed, coupled with improved insulin sensitivity and glucose homeostasis. It is worth noting that at baseline, the two groups presented with variable gut microbiota; on the one hand, Caucasian subjects had a higher concentration of *Collinsella*, Clostridiaceae, and *Ruminococcus* spp. than non-Caucasians, whereas *Streptococcus* and *Lactobacillales* spp. were over-represented in non-Caucasians. However, owing to the variability in the plasma levels of dihydroresveratrol, the primary resveratrol metabolite, the racial differences observed in the metabolic changes and relative *Akkermansia* abundance were not sustained [161], indicating that additional mechanisms other than microbial metabolism of phenolics could support the changes in *A. muciniphila* gut colonization and the improvements in glucose homeostasis.

In humans, it is difficult to conduct in-depth studies of the effect of polyphenols on the status of the mucosal colonic epithelium as an indirect outcome substantiating an improved niche for *A. muniniphila* to grow. However, the analysis of inflammatory markers in serum and fecal samples, such as C-reactive protein, IL-1, IL-6, and LBP, might support the amelioration of microbial LPS dislocation and the function of the gut epithelium barrier. However, these parameters were not studied in the above work. Likewise, extended dietary intervention coupled with metabolome analysis might yield broader outcomes with respect to the gut microbiota function associated with interindividual phenotypes beyond the gut microbiota composition.

Supplementation of Magnolia berry *Schisandra Chinensis* (SCF), a fruit popular in East Asian traditional medicine and rich in flavonoids, was studied in obese women for 12 weeks. The women were administered 100 mL of juice containing 12 mg total phenolic compounds and 3.34 mg total flavonoids twice a day (Table 7). At the genus level, the SCF induced an improved relative abundance of *Akkermansia, Roseburia, Prevotella, Bifidobacterium*, and *Bacteroides* spp. compared to placebo groups, as demonstrated by denaturing gradient gel electrophoresis analysis (DGGE) of the gut microbiota. Among the anthropometric and blood parameters, the juice decreased fat mass, fasting blood glucose, TG, AST, and ALT [173]. Fidélix and colleagues demonstrated a relative increase in *Akkermansia* spp., along with *Lactobacillus* spp. and *Ruminococcus* spp., after intervention with a flavanone-rich orange juice (300 mL d^{-1}) for 60 days in healthy female volunteers. In addition, an inverse correlation was detected between these bacteria and glycemia, insulin, HOMA-IR, TG, total cholesterol, and LDL-C [40].

In a recent trial, the effects of mango on the intestinal microbiota and inflammatory markers were evaluated in patients with IBD. Mango is a rich source of flavonoids, such as quercetin, kaempferol glycosides, gallic acid, and galloyl glycosides. Patients received a daily dose of 300–400 g of mango pulp for eight weeks. Changes in the gut microbiota composition were followed by qPCR analysis. Whereas *A. muciniphila* was one of the targeted taxa, it was not reported to be modulated in the participants' gut microbiota at the end of the intervention (Table 7). Mango mainly increased the relative levels of *Lactobacillus*, particularly *L. plantarum*, *Lactobacillus lactis*, and *L. reuteri*. These results are consistent with the capacity of *Lactobacillus* species to metabolize gallotannins. Daily intake of mango was shown to increase butyrate and decrease the plasma levels of proinflammatory cytokines, such as IL-8, growth-regulated oncogene (GRO), and granulocyte-macrophage colony-stimulating factor (GMCSF) [174].

Specific metabotypes might reflect an imprinted gut microbiota connected to the attenuation and progress of metabolic disorders and intestinal disruption upon the intake of polyphenol-rich foods. This feature has been reported with respect to the interaction between gut microbiota, isoflavones, ellagitannins, and ellagic acid [175,176]. For instance, previous research has underlined the role of certain gut bacteria, such as *Clostridium* spp. and two genera belonging to the Coriobacteriaceae family, *Gordonibacter* and *Ellagibacter* spp., in the conversion process of ET and ellagic acid into urolithin metabolites [177]. Metabotypes not only depend on the presence of polyphenol degraders in the gut microbiota but on the health status of the host. For example, the equol nonproducer metabotype (from isoflavone metabolism) and the urolithin metabotype B (formation of urolithin B and usourolithin A from ellagitannins metabolism) are more abundant in overweight and obese individuals [178–180]. Likewise, intraindividual variations in metabolite production in terms of stability over time are relevant in deciphering health benefits in the host [181]. Indeed, urolithin metabotypes were shown to be stable in humans. However, those individuals with the urolithin-nonproducing metabotype can acquire the benefits observed in urolithin producers after regular intake of ellagitannins and urolithin A [42], which can be transferred from breastfeeding mothers to their babies [182]. Notably, the latter observations indicate that dietary strategies rich in polyphenols may induce metabolic adaptions in the gut microbiota and re-establish the gut polyphenol-degrading enzymatic functions. In this context, Singh et al. (2022) showed that direct supplementation of urolithin A (500 mg) in healthy subjects with non-urolithin-producing features could provide the biological activities observed in urolithin A producers. Non-producers lacked the richness and diversity needed to transform the polyphenolic dietary precursors into urolithin A and had a lower Firmicutes-to-Bacteroidota ratio than high producers [42]. Furthermore, an increased abundance of *A. muciniphilia* was reported in the high urolithin A producers compared to the low and non-producers. This outcome was previously reported with a 47-fold higher proportion of this bacterium in urolithin A producers after a 4-week dietary intervention with pomegranate in healthy volunteers [52]. Analysis of *A. muciniphila* abundance was not tackled in subjects after the direct intake of urolithin A, given the short intervention with administering urolithin A mainly, with a focus on analyzing urolithin circulating levels after 24 h [42]. Here, urolithin A supplementation increased by sixfold relative to urolithin A glucuronide when compared with subjects receiving the pomegranate juice.

It is unclear whether the stimulation of this bacterium might stem from the intermediary byproducts generated by the initial hydrolysis of ellagitannins into ellagic acid and further urolithin metabolites. Trophic networking with other urolithin-producer microbes and the triggered homeostatic gut microenvironment might prevail more than the levels of urolithin A itself in the prebiotic effects of ellagitannins on *A. muciniphila.* Indeed, both ellagic acid and urolithin A could not directly prompt the growth of *A. muciniphila* in vitro [43]. The stimulation of this bacterium by the direct administration of urolithin A is yet to be confirmed in humans, although it has been demonstrated in animal models. Al Khalaf et al. (2021) [114] showed that oral supplementation of urolithin A and urolithin B (2.5 mg/kg each) induced the growth of *Akkermansia* spp. in rats fed a normal diet, with a higher proportion in urolithin-A-fed animals. In addition, only urolithin A significantly impacted the abundance of Firmicutes and Bacteroidota phyla. These latter findings might be associated with the antimicrobial potential of urolithin A as reported elsewhere [29]. It is also possible that inhibiting microbial competitors by polyphenol-derived metabolites might give *A. mucinihipla* an advantage in proliferating in the host gut.

The results of the above studies, including those performed in animal models, support the use of polyphenol-rich foods as therapeutic and dietary strategies to cope with gut inflammation, as well as liver and metabolic disturbances, boosting *A. muciniphila* as a critical intermediary.

6. Indirect Mechanism Favoring *A. muciniphila*: Polyphenol Signaling of the Host Gut Epithelium and Influencing the Gut Microbial Environment

The in vivo promoting effect of polyphenols on *A. muciniphila* embraces different routes that can directly or indirectly can enhance the colonization of this bacterium under steady and unsteady health conditions. It has been demonstrated that an HFD considerably perturbs the microbial composition, diversity, and functions of the gut microbiota and prompts mucosal inflammation and disturbed intestinal permeability. This is often associated with lower levels of *A. muciniphila* and a higher ratio of the major phyla Firmicutes and Bacteroidota in obese compared with lean subjects [183]. In this section, we tackle the impact of polyphenols on the restoration of the mucin-rich mucus layer and intestinal barrier function, as well as the role of polyphenols, such as antimicrobials, in reducing proinflammatory opportunistic microbes. These two factors provide *A. muciniphila* with a better gut microenvironment to proliferate and orchestrate regulatory anti-inflammatory and metabolic responses in the host.

6.1. Goblet Cells and Mucin Differentiation

Because *A. muciniphila* directly benefits from the increase in Muc2 secretion and improved mucus barrier [60,184], mucin modulation by polyphenols represents an indirect mechanism that prompts this bacterium. Polyphenols and microbial phenolic metabolites can actively signal the host epithelium, regulating mucin secretion and structure. A recent study mechanistically demonstrated that ellagitannins, ellagic acids, and urolithin metabolite can remodel the O-glycans profile by regulating the activities of N-acetyl-α-galactosaminyltransferases (ppGalNAc-T), preventing colorectal cancer cell migration and invasion [185]. Altered expression of mucin glycosylation has been documented in a variety of inflammatory, malignant, and metabolic diseases [23]. Crohn's disease, ulcerative colitis, polyps, and colon cancer are associated with qualitative and quantitative changes in secreted sialomucins and sulfomucins and a consequently disrupted gut microbiota. Supplementation with polyphenols enhanced the mucosal barrier function by regulating the proportion of sialomucin mucin in rodents with 1,2-dimethylhydrazine-induced colon carcinogenesis [45]. Significantly increased sulfated mucin has been correlated with enhanced protection against enteritis [186]. In line with this, dietary polyphenols might be promising targets for therapeutic strategies by interfering with the function of specific glycosylation processes and glycosylated receptors.

Polyphenols have also been shown to modulate mucin biosynthesis without the action of the resident gut microbiota in a germ-free mouse model. Forgie et al. (2022) [27] demonstrated that diets rich pea seed coat proanthocyanins and hydrolyzable red-osier dogwood (ROD) tannins increased the luminal and fecal mucin GalNAc content in both germ-free and conventional Swiss–Webster mice after fourteen days of intervention. In particular, in mice harboring a conventional gut microbiota, the ROD extract increased *A. muciniphila* and *Bacteroides thetaiotaomicron*, along with an unclassified member of the Muribaculaceae family, and reduced *Romboutsia*, Ruminococcaceae, and Oscillospiraceae members.

The reduced relative abundance of *A. muciniphila* in obesity might correlate with a significant increase in the ratio of sialo/sulfomucins. Intake of an HFD suppresses GC differentiation and increases ER stress; thus, non-glycosylated Muc2 precursors accumulate and defect the mucin glycosylation process [24]. These alterations could influence the bacterial composition by decreasing the *A. muciniphila* abundance and gut microbial diversity. In this sense, polyphenols have been demonstrated to attenuate the disrupted glycan-rich ecological niche by increasing the mucus-secreting GC and reducing iNOS expression on the surface and in crypts of the colonic epithelium [26]. The mucosa restored by dietary polyphenols counteracted gut inflammation and recovered the relative proportion of *A. muciniphila* in HFD-fed mice. Likewise, Wang et al. (2020) [51] demonstrated that 16-week dietary administration of resveratrol upregulated TJP expression and mucins in mouse colon. Polyphenols might function as signaling molecules, contributing to GC maturation, the development of the intestinal epithelium, and the foundation of a healthy gut microbiota, especially *A. muciniphila*.

Recently, Lu et al. (2021) [187] showed that early-life dietary supplementation with grape polyphenols for 2 weeks increased the amount of mucus and Muc2 and led to a significant bloom of *A. muciniphila* population in post-weaning mice under specific pathogen-free (SPF) conditions. Notably, GC number and expression of the Kruppel-like factor 4, a marker of GC differentiation, and Cdx2, a Muc2-related promoter, was increased by dietary polyphenols after two weeks of intervention. This period was long enough to both promote an enrich the mucous environment for *A. muciniphila* and increase the population of *Lactobacillus* spp. The latter was shown to be an essential player in early-life colonization of the intestine, contributing to a boost in lactate levels and, in this case, the generation of phenolic metabolites due to its polyphenol-degrading enzymatic abilities.

6.2. Aryl Hydrocarbon Receptor (AhR) and Gut Inflammation

The aryl hydrocarbon receptor (*AhR*) is a ligand-activated transcription factor that regulates physiological processes in health and disease. Emerging studies point to the relevance of this receptor in binding to food and microbial agonists, resulting in beneficial outcomes in the host. The activation of this multifunctional sensor has been associated with protective roles in modulating energy and lipid metabolism, cell differentiation, and the immune system [188–192]. *AhR* agonists have been found to be decreased in sera of humans and animals suffering from alcoholic liver inflammation [193,194], multiple sclerosis [195], and metabolic syndromes, including obesity and diabetes [196]. This altered *AhR* regulation considerably influences the host gut microbiota, particularly *A. muciniphila* abundance. For instance, *A. muciniphila* and the *AhR* are depleted in the liver and colon of mice with saccharin/sucralose-induced NAFLD [197]. Reduced *AhR* agonists in mice sera were correlated with a disrupted gut epithelium and an increase in inflammatory markers [197].

Phenolic molecules have been reported as the main source of *AhR* modulators found in the human diet [48]. Isoflavones, quercetin, urolithin A, luteolin, and baicalein have been shown to act as the *AhR* agonists or antagonists [198–201]. A recent study evaluated the roles of polyphenol–microbe interplay in *AhR*-mediated signaling using a simulator of the human intestinal microbial ecosystem (SHIME) coupled with HepG2 and Caco-2 cell assays [199]. Microbiota-derived metabolites of tryptophan and flavonoids were assessed. Koper et al. (2019) first demonstrated a structure-dependent *AhR* activation by flavonoids in HepG2 and Caco-2 cells. Interestingly, a planar structure and the number of hydroxyl groups of polyphenols were found to be essential features modulating *AhR* activation [199]. In this study, luteolin and baicalein were identified as compounds activating the most *AhR* signaling. In addition, the luteolin-fermented metabolites derived from the ascending (AC), transverse (TC), and descending (DC) colon of the SHIME system were shown to drive the *AhR* activation differently. Notably, AC metabolites led to *AhR* activation, whereas a greater response was induced by the TC- and DC-derived metabolites, suggesting that in addition to dietary substrates, the gut microbiota is an essential supplier of endogenous *AhR* agonists, such as tryptophan derivates.

Tryptophan catabolism by the resident gut microbiota has also been shown to be influenced by dietary polyphenols, yielding physiologically relevant *AhR* ligands in the host. For instance, Marques and colleagues [201] showed that the administration of blackberry anthocyanin-rich extracts (25 mg kg^{-1} BW per day) changes the gut microbiota composition and tryptophan metabolism, increasing the production of the neuroprotective metabolite kynurenic acid. Indeed, adding polyphenols, together with tryptophan derivatives, significantly boosted the *AhR* response compared to the tryptophan derivatives alone in both rat DR-H4IIE and human DR-HepG2 hepatoma cells. Isoflavones (e.g., daidzein and genistein) and resveratrol the *AhR* response in DR-H4IIE cells increased by up to fourfold compared to tryptophan-derived ligands alone, such as TCDD (2,3,7,8-tetrachlorodibenzo-p-dioxin) and FICZ (6-formylindolo [3,2-b]carbazole). Furthermore, quercetin and both flavones baicalin and chrysin were shown to act as *AhR* agonists in human DR-HepG2 cells, regardless of the presence of tryptophan derivatives in the medium.

The role of the polyphenol–*AhR* interactions has been further demonstrated in *AhR−/−* mouse models using urolithin A, a metabolite of microbial degradation of ellagitannins. This metabolite was implicated in protection against gut inflammation in experimental colitis via the *AhR–Nrf2* pathway [200]. The activation of *AhR* by polyphenols and microbial metabolites results in the downregulation of inflammatory cytokines and the upregulation of *IL-22* and TJPs [47,199]. In this sense, the *AhR/IL-22* axis has been shown to be crucial in maintaining mucosal barrier integrity and in protecting the gastrointestinal epithelium against pathogen colonization [49,202–205]. Mechanistic studies have also demonstrated that reduced secretion of *IL-22* was correlated with impaired glycocalyx- O-glycan in mice suffering from obesity [206] and that such glycosylation pattern could be restored by boosting the *IL-22* levels [24,207]. The immunomodulatory outcomes linked to the *AhR/IL-22* axis contribute to changes in the gut microbiome composition, stimulating beneficial species such as *A. muciniphila* and reducing the susceptibility to metabolic and inflammatory gastrointestinal disorders.

Bidirectional *AhR*-related effects might explain the *Akkermansia*-driven benefits in the host upon consumption of dietary polyphenols. On the one hand, polyphenols and derived microbial metabolites might activate *AhR*-dependent pathways, which restore the ecological niche of *A. muciniphila* (i.e., through IL-22, TJP regulation, and promotion of GC differentiation) [188,189,203,208]. On the other hand, phenolic compounds might potentiate the gut traits of *Akkermansia*, allowing this bacterium to orchestrate immunomodulatory markers and contribute to the maintenance of intestinal homeostasis. *A. muciniphila* and its membrane protein, Amuc_1100, can significantly increase gut indole loads and, consequently, *AhR* activation [71]. It is worth mentioning that Amuc_1100 is a well-known TLR2-inducer and substantiates the role of *A. muciniphila* in attenuating gut inflammation and metabolic disorders [14,17]. The mechanisms of action by polyphenols in sustaining an enhanced gut microecological niche promoting *A. muciniphila* colonization are illustrated in Figure 4.

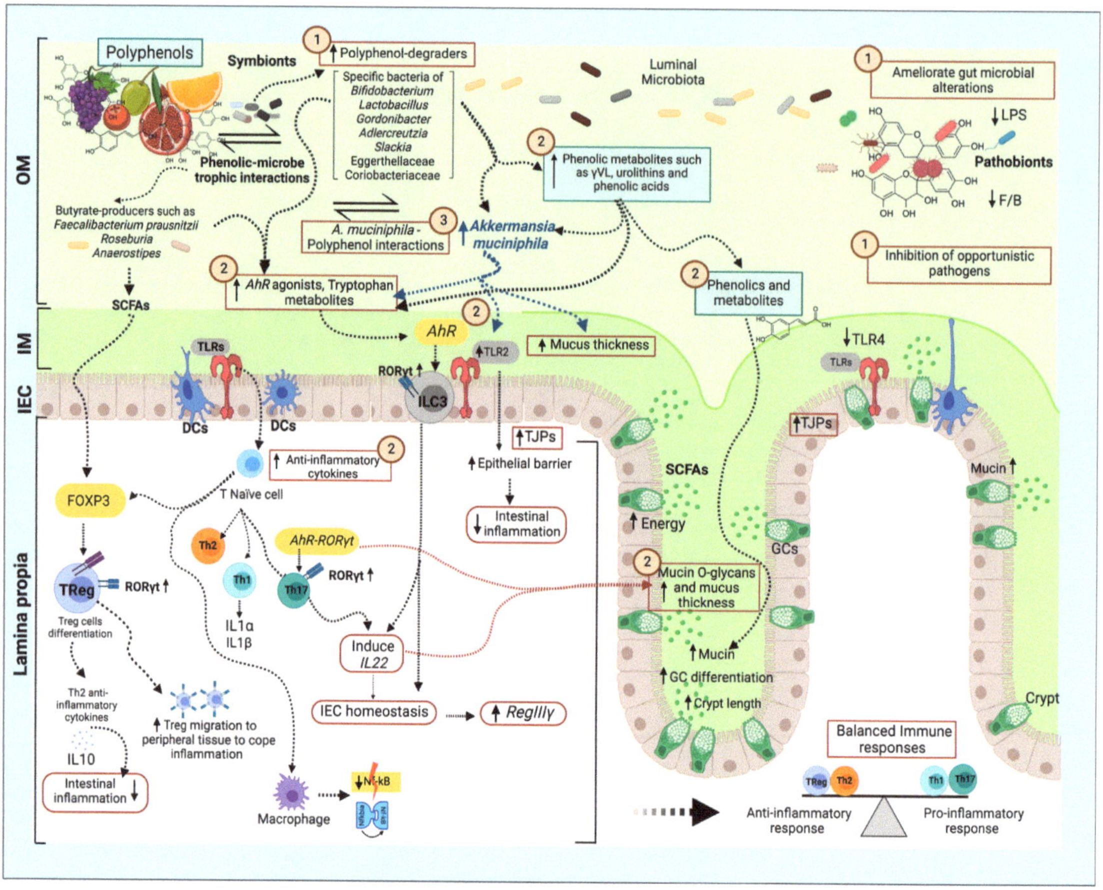

Figure 4. Pleiotropic polyphenol–microbiota interactions promoting an increase in *Akkermansia muciniphila* in the host gut. Metabolic diseases, such as obesity and type 2 diabetes, are associated with an altered intestinal ecological niche primarily associated with the onset of gut microbiota dysfunctions and a decrease in relative levels of *A. muciniphila*. The consumption of polyphenols, either pure or combined with other dietary susbstrates (e.g., fibers) shapes the gut microbiota composition, selectively promoting the relative abundance of *A. muciniphila* in association with improved metabolic outcomes in the host. Three main routes underline the stimulatory effects of polyphenols on *A. muciniphila* (represented by enumerate circles): (**1**) Interactions between polyphenols and the gut microbiota result in the inhibition of opportunistic pathogens, in addition to favoring symbiotics and the polyphenol-degrader microbes. Polyphenols reduce the level of lipopolysaccharides (LPS) and suppress the Toll-like receptor (TLR)-4, attenuating metabolic endotoxemia. Polyphenol-microbe trophic interactions yield phenolic metabolites, such as phenyl-γ-valerolactones (γVL), urolithins, and phenolic acids. De novo generated phenolics and the microbial metabolites trigger immunological pathways, restoring the gut epithelial function and homeostasis and positively affecting the abundance of *A. muciniphila*. (**2**) Notably, phenolics and tryptophan metabolites signal the Aryl hydrocarbon receptor (*AhR*). This activation controls cell recruitment, Th2 and Th17/IL-22 responses, and the remodeling of mucin O-glycans and intestinal epithelial cells (IEC) in an *IL-22* dependent manner. *AhR* activation in lymphocytes also influences cells recruitment and *IL-22* expression and improves intestinal inflammation. Then, the polyphenol-induced *AhR/IL-22* pathway provides *A. muciniphila* with a restored microecological niche. Polyphenols and *A. muciniphila* jointly act as activators of *AhR*, playing regulatory roles in intestinal cell function. *A. muciniphila* colonization contributes to a

reduction in proinflammatory markers via TLR2 induction, helping to cope with low-grade local inflammation and promote epithelial tight-junction proteins (TJPs). In addition, SCFAs, such as butyrate, can drive the activation of the transcriptional regulator Forkhead box P3 (Foxp3), which is also linked to the upregulation of the anti-inflammatory IL-10. (**3**) *A. muciniphila* can directly be affected by phenolics and derivatives, influencing the expression of its molecular and metabolic attributes; it can hydrolyze phenolics and liberate smaller molecules that are substrates for other microbial analogs to subsequently generate bioactive phenolic metabolites. Overall, polyphenols favor anti-inframammary reactions, thus re-establishing the equilibrium between Treg/Th2- and Th1/Th17-associated immune response. The IEC can sense and absorb phenolic metabolites with health effects in the gut barrier's permeability and distal organs and tissues. A hypothesis we cannot rule out is that some phenolics cause intestinal damage, resulting in increased mucus production and inducing *A. muciniphila* and improved barrier function as a kind of overshoot. *A. muciniphila* creates a positive feedback loop, thus renewing the mucus layer and maintaining its thickness. F/B: Firmicutes-to-Bacteroidota ratio; RAR: related orphan receptor gamma Rorγt; ILC3: innate lymphoid cell type; DC; dendritic cells; IM—inner mucus layer; OM: outer mucus layer. The graph was created with BioRender.

6.3. Inhibitory Action on Gut Microbial Competitors

Unabsorbed dietary polyphenols and their metabolites can behave as activators or inhibitors of bacterial growth, depending on their chemical structure and concentration [209]. Phenolic metabolites and parent compounds selectively inhibit pathogens and stimulate the growth of commensal bacteria [44,143]. It has also been indicated that the modulatory effect of polyphenols on gut bacteria relies on the alteration of the cell physiology [210] rather than the reduction in the bacterial quantity per se. The expression of stress response and protein chaperone genes is upregulated in gut bacteria exposed to polyphenols, indicating the adverse, species-dependent effects of polyphenols on the gut microbiota [211]. The inhibitory mechanisms of action of phenolics have been compared with those exhibited by antibiotics [99]. For instance, based on the inhibitory effects and intracellular metabolites induced by apigenin against pathogens, this phenolic was clustered together with rifampicin and norfloxacin [212]. These antibiotics target RNA polymerase, gyrase, and topoisomerase IV, and apigenin could affect nucleic acid processing enzymes. Polyphenols can affect cell membrane/wall synthesis, for example, by altering the D-alanine: D-alanine ligase and the type II fatty acid synthetic pathway [213,214]; therefore, most of the Gram-positive gut bacteria have been shown to be more susceptible to polyphenol–antimicrobial actions; this is the case of the Firmicutes phylum, as expounded above in the section discussing the outcomes of animal studies. *A. muciniphila* has been stimulated by administering vancomycin and other wide-spectrum antibiotics [215,216]. In analogy to the antimicrobial actions of phenolic phenolics, reducing microbial competitors and associated inflammation allows *Akkermansia* spp. to flourish in the gut. In addition, in vitro studies of *A. muciniphila* growth in media enriched with polyphenols validate the resistance and adaptation of this bacteria to be metabolically active and resilient upon polyphenol intake. The antimicrobial effects of dietary polyphenols on food and gut microbial pathobionts associated with metabolic and inflammatory diseases were recently reviewed by Makarewicz et al. (2021) [28].

7. Polyphenol Effects on *A. muciniphila* Fitness: Hints of Metabolism and Adaptation to Polyphenol-Rich Foods

The physiological properties of *A. muciniphila* and the potential to adapt to polyphenols might be closely related to its genetic composition [57]. It harbors genes not only encoding mucin-degrading enzymes, accounting for 14% of the genome, but also genes linked to capsular and slime polysaccharide production. These features might support the tolerance of *A. muciniphila* to the cell wall/membrane disturbances provoked by polyphenols in gut bacteria. Proteomics and transcriptomics revealed that *A. muciniphila* molecularly adapts when exposed to disadvantageous conditions, such as a medium rich in bile acids, or when its favorite nutritional substrate, mucin, is depleted [58,217]. Notably, markers linked to acid and oxidative stress responses, nucleotide excision repair, inorganic ion transport and motility, ABC and RND (Resistance-Nodulation-cell division) transporters, and polysaccharide biosynthesis systems are upregulated in *A. muciniphila*. Additionally, changes in the metabolic profile of *A. muciniphila* are reliant on the substrate input, which determines its versatility and resilience to adapt in a less favorable medium [58], such as that induced by polyphenols. Intriguingly, mucin-depleted conditions have been shown to trigger attributes in *A. muciniphila,* supporting antiobesity effects in mice [218]. Polyphenols have been shown to modulate similar molecular functions of adaptation in bacteria with and without polyphenol-degrading potential, such as *Salmonella enterica, Escherichia coli, L. plantarum, Enterococcus caccae, Ruminococcus gauvreauii,* and *B. catenulatum* [143,211,219–224]. Under polyphenols-induced pressure, *A. muciniphila* might acclimatize to phenolic antimicrobial actions by altering its metabolic pathways and membrane phenotypes. The expansion of the relative *A. muciniphila* proportion in the gut microbiota upon polyphenols intake might be encouraged by polyphenol degradation by other microbes rather than the direct polyphenol utilization by this bacterium. As discussed above, many studies have shown an increase in *A. muciniphila,* along with other well-known polyphenol-degrader and polysaccharide-degrader microbes. This is the case of specific species belonging to *Lactobacillus, Bifidobacterium, Slackia, Gordonibacter, Eggerthella, Adlercreutzia,* and *Bacteroides* genera, which have been demonstrated to possess enzymes involved in the breakdown and utilization of different phenolic classes. Likewise, given their polysaccharide-degrading potential, species such as *F. prausnitzii* and members of *Roseburia, Prevotella,* and Clostridiales are concomitantly prompted with *A. muciniphila* by the intake of fiber-bound polyphenols encountered in several polyphenol-rich fruits and vegetables. Given the inferred trophic interactions that occur among polyphenol and polysaccharide degraders in vivo, both polysaccharide fermentation and subsequent polyphenol-degrading activity by other microbes may support *A. muciniphila* in the gut. A comparable case of crossfeeding upon polyphenol administration was reported between the polysaccharide degrader *Bacteroides thetaiotaomicron* and the polyphenol degrader *E. ramulus* [225]. Here, the release of glucose and maltose upon the fiber fermentation by *B. thetaiotaomicrom* drove the degradation of the flavonol quercetin by *E. ramulus*. These polyphenol-degrading and fiber-degrading activities reduced significantly when both species were grown separately. Similarly, in another study, *E. caccae* was shown to be effectively inhibited by apigenin when cultured alone, but this genus was enhanced when tested within a gut microbial community [143]. It still remains to be determined whether *A. muciniphila* utilizes the ensuing phenolic metabolites and the SCFAs formed from polyphenol-rich extract fermentation. Coculturing *A. muciniphila* with polyphenol-degrader microbes would provide insights into the notion that this bacterium is powered by intimate crossfeeding with other microbes upon polyphenol-rich food intake.

The direct trophic utilization of phenolics by *A. muciniphila* has not yet been investigated, and the identification of novel enzymatic activity responsible for polyphenol degradation remains to be mechanistically studied in vitro. Conflicting findings have been reported in a few studies dealing with the ability of *A. muciniphila* to utilize polyphenols in polyphenol-enriched media (i.e., EGCG, gallic acid, ellagitannins, and ellagic acid) [43,54]. On the one hand, *A. muciniphila* could not succeed in a medium enriched with EGCG as a unique nutritional source. In addition, aside from gallic acid, the specific phenolic end metabolites, such as phenyl-γ-valerolactones, phenylvaleric, and propionic acid derivatives, were not identified when it was grown optimally in an EGCG-enriched medium containing mucin [54]. On the other hand, *A. muciniphila* growth was shown to be hampered by pomegranate ellagitannins and ellagic acid. However, the levels of ellagic acid were shown to increase when this bacterium was grown with pomegranate ellagitannins [43]. The fact that both gallic and ellagic acids derived from the initial hydrolysis of galloylated and gallated parent flavan-3-ols were found in the growing media indicates that *A. muciniphila* primes the hydrolysis of EGCG and ellagitannins but does not participate in the subsequent phenolic degradation. However, this initial hydrolytic step might contribute to empowering the polyphenol-degrading action of other microbes, in turn furnishing the gut with bioactive metabolites and contributing to the maintenance of intestinal homeostasis.

Because the main nutritional source of *A. muciniphila* is mucins, its growth is not strictly confined to metabolizing dietary polyphenol-rich sources; however, the latter modulates the abundance and activity of other microbes and promotes mucin-secreting GC, giving *A. muciniphila* a colonizing advantage compared to its counterparts. The fermentation of different substrates, such as galactose, fucose, glucose, Glc-NAc, and Gal-NAc, by *A. muciniphila* has been shown to be enhanced when this bacterium is grown in the presence of mucin [58]. In analogy to the above observation, we suggest that polyphenols elicit distinctive metabolic and molecular phenotypes in *A. muciniphila*, in addition to increasing the mucin supply in the host. By degrading the newly polyphenol-induced mucins, *A. muciniphila* creates a positive feedback loop, renewing the mucus layer and maintaining its thickness. Considering all these studies, *A. muciniphila* proliferation stems from the synergic effects between crossfeeding networking with polyphenol degraders and the local impacts of phenolic derivatives on the mucin-rich mucus layer in vivo.

8. Conclusions and Future Directions

In the present work, we revisited preclinical and clinical trials demonstrating the prebiotic-like effect of polyphenolic extracts and polyphenol-rich foods on *A. muciniphila*. Different phenolic classes, including phenolic acids, anthocyanins, flavan-3-ols, flavonols, flavones, and stilbenes, distinctly stimulate *A. muciniphila* in the gut. For example, high intra- and interindividual variability in the gut microbiota composition impacts the fate of polyphenols and health benefits in the host and, consequently, the relative abundance of *A. muciniphila*. We highlighted three main routes underlying the polyphenol-induced *A. muciniphila* increase in the gut: (1) modulation of gut microbiota ecology while inhibiting microbial competitors and promoting the polyphenol-degrader consortium, (2) attenuation of gut inflammation and restoration of mucin-rich barrier function, and (3) direct trophic utilization and crossfeeding networking between *A. muciniphila* and other gut microbes. A conglomerate of bacterial taxa with polyphenol-degrading abilities often stands out concomitantly with an increase in *A. muciniphila* resulting from the consumption of polyphenol-rich foods in animals and humans. However, we cannot rule out that some phenolics signal the intestine, which results in increased mucus production, inducing the growth of *A. muciniphila* and improving barrier function, in addition to other metabolic benefits.

Few gut bacteria are known to degrade polyphenols and significantly correlate with *A. muciniphila* abundance. This is the case of ellagitannin degraders, yielding smaller molecules of urolithin A. Inconclusive results have been reported with respect to how ellagitannins and ellagic acid directly promote *A. muciniphila* in vitro, as reduced growth was observed when these compounds were added to the growing media. In a real-life scenario, it is most likely that the effects would be due to the trophic utilization of other microbes of ellagitannins and the immunomodulation that the urolithin A and parent compounds trigger in the gut.

It is yet to be determined whether *A. muciniphila* can metabolize different phenolic classes as single or combined substrates in monoculture and a community setting. Specific questions remain to be answered, such as which intermediary enzymes and degradation pathways are involved in polyphenol metabolism by *A. muciniphila*, as well as the resulting metabolic products. Moreover, there is a need to move from relative abundancies towards absolute abundancies and correct for the impact of transit time on the relative abundance of *A. muciniphila* and other intestinal bacteria [226,227]. A potential crossfeeding phenomenon behind polyphenol degradation might occur, as the metabolites from one bacterium could empower substrate utilization by another. A combination of omics tools, including transcriptomics, lipidomics, proteomics, and metabolomics, would provide a whole picture of the prebiotic effect of polyphenols on *A. muciniphila* that is more relevant than the structure of the gut microbiota itself. Polyphenols continue to receive a considerable amount of attention, owing to their selectivity to sponsor *A. muciniphila*, representing a cost-effective therapeutic and preventive approach with no demonstrated adverse side effects to cope with metabolic diseases.

Emerging coencapsulation techniques for the development of new synbiotics using polyphenols and probiotics are promising strategies for further application in nutritional therapies to counteract gut microbiota disturbances and metabolic syndromes [228–230]. Research has begun on the coencapsulation of *A. muciniphila* with polyphenols. Cells of *A. muciniphila* were successfully coencapsulated in succinate-grafted alginate doped with EGCG by spray drying [230]. Moreover, a recently marketed European product contains tablets of freeze-dried pasteurized *A. muciniphila* and other components, including ECGC (www.theakkermansiacompany.com (accessed on 31 October 2022)). In this sense, the inclusion of recognized polyphenol-degrading probiotics in the encapsulation preparations represents a promising avenue for the delivery of the goods of polyphenol–*Akkermansia* synergy to individuals with less favorable metabotypes.

Author Contributions: Conceptualization, M.C.R.-D. and W.M.d.V.; investigation and resources, M.C.R.-D.; writing—original draft preparation, M.C.R.-D.; writing—review and editing, M.C.R.-D. and W.M.d.V.; supervision, W.M.d.V.; project administration, W.M.d.V.; funding acquisition, W.M.d.V. All authors have read and agreed to the published version of the manuscript.

Funding: This research was funded by SIAM Gravitation (024.002.002) of the Netherlands Organization for Scientific Research (NWO).

Institutional Review Board Statement: Not applicable.

Informed Consent Statement: Not applicable.

Conflicts of Interest: W.M.d.V. is founder of and scientific advisor to The Akkermansia Company.

References

1. Sender, R.; Fuchs, S.; Milo, R. Revised estimates for the number of human and bacteria cells in the body. *PLoS Biol.* **2016**, *14*, e1002533. [CrossRef] [PubMed]
2. Rajilic-Stojanovic, M.; de Vos, W.M. The first 1000 cultured species of the human gastrointestinal microbiota. *FEMS Microbiol. Rev.* **2014**, *38*, 996–1047. [CrossRef] [PubMed]
3. Lloyd-Price, J.; Arze, C.; Ananthakrishnan, A.N.; Schirmer, M.; Avila-Pacheco, J.; Poon, T.W.; Andrews, E.; Ajami, N.J.; Bonham, K.S.; Brislawn, C.J.; et al. Multi-omics of the gut microbial ecosystem in inflammatory bowel diseases. *Nature* **2019**, *569*, 655–662. [CrossRef] [PubMed]
4. Shin, A.; Preidis, G.A.; Shulman, R.; Kashyap, P.C. The gut microbiome in adult and pediatric functional gastrointestinal disorders. *Clin. Gastroenterol. Hepatol.* **2019**, *17*, 256–274. [CrossRef] [PubMed]
5. Sanchez-Rodriguez, E.; Egea-Zorrilla, A.; Plaza-Diaz, J.; Aragon-Vela, J.; Munoz-Quezada, S.; Tercedor-Sanchez, L.; Abadia-Molina, F. The gut microbiota and its implication in the development of atherosclerosis and related cardiovascular diseases. *Nutrients* **2020**, *12*, 605. [CrossRef]
6. De Vos, W.M.; Tilg, H.; Van Hul, M.; Cani, P.D. Gut microbiome and health: Mechanistic insights. *Gut* **2022**, *71*, 1020–1032. [CrossRef]
7. Tilg, H.; Cani, P.D.; Mayer, E.A. Gut microbiome and liver diseases. *Gut* **2016**, *65*, 2035–2044. [CrossRef]
8. Powell, N.; Walker, M.M.; Talley, N.J. The mucosal immune system: Master regulator of bidirectional gut-brain communications. *Nat. Rev. Gastroenterol. Hepatol.* **2017**, *14*, 143–159. [CrossRef]
9. Quigley, E.M.M. The gut-brain axis and the microbiome: Clues to pathophysiology and opportunities for novel management strategies in irritable bowel syndrome (IBS). *J. Clin. Med.* **2018**, *7*, 6. [CrossRef]
10. Cani, P.D.; Depommier, C.; Derrien, M.; Everard, A.; de Vos, W.M. *Akkermansia muciniphila*: Paradigm for next-generation beneficial microorganisms. *Nat. Rev. Gastroenterol. Hepatol.* **2022**, *19*, 625–637. [CrossRef]
11. Depommier, C.; Everard, A.; Druart, C.; Plovier, H.; Van Hul, M.; Vieira-Silva, S.; Falony, G.; Raes, J.; Maiter, D.; Delzenne, N.M.; et al. Supplementation with *Akkermansia muciniphila* in overweight and obese human volunteers: A proof-of-concept exploratory study. *Nat. Med.* **2019**, *25*, 1096–1103. [CrossRef]
12. Brahe, L.K.; Le Chatelier, E.; Prifti, E.; Pons, N.; Kennedy, S.; Hansen, T.; Pedersen, O.; Astrup, A.; Ehrlich, S.D.; Larsen, L.H. Specific gut microbiota features and metabolic markers in postmenopausal women with obesity. *Nutr. Diabetes* **2015**, *5*, e159. [CrossRef] [PubMed]
13. Yassour, M.; Lim, M.Y.; Yun, H.S.; Tickle, T.L.; Sung, J.; Song, Y.M.; Lee, K.; Franzosa, E.A.; Morgan, X.C.; Gevers, D.; et al. Sub-clinical detection of gut microbial biomarkers of obesity and type 2 diabetes. *Genome Med.* **2016**, *8*, 17. [CrossRef] [PubMed]

14. Plovier, H.; Everard, A.; Druart, C.; Depommier, C.; Hul, M.; Geurts, L.; Chilloux, J.; Ottman, N.; Duparc, T.; Lichtenstein, L.; et al. A purified membrane protein from *Akkermansia muciniphila* or the pasteurized bacterium improves metabolism in obese and diabetic mice. *Nat. Med.* **2016**, *23*, 107–113. [CrossRef]
15. Everard, A.; Plovier, H.; Rastelli, M.; Van Hul, M.; de Wouters d'Oplinter, A.; Geurts, L.; Druart, C.; Robine, S.; Delzenne, N.M.; Muccioli, G.G.; et al. Intestinal epithelial N-acylphosphatidylethanolamine phospholipase D links dietary fat to metabolic adaptations in obesity and steatosis. *Nat. Commun.* **2019**, *10*, 457. [CrossRef] [PubMed]
16. Zhao, S.; Liu, W.; Wang, J.; Shi, J.; Sun, Y.; Wang, W.; Ning, G.; Liu, R.; Hong, J. *Akkermansia muciniphila* improves metabolic profiles by reducing inflammation in chow diet-fed mice. *J. Mol. Endocrinol.* **2017**, *58*, 1–14. [CrossRef] [PubMed]
17. Ottman, N.; Reunanen, J.; Meijerink, M.; Pietilä, T.E.; Kainulainen, V.; Klievink, J.; Huuskonen, L.; Aalvink, S.; Skurnik, M.; Boeren, S.; et al. Pili-like proteins of *Akkermansia muciniphila* modulate host immune responses and gut barrier function. *PLoS ONE* **2017**, *12*, e0173004. [CrossRef] [PubMed]
18. De Moraes, A.C.; Fernandes, G.R.; da Silva, I.T.; Almeida-Pititto, B.; Gomes, E.P.; Pereira, A.D.; Ferreira, S.R. Enterotype may drive the dietary-associated cardiometabolic risk factors. *Front. Cell Infect. Microbiol.* **2017**, *7*, 47. [CrossRef]
19. Costea, P.I.; Hildebrand, F.; Arumugam, M.; Backhed, F.; Blaser, M.J.; Bushman, F.D.; de Vos, W.M.; Ehrlich, S.D.; Fraser, C.M.; Hattori, M.; et al. Enterotypes in the landscape of gut microbial community composition. *Nat. Microbiol.* **2018**, *3*, 8–16. [CrossRef]
20. Arumugam, M.; Raes, J.; Pelletier, E.; Le Paslier, D.; Yamada, T.; Mende, D.R.; Fernandes, G.R.; Tap, J.; Bruls, T.; Batto, J.M.; et al. Enterotypes of the human gut microbiome. *Nature* **2011**, *473*, 174–180. [CrossRef]
21. Cotillard, A.; Kennedy, S.P.; Kong, L.C.; Prifti, E.; Pons, N.; Le Chatelier, E.; Almeida, M.; Quinquis, B.; Levenez, F.; Galleron, N.; et al. Dietary intervention impact on gut microbial gene richness. *Nature* **2013**, *500*, 585–588. [CrossRef] [PubMed]
22. Mastrodonato, M.; Calamita, G.; Mentino, D.; Scillitani, G. High-fat diet alters the glycosylation patterns of duodenal mucins in a murine model. *J. Histochem. Cytochem.* **2020**, *68*, 279–294. [CrossRef] [PubMed]
23. Mastrodonato, M.; Mentino, D.; Portincasa, P.; Calamita, G.; Liquori, G.E.; Ferri, D. High-fat diet alters the oligosaccharide chains of colon mucins in mice. *Histochem. Cell Biol.* **2014**, *142*, 449–459. [CrossRef]
24. Gulhane, M.; Murray, L.; Lourie, R.; Tong, H.; Sheng, Y.H.; Wang, R.; Kang, A.; Schreiber, V.; Wong, K.Y.; Magor, G.; et al. High fat diets Iiduce colonic epithelial cell stress and inflammation that is reversed by IL-22. *Sci. Rep.* **2016**, *6*, 28990. [CrossRef] [PubMed]
25. Bergstrom, K.; Shan, X.; Casero, D.; Batushansky, A.; Lagishetty, V.; Jacobs, J.P.; Hoover, C.; Kondo, Y.; Shao, B.; Gao, L.; et al. Proximal colon-derived O-glycosylated mucus encapsulates and modulates the microbiota. *Science* **2020**, *370*, 467–472. [CrossRef]
26. Liu, H.Y.; Walden, T.B.; Ahl, D.; Nyman, M.; Bertilsson, S.; Phillipson, M.; Holm, L. High-fat diet enriched with bilberry modifies colonic mucus dynamics and restores marked alterations of gut microbiome in rats. *Mol. Nutr. Food Res.* **2019**, *63*, e1900117. [CrossRef]
27. Forgie, A.J.; Ju, T.; Tollenaar, S.L.; Willing, B.P. Phytochemical-induced mucin accumulation in the gastrointestinal lumen is independent of the microbiota. *bioRxiv* **2022**, bioRxiv:2022.03.11.483917.
28. Makarewicz, M.; Drozdz, I.; Tarko, T.; Duda-Chodak, A. The Interactions between polyphenols and microorganisms, especially gut microbiota. *Antioxidants* **2021**, *10*, 188. [CrossRef]
29. Giménez-Bastida, J.A.; Truchado, P.; Larrosa, M.; Espín, J.C.; Tomás-Barberán, F.A.; Allende, A.; García-Conesa, M.T. Urolithins, ellagitannin metabolites produced by colon microbiota, inhibit quorum sensing in *Yersinia Enterocolitica*: Phenotypic response and associated molecular changes. *Food Chem* **2012**, *132*, 1465–1474. [CrossRef]
30. Saha, P.; Yeoh, B.; Singh, R.; Chandrasekar, B.; Vemula, P.; Haribabu, B.; Vijay-Kumar, M.; Jala, V.R. Gut Microbiota Conversion of Dietary Ellagic Acid into Bioactive Phytoceutical Urolithin A Inhibits Heme Peroxidases. *PLoS ONE* **2016**, *11*, e0156811. [CrossRef]
31. Mele, L.; Carobbio, S.; Brindani, N.; Curti, C.; Rodriguez-Cuenca, S.; Bidault, G.; Mena, P.; Zanotti, I.; Vacca, M.; Vidal-Puig, A.; et al. Phenyl-gamma-valerolactones, flavan-3-ol colonic metabolites, protect brown adipocytes from oxidative stress without affecting their differentiation or function. *Mol. Nutr. Food Res.* **2017**, *61*, 1700074. [CrossRef] [PubMed]
32. Mena, P.; Bresciani, L.; Brindani, N.; Ludwig, I.A.; Pereira-Caro, G.; Angelino, D.; Llorach, R.; Calani, L.; Brighenti, F.; Clifford, M.N.; et al. Phenyl-γ-valerolactones and phenylvaleric acids, the main colonic metabolites of flavan-3-ols: Synthesis, analysis, bioavailability, and bioactivity. *Nat. Prod. Rep.* **2018**, *36*, 714–752. [CrossRef] [PubMed]
33. Olennikov, D.N.; Kashchenko, N.I.; Chirikova, N.K. In vitro bioaccessibility, human gut microbiota metabolites and hepatoprotective potential of Chebulic ellagitannins: A case of padma hepaten(R) formulation. *Nutrients* **2015**, *7*, 8456–8477. [CrossRef]
34. Karn, A.; Zhao, C.; Yang, F.; Cui, J.; Gao, Z.; Wang, M.; Wang, F.; Xiao, H.; Zheng, J. In-vivo biotransformation of citrus functional components and their effects on health. *Crit. Rev. Food Sci. Nutr.* **2021**, *61*, 756–776. [CrossRef] [PubMed]
35. Cardona, F.; Andres-Lacueva, C.; Tulipani, S.; Tinahones, F.J.; Queipo-Ortuno, M.I. Benefits of polyphenols on gut microbiota and implications in human health. *J. Nutr. Biochem.* **2013**, *24*, 1415–1422. [CrossRef] [PubMed]
36. Ottaviani, J.I.; Borges, G.; Momma, T.Y.; Spencer, J.P.; Keen, C.L.; Crozier, A.; Schroeter, H. The metabolome of [2-(14)C](-)-epicatechin in humans: Implications for the assessment of efficacy, safety, and mechanisms of action of polyphenolic bioactives. *Sci Rep.* **2016**, *6*, 29034. [CrossRef] [PubMed]

37. Nakano, H.; Wu, S.; Sakao, K.; Hara, T.; He, J.; Garcia, S.; Shetty, K.; Hou, D.X. Bilberry anthocyanins ameliorate NAFLD by improving dyslipidemia and gut microbiome dysbiosis. *Nutrients* **2020**, *12*, 3252. [CrossRef]
38. Cory, H.; Passarelli, S.; Szeto, J.; Tamez, M.; Mattei, J. The role of polyphenols in human health and food systems: A mini-review. *Front. Nutr.* **2018**, *5*, 87. [CrossRef]
39. Juárez-Fernández, M.; Porras, D.; Petrov, P.; Román-Sagüillo, S.; García-Mediavilla, M.V.; Soluyanova, P.; Martínez-Flórez, S.; González-Gallego, J.; Nistal, E.; Jover, R.; et al. The Synbiotic Combination of *Akkermansia muciniphila* and Quercetin Ameliorates Early Obesity and NAFLD through Gut Microbiota Reshaping and Bile Acid Metabolism Modulation. *Antioxidants* **2021**, *10*, 2001. [CrossRef]
40. Fidelix, M.; Milenkovic, D.; Sivieri, K.; Cesar, T. Microbiota modulation and effects on metabolic biomarkers by orange juice: A controlled clinical trial. *Food Funct.* **2020**, *11*, 1599–1610. [CrossRef]
41. Anhê, F.F.; Varin, T.V.; Barz, M.; Pilon, G.; Dudonné, S.; Trottier, J.; St-Pierre, P.; Harris, C.S.; Lucas, M.; Lemire, M.; et al. Arctic berry extracts target the gut–liver axis to alleviate metabolic endotoxaemia, insulin resistance and hepatic steatosis in diet-induced obese mice. *Diabetologia* **2018**, *61*, 919–931. [CrossRef] [PubMed]
42. Singh, A.; D'Amico, D.; Andreux, P.A.; Dunngalvin, G.; Kern, T.; Blanco-Bose, W.; Auwerx, J.; Aebischer, P.; Rinsch, C. Direct supplementation with Urolithin A overcomes limitations of dietary exposure and gut microbiome variability in healthy adults to achieve consistent levels across the population. *Eur. J. Clin. Nutr.* **2022**, *76*, 297–308. [CrossRef] [PubMed]
43. Henning, S.M.; Summanen, P.H.; Lee, R.-P.; Yang, J.; Finegold, S.M.; Heber, D.; Li, Z. Pomegranate ellagitannins stimulate the growth of *Akkermansia muciniphila* in vivo. *Anaerobe* **2017**, *43*, 56–60. [CrossRef] [PubMed]
44. Rodríguez-Daza, M.C.; Pulido-Mateos, E.C.; Lupien-Meilleur, J.; Guyonnet, D.; Desjardins, Y.; Roy, D. Polyphenol-mediated gut microbiota modulation: Toward prebiotics and further. *Front. Nutr.* **2021**, *8*, 689456. [CrossRef] [PubMed]
45. Lee, C.W.; Chen, H.J.; Xie, G.R.; Shih, C.K. Djulis (*Chenopodium Formosanum*) prevents colon carcinogenesis via regulating antioxidative and apoptotic pathways in rats. *Nutrients* **2019**, *11*, 2168. [CrossRef]
46. Van Dorsten, F.A.; Peters, S.; Gross, G.; Gomez-Roldan, V.; Klinkenberg, M.; de Vos, R.C.; Vaughan, E.E.; van Duynhoven, J.P.; Possemiers, S.; van de Wiele, T.; et al. Gut microbial metabolism of polyphenols from black tea and red wine/grape juice is source-specific and colon-region dependent. *J. Agric. Food Chem.* **2012**, *60*, 11331–11342. [CrossRef]
47. Yu, M.; Wang, Q.; Ma, Y.; Li, L.; Yu, K.; Zhang, Z.; Chen, G.; Li, X.; Xiao, W.; Xu, P.; et al. Aryl hydrocarbon receptor activation modulates intestinal epithelial barrier function by maintaining tight junction integrity. *Int. J. Biol. Sci.* **2018**, *14*, 69–77. [CrossRef]
48. Xue, Z.; Li, D.; Yu, W.; Zhang, Q.; Hou, X.; He, Y.; Kou, X. Mechanisms and therapeutic prospects of polyphenols as modulators of the aryl hydrocarbon receptor. *Food Funct.* **2017**, *8*, 1414–1437. [CrossRef]
49. Tang, J.S.; Cait, A.; Li, Y.; Abolins-Thompson, H.; Gell, K.; Herst, P.M.; O'Sullivan, D.; Gasser, O. Practical approach to explore the effects of polyphenols on aryl hydrocarbon receptor regulated immune function. *J. Agric. Food Chem.* **2021**, *69*, 8625–8633. [CrossRef]
50. Bachmann, R.; Van Hul, M.; Baldin, P.; Léonard, D.; Delzenne, N.M.; Belzer, C.; Ouwerkerk, J.P.; Repsilber, D.; Rangel, I.; Kartheuser, A.; et al. *Akkermansia muciniphila* reduces peritonitis and improves intestinal tissue wound healing after a colonic transmural defect by a MyD88-dependent mechanism. *Cells* **2022**, *11*, 2666. [CrossRef]
51. Wang, P.; Li, D.; Ke, W.; Liang, D.; Hu, X.; Chen, F. Resveratrol-induced gut microbiota reduces obesity in high-fat diet-fed mice. *Int. J. Obes.* **2020**, *44*, 213–225. [CrossRef] [PubMed]
52. Li, Z.; Henning, S.M.; Lee, R.P.; Lu, Q.Y.; Summanen, P.H.; Thames, G.; Corbett, K.; Downes, J.; Tseng, C.H.; Finegoldbce, S.M.; et al. Pomegranate extract induces ellagitannin metabolite formation and changes stool microbiota in healthy volunteers. *Food Funct.* **2015**, *6*, 2487–2495. [CrossRef] [PubMed]
53. Garcia-Villalba, R.; Vissenaekens, H.; Pitart, J.; Romo-Vaquero, M.; Espín, J.; Grootaert, C.; Selma, M.V.; Raes, K.; Smagghe, G.; Possemiers, S.; et al. The gastrointestinal simulation model TWIN-SHIME® Shows differences between human Urolithin-metabotypes in gut microbiota composition, pomegranate polyphenol metabolism, and transport along the intestinal tract. *J. Agric. Food Chem.* **2017**, *65*, 5480–5493. [CrossRef] [PubMed]
54. Xia, Y.; Zhang, X.; Jiang, M.; Zhang, H.; Wang, Y.; Zhang, Y.; Seviour, R.; Kong, Y. In vitro co-metabolism of epigallocatechin-3-gallate (EGCG) by the mucin-degrading bacterium *Akkermansia muciniphila*. *PLoS ONE* **2021**, *16*, e0260757. [CrossRef]
55. Johansson, M.E.; Larsson, J.M.; Hansson, G.C. The two mucus layers of colon are organized by the MUC2 mucin, whereas the outer layer is a legislator of host-microbial interactions. *Proc. Natl. Acad. Sci. USA* **2011**, *108* (Suppl. S1), 4659–4665. [CrossRef]
56. Bansil, R.; Turner, B.S. Mucin structure, aggregation, physiological functions and biomedical applications. *Curr. Opin. Colloid. Interface Sci.* **2006**, *11*, 164–170. [CrossRef]
57. van Passel, M.W.J.; Kant, R.; Zoetendal, E.G.; Plugge, C.M.; Derrien, M.; Malfatti, S.A.; Chain, P.S.G.; Woyke, T.; Palva, A.; de Vos, W.M.; et al. The genome of *Akkermansia muciniphila*, a dedicated intestinal mucin degrader, and its use in exploring intestinal metagenomes. *PLoS ONE* **2011**, *6*, e16876, 1-8. [CrossRef]
58. Ottman, N.; Davids, M.; Suarez-Diez, M.; Boeren, S.; Schaap, P.J.; Martins Dos Santos, V.A.P.; Smidt, H.; Belzer, C.; de Vos, W.M. Genome-scale model and omics analysis of metabolic capacities of *Akkermansia muciniphila* reveal a preferential mucin-degrading lifestyle. *Appl. Environ. Microbiol.* **2017**, *83*, e01014-17. [CrossRef]
59. Derrien, M.; Belzer, C.; de Vos, W.M. *Akkermansia muciniphila* and its role in regulating host functions. *Microb. Pathog.* **2017**, *106*, 171–181. [CrossRef]

60. Everard, A.; Belzer, C.; Geurts, L.; Ouwerkerk, J.P.; Druart, C.; Bindels, L.B.; Guiot, Y.; Derrien, M.; Muccioli, G.G.; Delzenne, N.M.; et al. Cross-talk between *Akkermansia muciniphila* and intestinal epithelium controls diet-induced obesity. *Proc. Natl. Acad. Sci. USA* **2013**, *110*, 9066–9071. [CrossRef]
61. Earley, H.; Lennon, G.; Balfe, A.; Coffey, J.C.; Winter, D.C.; O'Connell, P.R. The abundance of *Akkermansia muciniphila* and its relationship with sulphated colonic mucins in health and ulcerative colitis. *Sci. Rep.* **2019**, *9*, 15683. [CrossRef]
62. Yamada, T.; Hino, S.; Iijima, H.; Genda, T.; Aoki, R.; Nagata, R.; Han, K.H.; Hirota, M.; Kinashi, Y.; Oguchi, H.; et al. Mucin O-glycans facilitate symbiosynthesis to maintain gut immune homeostasis. *EBioMedicine* **2019**, *48*, 513–525. [CrossRef] [PubMed]
63. Larsson, J.M.; Karlsson, H.; Crespo, J.G.; Johansson, M.E.; Eklund, L.; Sjovall, H.; Hansson, G.C. Altered O-glycosylation profile of MUC2 mucin occurs in active ulcerative colitis and is associated with increased inflammation. *Inflamm. Bowel Dis.* **2011**, *17*, 2299–2307. [CrossRef] [PubMed]
64. Giron, L.B.; Tanes, C.E.; Schleimann, M.H.; Engen, P.A.; Mattei, L.M.; Anzurez, A.; Damra, M.; Zhang, H.; Bittinger, K.; Bushman, F.; et al. Sialylation and fucosylation modulate inflammasome-activating eIF2 Signaling and microbial translocation during HIV infection. *Mucosal. Immunol.* **2020**, *13*, 753–766. [CrossRef] [PubMed]
65. Pruss, K.M.; Marcobal, A.; Southwick, A.M.; Dahan, D.; Smits, S.A.; Ferreyra, J.A.; Higginbottom, S.K.; Sonnenburg, E.D.; Kashyap, P.C.; Choudhury, B.; et al. Mucin-derived O-glycans supplemented to diet mitigate diverse microbiota perturbations. *ISME J.* **2021**, *15*, 577–591. [CrossRef]
66. Rothhammer, V.; Quintana, F.J. The aryl hydrocarbon receptor: An environmental sensor integrating immune responses in health and disease. *Nat. Rev. Immunol.* **2019**, *19*, 184–197. [CrossRef]
67. Goto, Y.; Obata, T.; Kunisawa, J.; Sato, S.; Ivanov, I.I.; Lamichhane, A.; Takeyama, N.; Kamioka, M.; Sakamoto, M.; Matsuki, T.; et al. Innate lymphoid cells regulate intestinal epithelial cell glycosylation. *Science* **2014**, *345*, 1254009. [CrossRef]
68. Ehrlich, A.M.; Pacheco, A.R.; Henrick, B.M.; Taft, D.; Xu, G.; Huda, M.N.; Mishchuk, D.; Goodson, M.L.; Slupsky, C.; Barile, D.; et al. Indole-3-lactic acid associated with *Bifidobacterium*-dominated microbiota significantly decreases inflammation in intestinal epithelial cells. *BMC Microbiol.* **2020**, *20*, 357. [CrossRef]
69. Takamura, T.; Harama, D.; Fukumoto, S.; Nakamura, Y.; Shimokawa, N.; Ishimaru, K.; Ikegami, S.; Makino, S.; Kitamura, M.; Nakao, A. *Lactobacillus bulgaricus* OLL1181 activates the aryl hydrocarbon receptor pathway and inhibits colitis. *Immunol. Cell Biol.* **2011**, *89*, 817–822. [CrossRef]
70. Yin, J.; Song, Y.; Hu, Y.; Wang, Y.; Zhang, B.; Wang, J.; Ji, X.; Wang, S. Dose-dependent beneficial effects of tryptophan and its derived metabolites on *Akkermansia* In Vitro: A preliminary prospective study. *Microorganisms* **2021**, *9*, 1511. [CrossRef]
71. Gu, Z.; Pei, W.; Shen, Y.; Wang, L.; Zhu, J.; Zhang, Y.; Fan, S.; Wu, Q.; Li, L.; Zhang, Z. *Akkermansia muciniphila* and its outer protein Amuc_1100 regulates tryptophan metabolism in colitis. *Food Funct.* **2021**, *12*, 10184–10195. [CrossRef] [PubMed]
72. Manach, C.; Williamson, G.; Morand, C.; Scalbert, A.; Remesy, C. Bioavailability and bioefficacy of polyphenols in humans. I. Review of 97 bioavailability studies. *Am. J. Clin. Nutr.* **2005**, *81* (Suppl. S1), 230S–242S. [CrossRef] [PubMed]
73. Serra, D.; Almeida, L.M.; Dinis, T.C.P. Dietary polyphenols: A novel strategy to modulate microbiota-gut-brain axis. *Trends Food Sci. Technol.* **2018**, *78*, 224–233. [CrossRef]
74. Gao, X.; Xie, Q.; Kong, P.; Liu, L.; Sun, S.; Xiong, B.; Huang, B.; Yan, L.; Sheng, J.; Xiang, H. Polyphenol- and caffeine-rich postfermented Pu-erh tea improves diet-induced metabolic syndrome by remodeling intestinal homeostasis in mice. *Infect. Immun.* **2018**, *86*, e00601-17. [CrossRef] [PubMed]
75. Zhang, S.; Zhao, Y.; Ohland, C.; Jobin, C.; Sang, S. Microbiota facilitates the formation of the aminated metabolite of green tea polyphenol (-)-epigallocatechin-3-gallate which trap deleterious reactive endogenous metabolites. *Free Radic. Biol. Med.* **2019**, *131*, 332–344. [CrossRef]
76. Hara-Terawaki, A.; Takagaki, A.; Kobayashi, H.; Nanjo, F. Inhibitory activity of catechin metabolites produced by intestinal microbiota on proliferation of HeLa cells. *Biol. Pharm. Bull.* **2017**, *40*, 1331–1335. [CrossRef]
77. Takagaki, A.; Nanjo, F. Effects of metabolites produced from (-)-epigallocatechin gallate by rat intestinal bacteria on angiotensin I-converting enzyme activity and blood pressure in spontaneously hypertensive rats. *J. Agric. Food Chem* **2015**, *63*, 8262–8266. [CrossRef]
78. Rodriguez-Mateos, A.; Feliciano, R.P.; Boeres, A.; Weber, T.; Dos Santos, C.N.; Ventura, M.R.; Heiss, C. Cranberry (poly)phenol metabolites correlate with improvements in vascular function: A double-blind, randomized, controlled, dose-response, crossover study. *Mol. Nutr. Food Res.* **2016**, *60*, 2130–2140. [CrossRef]
79. Mi, Y.; Qi, G.; Gao, Y.; Li, R.; Wang, Y.; Li, X.; Huang, S.; Liu, X. (-)-Epigallocatechin-3-gallate ameliorates insulin resistance and mitochondrial dysfunction in HepG2 cells: Involvement of Bmal1. *Mol. Nutr. Food Res.* **2017**, *61*, 1700440. [CrossRef]
80. Man, A.W.C.; Zhou, Y.; Xia, N.; Li, H. Involvement of gut microbiota, microbial metabolites and interaction with polyphenol in host immunometabolism. *Nutrients* **2020**, *12*, 3054. [CrossRef]
81. Gibson, G.R.; Hutkins, R.; Sanders, M.; Prescott, S.L.; Reimer, R.A.; Salminen, S.J.; Scott, K.; Stanton, C.; Swanson, K.S.; Cani, P.D.; et al. Expert consensus document: The International Scientific Association for Probiotics and Prebiotics (ISAPP) consensus statement on the definition and scope of prebiotics. *Nat. Rev. Gastroenterol. Hepatol.* **2017**, *14*, 491–502. [CrossRef] [PubMed]
82. Hair, R.; Sakaki, J.R.; Chun, O.K. Anthocyanins, microbiome and health benefits in aging. *Molecules* **2021**, *26*, 537. [CrossRef] [PubMed]
83. Lavefve, L.; Howard, L.R.; Carbonero, F. Berry polyphenols metabolism and impact on human gut microbiota and health. *Food Funct.* **2020**, *11*, 45–65. [CrossRef] [PubMed]

84. Cheng, J.R.; Liu, X.M.; Chen, Z.Y.; Zhang, Y.S.; Zhang, Y.H. Mulberry anthocyanin biotransformation by intestinal probiotics. *Food Chem.* **2016**, *213*, 721–727. [CrossRef] [PubMed]
85. Dabek, M.; McCrae, S.I.; Stevens, V.J.; Duncan, S.H.; Louis, P. Distribution of beta-glucosidase and beta-glucuronidase activity and of beta-glucuronidase gene gus in human colonic bacteria. *FEMS Microbiol. Ecol.* **2008**, *66*, 487–495. [CrossRef] [PubMed]
86. Song, H.; Shen, X.; Deng, R.; Zhang, Y.; Zheng, X. Dietary anthocyanin-rich extract of acai protects from diet-induced obesity, liver steatosis, and insulin resistance with modulation of gut microbiota in mice. *Nutrition* **2021**, *86*, 111176. [CrossRef]
87. John, O.D.; Mouatt, P.; Prasadam, I.; Xiao, Y.; Panchal, S.K.; Brown, L. The edible native Australian fruit, Davidson's plum (*Davidsonia pruriens*), reduces symptoms in rats with diet-induced metabolic syndrome. *J. Funct. Foods* **2019**, *56*, 204–215. [CrossRef]
88. Pan, P.; Lam, V.; Salzman, N.; Huang, Y.-W.; Yu, J.; Zhang, J.; Wang, L.-S. Black raspberries and their anthocyanin and fiber fractions alter the composition and diversity of gut microbiota in F-344 rats. *Nutr. Cancer.* **2017**, *69*, 943–951. [CrossRef]
89. Tan, J.; Li, Y.; Hou, D.X.; Wu, S. The effects and mechanisms of cyanidin-3-glucoside and its phenolic metabolites in maintaining intestinal integrity. *Antioxidants* **2019**, *8*, 479. [CrossRef]
90. Ajiboye, T.O.; Habibu, R.S.; Saidu, K.; Haliru, F.Z.; Ajiboye, H.O.; Aliyu, N.O.; Ibitoye, O.B.; Uwazie, J.N.; Muritala, H.F.; Bello, S.A.; et al. Involvement of oxidative stress in protocatechuic acid-mediated bacterial lethality. *Microbiologyopen* **2017**, *6*, e00472. [CrossRef]
91. Esteban-Torres, M.; Santamaria, L.; Cabrera-Rubio, R.; Plaza-Vinuesa, L.; Crispie, F.; de Las Rivas, B.; Cotter, P.; Munoz, R. A diverse range of human gut bacteria have the potential to metabolize the dietary component gallic acid. *Appl. Environ. Microbiol.* **2018**, *84*, e01558-18. [CrossRef]
92. Chia, L.; Hornung, B.V.H.; Aalvink, S.; Schaap, P.J.; de Vos, W.M.; Knol, J.; Belzer, C. Deciphering the trophic interaction between *Akkermansia muciniphila* and the butyrogenic gut commensal Anaerostipes caccae using a metatranscriptomic approach. *Antonie Van Leeuwenhoek* **2018**, *111*, 859–873. [CrossRef] [PubMed]
93. Luo, Y.; Jian, Y.; Liu, Y.; Jiang, S.; Muhammad, D.; Wang, W. Flavanols from nature: A phytochemistry and biological activity review. *Molecules* **2022**, *27*, 719. [CrossRef] [PubMed]
94. Kuhnle, G.G.C. Nutrition epidemiology of flavan-3-ols: The known unknowns. *Mol. Asp. Med.* **2018**, *61*, 2–11. [CrossRef] [PubMed]
95. Chen, H.; Sang, S. Biotransformation of tea polyphenols by gut microbiota. *J. Funct. Foods* **2014**, *7*, 26–42. [CrossRef]
96. Kan, J.; Chen, C.; Huo, T.; Xie, W.; Hui, Y.; Liu, J.; Jin, C. Polyphenolic-enriched peach peels extract regulates lipid metabolism and improves the gut microbiota composition in high fat diet-fed mice. *J. Funct. Foods* **2020**, *72*, 104082. [CrossRef]
97. Xie, Y.; Yang, W.; Tang, F.; Chen, X.; Ren, L. Antibacterial activities of flavonoids: Structure-activity relationship and mechanism. *Curr. Med. Chem.* **2014**, *22*, 132–149. [CrossRef]
98. Smeriglio, A.; Barreca, D.; Bellocco, E.; Trombetta, D. Proanthocyanidins and hydrolysable tannins: Occurrence, dietary intake and pharmacological effects. *Br. J. Pharmacol.* **2017**, *174*, 1244–1262. [CrossRef]
99. Farha, A.K.; Yang, Q.-Q.; Kim, G.; Li, H.-B.; Zhu, F.; Liu, H.-Y.; Gan, R.-Y.; Corke, H. Tannins as an alternative to antibiotics. *Food Biosci.* **2020**, *38*, 100751. [CrossRef]
100. Rodriguez-Daza, M.C.; Daoust, L.; Boutkrabt, L.; Pilon, G.; Varin, T.; Dudonne, S.; Levy, E.; Marette, A.; Roy, D.; Desjardins, Y. Wild blueberry proanthocyanidins shape distinct gut microbiota profile and influence glucose homeostasis and intestinal phenotypes in high-fat high-sucrose fed mice. *Sci. Rep.* **2020**, *10*, 2217. [CrossRef]
101. Neto, C.C.; Mortzfeld, B.M.; Turbitt, J.R.; Bhattarai, S.K.; Yeliseyev, V.; DiBenedetto, N.; Bry, L.; Bucci, V. Proanthocyanidin-enriched cranberry extract induces resilient bacterial community dynamics in a gnotobiotic mouse model. *Microb. Cell* **2021**, *8*, 131–142. [CrossRef] [PubMed]
102. Rodriguez-Daza, M.C.; Roquim, M.; Dudonne, S.; Pilon, G.; Levy, E.; Marette, A.; Roy, D.; Desjardins, Y. Berry polyphenols and fibers modulate distinct microbial metabolic functions and gut microbiota enterotype-like clustering in obese mice. *Front. Microbiol* **2020**, *11*, 2032. [CrossRef] [PubMed]
103. Medina-Larque, A.S.; Rodriguez-Daza, M.C.; Roquim, M.; Dudonne, S.; Pilon, G.; Levy, E.; Marette, A.; Roy, D.; Jacques, H.; Desjardins, Y. Cranberry polyphenols and agave agavins impact gut immune response and microbiota composition while improving gut barrier function, inflammation, and glucose metabolism in mice fed an obesogenic diet. *Front. Immunol.* **2022**, *13*, 871080. [CrossRef] [PubMed]
104. Casanova-Martí, À.; Serrano, J.; Portune, K.J.; Sanz, Y.; Blay, T.M.; Terra, X.; Ardévol, A.; Pinent, M. Grape seed proanthocyanidins influence gut microbiota and enteroendocrine secretions in female rats. *Food Funct.* **2018**, *9*, 1672–1682. [CrossRef] [PubMed]
105. Roopchand, D.E.; Kuhn, P.; Rojo, L.E.; Lila, M.; Raskin, I. Blueberry polyphenol-enriched soybean flour reduces hyperglycemia, body weight gain and serum cholesterol in mice. *Pharmacol. Res.* **2013**, *68*, 59–67. [CrossRef]
106. Zhang, L.; Carmody, R.N.; Kalariya, H.M.; Duran, R.M.; Moskal, K.; Poulev, A.; Kuhn, P.; Tveter, K.M.; Turnbaugh, P.J.; Raskin, I.; et al. Grape proanthocyanidin-induced intestinal bloom of *Akkermansia muciniphila* is dependent on its baseline abundance and precedes activation of host genes related to metabolic health. *J. Nutr. Biochem.* **2018**, *56*, 142–151. [CrossRef]
107. Masumoto, S.; Terao, A.; Yamamoto, Y.; Mukai, T.; Miura, T.; Shoji, T. Non-absorbable apple procyanidins prevent obesity associated with gut microbial and metabolomic changes. *Sci. Rep.* **2016**, *6*, 31208. [CrossRef]

108. Anhe, F.F.; Nachbar, R.T.; Varin, T.V.; Trottier, J.; Dudonne, S.; Le Barz, M.; Feutry, P.; Pilon, G.; Barbier, O.; Desjardins, Y.; et al. Treatment with camu camu (*Myrciaria dubia*) prevents obesity by altering the gut microbiota and increasing energy expenditure in diet-induced obese mice. *Gut* **2019**, *68*, 453–464. [CrossRef]
109. Abot, A.; Brochot, A.; Pomie, N.; Wemelle, E.; Druart, C.; Regnier, M.; Delzenne, N.M.; de Vos, W.M.; Knauf, C.; Cani, P.D. Camu-camu reduces obesity and improves diabetic profiles of obese and diabetic mice: A dose-ranging study. *Metabolites* **2022**, *12*, 301. [CrossRef]
110. Xia, T.; Duan, W.; Zhang, Z.; Li, S.; Zhao, Y.; Geng, B.; Zheng, Y.; Yu, J.; Wang, M. Polyphenol-rich vinegar extract regulates intestinal microbiota and immunity and prevents alcohol-induced inflammation in mice. *Food Res. Int.* **2021**, *140*, 110064. [CrossRef]
111. Roopchand, D.E.; Carmody, R.N.; Kuhn, P.; Moskal, K.; Rojas-Silva, P.; Turnbaugh, P.J.; Raskin, I. Dietary polyphenols promote growth of the gut bacterium *Akkermansia muciniphila* and attenuate high-fat diet-induced metabolic syndrome. *Diabetes* **2015**, *64*, 2847–2858. [CrossRef] [PubMed]
112. Dey, P.; Sasaki, G.Y.; Wei, P.; Li, J.; Wang, L.; Zhu, J.; McTigue, D.; Yu, Z.; Bruno, R.S. Green tea extract prevents obesity in male mice by alleviating gut dysbiosis in association with improved intestinal barrier function that limits endotoxin translocation and adipose inflammation. *J. Nutr. Biochem.* **2019**, *67*, 78–89. [CrossRef]
113. Henning, S.M.; Choo, J.J.; Heber, D. Nongallated compared with gallated flavan-3-ols in green and black tea are more bioavailable. *J. Nutr.* **2008**, *138*, 1529S–1534S. [CrossRef] [PubMed]
114. Mueller, M.; Zartl, B.; Schleritzko, A.; Stenzl, M.; Viernstein, H.; Unger, F.M. Rhamnosidase activity of selected probiotics and their ability to hydrolyse flavonoid rhamnoglucosides. *Bioprocess. Biosyst. Eng.* **2018**, *41*, 221–228. [CrossRef] [PubMed]
115. Gwiazdowska, D.; Jus, K.; Jasnowska-Malecka, J.; Kluczynska, K. The impact of polyphenols on *Bifidobacterium* growth. *Acta Biochim. Pol.* **2015**, *62*, 895–901. [CrossRef]
116. Liu, Z.; Chen, Z.; Guo, H.; He, D.; Zhao, H.; Wang, Z.; Zhang, W.; Liao, L.; Zhang, C.; Ni, L. The modulatory effect of infusions of green tea, oolong tea, and black tea on gut microbiota in high-fat-induced obese mice. *Food Funct.* **2016**, *7*, 4869–4879. [CrossRef]
117. Ushiroda, C.; Naito, Y.; Takagi, T.; Uchiyama, K.; Mizushima, K.; Higashimura, Y.; Yasukawa, Z.; Okubo, T.; Inoue, R.; Honda, A.; et al. Green tea polyphenol (epigallocatechin-3-gallate) improves gut dysbiosis and serum bile acids dysregulation in high-fat diet-fed mice. *J. Clin. Biochem. Nutr.* **2019**, *65*, 34–46. [CrossRef]
118. Jeong, H.W.; Kim, J.K.; Kim, A.Y.; Cho, D.; Lee, J.H.; Choi, J.K.; Park, M.; Kim, W. Green tea encourages growth of *Akkermansia muciniphila*. *J. Med. Food* **2020**, *23*, 841–851. [CrossRef]
119. Takagaki, A.; Nanjo, F. Biotransformation of (-)-epicatechin, (+)-epicatechin, (-)-catechin, and (+)-catechin by intestinal bacteria involved in isoflavone metabolism. *Biosci. Biotechnol. Biochem.* **2016**, *80*, 199–202. [CrossRef]
120. Gaya, P.; Peirotén, Á.; Medina, M.; Álvarez, I.; Landete, J.M. *Bifidobacterium pseudocatenulatum* INIA P815: The first bacterium able to produce urolithins A and B from ellagic acid. *J. Funct. Foods* **2018**, *45*, 95–99. [CrossRef]
121. Kelly, S.M.; O'Callaghan, J.; Kinsella, M.; van Sinderen, D. Characterisation of a hydroxycinnamic acid esterase from the *Bifidobacterium longum* subsp. *longum taxon. Front. Microbiol.* **2018**, *9*, 2690. [CrossRef] [PubMed]
122. Feng, G.; Han, K.; Yang, Q.; Feng, W.; Guo, J.; Wang, J.; Yang, X. Interaction of pyrogallol-containing polyphenols with mucin reinforces intestinal mucus barrier properties. *J. Agric. Food Chem.* **2022**, *70*, 9536–9546. [CrossRef]
123. Al Khalaf, A.K.; Abdulrahman, A.O.; Kaleem, M.; Nur, S.M.; Asseri, A.H.; Choudhry, H.; Khan, M.I. Comparative analysis of the impact of urolithins on the composition of the gut microbiota in normal-diet fed rats. *Nutrients* **2021**, *13*, 3885. [CrossRef] [PubMed]
124. Era, M.; Matsuo, Y.; Saito, Y.; Tanaka, T. Production of.f ellagitannin hexahydroxydiphenoyl ester by spontaneous reduction of dehydrohexa-hydroxydiphenoyl ester. *Molecules* **2020**, *25*, 1051. [CrossRef] [PubMed]
125. Ozdal, T.; Sela, D.A.; Xiao, J.; Boyacıoglu, D.; Chen, F.; Capanoglu, E. The Reciprocal Interactions between Polyphenols and Gut Microbiota and Effects on Bioaccessibility. *Nutrients* **2016**, *8*, 78. [CrossRef] [PubMed]
126. Zhang, Z.; Peng, X.; Li, S.; Zhang, N.; Wang, Y.; Wei, H. Isolation and identification of quercetin degrading bacteria from human fecal microbes. *PLoS ONE* **2014**, *9*, e90531. [CrossRef]
127. Cao, H.; Chen, X.; Jassbi, A.R.; Xiao, J. Microbial biotransformation of bioactive flavonoids. *Biotechnol. Adv.* **2015**, *33*, 214–223. [CrossRef]
128. Rodriguez-Castano, G.P.; Dorris, M.R.; Liu, X.; Bolling, B.W.; Acosta-Gonzalez, A.; Rey, F.E. *Bacteroides thetaiotaomicron* starch utilization promotes quercetin degradation and butyrate production by *Eubacterium ramulus. Front. Microbiol.* **2019**, *10*, 1145. [CrossRef]
129. Fetzner, S. Ring-Cleaving Dioxygenases with a Cupin Fold. *Appl. Environ. Microbiol.* **2012**, *78*, 2505–2514. [CrossRef]
130. Braune, A.; Gutschow, M.; Engst, W.; Blaut, M. Degradation of quercetin and luteolin by *Eubacterium ramulus. Appl. Environ. Microbiol.* **2001**, *67*, 5558–5567. [CrossRef]
131. Nie, J.; Zhang, L.; Zhao, G.; Du, X. Quercetin reduces atherosclerotic lesions by altering the gut microbiota and reducing atherogenic lipid metabolites. *J. Appl. Microbiol.* **2019**, *127*, 1824–1834. [CrossRef] [PubMed]
132. Etxeberria, U.; Arias, N.; Boqué, N.; Macarulla, M.T.; Portillo, M.P.; Martínez, J.A.; Milagro, F.I. Reshaping faecal gut microbiota composition by the intake of trans-resveratrol and quercetin in high-fat sucrose diet-fed rats. *J. Nutr. Biochem.* **2015**, *26*, 651–660. [CrossRef] [PubMed]
133. Zhao, L.; Zhang, Q.; Ma, W.; Tian, F.; Shen, H.; Zhou, M. A combination of quercetin and resveratrol reduces obesity in high-fat diet-fed rats by modulation of gut microbiota. *Food. Funct.* **2017**, *8*, 4644–4656. [CrossRef] [PubMed]

134. Li, X.; Yang, L.; Xu, M.; Qiao, G.; Li, C.; Lin, L.; Zheng, G. *Smilax china L.* polyphenols alleviates obesity and inflammation by modulating gut microbiota in high fat/high sucrose diet-fed C57BL/6J mice. *J. Funct. Foods* **2021**, *77*, 104332. [CrossRef]
135. Amaretti, A.; Raimondi, S.; Leonardi, A.; Quartieri, A.; Rossi, M. Hydrolysis of the rutinose-conjugates flavonoids rutin and hesperidin by the gut microbiota and bifidobacteria. *Nutrients* **2015**, *7*, 2788–2800. [CrossRef]
136. Riva, A.; Kolimar, D.; Spittler, A.; Wisgrill, L.; Herbold, C.W.; Abranko, L.; Berry, D. Conversion of rutin, a prevalent dietary flavonol, by the human gut microbiota. *Front. Microbiol.* **2020**, *11*, 585428. [CrossRef]
137. Bu, F.; Ding, Y.; Chen, T.; Wang, Q.; Wang, R.; Zhou, J.Y.; Jiang, F.; Zhang, D.; Xu, M.; Shi, G.; et al. Total flavone of *Abelmoschus Manihot* improves colitis by promoting the growth of *Akkermansia* in mice. *Sci. Rep.* **2021**, *11*, 20787. [CrossRef]
138. Stevens, Y.; Rymenant, E.V.; Grootaert, C.; Camp, J.V.; Possemiers, S.; Masclee, A.; Jonkers, D. The Intestinal fate of citrus flavanones and their effects on gastrointestinal health. *Nutrients* **2019**, *11*, 1464. [CrossRef]
139. Mas-Capdevila, A.; Teichenne, J.; Domenech-Coca, C.; Caimari, A.; Del Bas, J.M.; Escote, X.; Crescenti, A. Effect of hesperidin on cardiovascular disease risk factors: The role of intestinal microbiota on hesperidin bioavailability. *Nutrients* **2020**, *12*, 1488. [CrossRef]
140. Liu, Z.; Liu, T.; Lei, C.; Song, W.; Fang, R.; Chen, H.; Li, C.; Li, X.; Liang, X.; Huang, Q. Novel role of hesperidin improve obesity in HFD mice by modulating the composition of the gut microbiota. *Res. Sq.* **2020**, preprint. [CrossRef]
141. Blaut, M.; Schoefer, L.; Braune, A. Transformation of flavonoids by intestinal microorganisms. *Int. J. Vitam. Nutr. Res.* **2003**, *73*, 79–87. [CrossRef] [PubMed]
142. Schoefer, L.; Braune, A.; Blaut, M. Cloning and expression of a phloretin hydrolase gene from *Eubacterium ramulus* and characterization of the recombinant enzyme. *Appl. Environ. Microbiol.* **2004**, *70*, 6131–6137. [CrossRef]
143. Wang, M.; Firrman, J.; Zhang, L.; Arango-Argoty, G.; Tomasula, P.; Liu, L.; Xiao, W.; Yam, K. Apigenin impacts the growth of the gut microbiota and alters the gene expression of *Enterococcus*. *Molecules* **2017**, *22*, 1292. [CrossRef] [PubMed]
144. Qiao, Y.; Zhang, Z.; Zhai, Y.; Yan, X.; Zhou, W.; Liu, H.; Guan, L.; Peng, L. Apigenin alleviates obesity-associated metabolic syndrome by regulating the composition of the gut microbiome. *Front. Microbiol.* **2021**, *12*, 805827. [CrossRef] [PubMed]
145. Zhang, H.; Troise, A.D.; Qi, Y.; Wu, G.; Zhang, H.; Fogliano, V. Insoluble dietary fibre scavenges reactive carbonyl species under simulated physiological conditions: The key role of fibre-bound polyphenols. *Food Chem.* **2021**, *349*, 129018. [CrossRef] [PubMed]
146. Williamson, G.; Dionisi, F.; Renouf, M. Flavanols from green tea and phenolic acids from coffee: Critical quantitative evaluation of the pharmacokinetic data in humans after consumption of single doses of beverages. *Mol. Nutr. Food Res.* **2011**, *55*, 864–873. [CrossRef] [PubMed]
147. Liang, N.; Kitts, D.D. Role of chlorogenic acids in controlling oxidative and inflammatory stress conditions. *Nutrients* **2015**, *8*, 16. [CrossRef]
148. Zhang, Z.; Wu, X.; Cao, S.; Cromie, M.; Shen, Y.; Feng, Y.; Yang, H.; Li, L. Chlorogenic acid ameliorates experimental colitis by promoting growth of *Akkermansia* in mice. *Nutrients* **2017**, *9*, 677. [CrossRef]
149. Zhang, Z.; Wu, X.; Cao, S.; Wang, L.; Wang, D.; Yang, H.; Feng, Y.; Wang, S.; Li, L. Caffeic acid ameliorates colitis in association with increased *Akkermansia* population in the gut microbiota of mice. *Oncotarget* **2016**, *7*, 31790–31799. [CrossRef]
150. Zhang, Z.; Wang, D.; Qiao, S.; Wu, X.; Cao, S.; Wang, L.; Su, X.; Li, L. Metabolic and microbial signatures in rat hepatocellular carcinoma treated with caffeic acid and chlorogenic acid. *Sci. Rep.* **2017**, *7*, 4508. [CrossRef]
151. Zhang, X.; Zhang, M.; Dong, L.; Jia, X.; Liu, L.; Ma, Y.; Huang, F.; Zhang, R. Phytochemical profile, bioactivity, and prebiotic potential of bound phenolics released from rice bran dietary fiber during in vitro gastrointestinal digestion and colonic fermentation. *J. Agric. Food Chem.* **2019**, *67*, 12796–12805. [CrossRef] [PubMed]
152. Fritsch, C.; Jansch, A.; Ehrmann, M.A.; Toelstede, S.; Vogel, R.F. Characterization of cinnamoyl esterases from different Lactobacilli and Bifidobacteria. *Curr. Microbiol.* **2017**, *74*, 247–256. [CrossRef] [PubMed]
153. Esteban-Torres, M.; Landete, J.M.; Reveron, I.; Santamaria, L.; de las Rivas, B.; Munoz, R. A *Lactobacillus plantarum* esterase active on a broad range of phenolic esters. *Appl. Environ. Microbiol.* **2015**, *81*, 3235–3242. [CrossRef] [PubMed]
154. Santamaria, L.; Reveron, I.; Lopez de Felipe, F.; de Las Rivas, B.; Munoz, R. Unravelling the reduction pathway as an alternative metabolic route to hydroxycinnamate decarboxylation in *Lactobacillus plantarum*. *Appl. Environ. Microbiol.* **2018**, *84*, e01123-18. [CrossRef] [PubMed]
155. Poti, F.; Santi, D.; Spaggiari, G.; Zimetti, F.; Zanotti, I. Polyphenol health effects on cardiovascular and neurodegenerative disorders: A review and meta-analysis. *Int. J. Mol. Sci.* **2019**, *20*, 351. [CrossRef] [PubMed]
156. Iglesias-Aguirre, C.E.; Vallejo, F.; Beltran, D.; Berna, J.; Puigcerver, J.; Alajarin, M.; Selma, M.V.; Espin, J.C. 4-Hydroxydibenzyl: A novel metabolite from the human gut microbiota after consuming resveratrol. *Food Funct* **2022**, *13*, 7487–7493. [CrossRef]
157. Bode, L.M.; Bunzel, D.; Huch, M.; Cho, G.S.; Ruhland, D.; Bunzel, M.; Bub, A.; Franz, C.M.; Kulling, S.E. In vivo and in vitro metabolism of trans-resveratrol by human gut microbiota. *Am. J. Clin. Nutr.* **2013**, *97*, 295–309. [CrossRef]
158. Chen, M.L.; Yi, L.; Zhang, Y.; Zhou, X.; Ran, L.; Yang, J.; Zhu, J.D.; Zhang, Q.Y.; Mi, M.T. Resveratrol attenuates trimethylamine-N-oxide (TMAO)-induced atherosclerosis by regulating TMAO synthesis and bile acid metabolism via remodeling of the gut microbiota. *mBio* **2016**, *7*, e02210-15. [CrossRef]
159. Chen, M.; Hou, P.; Zhou, M.; Ren, Q.; Wang, X.; Huang, L.; Hui, S.; Yi, L.; Mi, M. Resveratrol attenuates high-fat diet-induced non-alcoholic steatohepatitis by maintaining gut barrier integrity and inhibiting gut inflammation through regulation of the endocannabinoid system. *Clin. Nutr.* **2020**, *39*, 1264–1275. [CrossRef]

160. Alrafas, H.R.; Busbee, P.B.; Nagarkatti, M.; Nagarkatti, P.S. Resveratrol modulates the gut microbiota to prevent murine colitis development through induction of Tregs and suppression of Th17 cells. *J. Leukoc. Biol.* **2019**, *106*, 467–480. [CrossRef]
161. Walker, J.M.; Eckardt, P.; Aleman, J.O.; Correa da Rosa, J.; Liang, Y.; Lizumi, T.; Etheve, S.; Blaser, M.J.; Breslow, J.L.; Holt, P.R. The effects of trans-resveratrol on insulin resistance, inflammation, and microbiota in men with the metabolic syndrome: A pilot randomized, placebo controlled clinical trial. *J. Clin. Transl. Res.* **2018**, *4*, 122–135. [PubMed]
162. Etxeberria, U.; Hijona, E.; Aguirre, L.; Milagro, F.I.; Bujanda, L.; Rimando, A.M.; Martinez, J.A.; Portillo, M.P. Pterostilbene-induced changes in gut microbiota composition in relation to obesity. *Mol. Nutr. Food Res.* **2017**, *61*, 1500906. [CrossRef] [PubMed]
163. Kala, R.; Tollefsbol, T.O.; Li, Y. Potential of resveratrol in inhibiting cancer and slowing aging. *J. Nutr. Sci.* **2012**, *5*, 2. [CrossRef]
164. Quartieri, A.; Garcia-Villalba, R.; Amaretti, A.; Raimondi, S.; Leonardi, A.; Rossi, M.; Tomas-Barberan, F. Detection of novel metabolites of flaxseed lignans in vitro and in vivo. *Mol. Nutr. Food Res.* **2016**, *60*, 1590–1601. [CrossRef]
165. Clavel, T.; Borrmann, D.; Braune, A.; Dore, J.; Blaut, M. Occurrence and activity of human intestinal bacteria involved in the conversion of dietary lignans. *Anaerobe* **2006**, *12*, 140–147. [CrossRef]
166. Mabrok, H.B.; Klopfleisch, R.; Ghanem, K.Z.; Clavel, T.; Blaut, M.; Loh, G. Lignan transformation by gut bacteria lowers tumor burden in a gnotobiotic rat model of breast cancer. *Carcinogenesis* **2012**, *33*, 203–208. [CrossRef]
167. Power, K.A.; Lepp, D.; Zarepoor, L.; Monk, J.M.; Wu, W.; Tsao, R.; Liu, R. Dietary flaxseed modulates the colonic microenvironment in healthy C57Bl/6 male mice which may alter susceptibility to gut-associated diseases. *J. Nutr. Biochem.* **2016**, *28*, 61–69. [CrossRef]
168. Cho, S.Y.; Kim, J.; Lee, J.H.; Sim, J.H.; Cho, D.H.; Bae, I.H.; Lee, H.; Seol, M.A.; Shin, H.M.; Kim, T.J.; et al. Modulation of gut microbiota and delayed immunosenescence as a result of syringaresinol consumption in middle-aged mice. *Sci. Rep.* **2016**, *6*, 39026. [CrossRef]
169. Timmers, S.; Konings, E.; Bilet, L.; Houtkooper, R.H.; van de Weijer, T.; Goossens, G.H.; Hoeks, J.; van der Krieken, S.; Ryu, D.; Kersten, S.; et al. Calorie restriction-like effects of 30 days of resveratrol supplementation on energy metabolism and metabolic profile in obese humans. *Cell Metab.* **2011**, *14*, 612–622. [CrossRef]
170. Korsholm, A.S.; Kjaer, T.N.; Ornstrup, M.J.; Pedersen, S.B. Comprehensive metabolomic analysis in blood, urine, fat, and muscle in men with metabolic syndrome: A randomized, placebo-controlled clinical trial on the effects of resveratrol after four months' treatment. *Int. J. Mol. Sci.* **2017**, *18*, 544. [CrossRef]
171. Xu, D.; Feng, M.; Chu, Y.; Wang, S.; Shete, V.; Tuohy, K.M.; Liu, F.; Zhou, X.; Kamil, A.; Pan, D.; et al. The prebiotic effects of oats on blood lipids, gut microbiota, and short-chain fatty acids in mildly hypercholesterolemic subjects compared with rice: A randomized, controlled trial. *Front. Immunol.* **2021**, *12*, 787797. [CrossRef] [PubMed]
172. Most, J.; Penders, J.; Lucchesi, M.; Goossens, G.H.; Blaak, E.E. Gut microbiota composition in relation to the metabolic response to 12-week combined polyphenol supplementation in overweight men and women. *Eur. J. Clin. Nutr.* **2017**, *71*, 1040–1045. [CrossRef] [PubMed]
173. Song, M.-y.; Wang, J.-h.; Eom, T.; Kim, H. *Schisandra chinensis* fruit modulates the gut microbiota composition in association with metabolic markers in obese women: A randomized, double-blind placebo-controlled study. *Nutr. Res.* **2015**, *35*, 655–663. [CrossRef] [PubMed]
174. Kim, H.; Venancio, V.P.; Fang, C.; Dupont, A.W.; Talcott, S.T.; Mertens-Talcott, S.U. Mango (*Mangifera indica* L.) polyphenols reduce IL-8, GRO, and GM-SCF plasma levels and increase *Lactobacillus* species in a pilot study in patients with inflammatory bowel disease. *Nutr. Res.* **2020**, *75*, 85–94. [CrossRef]
175. Tomás-Barberán, F.A.; González-Sarrías, A.; García-Villalba, R.; Núñez-Sánchez, M.A.; Selma, M.V.; García-Conesa, M.T.; Espín, J. Urolithins, the rescue of "old" metabolites to understand a "new" concept: Metabotypes as a nexus among phenolic metabolism, microbiota dysbiosis, and host health status. *Mol. Nutr. Food. Res.* **2017**, *61*, 1500901. [CrossRef]
176. Atkinson, C.; Newton, K.M.; Bowles, E.J.; Yong, M.; Lampe, J.W. Demographic, anthropometric, and lifestyle factors and dietary intakes in relation to daidzein-metabolizing phenotypes among premenopausal women in the United States. *Am. J. Clin. Nutr.* **2008**, *87*, 679–687. [CrossRef]
177. Selma, M.V.; Beltran, D.; Garcia-Villalba, R.; Espin, J.C.; Tomas-Barberan, F.A. Description of urolithin production capacity from ellagic acid of two human intestinal *Gordonibacter* species. *Food Funct.* **2014**, *5*, 1779–1784. [CrossRef]
178. Cortes-Martin, A.; Romo-Vaquero, M.; Garcia-Mantrana, I.; Rodriguez-Varela, A.; Collado, M.C.; Espin, J.C.; Selma, M.V. Urolithin metabotypes can anticipate the different restoration of the gut microbiota and anthropometric profiles during the first year postpartum. *Nutrients* **2019**, *11*, 2079. [CrossRef]
179. Tomas-Barberan, F.A.; Garcia-Villalba, R.; Gonzalez-Sarrias, A.; Selma, M.V.; Espin, J.C. Ellagic acid metabolism by human gut microbiota: Consistent observation of three urolithin phenotypes in intervention trials, independent of food source, age, and health status. *J. Agric. Food Chem.* **2014**, *62*, 6535–6538. [CrossRef]
180. Tomas-Barberan, F.A.; Selma, M.V.; Espin, J.C. Polyphenols' gut microbiota metabolites: Bioactives or biomarkers? *J. Agric. Food Chem.* **2018**, *66*, 3593–3594. [CrossRef]
181. Tomas-Barberan, F.A.; Selma, M.V.; Espin, J.C. Interactions of gut microbiota with dietary polyphenols and consequences to human health. *Curr. Opin. Clin. Nutr. Metab. Care* **2016**, *19*, 471–476. [CrossRef] [PubMed]

182. Cortes-Martin, A.; Garcia-Villalba, R.; Garcia-Mantrana, I.; Rodriguez-Varela, A.; Romo-Vaquero, M.; Collado, M.C.; Tomas-Barberan, F.A.; Espin, J.C.; Selma, M.V. Urolithins in human breast milk after walnut intake and kinetics of *Gordonibacter* colonization in newly born: The role of mothers' urolithin metabotypes. *J. Agric. Food Chem.* **2020**, *68*, 12606–12616. [CrossRef] [PubMed]
183. Li, X.; Watanabe, K.; Kimura, I. Gut microbiota dysbiosis drives and implies novel therapeutic strategies for diabetes mellitus and related metabolic diseases. *Front. Immunol.* **2017**, *8*, 1882. [CrossRef] [PubMed]
184. Van Herreweghen, F.; De Paepe, K.; Marzorati, M.; Van de Wiele, T. Mucin as a functional niche is a more important driver of in vitro gut microbiota composition and functionality than supplementation of *Akkermansia muciniphila*. *Appl Environ. Microbiol.* **2020**, *87*, e02647-20. [CrossRef]
185. Liu, F.; Cui, Y.; Yang, F.; Xu, Z.; Da, L.T.; Zhang, Y. Inhibition of polypeptide N-acetyl-alpha-galactosaminyltransferases is an underlying mechanism of dietary polyphenols preventing colorectal tumorigenesis. *Bioorg. Med. Chem.* **2019**, *27*, 3372–3382. [CrossRef]
186. Gabrielli, M.G.; Tomassoni, D. Starch-enriched diet modulates the glucidic profile in the rat colonic mucosa. *Eur. J. Nutr.* **2018**, *57*, 1109–1121. [CrossRef]
187. Lu, F.; Li, Y.; Wang, X.; Hu, X.; Liao, X.; Zhang, Y. Early-life polyphenol intake promotes *Akkermansia* growth and increase of host goblet cells in association with the potential synergistic effect of *Lactobacillus*. *Food Res. Int.* **2021**, *149*, 110648. [CrossRef]
188. Alvarado, D.M.; Chen, B.; Iticovici, M.; Thaker, A.I.; Dai, N.; VanDussen, K.L.; Shaikh, N.; Lim, C.K.; Guillemin, G.J.; Tarr, P.I.; et al. Epithelial indoleamine 2,3-dioxygenase 1 modulates aryl hydrocarbon receptor and notch signaling to increase differentiation of secretory cells and alter mucus-associated microbiota. *Gastroenterology* **2019**, *157*, 1093–1108.e11. [CrossRef]
189. Powell, D.N.; Swimm, A.; Sonowal, R.; Bretin, A.; Gewirtz, A.T.; Jones, R.M.; Kalman, D. Indoles from the commensal microbiota act via the AHR and IL-10 to tune the cellular composition of the colonic epithelium during aging. *Proc. Natl. Acad. Sci. USA* **2020**, *117*, 21519–21526. [CrossRef]
190. Stockinger, B.; Shah, K.; Wincent, E. AHR in the intestinal microenvironment: Safeguarding barrier function. *Nat. Rev. Gastroenterol. Hepatol.* **2021**, *18*, 559–570. [CrossRef]
191. Wang, J.; Wang, P.; Tian, H.; Tian, F.; Zhang, Y.; Zhang, L.; Gao, X.; Wang, X. Aryl hydrocarbon receptor/IL-22/Stat3 signaling pathway is involved in the modulation of intestinal mucosa antimicrobial molecules by commensal microbiota in mice. *Innate Immun.* **2018**, *24*, 297–306. [CrossRef] [PubMed]
192. Goya-Jorge, E.; Jorge Rodriguez, M.E.; Veitia, M.S.; Giner, R.M. Plant occurring flavonoids as modulators of the aryl hydrocarbon receptor. *Molecules* **2021**, *26*, 2315. [CrossRef] [PubMed]
193. Wrzosek, L.; Ciocan, D.; Hugot, C.; Spatz, M.; Dupeux, M.; Houron, C.; Lievin-Le Moal, V.; Puchois, V.; Ferrere, G.; Trainel, N.; et al. Microbiota tryptophan metabolism induces aryl hydrocarbon receptor activation and improves alcohol-induced liver injury. *Gut* **2021**, *70*, 1299–1308. [CrossRef]
194. Carambia, A.; Schuran, F.A. The aryl hydrocarbon receptor in liver inflammation. *Semin. Immunopathol.* **2021**, *43*, 563–575. [CrossRef] [PubMed]
195. Rothhammer, V.; Borucki, D.M.; Kenison, J.E.; Hewson, P.; Wang, Z.; Bakshi, R.; Sherr, D.H.; Quintana, F.J. Detection of aryl hydrocarbon receptor agonists in human samples. *Sci. Rep.* **2018**, *8*, 4970. [CrossRef]
196. Natividad, J.M.; Agus, A.; Planchais, J.; Lamas, B.; Jarry, A.C.; Martin, R.; Michel, M.L.; Chong-Nguyen, C.; Roussel, R.; Straube, M.; et al. Impaired aryl hydrocarbon receptor ligand production by the gut microbiota is a key factor in metabolic syndrome. *Cell Metab.* **2018**, *28*, 737–749.e4. [CrossRef]
197. Shi, Z.; Lei, H.; Chen, G.; Yuan, P.; Cao, Z.; Ser, H.L.; Zhu, X.; Wu, F.; Liu, C.; Dong, M.; et al. Impaired Intestinal *Akkermansia muciniphila* and aryl hydrocarbon receptor ligands contribute to nonalcoholic fatty liver disease in mice. *mSystems* **2021**, *6*, e00985-20. [CrossRef]
198. Jin, U.H.; Park, H.; Li, X.; Davidson, L.A.; Allred, C.; Patil, B.; Jayaprakasha, G.; Orr, A.A.; Mao, L.; Chapkin, R.S.; et al. Structure-dependent modulation of aryl hydrocarbon receptor-mediated activities by flavonoids. *Toxicol. Sci.* **2018**, *164*, 205–217. [CrossRef]
199. Koper, J.E.B.; Loonen, L.M.P.; Wells, J.M.; Troise, A.D.; Capuano, E.; Fogliano, V. Polyphenols and tryptophan metabolites activate the aryl hydrocarbon receptor in an in vitro model of colonic fermentation. *Mol. Nutr. Food Res.* **2019**, *63*, e1800722. [CrossRef]
200. Singh, R.; Chandrashekharappa, S.; Bodduluri, S.R.; Baby, B.V.; Hegde, B.; Kotla, N.G.; Hiwale, A.A.; Saiyed, T.; Patel, P.; Vijay-Kumar, M.; et al. Enhancement of the gut barrier integrity by a microbial metabolite through the Nrf2 pathway. *Nat. Commun.* **2019**, *10*, 89. [CrossRef]
201. Marques, C.; Fernandes, I.; Meireles, M.; Faria, A.; Spencer, J.P.E.; Mateus, N.; Calhau, C. Gut microbiota modulation accounts for the neuroprotective properties of anthocyanins. *Sci. Rep.* **2018**, *8*, 11341. [CrossRef] [PubMed]
202. Rannug, A. How the AhR became important in intestinal homeostasis- A Diurnal FICZ/AhR/CYP1A1 feedback controls both Iimunity and immunopathology. *Int. J. Mol. Sci.* **2020**, *21*, 5681. [CrossRef] [PubMed]
203. Monteleone, I.; Rizzo, A.; Sarra, M.; Sica, G.; Sileri, P.; Biancone, L.; MacDonald, T.T.; Pallone, F.; Monteleone, G. Aryl hydrocarbon receptor-induced signals up-regulate IL-22 production and inhibit inflammation in the gastrointestinal tract. *Gastroenterology* **2011**, *141*, 237–248.e1. [CrossRef] [PubMed]

204. Meynier, M.; Baudu, E.; Rolhion, N.; Defaye, M.; Straube, M.; Daugey, V.; Modoux, M.; Wawrzyniak, I.; Delbac, F.; Villeger, R.; et al. AhR/IL-22 pathway as new target for the treatment of post-infectious irritable bowel syndrome symptoms. *Gut Microbes* **2022**, *14*, 2022997. [CrossRef]
205. Han, H.; Davidson, L.A.; Fan, Y.Y.; Landrock, K.K.; Jayaraman, A.; Safe, S.H.; Chapkin, R.S. Loss of aryl hydrocarbon receptor suppresses the response of colonic epithelial cells to IL22 signaling by upregulating SOCS3. *Am. J. Physiol. Gastrointest. Liver Physiol.* **2022**, *322*, G93–G106. [CrossRef]
206. Rohr, M.W.; Narasimhulu, C.A.; Rudeski-Rohr, T.A.; Parthasarathy, S. Negative effects of a high-fat diet on intestinal permeability: A review. *Adv. Nutr.* **2020**, *11*, 77–91. [CrossRef]
207. Roth, S.; Spalinger, M.R.; Gottier, C.; Biedermann, L.; Zeitz, J.; Lang, S.; Weber, A.; Rogler, G.; Scharl, M. Bilberry-derived anthocyanins modulate cytokine expression in the intestine of patients with ulcerative colitis. *PLoS ONE* **2016**, *11*, e0154817. [CrossRef]
208. Zelante, T.; Iannitti, R.G.; Cunha, C.; De Luca, A.; Giovannini, G.; Pieraccini, G.; Zecchi, R.; D'Angelo, C.; Massi-Benedetti, C.; Fallarino, F.; et al. Tryptophan catabolites from microbiota engage aryl hydrocarbon receptor and balance mucosal reactivity via interleukin-22. *Immunity* **2013**, *39*, 372–385. [CrossRef]
209. Tabasco, R.; Sánchez-Patán, F.; Monagas, M.; Bartolomé, B.; Moreno-Arribas, V.M.; Peláez, C.; Requena, T. Effect of grape polyphenols on lactic acid bacteria and bifidobacteria growth: Resistance and metabolism. *Food Microbiol.* **2011**, *28*, 1345–1352. [CrossRef]
210. Dong, J.; Qiu, J.; Wang, J.; Li, H.; Dai, X.; Zhang, Y.; Wang, X.; Tan, W.; Niu, X.; Deng, X.; et al. Apigenin alleviates the symptoms of *Staphylococcus aureus* pneumonia by inhibiting the production of alpha-hemolysin. *FEMS Microbiol. Lett.* **2013**, *338*, 124–131. [CrossRef]
211. Lopez de Felipe, F.; de Las Rivas, B.; Munoz, R. Molecular responses of Lactobacilli to plant phenolic compounds: A comparative review of the mechanisms involved. *Antioxidants* **2021**, *11*, 18. [CrossRef] [PubMed]
212. Yu, Y.; Yi, Z.B.; Liang, Y.Z. Validate antibacterial mode and find main bioactive components of traditional Chinese medicine Aquilegia oxysepala. *Bioorg. Med. Chem. Lett.* **2007**, *17*, 1855–1859. [CrossRef] [PubMed]
213. Wu, D.; Kong, Y.; Han, C.; Chen, J.; Hu, L.; Jiang, H.; Shen, X. D-Alanine:D-alanine ligase as a new target for the flavonoids quercetin and apigenin. *Int. J. Antimicrob. Agents* **2008**, *32*, 421–426. [CrossRef] [PubMed]
214. Alvarez-Martinez, F.J.; Barrajon-Catalan, E.; Encinar, J.A.; Rodriguez-Diaz, J.C.; Micol, V. Antimicrobial capacity of plant polyphenols against Gram-positive bacteria: A comprehensive review. *Curr. Med. Chem.* **2020**, *27*, 2576–2606. [CrossRef]
215. Hansen, C.H.F.; Krych, L.; Nielsen, D.S.; Diabetologia, V.-F.K. Early life treatment with vancomycin propagates *Akkermansia muciniphila* and reduces diabetes incidence in the NOD mouse. *Diabetologia* **2012**, *55*, 2285–2294. [CrossRef]
216. Dubourg, G.; Lagier, J.-C.; Armougom, F.; Robert, C.; Audoly, G.; Papazian, L.; Raoult, D. High-level colonisation of the human gut by Verrucomicrobia following broad-spectrum antibiotic treatment. *Int. J. Antimicrob. Agents* **2013**, *41*, 149–155. [CrossRef]
217. Hagi, T.; Geerlings, S.Y.; Nijsse, B.; Belzer, C. The effect of bile acids on the growth and global gene expression profiles in *Akkermansia muciniphila*. *Appl. Microbiol. Biotechnol.* **2020**, *104*, 10641–10653. [CrossRef]
218. Shin, J.; Noh, J.-R.; Chang, D.-H.; Kim, Y.-H.; Kim, M.; Lee, E.; Cho, S.; Ku, B.; Rhee, M.-S.; Kim, B.-C.; et al. Elucidation of *Akkermansia muciniphila* probiotic traits driven by mucin depletion. *Front. Microbiol.* **2019**, *10*, 1137. [CrossRef]
219. Das, Q.; Lepp, D.; Yin, X.; Ross, K.; McCallum, J.L.; Warriner, K.; Marcone, M.F.; Diarra, M.S. Transcriptional profiling of *Salmonella enterica* serovar Enteritidis exposed to ethanolic extract of organic cranberry pomace. *PLoS ONE* **2019**, *14*, e0219163. [CrossRef]
220. Curiel, J.A.; Rodríguez, H.; De la Rivas, B.; Anglade, O.; Baraige, F.; Zagorec, M.; Champomier-Vergès, M.; Muñoz, R.; de Felipe, F. Response of a *Lactobacillus plantarum* human isolate to tannic acid challenge assessed by proteomic analyses. *Mol. Nutr. Food Res.* **2011**, *55*, 1454–1465. [CrossRef]
221. Reveron, I.; de las Rivas, B.; Matesanz, R.; Munoz, R.; Lopez de Felipe, F. Molecular adaptation of *Lactobacillus plantarum* WCFS1 to gallic acid revealed by genome-scale transcriptomic signature and physiological analysis. *Microb. Cell Fact.* **2015**, *14*, 160. [CrossRef] [PubMed]
222. Khalil, R.K.S. Influence of gallic acid and catechin polyphenols on probiotic properties of *Streptococcus thermophilus* CHCC 3534 strain. *World J. Microbiol. Biotechnol.* **2010**, *26*, 2069–2079. [CrossRef]
223. Firrman, J.; Liu, L.; Zhang, L.; Argoty, G.; Wang, M.; Tomasula, P.; Kobori, M.; Pontious, S.; Xiao, W. The effect of quercetin on genetic expression of the commensal gut microbes *Bifidobacterium catenulatum, Enterococcus caccae* and *Ruminococcus gauvreauii*. *Anaerobe* **2016**, *42*, 130–141. [CrossRef] [PubMed]
224. Cho, Y.S.; Schiller, N.L.; Kahng, H.Y.; Oh, K.H. Cellular responses and proteomic analysis of *Escherichia coli* exposed to green tea polyphenols. *Curr. Microbiol.* **2007**, *55*, 501–506. [CrossRef] [PubMed]
225. Simmering, R.; Pforte, H.; Jacobasch, G.; Blaut, M. The growth of the flavonoid-degrading intestinal bacterium, *Eubacterium ramulus*, is stimulated by dietary flavonoids in vivo. *FEMS Microbiol. Ecol.* **2002**, *40*, 243–248. [CrossRef] [PubMed]
226. Vandeputte, D.; Kathagen, G.; D'Hoe, K.; Vieira-Silva, S.; Valles-Colomer, M.; Sabino, J.; Wang, J.; Tito, R.Y.; De Commer, L.; Darzi, Y.; et al. Quantitative microbiome profiling links gut community variation to microbial load. *Nature* **2017**, *551*, 507–511. [CrossRef] [PubMed]
227. Vandeputte, D.; Falony, G.; Vieira-Silva, S.; Tito, R.Y.; Joossens, M.; Raes, J. Stool consistency is strongly associated with gut microbiota richness and composition, enterotypes and bacterial growth rates. *Gut* **2016**, *65*, 57–62. [CrossRef]

228. Silva, M.P.; Martelli-Tosi, M.; Massarioli, A.P.; Melo, P.S.; Alencar, S.M.; Favaro-Trindade, C.S. Co-encapsulation of guaraná extracts and probiotics increases probiotic survivability and simultaneously delivers bioactive compounds in simulated gastrointestinal fluids. *LWT* **2022**, *161*, 113351. [CrossRef]
229. Chang, Y.; Yang, Y.; Xu, N.; Mu, H.; Zhang, H.; Duan, J. Improved viability of *Akkermansia muciniphila* by encapsulation in spray dried succinate-grafted alginate doped with epigallocatechin-3-gallate. *Int. J. Biol. Macromol.* **2020**, *159*, 373–382. [CrossRef]
230. Kvakova, M.; Bertkova, I.; Stofilova, J.; Savidge, T.C. Co-encapsulated synbiotics and immobilized probiotics in human health and gut microbiota modulation. *Foods* **2021**, *10*, 1297. [CrossRef]

International Journal of *Molecular Sciences*

MDPI

Article

Integrated Multi-Omics Analysis Reveals Differential Effects of Fructo-Oligosaccharides (FOS) Supplementation on the Human Gut Ecosystem

Tamotsu Kato [1,2,†], Masaharu Kagawa [3,†], Wataru Suda [4], Yuuri Tsuboi [5], Sayo Inoue-Suzuki [1,2], Jun Kikuchi [5,6], Masahira Hattori [4,7], Toshiko Ohta [1,3,8] and Hiroshi Ohno [1,2,*]

1 Laboratory for Intestinal Ecosystem, RIKEN Center for Integrative Medical Sciences (IMS), Kanagawa 230-0045, Japan
2 Immunobiology Laboratory, Graduate School of Medical Life Science, Yokohama City University, Kanagawa 230-0045, Japan
3 Institute of Nutrition Sciences, Kagawa Nutrition University, Saitama 350-0288, Japan
4 Laboratory for Microbiome Sciences, RIKEN Center for Integrative Medical Sciences (IMS), Kanagawa 230-0045, Japan
5 Environmental Metabolic Analysis Research Team, RIKEN Center for Sustainable Resource Sciences (CSRS), Kanagawa 230-0045, Japan
6 Environmental Factor Analysis Laboratory, Graduate School of Medical Life Science, Yokohama City University, Kanagawa 230-0045, Japan
7 Graduate School of Advanced Science and Engineering, Waseda University, Tokyo 169-8555, Japan
8 Japan Aerospace Exploration Agency (JAXA), Ibaraki 305-8505, Japan
* Correspondence: hiroshi.ohno@riken.jp; Tel.:+81-45-503-7031
† These authors contributed equally to this work.

Citation: Kato, T.; Kagawa, M.; Suda, W.; Tsuboi, Y.; Inoue-Suzuki, S.; Kikuchi, J.; Hattori, M.; Ohta, T.; Ohno, H. Integrated Multi-Omics Analysis Reveals Differential Effects of Fructo-Oligosaccharides (FOS) Supplementation on the Human Gut Ecosystem. *Int. J. Mol. Sci.* **2022**, *23*, 11728. https://doi.org/10.3390/ijms231911728

Academic Editors: Baojun Xu and Matteo Bordiga

Received: 30 August 2022
Accepted: 26 September 2022
Published: 3 October 2022

Abstract: Changes in the gut ecosystem, including the microbiome and the metabolome, and the host immune system after fructo-oligosaccharide (FOS) supplementation were evaluated. The supplementation of FOS showed large inter-individual variability in the absolute numbers of fecal bacteria and an increase in *Bifidobacterium*. The fecal metabolome analysis revealed individual variability in fructose utilization in response to FOS supplementation. In addition, immunoglobulin A(IgA) tended to increase upon FOS intake, and peripheral blood monocytes significantly decreased upon FOS intake and kept decreasing in the post-FOS phase. Further analysis using a metagenomic approach showed that the differences could be at least in part due to the differences in gene expressions of enzymes that are involved in the fructose metabolism pathway. While the study showed individual differences in the expected health benefits of FOS supplementation, the accumulation of "personalized" knowledge of the gut ecosystem with its genetic expression may enable effective instructions on prebiotic consumption to optimize health benefits for individuals in the future.

Keywords: fructo-oligosaccharides; microbiome; gut ecosystem; omics analysis; metabolome

1. Introduction

The commensal microbiota is a complex microbial community that consists of huge numbers of microbes that reside in both mucosal and external surfaces of the human body and their metabolites [1]. Commensal microbiota on the skin, mouth, digestive tract, and vagina are known to have different compositions [2]. With a comparable number to the eukaryotic cells constituting a host human body, gut commensal microbes construct a symbiotic ecosystem with the host [3]. Along with other various functions, the microbiome has been reported to affect the immune function of the host and has also been considered to play an important role in the maintenance of our health [2,4,5].

In the gastrointestinal tract, a vast majority of approximately 1000 bacterial species reside in the colon [6–8]. Gut microbiota has been reported to be influenced by a number of factors, including the diet of the host [9,10]. Among dietary components, substrates selectively utilized by host microorganisms to confer a health benefit are defined as prebiotics [11]. Non-digestible oligosaccharides such as fructo-oligosaccharides (FOS) are common prebiotics that particularly assist the growth of lactic acid-producing bacteria species (e.g., *Bifidobacterium* and *Lactobacillus*) [12,13]. FOS are also known to have a number of beneficial effects on the immune system, including increased production of short-chain fatty acids (SCFAs), enhancement of immune responses to lipopolysaccharides (LPS) [14], and also augmented secretion of Immunoglobulin A (IgA) from Peyer's patch cells [15,16]. However, a vast majority of these findings on FOS supplementation were from animal studies, and results from human subjects are limited. Considering a large inter-individual variability in the human gut microbial composition [2,17] and the presence of bacterial strains that differ in carbohydrate metabolism [18–20], it is hypothesized that the effects of FOS supplementation on the human gut ecosystem and its health benefits may vary between individuals. Additionally, there have been scarce studies that use comprehensive approaches to explore the effects of FOS supplementation on the human gut ecosystem, including metabolites and health benefits to the host. Therefore, the present study aimed to evaluate the effects of FOS supplementation on the gut microbiome and immune cells of the host using an integrated multi-omics analysis.

2. Results

A total of 11 Japanese males underwent a nine-week longitudinal intervention study with a four-week FOS supplementation. The study period was divided into three distinct phases (i.e., pre, FOS, and post phases). At each phase, samples of feces and blood were collected twice from all participants (Supplementary Figure S1).

2.1. Increase of *Bifidobacterium* in Feces by FOS Supplementation

From 16S rRNA gene amplicon sequencing of extracted bacterial DNAs, fecal samples showed large inter-individual variability in the total number of bacteria and their response to the FOS supplementation that resulted in no significant differences in the total number of bacteria in the feces (Figure 1a). While major bacterial genera that constituted the composition of the microbiome were consistent among participants, the number of bacteria in each genus and their response to the FOS supplementation varied individually (Figure 1b). PCoA using the Bray–Curtis distance based on the 16S rRNA gene amplicon sequencing confirmed the significant inter-individual variabilities of microbiome in the fecal samples (Figure 1c). Similar results were evident when the bacterial composition of fecal microbiota was analyzed based on the relative abundance (Supplementary Figure S3a,b). Among the major bacteria that constitute intestinal microbiota, the number of *Bifidobacterium* was found to be significantly ($p < 0.05$) increased during the FOS supplementation period (Figure 1d). By contrast, when the samples were analyzed based on a relative abundance using the Wilcoxon test with the Benjamini-Hochberg adjustment, although the difference was smaller, a significant increase in *Bifidobacterium* and a significant decrease in *Blautia* were observed during the FOS phase ($p < 0.05$ or $p < 0.01$, Supplementary Figure S3c). A significantly ($p < 0.05$) lower Shannon index at the FOS phase compared with both pre and post phases may indicate an occurrence of decreased fecal microbial diversity as a result of FOS intake (Supplementary Figure S3d).

2.2. Increase of Fecal Fructose during the FOS Intake in Some Individuals

Next, we evaluated the effects of FOS intake on the gut metabolome using nuclear magnetic resonance (NMR)-based measurement of fecal metabolites. We detected 32 metabolites as listed in Table S3. Consistent with the results observed from PCoA on the gut microbiota, principal component analysis (PCA) of NMR-based fecal metabolites analysis showed a large inter-individual variability of gut metabolites and no particular clustering or distribution among different phases (Figure 2a). A detailed examination of PCA on the fecal metabolome data revealed peaks corresponding to carbohydrates toward the PC2 negative direction (Figure 2b, rectangle).

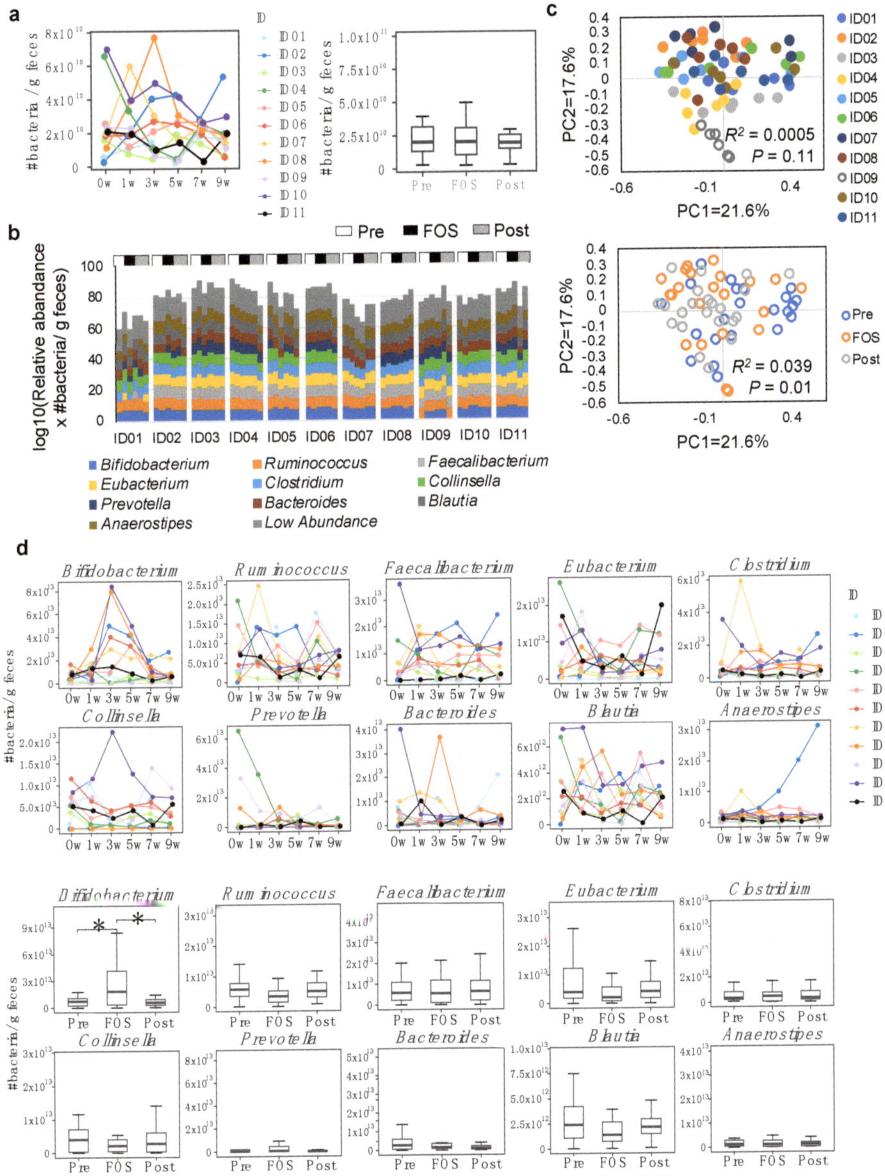

Figure 1. Changes in the total number of bacteria in the fecal microbiota upon FOS supplementation. (**a**) The absolute number of bacteria in each individual by real-time 16S rRNA gene qPCR. (**b**) The composition of microbiota at the genus level in fecal samples using 16S rRNA gene amplicon sequencing. These figures were constructed by the top 10 bacteria at the genus level, respectively. (**c**) PCoA analysis using Bray–Curtis distance of fecal microbiome colored by individuals (**Upper**) and sampling periods (**Lower**). (**d**) The total number of each bacterium in feces at the genus level. The upper figure is illustrated for each individual and the lower figure is summarized for the sample period. * = padj < 0.05.

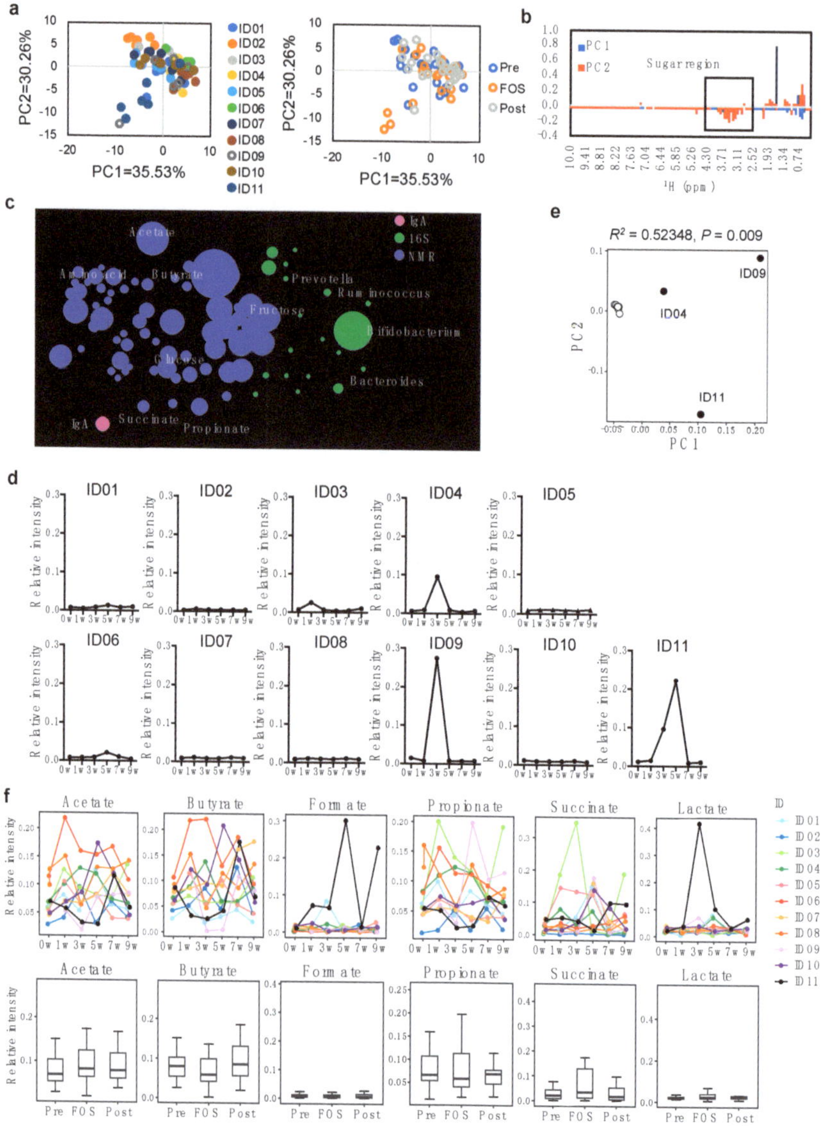

Figure 2. Differences of fecal metabolites upon intake of FOS. (**a**) Score plots of PCA for metabolic profiles of feces colored by individuals (**Left**) and sampling periods (**Right**). (**b**) Loading plot of PCA for fecal metabolome. Blue bars indicate PC1 direction and red bars indicate PC2 direction in the PCA plot in a. PC2 negative direction was influenced by the sugar region (rectangle). (**c**) The correlation network of feces was constructed as described in the Materials and Methods. The line between each element was connected to the calculated correlation coefficient and elements with $p < 0.05$. The size of each element reflects the relative value within each measurement. The amount of fecal IgA was shown in magenta, bacteria in green, and metabolites in blue. (**d**) Relative intensity of fructose in individual fecal samples. (**e**) PCoA based on Euclidean distance of the changes in fructose concentration over time in feces from individuals shown in (**b**). ADONIS revealed that ID04, ID09 and ID11 were significantly different from the others (R2 = 0.52348, p = 0.009). (**f**) Relative intensity of fecal SCFAs measured by NMR.

A network analysis of the microbiome and metabolites indicated a relatively large amount of *Bifidobacterium*, fructose, and glucose in the samples (Figure 2c). The network analysis of individuals revealed that some participants showed distinctively increased *Bifidobacterium* and fructose during the FOS phase compared with the pre and post phases (Supplementary Figure S4). The analysis specifically focused on fructose throughout the study period revealed that three participants (i.e., ID04, ID09, and ID11) clearly showed increased fructose in their fecal samples collected in the FOS phase (Figure 2d and Supplementary Figure S4). Results from PCA also confirmed that these three participants had a significantly ($p < 0.01$) distinct amount of fructose compared to the rest of the participants (Figure 2e). In addition to carbohydrates, the presence of other metabolites in fecal samples, including SCFAs, was analyzed throughout the study period (Figure 2f). There were large differences in the metabolites produced throughout the study period among individuals that resulted in no significant differences in these metabolites in fecal samples at different phases.

2.3. Genetic Differences in Fructose Metabolism from Metagenomic Analysis

In order to gain mechanistic insights into the individual variation in the amount of fructose in their feces, a metagenomic sequencing that focused on the genetic pathways categorized as carbohydrate metabolism in the Kyoto Encyclopedia of Genes and Genomes (KEGG) pathways was conducted. A relative abundance of genes associated with carbohydrate metabolism showed differences in response to FOS supplementation (Figure 3a). This KEGG analysis showed individual differences in gene expressions were particularly eminent in those of at least 16 carbohydrate metabolic enzymes (Figure 3b, green rectangle), including fructose-bisphosphate aldolase, ATP-dependent phosphofructokinase/diphosphate-dependent phosphofructokinase, and fructokinase (Figure 3b,c). The majority of these enzymes are involved in fructose metabolism (Figure 3d). These analyses indicate that there is individual variability in the fructose metabolism that could result in different fructose consumptions in response to FOS supplementation. The gut microbiome of ID02, ID05, ID06, ID8, and ID10 had a relatively higher possession of genes encoding these enzymes related to fructose metabolism (Figure 3b, red rectangle) which likely results in no accumulation of fructose in their feces collected during the FOS phase, while the gut microbiomes of the other participants (i.e., ID01, ID04, ID09, and ID11) were almost devoid of genes encoding these enzymes, supposedly leading to the higher amounts of fructose in their feces upon FOS supplementation.

2.4. Impact of FOS Intake on the Immune Compartment

FOS supplementation has been reported to affect the immune system including the augmented secretion of IgA [15,16]. Therefore, we also examined the effect of FOS intake on the immune compartment. There was again a large individual variation and no significant increase in the fecal IgA concentration during FOS intake, although it tended to increase in some individuals (Figure 4a).

Investigation of the composition of PBMCs using FACS also showed individual differences throughout the study period (Figure 4b). While most PBMC subsets did not show significant differences in response to the FOS supplementation, the number of monocytes during the FOS phase was significantly ($p < 0.01$) less than in the pre phase, and the number in the post phase was significantly ($p < 0.05$) less than the FOS period. In addition to monocytes, a number of conventional dendritic cells (pDC) showed a significant ($p < 0.05$) reduction in the post phase compared to the number in the FOS phase. However, the observed results may be partly due to a considerable increase of cDC by one participant (i.e., ID06) during the FOS phase.

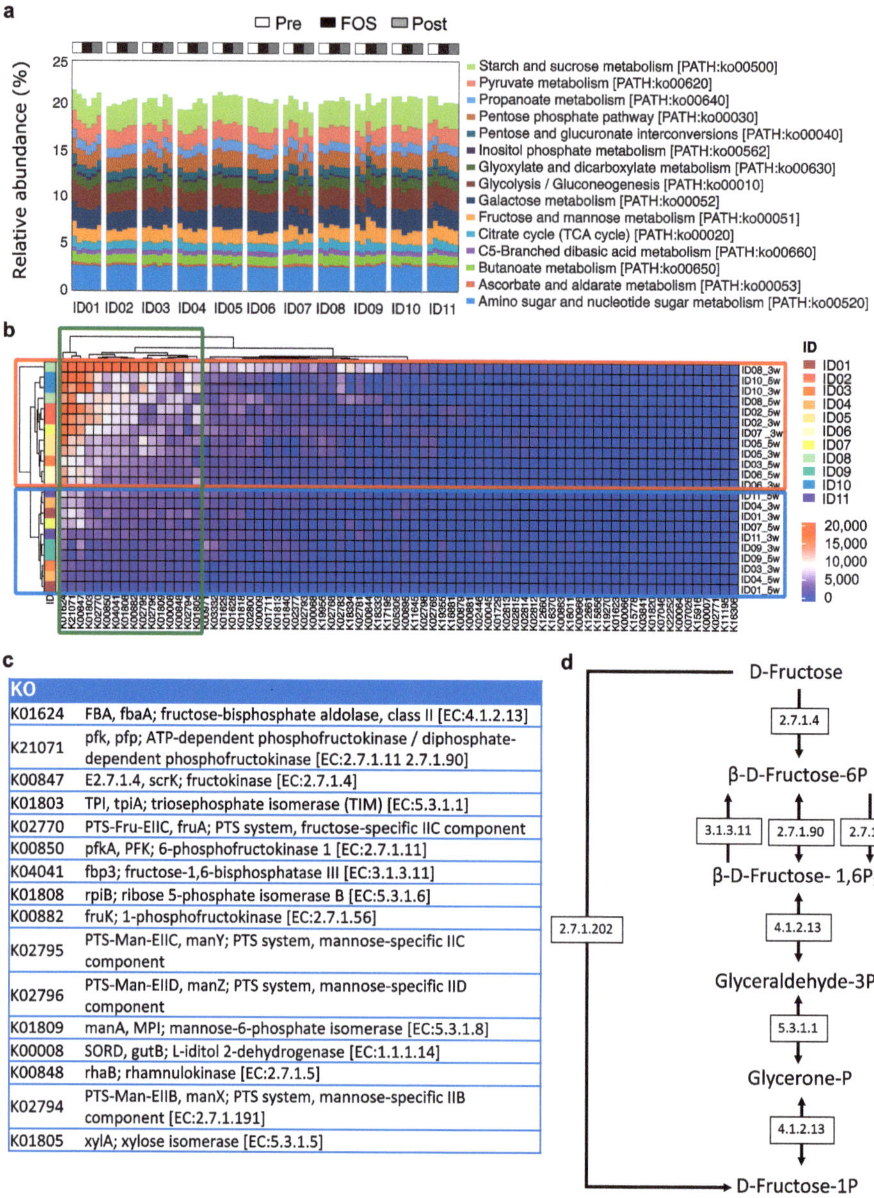

Figure 3. KEGG pathways in carbohydrate metabolism with the metagenomic analysis. (**a**) Relative abundance of the KEGG orthologies (KOs) associated with carbohydrate metabolism in fecal samples was inferred based on the metagenome analysis. (**b**) A comparison of KOs of the fructose and mannose metabolism pathways in individuals obtained from the metagenome analysis of fecal samples using a heatmap calculated by multiplying individual metagenomic genes and a number of bacteria. The green rectangle indicates the gut microbial fructose and mannose metabolism KOs with high variability, and red and blue rectangles indicate groups of individuals having relatively higher and lower expressions of these highly variable KOs, respectively. (**c**) A list of 16 KOs with high individual variability in gene expression that was identified from (**b**). (**d**) Summary of a metabolic pathway that involves many KOs listed in (**c**).

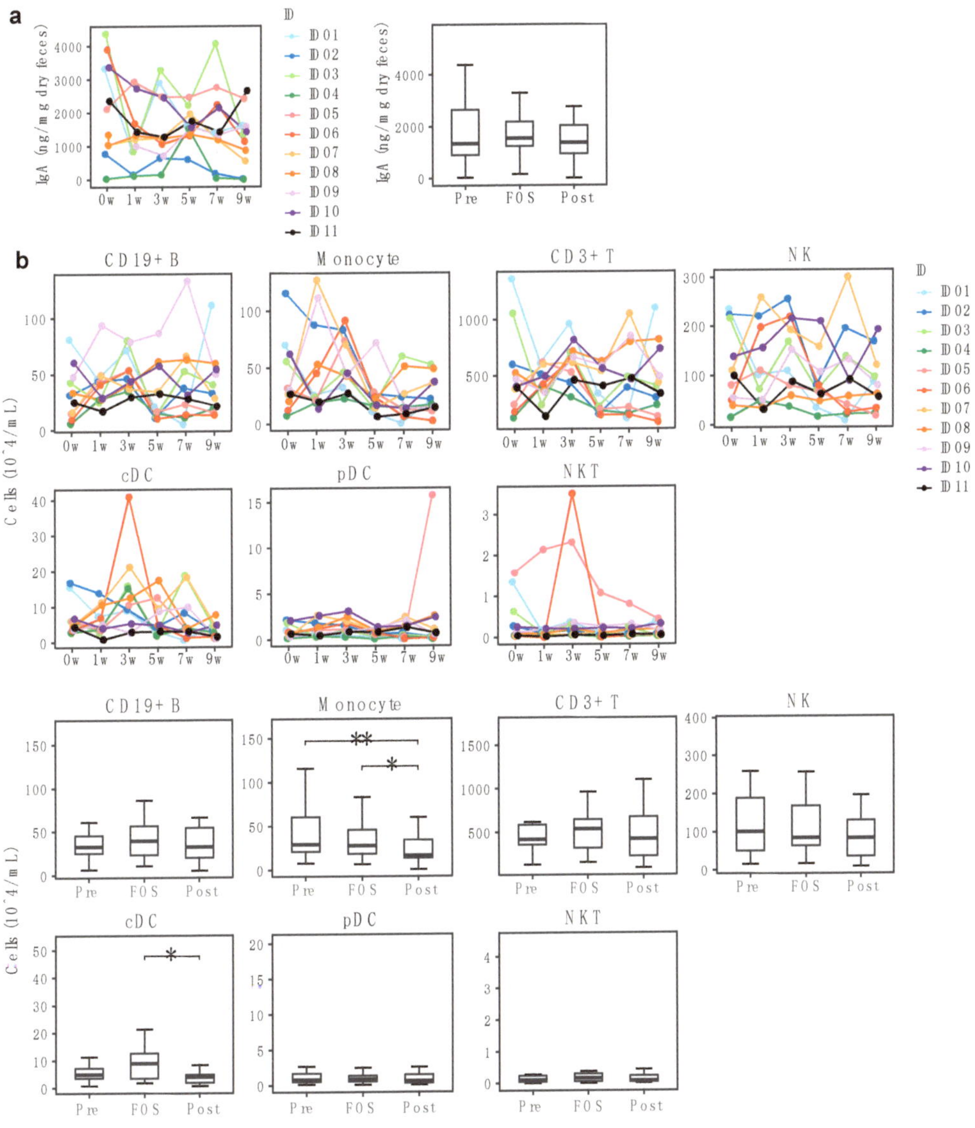

Figure 4. FOS intake and the immune system. (**a**) Fecal IgA at different time points. (**b**) The absolute number of immune cells at each time point. Proportion of immune cells was analyzed using FACS with antibodies listed in Table S2. CD19+ B, B cells; CD3+ T, T cells; NK, natural killer cells; cDC, conventional dendritic cells; pDC, plasmacytoid dendritic cells; NKT, natural killer T cells. * = padj < 0.05, ** = padj < 0.01.

3. Discussion

It is well documented that microbiota residing in the host vary among individuals [17,21–24]. In the present study, responses of microbiota and resultant metabolites, as well as the host immune capacity to FOS supplementation, were investigated. While it is common to report changes in the microbiome in relative abundance, the present study focused on absolute numbers of bacteria, as recent studies reported the importance of analyzing microbes using absolute numbers for understanding health risks and observing inter-microbial relationships [25–27]. Overall, the study confirmed a high variability of gut microbiomes among individuals, which was consistent with previous studies [17,21,22]. Because of such variability and the small sample size, we were unable to observe the presence of significant changes in the total number as well as the relative abundance of gut bacteria in response to FOS intake as a group. However, when the collected fecal samples were analyzed intra-individually, FOS intake drastically impacted the gut environment and resulted in an increase in *Bifidobacterium*. The observation is comparable to a previous study [13], and, therefore, it can be considered that FOS has a positive impact on the inhabitation of *Bifidobacterium* in the gut. Interestingly, although there was no significant difference in the absolute number of *Blautia* throughout the study period, we observed a small but significant reduction in its relative abundance during FOS intake. This suggests that while *Bifidobacterium* increased both the actual number and the proportion relative to the total number of bacteria, *Blautia* did not change its number but reduced its proportion during the FOS supplementation period. The results also suggest that the FOS supplementation decreases microbial diversity in fecal samples as confirmed by the Shannon index, probably due to the large increase of *Bifidobacterium*.

Loss of bacterial diversity in feces has been reported to be associated with disorders such as inflammatory bowel disease [6,28–30]. However, to the best of our knowledge, there is no report of the influence of dietary components on gut microbial diversity. Therefore, the observed reduction of gut microbial diversity by FOS supplementation is surprising. Since the intervention period in the present study was relatively short (i.e., four weeks), further investigation into the long-term effects, as well as the associated physiological significance, is warranted.

It Is hypothesized that supplemented FOS should be easily digested by gut microbes and utilized to produce microbial metabolites such as SCFAs. Hence, participants were assumed to show an increase in carbohydrates, especially an amount of fructose during the FOS supplementation period. However, the present study revealed considerable individual variability in the amount of fructose produced during the FOS phase and only three participants (ID04, ID09, and ID11) showed a significant increase of fructose in fecal samples collected during the FOS phase. Further analysis using a metagenomic approach showed that all three participants showed marginal gene expressions of the enzymes involved in the fructose metabolism pathway, including fructose-bisphosphate aldolase, fructokinase, and fructose-1,6-bisphosphatase III. Therefore, these three participants may be considered individuals with poor metabolic capacity in fructose catabolism. This suggests that FOS supplementation does not provide the same degree of health benefits to every individual, and caution is required for FOS supplementation. In fact, fructose consumption is linked to the rising incidence of obesity and cancer, which are two of the leading causes of morbidity and mortality globally [31,32]. Although monosaccharides including fructose are generally thought to be absorbed in the small intestine, fructose could be possibly absorbed in the colon as well since a fructose transporter GLUT7 is also expressed in the colon [33,34]. In this regard, FOS administration should be cautiously administered to individuals whose gut microbiota is insufficient in fructose catabolism. In the present study, ID01 also showed low gene expression of fructose metabolism-related enzymes but did not show a significant increase in fructose during the FOS supplementation period. Therefore, it should be considered that other metabolic pathways may also be involved in fructose degradation.

Previous studies reported that at least 21 species in the gut can metabolize FOS [35], including lactic acid-producing bacteria such as *Bifidobacterium* and Lactobacillus [12,13]. Therefore, the total number and relative abundance of these species within microbiota may affect the efficiency of FOS degradation. In addition, it has been reported that a strain of *Bifidobacterium* with specific ATP-binding cassette (ABC) carbohydrate transporters [20] can contribute to the efficiency of the clearance of fructose.

In the present study, no clear changes in metabolites were observed. This is likely due to large inter-individual variability in the gut microbiome and the efficiency of FOS degradation. Fermentation of FOS by *Bifidobacterium* is known to produce acetate and lactate and is further utilized by other gut microbes such as *Eubacterium* and *Roseburia* to produce butyrate [36,37]. In addition, *Bacteroides, Faecalibacterium, Clostridium,* and *Ruminococcus*, may increase the production of lactate, propionate, and butylate [21,37,38]. Although FOS supplementation increased the total number and a relative abundance of *Bifidbacterium*, we were unable to observe a clear pattern of increase in *Bifidobacterium*-related metabolites (e.g., acetate and lactate) and the reduction of other metabolites upon FOS intake. Considering the presence of individuals with efficient fructose metabolism, it may be possible to investigate the effects of FOS intake by a careful selection of participants with influencing factors (e.g., an absolute and relative abundance of *Bifidobacterium*, a presence of *Bifidobacterium* strains with ABC fructose transporters, and a low expression of enzymes involved in the fructose metabolism pathway).

Consumption of FOS has been reported to have beneficial effects on the immune functions of the host [14–16]. Due to the large inter-individual variability of the gut microbiome and small sample size, the present study was unable to indicate statistical significance for changes in immune functions from FOS supplementation. Nevertheless, the fecal samples showed a tendency for increasing IgA in response to FOS intake. Previous studies have reported the increase in fecal IgA upon FOS intake in experimental animals such as mice and rats [39]. By contrast, fecal IgA tended to increase, but there was no statistical significance from FOS intake in human studies [17,39], consistent with the present study. This discrepancy could at least partly be because human studies reflect large inter-individual differences in factors such as genetic backgrounds and diets, whereas these factors are much more uniform in experimental animals. In addition to IgA, FACS analysis of PMBC showed a significant decrease in monocytes and an increase in cDC during FOS supplementation. This may suggest that short-term supplementation of FOS could cause an acute change in the systemic immune status of these cells regardless of differences in the gut microbiome.

4. Materials and Methods

4.1. Participants

Eleven Japanese males aged between 30 and 50 years were recruited. Participants were excluded if they were diagnosed or susceptible to immune dysfunctions including human immunodeficiency virus (HIV) infection; had undergone any operations on the gastrointestinal tract excluding the gallbladder and appendix in the last five years; had any prescription of medications, or were under treatment on the gastrointestinal tract; had chronic constipation; or were under treatment or susceptible to toxic shock syndrome. In order to remove the influence of gender on the composition of the gut microbiome and its response to FOS supplementation, females were also excluded from the study. All participants were given information documents and a verbal explanation about the study and written informed consent was obtained prior to their participation in the study. Ethical approval was obtained from the human research ethics committee of relevant institutions prior to the commencement of the study, and all methods were performed in accordance with the relevant guidelines and regulations. Participants' self-reported age, height, weight, and body mass index (BMI) were 37.4 $\pm$ 4.6 years (range: 30.0–44.0 years), 172.7 $\pm$ 6.0 cm (range: 159.0–180.0 cm), 75.4 $\pm$ 15.4 kg (range: 58.0–110.0 kg), and 25.2 $\pm$ 4.5 kg/m^2 (range: 20.1–33.9 kg/m^2), respectively (Supplementary Table S1).

4.2. Study Design

The study was conducted as a nine-week longitudinal study. After two weeks of the baseline period (the "Pre" phase), participants were instructed to consume FOS supplements as described below for four weeks (the "FOS" phase). After this intervention period, changes in the participants' gut microbiome and its metabolites were monitored for three weeks (the "Post" phase) (Supplementary Figure S1). Participants were instructed to maintain an ordinary lifestyle during the entire period of the study.

4.3. FOS Supplementation

Participants were instructed to consume 12 g of a FOS syrup (MeioligoP®, Meiji Food Materia Co., Ltd., Tokyo, Japan) for 28 consecutive days. The syrup consists of more than 95% FOS with less than 5% glucose, fructose, and sucrose. The syrup was prepared into two packages each containing 6 g of FOS. It was recommended to participants to consume each package at different times of the day in order to prevent the possibility of having a loose stool.

4.4. Sample Collection

Samples of blood and feces were collected at the Pre-, FOS, and Post-phases of the study. Two samples were collected at each phase, and, therefore, a total of six samples were collected (Supplementary Figure S1). Blood samples were collected from the median cubital vein by medical personnel. A total of 5 mL of peripheral blood was collected into a blood collection tube containing EDTA-2Na (Venoject II VP-NA050K, Terumo Corp., Tokyo, Japan). Fecal samples were also collected using a stool collection kit consisting of collection sheets, plastic spoons, and plastic containers. Participants were instructed to wear sterile gloves and collect about 3 g of feces without any contamination of toilet water or urine within 24 h prior to blood sampling, and the samples were stored in a fridge. Samples submitted to the research group were immediately stored in a freezer at −80 °C until analysis.

4.5. Measurement of Fecal IgA

Fecal samples were lyophilized using a VD-800R lyophilizer (TAITEC Co., Ltd., Saitama, Japan) for 24 h. Freeze-dried feces were then ground with 3.0 mm Zirconia Beads (BioSpec Products, OK, USA) using a Shake Master (Biomedical Science Corp., Tokyo, Japan) for 10 min. Approximately 10 mg of each fecal sample was suspended in 300 μL of phosphate-buffered saline (PBS) containing proteinase inhibitor cocktail tablets (Complete EDTA-free, Roche Diagnostics GmbH, Mannheim, Germany) and homogenized for 1 min. After being centrifuged at 17,800× *g* for 15 min at 4 °C, the supernatant was collected from the homogenate sample and frozen at −20 °C. The measurement of fecal IgA was performed according to the instruction of the ELISA kit (E80-102, Bethyl Laboratories Inc., Montgomery, TX, USA).

4.6. Bacterial DNA Extraction

DNA extraction from feces was performed according to the literature with minor modifications [40,41]. Bacterial pellets from feces were suspended with a 10 mM Tris-HCl/10 mM EDTA solution. The sample suspension was incubated with 15 mg/mL lysozyme (FUJIFILM Wako Pure Chemical Corp., Osaka, Japan) at 37 °C for 1 h. A final concentration of 2000 units/mL of purified achromopeptidase (FUJIFILM Wako Pure Chemical Corp., Osaka, Japan) was added and then incubated at 37 °C for 30 min. The suspension was added at 1% (*w/v*) sodium dodecyl sulfate and 1 mg/mL proteinase K (Merck KGaA) and incubated at 55 °C for 1 h. After centrifugation, the bacterial DNA was purified using a phenol/chloroform/isoamyl alcohol (25:24:1) solution. The DNA was precipitated by adding ethanol and sodium acetate, and RNase treatment and polyethylene glycol (PEG) precipitation were performed.

4.7. Bacterial 16S rRNA Gene Amplicon Sequencing and Processing

The 16S rRNA gene V1-V2 regions were amplified by PCR with barcoded bacterial universal primers (27Fmod and 338R) [40]. Amplification of V1-V2 regions and preparation of the sequencing library were performed according to the literature with minor modifications [42,43]. Equal amounts of multiplexed PCR amplicon were mixed and then sequenced using either 454 GS FLX Titanium or 454 GS JUNIOR (Roche Diagnostics GmbH, Mannheim, Germany).

Analysis of the 16S rRNA gene sequencing data was conducted using an analysis pipeline established by the research group. Each sample was demultiplexed based on sample-specific barcodes followed by the removal of reads lacking both forward and reverse primer sequences. The sequence data were denoised by removing sequences with an average quality of <25 and possible chimeric sequences. We performed chimera checking and taxonomy assignment using the database constructed from public databases (Ribosomal Database Project (RDP) v. 10.27, CORE (http://microbiome.osu.edu/, accessed on 1 April 2020) and a reference genome sequence database obtained from the NCBI FTP site (ftp://ftp.ncbi.nih.gov/genbank/, accessed on 1 December 2011)). Reads removed in these processes accounted for about 64.6% of all reads, most of which represented reads lacking PCR primer sequences (63.4% of raw reads). Three thousand reads per sample were randomly selected from each filter-passed read and were used for further analysis since this read number represents more than 90% of the total reads according to Good's coverage estimator [44]. These high-quality reads were clustered into operational taxonomic units (OTUs) using a 96% pairwise-identity using UCLUST (http://www.drive5.com/, accessed on 29 August 2022). Taxonomic assignments of OTUs were performed against the database using the GLsearch program accessed on 29 August 2022. The weighted UniFrac distance [45] was calculated to assess the distance between each sample. Shannon's diversity index (Shannon Index) was calculated to evaluate the diversity of microbial communities in a sample using the vegan package of R software (version 3.6.1). All of the 16S rRNA sequence data used in this study were deposited in DDBJ/GenBank/EMBL under accession numbers: DRA014786.

4.8. Metagenomic Sequencing

Metagenome shotgun libraries (insert size of ~500 bp) were prepared using the TruSeq Nano DNA kit (Illumina) and sequenced by the Illumina NovaSeq platform. After quality filtering, reads mapped to the human genome (HG19) and the phiX bacteriophage genome were removed.

4.9. Assembly of Metagenomic Sequences and Gene Prediction

For each individual, the filter-passed NovaSeq reads were assembled using MEGAHIT (v1.2.4). Prodigal (v2.6.3) was used to predict protein-coding genes (≥100 bp) in the contigs (≥500 bp) and singletons (≥300 bp). Finally, 2,234,201 non-redundant genes were identified in the 290 samples by clustering the predicted genes using CD-HIT with a 95% nucleotide identity and 90% length coverage cut-off.

4.10. Functional Assignment of Non-Redundant Genes in Human Gut Microbiomes

Functional assignment of the non-redundant genes was performed using DIAMOND (*e*-value $\leq 1.0 \times 10^{-5}$) against the Kyoto Encyclopedia of Genes and Genomes (KEGG) database (release 7 October 2019) to obtain the KEGG orthologies (KOs). The genes with the best hit correlating to eukaryotic genes were excluded from further analysis.

4.11. Quantification of the Annotated Genes in Human Gut Microbiomes

A million metagenomic reads per individual were mapped to the JPGM [7] and IGC [46] merged reference gene set using Bowtie2 with a 95% identity cut-off. The number of reads that mapped equally to more than one gene was normalized by the proportion of the number of reads uniquely mapped to the genes as was conducted for the mapping analysis to the reference genomes. The proportion of KOs was calculated from the number of reads mapped to them.

4.12. Metabolic Profiling of Human Fecal Samples

Lyophilized fecal samples were extracted in a 100 mmol/L potassium phosphate buffer (in deuterium oxide containing 1 mmol/L sodium 2,2-dimethyl-2-silapentane-5-sulfonate, pH = 7.0) as previously described [47]. In total, 0.8 mL of the above samples was inserted into a 5 mmϕ-Nuclear Magnetic Resonance (NMR) tube, then NMR experiments were conducted using a 700 MHz- NMR spectrometer (Bruker AVANCE II 700, Bruker Biospin GmbH, Rheinstetten, Germany). All ^{1}H NMR spectra were acquired using a Bruker standard pulse program "p3919gp" with 32 K data points, 32 scans, 16 dummy scans, 14 ppm spectral width, and 3 s relaxation delay [48]. For the annotation of the signals detected in the ^{1}H NMR spectra, two-dimensional J-resolved NMR measurements were performed using a Bruker standard pulse program "jresgpprqf" with 32 data points for F1 and 16 K data points for F2, 8 scans, 16 dummy scans, a 50 Hz spectral width for F1 and 16 ppm spectral width for F2, and 1.5 s relaxation delay, as described previously [49]. The detected signals were annotated using the SpinCouple program (http://dmar.riken.jp/spincouple/, accessed on 29 August 2022) [50] with reference to the Human Metabolome Database (http://www.hmdb.ca/, accessed on 29 August 2022) [51].

4.13. Quantification of Bacterial Number and Gene Expression Using Quantitative Polymerase Chain Reaction (qPCR)

The universal 16S rRNA primers (27Fmod: 5′-AGRGTTTGATYMTGGCTCAG-3′ and 338R: 5′-TGCTGCCTCCCGTAGGAGT-3′) were used to estimate the microbial cell number by real-time qPCR. DNA of *Escherichia coli* that had been counted by the Colony Forming Unit (CFU) assay was used as a standard. Fecal DNA was assayed in 20 μL PCR reactions according to the protocol for TB Green Premix Ex Taq II (Tli RNaseH Plus) (TaKaRa Bio, Tokyo, Japan). Each reaction mixture contained 10 pmol of each primer, 10 μL of TB Green Premix Ex Taq II (Tli RNase H Plus) (Takara Bio, Tokyo, Japan), extracted DNA, and UltraPure DNase/RNase-Free Distilled Water (Thermo Fisher Scientific, Waltham, MA, USA) to reach a final volume of 20 μL. PCR conditions were as follows: 95 °C for 30 s and then 40 cycles of 95 °C for 5 s and 60 °C for 30 s. The threshold cycle (Ct) value of qPCR was calculated by the 2nd Derivative Maximum (SDM) method using the Thermal Cycler Dice Real Time System TP800 (Takara Bio, Tokyo, Japan). To estimate the absolute abundance of bacteria in the fecal sample, the log10-fold standard curves ranging from 10^4 to 10^9 copies were produced using the DNA of *E. coli*. The threshold cycle values were converted into the estimates of the absolute abundance of bacteria in the fecal samples (copy numbers/g of dry feces).

4.14. Correlation Network Construction

A correlation network was built from the relative intensities of metabolites assigned by metabolome data measured by NMR, and the relative abundance of the microbiome at the genus level by 16S rRNA amplicon sequencing. Edges of the network show significant pairwise interactions between measurements using Spearman's rank correlation coefficient. Nodes' sizes represent the proportion to relative scores computed for each measurement. The two-dimensional correlation network was drawn by the Cytoscape software v3.9.1 [52].

4.15. Peripheral Blood Mononuclear Cell (PBMC) Isolation and Analysis

PBMCs were isolated by Ficoll-Paque (GE Healthcare Life Sciences, Tokyo, Japan) density gradient centrifugation and suspended in 2% FCS/D-PBS. Specific volumes of antibodies that are listed in Supplementary Table S3 were added to 50 μL of the cell suspensions for 30 min at 4 °C. In total, 1 mL of 2% FCS/D-PBS was added to the stained cell suspensions, centrifuged at 1200 rpm for 5 min at 4 °C, washed with 2% FCS/D-PBS twice, resuspended in 300 μL of 2% FCS/D-PBS, and analyzed using a FACS Canto II (BD Bioscience, Franklin Lakes, NJ, USA). The gating strategy is shown in Supplementary Figure S2. A fraction of cells within each gating scheme was used for subsequent analysis.

4.16. Multivariate Statistical Analysis

Statistical analyses were conducted with R (version 3.6.1). Depending on the normality of the data, Mann–Whitney's U-test or Kruskal–Wallis test was used to perform statistical analysis among a comparison of the 3 groups. Adonis was performed for the principal coordinates analysis (PCoA) of microbiomes and the changes in fructose concentration over time in feces from the individuals. For time-course analysis of SCFAs, repeated one-way ANOVA was applied. The *p*-values were corrected for multiple testing using the Benjamini-Hochberg (BH) method.

5. Conclusions

The present study confirmed large inter-individual variability in the gut ecosystem and its response to FOS supplementation. This study also suggests that differences in the gene expression of enzymes that are involved in the fructose metabolism pathway contribute to individual differences in fructose catabolism upon their FOS intake. Furthermore, the study shows an acute response in the immune system regardless of individual variability in the composition of the gut microbiome. Since the study was conducted with only 11 male participants, the results cannot be generalized, and further studies are necessary with a larger sample size, and the inclusion of females in future research, to obtain more comprehensive conclusions. Nevertheless, the present study provided insight into the potential factors that influence the effects of FOS supplementation and emphasizes the importance of understanding individual variability. The accumulation of "personalized" knowledge on gut ecosystems may enable effective instructions on prebiotic consumption to optimize health benefits for individuals in the future.

Supplementary Materials: The following supporting information can be downloaded at: https://www.mdpi.com/article/10.3390/ijms231911728/s1.

Author Contributions: T.K., H.O. and S.I.-S. performed all experiments for DNA extraction of samples and for immune factor analysis. M.K., T.K., S.I.-S., Y.T., M.H., W.S. and T.O. collected samples. J.K. and Y.T. contributed to NMR experiments and analysis. M.H. and W.S. performed 16S rRNA sequencing and bioinformatics analysis. T.K., M.K., J.K., M.H., T.O. and H.O. contributed to interpreting the results. H.O. and T.O. designed the research. All authors contributed to the preparation of the manuscript. All authors have read and agreed to the published version of the manuscript.

Funding: This study was supported by funding provided by Japan Aerospace Exploration Agency (JAXA) for research selected for life science to O.H. (14YPTK-009179).

Institutional Review Board Statement: Ethical approval was obtained from the Human Research Ethics Committees of Japan Aerospace Exploration Agency (JAXA)(JX-IRBA-20-008, 9/3/2022), RIKEN(RIKEN-Y-2021-068, 21/2/2022), and Kagawa Nutrition University (No418, 20/7/2022).

Informed Consent Statement: Informed consent was obtained from all subjects involved in the study.

Data Availability Statement: All sequence data are available from DDBJ DRA under DRA014786.

Acknowledgments: The authors wish to thank Amiu Shino (RIKEN, CSRS) for the NMR experiments at the initial stage of this study. We also thank JAXA members, Astronaut Chiaki Mukai, and Tatsuya Aiba for the management of the research planning. The authors also express appreciation to Kazuhiko Yamada, Kyoko Ishii, Professor Yasuo Kagawa, and Eiji Araki for their assistance in data collection. This study was supported in part by the funding provided by the Japan Aerospace Exploration Agency (JAXA) for the Japan Experiment Module (JEM) utilization program.

Conflicts of Interest: The authors declare no conflict of interest.

References

1. Rooks, M.G.; Garrett, W.S. Gut microbiota, metabolites and host immunity. *Nat. Rev. Immunol.* **2016**, *16*, 341–352. [CrossRef] [PubMed]
2. Human Microbiome Project, C. Structure, function and diversity of the healthy human microbiome. *Nature* **2012**, *486*, 207–214. [CrossRef] [PubMed]
3. Bianconi, E.; Piovesan, A.; Facchin, F.; Beraudi, A.; Casadei, R.; Frabetti, F.; Vitale, L.; Pelleri, M.C.; Tassani, S.; Piva, F.; et al. An estimation of the number of cells in the human body. *Ann. Hum. Biol.* **2013**, *40*, 463–471. [CrossRef] [PubMed]
4. Qin, J.; Li, Y.; Cai, Z.; Li, S.; Zhu, J.; Zhang, F.; Liang, S.; Zhang, W.; Guan, Y.; Shen, D.; et al. A metagenome-wide association study of gut microbiota in type 2 diabetes. *Nature* **2012**, *490*, 55–60. [CrossRef] [PubMed]

5. Karlsson, F.H.; Tremaroli, V.; Nookaew, I.; Bergstrom, G.; Behre, C.J.; Fagerberg, B.; Nielsen, J.; Backhed, F. Gut metagenome in European women with normal, impaired and diabetic glucose control. *Nature* **2013**, *498*, 99–103. [CrossRef]
6. Qin, J.; Li, R.; Raes, J.; Arumugam, M.; Burgdorf, K.S.; Manichanh, C.; Nielsen, T.; Pons, N.; Levenez, F.; Yamada, T.; et al. A human gut microbial gene catalogue established by metagenomic sequencing. *Nature* **2010**, *464*, 59–65. [CrossRef]
7. Nishijima, S.; Suda, W.; Oshima, K.; Kim, S.W.; Hirose, Y.; Morita, H.; Hattori, M. The gut microbiome of healthy Japanese and its microbial and functional uniqueness. *DNA Res.* **2016**, *23*, 125–133. [CrossRef]
8. Human Microbiome Project, C. A framework for human microbiome research. *Nature* **2012**, *486*, 215–221. [CrossRef]
9. Lozupone, C.A.; Stombaugh, J.I.; Gordon, J.I.; Jansson, J.K.; Knight, R. Diversity, stability and resilience of the human gut microbiota. *Nature* **2012**, *489*, 220–230. [CrossRef]
10. Fan, C.; Zhang, L.; Jia, S.; Tang, X.; Fu, H.; Li, W.; Liu, C.; Zhang, H.; Cheng, Q.; Zhang, Y. Seasonal variations in the composition and functional profiles of gut microbiota reflect dietary changes in plateau pikas. *Integr. Zool.* **2022**, *17*, 379–395. [CrossRef]
11. Gibson, G.R.; Hutkins, R.; Sanders, M.E.; Prescott, S.L.; Reimer, R.A.; Salminen, S.J.; Scott, K.; Stanton, C.; Swanson, K.S.; Cani, P.D.; et al. Expert consensus document: The International Scientific Association for Probiotics and Prebiotics (ISAPP) consensus statement on the definition and scope of prebiotics. *Nat. Rev. Gastroenterol. Hepatol.* **2017**, *14*, 491–502. [CrossRef] [PubMed]
12. Macfarlane, G.T.; Macfarlane, S. Fermentation in the human large intestine: Its physiologic consequences and the potential contribution of prebiotics. *J. Clin. Gastroenterol.* **2011**, *45*, S120–S127. [CrossRef] [PubMed]
13. Oozeer, R.; van Limpt, K.; Ludwig, T.; Ben Amor, K.; Martin, R.; Wind, R.D.; Boehm, G.; Knol, J. Intestinal microbiology in early life: Specific prebiotics can have similar functionalities as human-milk oligosaccharides. *Am. J. Clin. Nutr.* **2013**, *98*, 561S–571S. [CrossRef] [PubMed]
14. Ortega-González, M.; Ocón, B.; Romero-Calvo, I.; Anzola, A.; Guadix, E.; Zarzuelo, A.; Suárez, M.D.; Sánchez de Medina, F.; Martínez-Augustin, O. Nondigestible oligosaccharides exert nonprebiotic effects on intestinal epithelial cells enhancing the immune response via activation of TLR4-NFκB. *Mol. Nutr. Food Res.* **2014**, *58*, 384–393. [CrossRef] [PubMed]
15. Hosono, A.; Ozawa, A.; Kato, R.; Ohnishi, Y.; Nakanishi, Y.; Kimura, T.; Nakamura, R. Dietary fructooligosaccharides induce immunoregulation of intestinal IgA secretion by murine Peyer's patch cells. *Biosci. Biotechnol. Biochem.* **2003**, *67*, 758–764. [CrossRef] [PubMed]
16. Nakamura, Y.; Nosaka, S.; Suzuki, M.; Nagafuchi, S.; Takahashi, T.; Yajima, T.; Takenouchi-Ohkubo, N.; Iwase, T.; Moro, I. Dietary fructooligosaccharides up-regulate immunoglobulin A response and polymeric immunoglobulin receptor expression in intestines of infant mice. *Clin. Exp. Immunol.* **2004**, *137*, 52–58. [CrossRef]
17. Kato, T.; Fukuda, S.; Fujiwara, A.; Suda, W.; Hattori, M.; Kikuchi, J.; Ohno, H. Multiple omics uncovers host-gut microbial mutualism during prebiotic fructooligosaccharide supplementation. *DNA Res.* **2014**, *21*, 469–480. [CrossRef]
18. Sonnenburg, E.D.; Sonnenburg, J.L.; Manchester, J.K.; Hansen, E.E.; Chiang, H.C.; Gordon, J.I. A hybrid two-component system protein of a prominent human gut symbiont couples glycan sensing in vivo to carbohydrate metabolism. *Proc. Natl. Acad. Sci. USA* **2006**, *103*, 8834–8839. [CrossRef]
19. McNulty, N.P.; Wu, M.; Erickson, A.R.; Pan, C.; Erickson, B.K.; Martens, E.C.; Pudlo, N.A.; Muegge, B.D.; Henrissat, B.; Hettich, R.L.; et al. Effects of diet on resource utilization by a model human gut microbiota containing Bacteroides cellulosilyticus WH2, a symbiont with an extensive glycobiome. *PLoS Biol.* **2013**, *11*, e1001637. [CrossRef]
20. Fukuda, S.; Toh, H.; Hase, K.; Oshima, K.; Nakanishi, Y.; Yoshimura, K.; Tobe, T.; Clarke, J.M.; Topping, D.L.; Suzuki, T.; et al. Bifidobacteria can protect from enteropathogenic infection through production of acetate. *Nature* **2011**, *469*, 543–547. [CrossRef]
21. Lazar, V.; Ditu, L.M.; Pircalabioru, G.G.; Picu, A.; Petcu, L.; Cucu, N.; Chifiriuc, M.C. Gut Microbiota, Host Organism, and Diet Trialogue in Diabetes and Obesity. *Front. Nutr.* **2019**, *6*, 21. [CrossRef] [PubMed]
22. Mills, S.; Lane, J.A.; Smith, G.J.; Grimaldi, K.A.; Ross, R.P.; Stanton, C. Precision Nutrition and the Microbiome Part II: Potential Opportunities and Pathways to Commercialisation. *Nutrients* **2019**, *11*, 1468. [CrossRef] [PubMed]
23. Nasidze, I.; Li, J.; Quinque, D.; Tang, K.; Stoneking, M. Global diversity in the human salivary microbiome. *Genome Res.* **2009**, *19*, 636–643. [CrossRef] [PubMed]
24. Takeshita, T.; Matsuo, K.; Furuta, M.; Shibata, Y.; Fukami, K.; Shimazaki, Y.; Akifusa, S.; Han, D.H.; Kim, H.D.; Yokoyama, T.; et al. Distinct composition of the oral indigenous microbiota in South Korean and Japanese adults. *Sci. Rep.* **2014**, *4*, 6990. [CrossRef] [PubMed]
25. Gifford, S.M.; Sharma, S.; Rinta-Kanto, J.M.; Moran, M.A. Quantitative analysis of a deeply sequenced marine microbial metatoranscriptome. *ISME J.* **2011**, *5*, 461–472. [CrossRef] [PubMed]
26. Vandeputte, D.; Kathagen, G.; D'Hoe, K.; Vieira-Silva, S.; Valles-Colomer, M.; Sabino, J.; Wang, J.; Tito, R.Y.; De Commer, L.; Darzi, Y.; et al. Quantitative microbiome profiling links gut community variation to microbial load. *Nature* **2017**, *551*, 507–511. [CrossRef]
27. Rao, C.; Coyte, K.Z.; Bainter, W.; Geha, R.S.; Martin, C.R.; Rakoff-Nahoum, S. Multi-kingdom ecological drivers of microbiota assembly in preterm infants. *Nature* **2021**, *591*, 633–638. [CrossRef]
28. Kowalska-Duplaga, K.; Gosiewski, T.; Kapusta, P.; Sroka-Oleksiak, A.; Wedrychowicz, A.; Pieczarkowski, S.; Ludwig-Slomczynska, A.H.; Wolkow, P.P.; Fyderek, K. Differences in the intestinal microbiome of healthy children and patients with newly diagnosed Crohn's disease. *Sci. Rep.* **2019**, *9*, 18880. [CrossRef]
29. Quince, C.; Ijaz, U.Z.; Loman, N.; Eren, A.M.; Saulnier, D.; Russell, J.; Haig, S.J.; Calus, S.T.; Quick, J.; Barclay, A.; et al. Extensive Modulation of the Fecal Metagenome in Children With Crohn's Disease During Exclusive Enteral Nutrition. *Am. J. Gastroenterol.* **2015**, *110*, 1718–1729. [CrossRef]

30. Manichanh, C.; Rigottier-Gois, L.; Bonnaud, E.; Gloux, K.; Pelletier, E.; Frangeul, L.; Nalin, R.; Jarrin, C.; Chardon, P.; Marteau, P.; et al. Reduced diversity of faecal microbiota in Crohn's disease revealed by a metagenomic approach. *Gut* **2006**, *55*, 205–211. [CrossRef]
31. Bray, G.A.; Nielsen, S.J.; Popkin, B.M. Consumption of high-fructose corn syrup in beverages may play a role in the epidemic of obesity. *Am. J. Clin. Nutr.* **2004**, *79*, 537–543. [CrossRef] [PubMed]
32. Joh, H.-K.; Lee, D.H.; Hur, J.; Nimptsch, K.; Chang, Y.; Joung, H.; Zhang, X.; Rezende, L.F.M.; Lee, J.E.; Ng, K.; et al. Simple sugar and sugar-sweetened beverage intake during adolescence and risk of colorectal cancer precursors. *Gastroenterology* **2021**, *161*, 128–142. [CrossRef] [PubMed]
33. Li, Q.; Manolescu, A.; Ritzel, M.; Yao, S.; Slugoski, M.; Young, J.D.; Chen, X.-Z.; Cheeseman, C.I. Cloning and functional characterization of the human GLUT7 isoform SLC2A7 from the small intestine. *Am. J. Physiol. Gastrointest. Liver Physiol.* **2004**, *287*, G236–G242. [CrossRef] [PubMed]
34. Cheeseman, C. GLUT7: A new intestinal facilitated hexose transporter. *Am. J. Physiol. Endocrinol. Metab.* **2008**, *295*, E238–E241. [CrossRef]
35. Mao, B.; Li, D.; Zhao, J.; Liu, X.; Gu, Z.; Chen, Y.Q.; Zhang, H.; Chen, W. In vitro fermentation of fructooligosaccharides with human gut bacteria. *Food Funct.* **2015**, *6*, 947–954. [CrossRef]
36. Belenguer, A.; Duncan, S.H.; Calder, A.G.; Holtrop, G.; Louis, P.; Lobley, G.E.; Flint, H.J. Two routes of metabolic cross-feeding between Bifidobacterium adolescentis and butyrate-producing anaerobes from the human gut. *Appl. Environ. Microbiol.* **2006**, *72*, 3593–3599. [CrossRef]
37. Rowland, I.; Gibson, G.; Heinken, A.; Scott, K.; Swann, J.; Thiele, I.; Tuohy, K. Gut microbiota functions: Metabolism of nutrients and other food components. *Eur. J. Nutr.* **2018**, *57*, 1–24. [CrossRef]
38. Adamberg, S.; Tomson, K.; Vija, H.; Puurand, M.; Kabanova, N.; Visnapuu, T.; Jogi, E.; Alamae, T.; Adamberg, K. Degradation of Fructans and Production of Propionic Acid by Bacteroides thetaiotaomicron are Enhanced by the Shortage of Amino Acids. *Front. Nutr.* **2014**, *1*, 21. [CrossRef]
39. Komura, M.; Fukuta, T.; Genda, T.; Hino, S.; Aoe, S.; Kawagishi, H.; Morita, T. A short-term ingestion of fructo-oligosaccharides increases immunoglobulin A and mucin concentrations in the rat cecum, but the effects are attenuated with the prolonged ingestion. *Biosci. Biotechnol. Biochem.* **2014**, *78*, 1592–1602. [CrossRef]
40. Kim, S.W.; Suda, W.; Kim, S.; Oshima, K.; Fukuda, S.; Ohno, H.; Morita, H.; Hattori, M. Robustness of gut microbiota of healthy adults in response to probiotic intervention revealed by high-throughput pyrosequencing. *DNA Res.* **2013**, *20*, 241–253. [CrossRef]
41. Morita, H.; Kuwahara, T.; Ohshima, K.; Sasamoto, H.; Itoh, K.; Hattori, M.; Hayashi, T.; Takami, H. An improved DNA isolation method for metagenomic analysis of the microbial flora of the human intestine. *Microbes Environ.* **2007**, *22*, 214–222. [CrossRef]
42. Said, H.S.; Suda, W.; Nakagome, S.; Chinen, H.; Oshima, K.; Kim, S.; Kimura, R.; Iraha, A.; Ishida, H.; Fujita, J.; et al. Dysbiosis of salivary microbiota in inflammatory bowel disease and its association with oral immunological biomarkers. *DNA Res.* **2014**, *21*, 15–25. [CrossRef] [PubMed]
43. Takayasu, L.; Suda, W.; Takanashi, K.; Iioka, E.; Kurokawa, R.; Shindo, C.; Hattori, Y.; Yamashita, N.; Nishijima, S.; Oshima, K.; et al. Circadian oscillations of microbial and functional composition in the human salivary microbiome. *DNA Res.* **2017**, *24*, 261–270. [CrossRef] [PubMed]
44. Good, I.J.; Toulmin, G.H. The number of new species, and the increase in population coverage, when a sample is increased. *Biometrika* **1956**, *43*, 45–63. [CrossRef]
45. Lozupone, C.; Hamady, M.; Knight, R. UniFrac -an online tool for comparing microbial community diversity in a phylogenetic context. *BMC Bioinform.* **2006**, *7*, 371. [CrossRef]
46. Li, J.; Jia, H.; Cai, X.; Zhong, H.; Feng, Q.; Sunagawa, S.; Arumugam, M.; Kultima, J.R.; Prifti, E.; Nielsen, T.; et al. An integrated catalog of reference genes in the human gut microbiome. *Nat. Biotechnol.* **2014**, *32*, 834–841. [CrossRef]
47. Misawa, T.; Date, Y.; Kikuchi, J. Human metabolic, mineral, and microbiota fluctuations across daily nutritional intake visualized by a data-driven approach. *J. Proteome Res.* **2015**, *14*, 1526–1534. [CrossRef]
48. Motegi, H.; Tsuboi, Y.; Saga, A.; Kagami, T.; Inoue, M.; Toki, H.; Minowa, O.; Noda, T.; Kikuchi, J. Identification of Reliable Components in Multivariate Curve Resolution-Alternating Least Squares (MCR-ALS): A data-driven approach across metabolic processes. *Sci. Rep.* **2015**, *5*, 15710. [CrossRef]
49. Misawa, T.; Wei, F.; Kikuchi, J. Application of two-dimensional nuclear magnetic resonance for signal enhancement by spectral integration using a large data set of metabolic mixtures. *Anal. Chem.* **2016**, *88*, 6130–6134. [CrossRef]
50. Kikuchi, J.; Tsuboi, Y.; Komatsu, K.; Gomi, M.; Chikayama, E.; Date, Y. SpinCouple: Development of a web tool for analyzing metabolite mixtures via two-dimensional J-resolved NMR database. *Anal. Chem.* **2016**, *88*, 659–665. [CrossRef]
51. Wishart, D.S.; Jewison, T.; Guo, A.C.; Wilson, M.; Knox, C.; Liu, Y.; Djoumbou, Y.; Mandal, R.; Aziat, F.; Dong, E.; et al. HMDB 3.0-The Human Metabolome Database in 2013. *Nucleic Acids Res.* **2013**, *41*, D801–D807. [CrossRef] [PubMed]
52. Shannon, P.; Markiel, A.; Ozier, O.; Baliga, N.S.; Wang, J.T.; Ramage, D.; Amin, N.; Schwikowski, B.; Ideker, T. Cytoscape: A software environment for integrated models of biomolecular interaction networks. *Genome Res.* **2003**, *13*, 2498–2504. [CrossRef] [PubMed]

International Journal of *Molecular Sciences*

MDPI

Article

Egg Protein Transferrin-Derived Peptides Irw (Lle-Arg-Trp) and Iqw (Lle-Gln-Trp) Prevent Obesity Mouse Model Induced by a High-Fat Diet via Reducing Lipid Deposition and Reprogramming Gut Microbiota

Zhuangzhuang Liu [1], Sujuan Ding [2], Hongmei Jiang [1] and Jun Fang [1,*]

[1] College of Bioscience and Biotechnology, Hunan Agricultural University, Changsha 410128, China
[2] Hunan Provincial Key Laboratory of Animal Nutritional Physiology and Metabolic Process, CAS Key Laboratory of Agro-Ecological Processes in Subtropical Region, Institute of Subtropical Agriculture, Chinese Academy of Sciences, National Engineering Laboratory for Pollution Control and Waste Utilization in Livestock and Poultry Production, Changsha 410125, China
* Correspondence: fangjun1973@hunau.edu.cn; Tel.: +86-731-8461-3600

Abstract: Egg-derived peptides play important roles in insulin secretion and sensitivity, oxidative stress, and inflammation, suggesting their possible involvement in obesity management. Hence, the aim of this study is to explore the alleviating effects of IRW (lle-Arg-Trp) and IQW (lle-Gln-Trp) on obesity via the mouse model induced by a high-fat diet. The entire experimental period lasted eight weeks. The results demonstrated that IQW prevented weight gain (6.52%), decreased the glucose, low-density lipoprotein (LDL), malonaldehyde, triglycerides, total cholesterol (TC), and leptin levels, and increased the concentration of adiponectin ($p < 0.05$, $n = 8$). Although IRW failed to prevent weight gain, it reduced the concentration of glucose, high-density lipoprotein (HDL), LDL, and leptin, and increased the concentration of adiponectin ($p < 0.05$, $n = 8$). Moreover, IRW and IQW increased glucose tolerance and insulin resistance based on the results of the intraperitoneal glucose test and insulin tolerance test ($p < 0.05$, $n = 8$). The quantitative polymerase chain reaction results revealed that IRW and IQW downregulated the mRNA expression of DGAT1 (Diacylglycerol O-Acyltransferase 1), DGAT2 (Diacylglycerol O-Acyltransferase 2), TNF-α, IL-6, and IL-1β of liver tissue ($p < 0.05$, $n = 8$). The results of the 16S ribosomal RNA amplicon sequencing showed that IQW and IRW tended to reduce the relative abundance of Firmicutes and Parabacteroides, and that IRW enhanced the abundance of Bacteroides ($p < 0.05$, $n = 8$). Collectively, IRW and IQW supplementation could alleviate the progression of obesity due to the fact that the supplementation reduced lipid deposition, maintained energy balance, and reprogrammed gut microbiota.

Keywords: peptides; obesity; lipid determination; gut microbiota; short chain fatty acids

check for updates

Citation: Liu, Z.; Ding, S.; Jiang, H.; Fang, J. Egg Protein Transferrin-Derived Peptides Irw (Lle-Arg-Trp) and Iqw (Lle-Gln-Trp) Prevent Obesity Mouse Model Induced by a High-Fat Diet via Reducing Lipid Deposition and Reprogramming Gut Microbiota. *Int. J. Mol. Sci.* **2022**, *23*, 11227. https://doi.org/10.3390/ijms231911227

Academic Editors: Baojun Xu and Matteo Bordiga

Received: 26 August 2022
Accepted: 19 September 2022
Published: 23 September 2022

1. Introduction

Obesity, as a major health hazard, has attracted widespread attention due to its association with an elevated risk of death and a substantially increased risk of costly chronic diseases. A study calculated the weight and height data of 128.9 million children, adolescents, and adults from 1975 to 2018 and provided the body mass index trends for all countries in the world [1]. It warns of the rise in the prevalence of obesity in each country. Obesity is accompanied by health-threatening complications such as diabetes, hypertension, cardiovascular diseases, and stroke, thereby leading to a diminished quality of life and a shorter life expectancy [2]. The underlying cause of obesity is the energy imbalance between long-term excessive calorie intake and low calorie expenditure [3]. Enlarged adipose tissue is one of the characteristics of obesity, as it stores excess energy intake during the development of obesity, while adipocyte hypertrophy may damage adipose tissue function by inducing metabolic changes, mechanical stress, and local inflammation [4–6].

Although the liver is the main organ of lipid distribution, when it reaches the upper limit of lipid storage capacity, it leads to fatty liver [7]. Lipid deposition in the liver is a complex process, and the lipid secreted by the liver is essential for preventing it. Studies in rat models have shown that the main mechanism of dietary-induced hepatic steatosis is the inhibition of lipoprotein secretion [8].

Gut microbiota is a complex microbial ecosystem in the gut that plays a vital role in metabolism, immune response, and other key physiological pathways of the host [9–11]. The influence of gut microbiota on the host is multifaceted, including the provision of nutrients, the regulation of metabolism, and the regulation of immunity [12]. Environmental disruption of gut microbiota composition and function may cause mild inflammation, leading to obesity-related diseases [13–15]. Another contribution of gut microbes is the decomposition of non-digestible nutrients, including pectin, fiber, and resistant starches, which are produced by the fermentation of short-chain fatty acids (SCFAs), such as acetate, propionate, and butyrate, in the distal intestine [16]. SCFAs are an important energy source for the intestinal epithelium and liver, and they affect many important metabolic processes, including intestinal barrier function [17,18], intestinal immunity [19,20], hepatic gluconeogenesis, and adipogenesis [21,22].

Bioactive peptides play a vital role in metabolic regulation. Substantial evidence suggests that the small peptides released from the hydrolysis of animal, plant, and microbial proteins have many beneficial health properties, such as anti-hypertension [23], anti-oxidation [24], anti-obesity [25], hypocholesterolemic, and immune regulation [26,27]. IRW and IQW are identified as novel ACE inhibitory peptides that can alleviate the inflammatory response and oxidative stress of endothelial cells induced by tumor necrosis factor (TNF) [28]. These two egg protein transferrin-derived peptides were characterized from an integrated in silico digestion and comprehensive quantitative structure–activity relationship (QSAR) prediction and bioinformatics methods [29]. A study on a spontaneous hypertensive rodent model showed that IRW reduced blood pressure through the angiotensin-converting enzyme (ACE)2/Ang (1–7)/MasR axis, suggesting that IRW was the activator of ACE2 in vivo and that the activation of ACE2 was beneficial in enhancing endothelium-dependent vasodilation and reducing vascular inflammation [23]. Majumder et al. also demonstrated that IQW could not only reduce blood pressure by suppressing the generation of plasma Angiontein II, but also exert protective effects against inflammation in spontaneously hypertensive rats [30]. The purpose of the current study is to investigate whether IRW and IQW have preventive effects on the development of obesity by establishing a rodent obesity model on a high-fat (HF) diet. We explored the effects of IRW and IQW on lipid deposition, energy homeostasis, gut microbiota, and the metabolites of SCFAs, as well as the potential effect of gut microbiota on the development of obesity. The results showed that IRW and IQW could prevent obesity induced by an HF diet by reducing lipid deposition and regulating intestinal microbial composition.

2. Results

2.1. Body Weight and Organ Indexes of the Mice

The body weight and organ indexes of the mice are shown in Figure 1A–D. The results demonstrated that the body weight of the HF group remarkably increased compared to the CON group from the second week and the HF-IQW group from the fourth week ($p < 0.05$). No significant difference was observed in the HF-IRW group ($p > 0.05$). The HF diet caused a greater increase in body weight of the HF and HF-IRW groups from the second week in comparison with CON and HF-IQW groups ($p < 0.05$). IQW reduced the weight of the liver compared to the HF group ($p < 0.05$). IRW tended to decrease the weight of the liver, but no significant difference was found ($p > 0.05$). Moreover, the weight of perirenal adipose tissue showed no significant difference ($p > 0.05$). Collectively, IRW and IQW supplementation retarded weight gain.

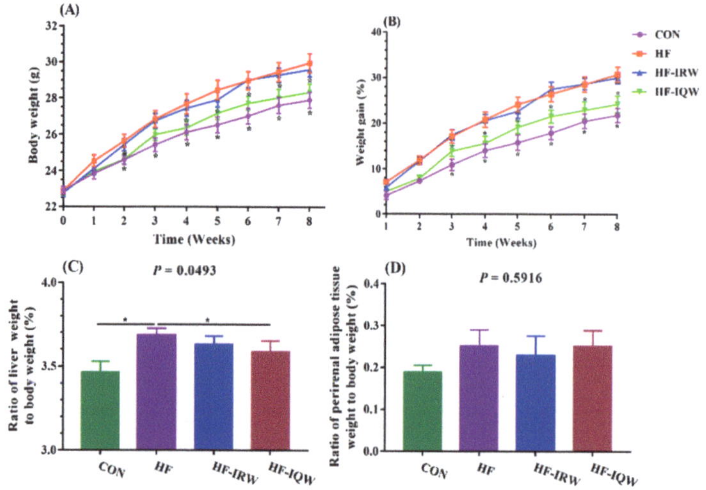

Figure 1. Egg protein transferrin-derived peptides IRW (Ile-Arg-Trp) and IQW (Ile-Gln-Trp) slowed weight gain, and IQW decreased the liver index of mice fed an HF (high-fat) diet. (**A**) Body weight of mice from week 0 to week 8; (**B**) body gain of mice from week 1 to week 8; (**C**) liver index of the mice; (**D**) perirenal adipose index of the mice. CON: the mice were fed the control diet; HF: the mice were fed an HF diet; HF-IRW: the mice were fed an HF diet and drank water supplemented with 0.03 g/L IRW starting from the fifth week; HF-IQW: the mice were fed an HF diet and drank water supplemented with 0.03 g/L IQW starting from the fifth week. * indicates $p < 0.05$ between the HF group and other groups.

2.2. Biochemical Assays, Glucose Tolerance and Insulin Aensitivity of the Mice

The lipid and GLU levels in the serum were analyzed (Figure 2C–F). The results revealed that the HF diet increased the GLU, HDL, and LDL levels ($p < 0.05$), but it did not significantly affect the serum TG ($p > 0.05$). The GLU, HDL, and LDL levels decreased with IRW, and IQW decreased the concentrations of GLU and LDL. The serum ALT and AST levels were also measured in this study, but no significant difference was found after administration of IRW and IQW ($p > 0.05$). The IGTT and ITT results showed that the HF diet increased the GLU level ($p < 0.05$). The administration of IRW and IQW reduced the GLU level after intraperitoneal GLU and insulin. In summary, IRW and IQW improved lipid accumulation and metabolic disorders induced by an HF diet.

2.3. Pathological Observation and Lipid Deposition of Hepatic Tissue

In order to investigate the molecular mechanism of IRW and IQW in triglyceride synthesis and β-oxidation, we determined the expression of several genes in the liver tissue, such as DGAT1, DGAT2, Acadm, and Cpt1a (Figure 3A–D). The results showed that IRW and IQW decreased the mRNA expression of DGAT1 and DGAT2 ($p < 0.05$) but had no effect on the expression of Acadm and Cpt1a ($p > 0.05$). In order to analyze the hepatic inflammation response induced by an HF diet, inflammation-related genes (TNF-α, interleukin (IL)-6, and IL-1β) were measured (Figure 3E–G). The results revealed that IRW and IQW reduced the mRNA expression of TNF-α, IL-6, and IL-1β induced by an HF diet ($p < 0.05$). Moreover, we determined the oxidation products and hepatic lipid level (Figure 3H–J). The results showed that IRW and IQW reduced the hepatic MDA level ($p < 0.05$) and that IQW supplementation decreased the concentration of TG and TC ($p < 0.05$). Likewise, the results of hepatic lipid deposition measured by oil red O staining showed that IRW and IQW alleviated the lipid deposition of hepatic tissue (Figure 3K–N). These results indicate that IRW and IQW prevent the liver from undergoing an inflammation response or oxidative stress, and alleviate hepatic lipid deposition.

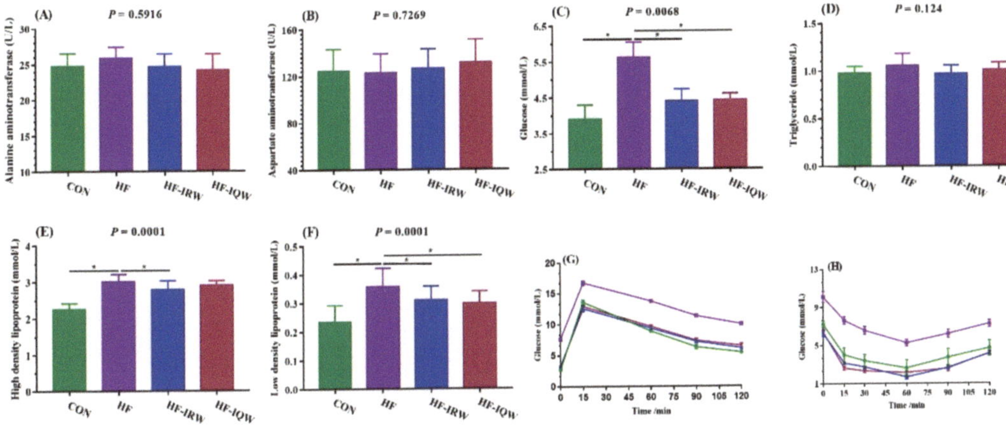

Figure 2. Effects of egg protein transferrin-derived peptides IRW (lle-Arg-Trp) and IQW (lle-Gln-Trp) on the biochemical serum biochemistry, GLU tolerance, and insulin sensitivity of mice fed an HF diet. (**A–F**) The biochemical analyses included alanine aminotransferase (ALT), aspartate aminotransferase (AST), glucose (GLU), triglyceride (TG), high-density lipoprotein (HDL), and low-density lipoprotein (LDL). GLU tolerance was indicated by the changing tendency of GLU after intraperitoneal glucose (**G**), and insulin sensitivity was indicated by the changing tendency of GLU after intraperitoneal insulin (**H**). * indicates $p < 0.05$ between the HF group and other groups.

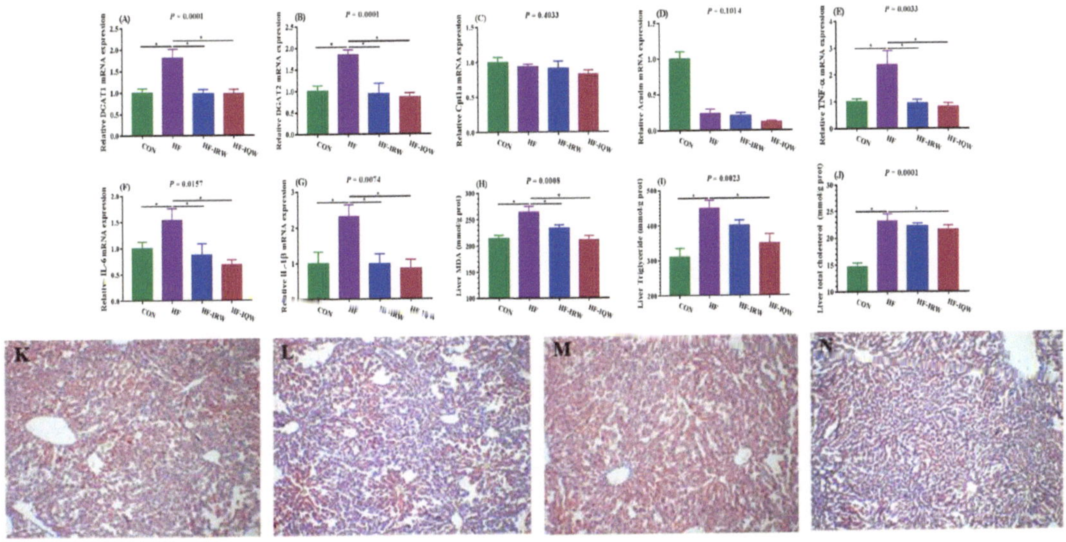

Figure 3. Effects of egg protein transferrin-derived peptides IRW (lle-Arg-Trp) and IQW (lle-Gln-Trp) on lipid deposition and inflammation of the liver of mice fed an HF diet. (**A–G**) mRNA expression of DGAT1 (Diacylglycerol O-Acyltransferase (1), DGAT2 (Diacylglycerol O-Acyltransferase (2), Cpt1a (carnitine palmitoyltransferase 1A), Acadm (acyl-CoA dehydrogenase medium chain), TNF-α, IL-6, and IL-1β. (**H–J**) MDA, triglyceride, and total cholesterol levels. (**K–N**) Oil red O staining (microscope's magnification= 100 times) of the CON, HF, HF-IRW, and HF-IQW groups. MDA: malondialdehyde. * indicates $p < 0.05$ between the HF group and other groups.

2.4. Pathological Observation and Lipid Deposition of Adipose Tissue

In order to explore the energy status of adipose tissue induced by an HF diet, the weight of adipose tissue was analyzed, and the concentrations of leptin and adiponectin were measured (Figure 4A–D). The results revealed that, compared to the CON group, the HF diet increased the weight of eWAT and ingWAT as well as the leptin level, and decreased the adiponectin level ($p < 0.05$). IRW and IQW supplementation mitigated the weight of eWAT and ingWAT, decreased the concentration of leptin, and increased adiponectin induced by the HF diet ($p < 0.05$). Moreover, the HE staining results of adipose tissue demonstrated that the HF diet increased eWAT and ingWAT hypertrophy relative to the tissue in the CON group, and that IRW and IQW supplementation could retard this trend (Figure 4E). Thus, IRW and IQW could protect adipose tissue from energy disorders and tissue hypertrophy induced by an HF diet.

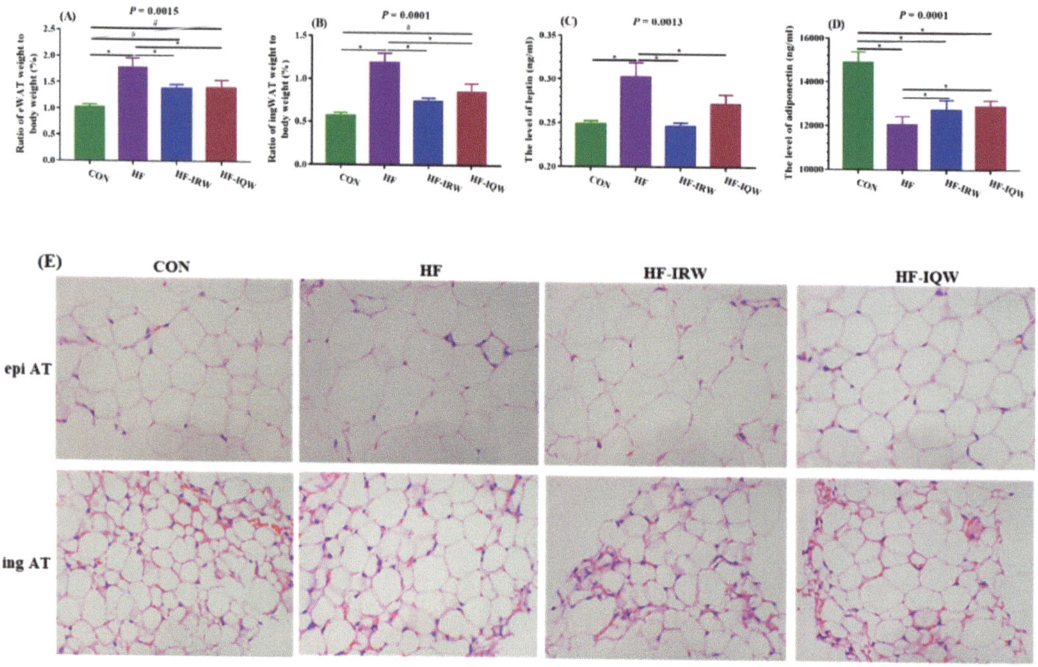

Figure 4. Effects of egg protein transferrin-derived peptides IRW (Ile-Arg-Trp) and IQW (Ile-Gln-Trp) on the lipid deposition of adipose tissue of mice fed an HF diet. (**A**,**B**) Index of ingWAT and eWAT, respectively. (**C**,**D**) Leptin and adiponectin levels. (**E**) ingWAT: inguinal white adipose tissue; eWAT: epididymal white adipose tissue (microscope's magnification= 200 times). * indicates $p < 0.05$ between the HF group and other groups. # indicates $p < 0.05$ between the CON group and other groups.

2.5. Egg Protein Transferrin-Derived Peptides IRW and IQW Can Reprogram Gut Microbiota

Gut microbiota is closely related to obesity. Thus, the colonic microbiota was examined using 16S ribosomal RNA amplicon sequencing. The results showed that IRW and IQW increased the ACE and Shannon indexes compared to the CON group, but decreased the Simpson index (Figure 5A–D, $p < 0.05$). We further analyzed the composition of microbiota at the phylum level (Figure 5E–K) and found that IQW supplementation inhibited the growth of Bacteroidetes compared to the HF group and reduced the growth of *Verrucomicrobia* compared to the CON group. Moreover, compared to the CON group, an increased relative abundance of Firmicutes was observed in the HF diet. IRW and IQW tended to decrease the relative abundance of Firmicutes ($p > 0.05$). Moreover, IRW and IQW administration had no effect on the relative abundance of Actinobacteria, Proteobacteria, and Deferribacteres ($p > 0.05$). At the genus level, the top 10 microbiota are illustrated in Figure 5L. The results showed that *Bacteroides, Akkermansia, Parabacteroides*, and *Ruminococcus* are the main bacteria at the genus level. The HF diet decreased the relative abundance of *Bacteroidales, Akkermansia*, and *Parabacteroides* compared to the CON group ($p < 0.05$). IRW supplementation decreased the abundance of *Bacteroidales* and increased that of *Parabacteroides* compared to the HF group ($p < 0.05$). IQW supplementation tended to increase *Parabacteroides*, but no significant difference ($p > 0.05$) was observed. Thus, IRW and IQW supplementation could reprogram gut microbiota to maintain gut health.

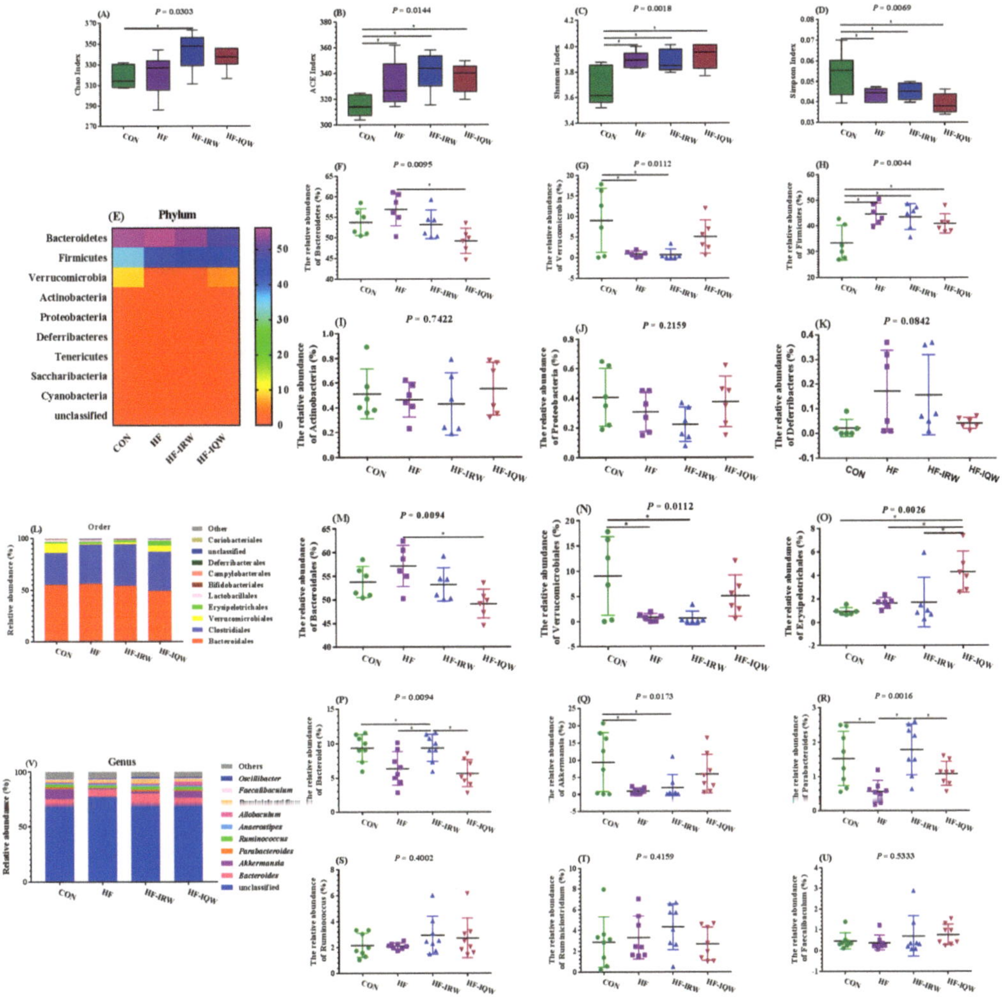

Figure 5. Effects of egg protein transferrin-derived peptides IRW (Ile-Arg-Trp) and IQW (Ile-Gln-Trp) on the gut microbiota of mice fed an HF diet. (**A–D**) The diverse indexes include chao index, ACE, Shannon index, and Simpson index. (**E**) Heat map of the top 10 microbiota at the phylum level. (**F–K**) Relative abundance of *Bacteroidetes, Verrucomicrobia, Firmicutes, Actinobacteria, Proteobacteria, and Deferribacteres*, respectively. (**L**) Microbiota composition at the order level. (**M–U**) Relative abundance of *Bacteroidales, Verrucomicrobiales, Erysipelotrichales, Bacteroides, Akkermansia, Parabacteroides, Ruminococcus, Ruminiclostridium*, and *Faecalibaculum*, respectively. * indicates $p < 0.05$. (**V**) Microbiota composition at the genus level.

The LEfSe analysis of colonic microorganisms among CON, HFD, and HFD-IRW groups is shown in Figure 6A. The result revealed that four bacterial biomarkers were detected in the CON group, including Porphyromonadaceae, Lactobacillaceae, Lactobacillales and Bacilli; two bacterial biomarkers were detected in the HFD group, including Mollicutes and Tenericutes; and two bacterial biomarkers were detected in the HFD-IRW group, including Prevotellaceae and Ruminococcaceae. The LEfSe analysis of colonic microorganisms among CON, HFD, and HFD-IQW groups is shown in Figure 6B. The result revealed that one bacterial biomarker was detected in the CON group (Porphyromonadaceae); six bacterial biomarkers were detected in the HFD group, including Bacteroidales, Bacteroidia, Bacteroidetes, Peptostreptococcaceae, Mollicutes and Tenericutes; and nine bacterial biomarkers were detected in the HFD-IQW group, including Bacillales, Carnobacteriaceae, Clostridiaceae, Erysipelotrichaceae, Erysipelotrichales, Erysipelotrichia, Moraxellaceae, Pseudomonadales, Gammaproteobacteria.

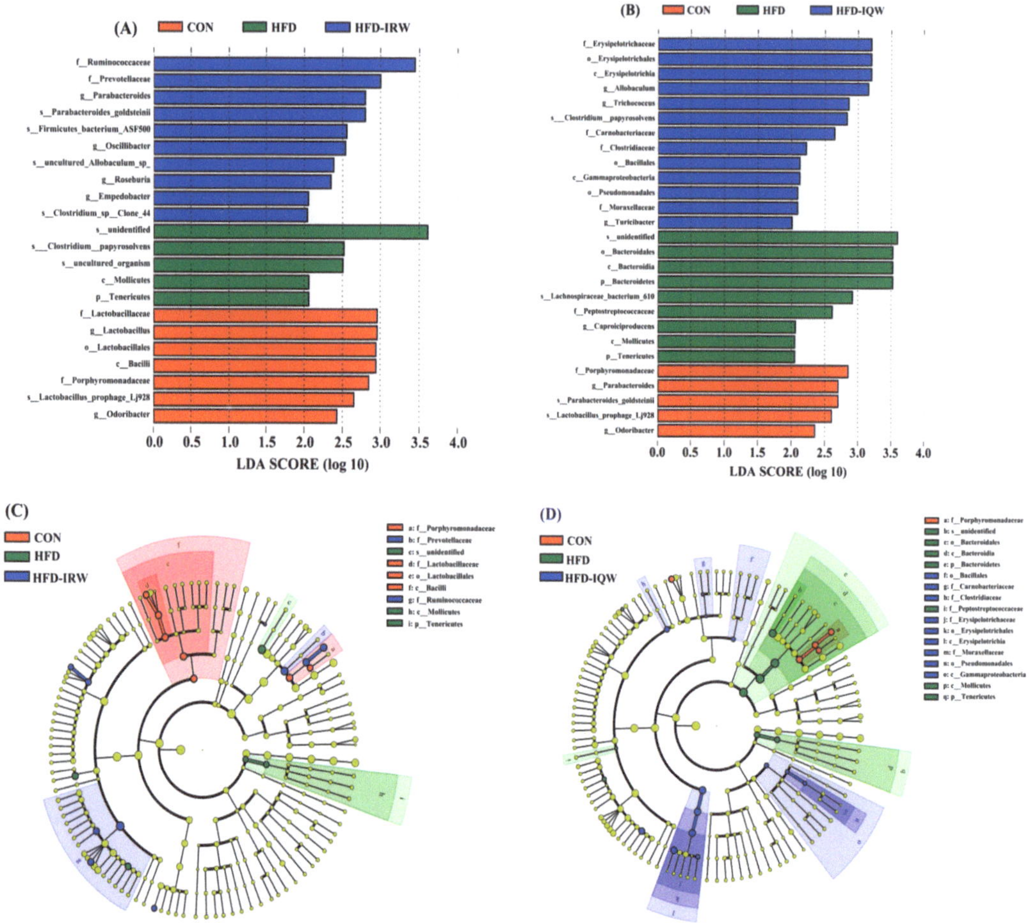

Figure 6. LEfSe analysis of microbiota in colon content. Linear discriminant analysis among (**A**) and (**C**) CON, HFD, and HFD-IRW, (**B**) and (**D**) among CON, HFD, and HFD-IQW. The red, green, and blue nodes in the phylogenetic tree signify the microbial species which perform a vital role in the groups. The yellow nodes signify the species with no significant difference.

2.6. SCFA Concentration in Feces

SCFAs have many beneficial effects on the mammalian energy metabolism. To examine the change in SCFAs in feces, we measured their concentrations (Figure 7A–F). The HF diet was found to decrease the level of propionic acid and increase the level of valeric acid compared to the CON group ($p < 0.05$). IRW and IQW supplementation decreased the concentration of valeric acid and isobutyric acid relative to the HF group ($p < 0.05$). Moreover, IQW supplementation decreased the acetic acid level compared to the HF group, and IRW supplementation increased the propionic acid level compared to the HF group ($p < 0.05$).

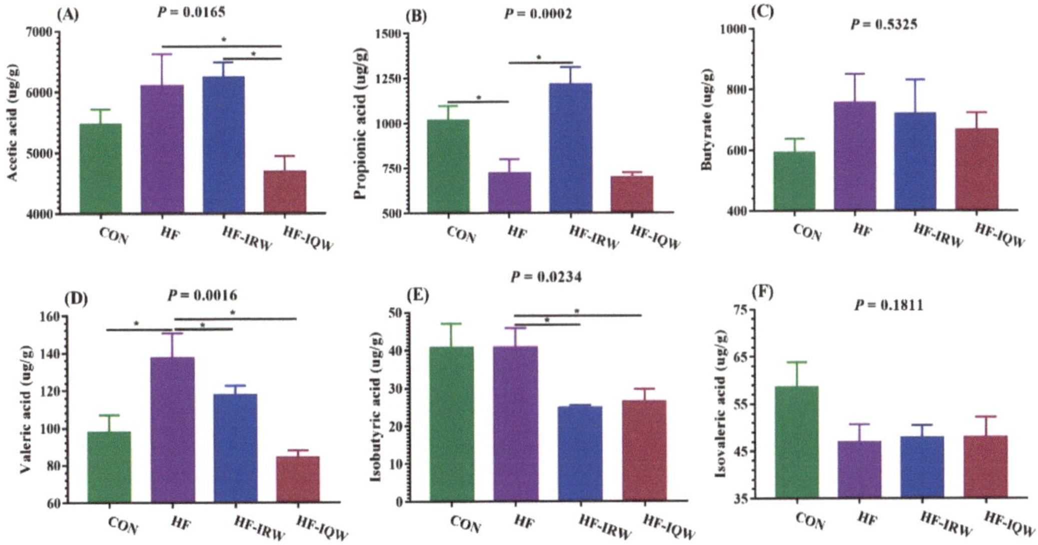

Figure 7. Effects of egg protein transferrin-derived peptides IRW (lle-Arg-Trp) and IQW (lle-Gln-Trp) on SCFAs (short-chain fatty acid) in the feces of mice fed an HF diet. (**A**) Concentration of acetic acid; (**B**) concentration of propionic acid; (**C**) concentration of butyrate; (**D**) concentration of Valeric acid; (**E**) concentration of Isobutyric acid; (**F**) concentration of Isovaleric acid. * indicates $p < 0.05$ between the HF (high-fat diet) group and other groups.

2.7. Correlation Analysis between Differential Gut Microbe with SCFA or Lipid Deposition-Related Factors in Serum

Gut microbiota affects host nutrition and energy regulation, as well as the development of obesity. In order to explore the relationship between gut microbiota and other chemical substances in the development of obesity, we investigated the correlation between gut microbiota with SCFA and lipid deposition-related factors (Figure 8A–P). The results showed a positive correlation between acetic acid and *Bacteroides;* a positive correlation between propionic acid and *Parabacteroides;* a negative correlation between GLU and *Parabacteroides / Akkermansia / Verrucomicrobia;* a negative correlation between LDL and *Parabacteroides,* but a positive correlation with *Firmicutes;* a negative correlation between HLD and *Parabacteroides/Akkermansia/Verrucomicrobia,* but a positive correlation with Firmicutes; a negative correlation between leptin and *Parabacteroides;* and a positive correlation between adiponectin and *Akkermansia / Verrucomicrobia,* but a negative correlation with *Firmicutes* and *Erysipelotrichales.*

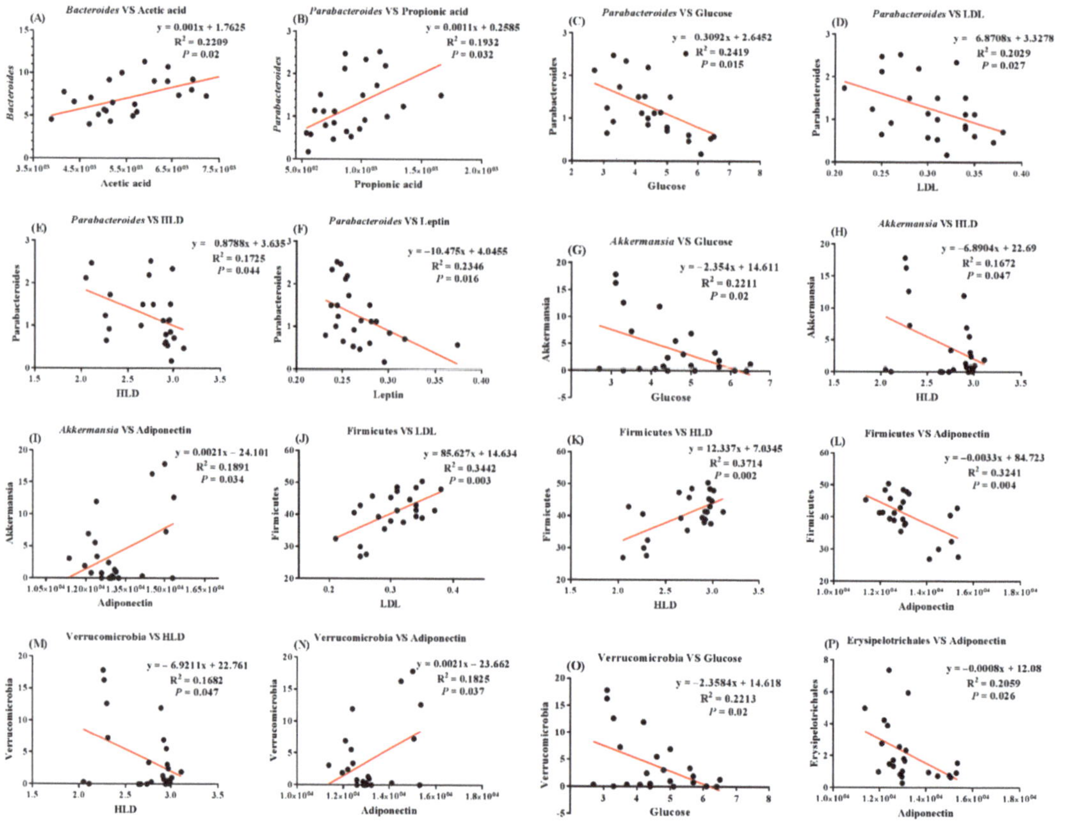

Figure 8. Correlation analysis between the gut microbe with SCFAs (short-chain fatty acids) and the lipid deposition-related factor. (**A**) *Bacteroides* vs. acetic acid; (**B**) *Parabacteroides* vs. propionic acid; (**C**) *Parabacteroides* vs. GLU; (**D**) *Parabacteroides* vs. LDL; (**E**) *Parabacteroides* vs. HLD; (**F**) *Parabacteroides* vs. leptin; (**G**) *Akkermansia* vs. GLU; (**H**) *Akkermansia* vs. HLD; (**I**) *Akkermansia* vs. adiponectin; (**J**) *Firmicutes* vs. LDL; (**K**) *Firmicutes* vs. HLD; (**L**) *Firmicutes* vs. adiponectin; (**M**) *Verrucomicrobia* vs. HLD; (**N**) *Verrucomicrobia* vs. adiponectin; (**O**) *Verrucomicrobia* vs. GLU; (**P**) *Erysipelotrichales* vs. adiponectin.

3. Discussion

As multifunctional compounds derived from proteins, bioactive peptides have a positive effect on body functions and may ultimately affect health [31]. Most natural processes in the body are signaled or regulated by the interaction of specific amino acid sequences, which can be peptide or protein segments [32]. The amino acid sequence of bioactive peptides may affect major body systems, such as hypertension, diabetes, cardiovascular, antimicrobial, and immune, and therefore have broad therapeutic applications in the future [23,33]. The milk-derived peptide Val-Pro-Pro was found to prevent a fatty inflammation response between fat cells and macrophages and to act as an ACE inhibitor to improve obesity-related insulin resistance, thereby improving the development of obesity [34]. In addition, studies have shown that IRW reduces blood pressure in spontaneously hypertensive rats, acts as an ACE2 activator to enhance endothelium-dependent vasodilation, and reduces vascular inflammation [23]. Our study demonstrated that IRW and IQW could reduce lipid deposition, increase GLU tolerance and insulin resistance, and reprogram gut microbiota.

WAT is a loose connective tissue with a highly organized vascular system. In addition to the functions of structural buffering, passive insulation, and GLU and lipid homeostasis, WAT also has other functions, such as regulating the metabolism and immune function of endocrine signaling organs [35,36]. Lipid homeostasis and immune function in the body may be out of order after excess adiposity caused by adipocyte hypertrophy and hyperplasia. Excess lipids and impaired lipid storage are partly responsible for obesity; thus, fatty acids affect insulin signaling pathways [37]. Adipose tissue is the main source of energy for the human body, along with adipocytokine adiponectin and leptin [38]. Leptin is an anti-obesity hormone secreted by adipocytes discovered through positional cloning, and it has pleiotropic effects on energy homeostasis as well as endocrine and metabolic physiology and pathology [39]. Leptin levels in human plasma are usually associated with fat mass and changes in energy [40]. Adiponectin, the protein most abundant in WAT, has been implicated in the regulation of insulin resistance, type 2 diabetes, atherosclerosis, and other diseases induced by obesity [41]. Our results showed that IRW and IQW supplementation inhibited the hypertrophy of eWAT and ingWAT and reduced the leptin and adiponectin levels induced by the HF diet (Figure 4A–D).

Hepatic lipid accumulation may lead to fatty liver and contribute to systemic metabolic dysfunction. The severity of fatty liver disease is directly relevant to the characteristics of metabolic syndrome, including hyperglycemia, insulin resistance, hypertriglyceridemia, and hyperinsulinemia [42,43]. An HF diet has been found to induce hepatomegaly and lipid deposition and to increase triglyceride levels in rodent models [44]. The liver plays an important role in maintaining GLU homeostasis. In insulin resistance, an increase in gluconeogenesis leads to an increase in hepatic glucose production [45,46]. Moreover, the disorder of glucose metabolism regulation is one of the characteristics of obesity, and is often accompanied by elevated levels of chronic inflammatory markers in the liver and obese tissues, such as TNF-α, IL-6, and IL-1 β [47]. An uncontrolled proinflammatory response may contribute to a chronic inflammatory state, promote a favorable tumor microenvironment, or promote immune transition activation and cancer growth [48]. Our results suggest that IRW and IQW reduced the levels of glucose, HDL, and LDL, improved GLU tolerance and insulin resistance induced by the HF diet, and decreased the mRNA expression of TNF-α, IL-6, and IL-1β (Figure 3E–G). Moreover, we examined the expression of Cpt1a and Acadm, which participated in β-oxidation, as well as the expression of DGAT1 and DGAT2, which participate in triglyceride synthesis. IRW and IQW promoted the expression of DGAT1 and DGAT2 in liver tissue and reduced the concentration of serum MDA, but they had no significant effect on Cpt1a and Acadm (Figure 3A–G).

Unhealthy eating habits have an adverse effect on gut microbiota homeostasis, which can lead to low levels of inflammation, thereby inducing diseases associated with obesity [12]. The variation in gut microbiota controls metabolic endotoxemia, inflammation, and related diseases by increasing intestinal permeability. The level of LPS in the cecum was reduced in *ob/ob* mice, and the mice were fed an HF diet after administering antibiotics. This effect has also been found to be associated with reduced GLU intolerance, weight gain, and inflammation. In addition, better intestinal permeability was shown in the HF diet, but the expression of genes related to the tight junction protein in intestinal tissue was inhibited [49]. Generally, obese individuals had decreased Bacteroidetes and increased Firmicutes levels (Figure 5F–K) [50]. However, changes at the smaller microbial community level, rather than at the phylum level, have been found to be associated with the development of obesity [51,52]. *Parabacteroides distasonis* (PD) is a probiotic in the human body that treats diarrhea and constipation. Its content is negatively correlated with obesity, non-alcoholic fatty liver disease, and diabetes. One study revealed that PD improved the symptoms of obesity, insulin resistance, lipid metabolism disorder, and non-alcoholic fatty liver in HF diet-fed mice and *ob/ob* model mice, which could be related to the extensive bile acid conversion and intestinal gluconeogenesis function of PD [53]. In addition, Bacteroides, as the main microorganism in the intestine, has anti-obesity effects in animal models [54]. Our results show that IQW and IRW tended to decrease the relative abundance of *Firmicutes* and *Parabacteroides*, and that IRW increased the abundance of *Bacteroides*. The mechanism of egg protein transferrin-derived peptides regulating fat deposition based on microorganisms needs further exploration.

4. Methods and Materials

4.1. Animal and Experimental Design

An animal experiment was conducted in accordance with Hunan Agricultural University's rules for the care and use of laboratory animals, with approval from the animal care committee of Hunan Agricultural University. A total of 32 nine-week-old C57BL/6J male mice were obtained from SLAC Laboratory Animal Central (Changsha, China).

Egg protein transferrin-derived peptides were synthesized by ChinaPeptides (Suzhou, Jiangsu, China). Purity detection of freeze-dried peptides was conducted by high performance liquid chromatography. The purity of IRW (Molecular Weight: 473.58) and IQW (Molecular Weight: 445.52) was 87.91% and 93.04%, respectively, and impurity includes moisture and water-soluble salt.

All animals were bred in a sterile environment at a room temperature of 22 ± 2 °C, relative humidity of 50 ± 5%, light cycle of 12 h/d (i.e., 6:30 a.m. to 6:30 p.m.), and with ad libitum food and water. After seven days of acclimation, all animals were randomly divided into four groups (Table 1, $n = 8$). The HF diet consisted of 54.9% corn, 5.6% casein, 18% soybean, 6.5% brewer's yeast, 11.3% lard, 0.8% soybean oil, 0.5% salt, 1.4% fish meal, and 1% premix. In accordance with Research Diets, the lard content in the control diet was 0.7%. The entire experimental period lasted eight weeks. During the last week of the experiment, after collecting blood from the retroorbital sinus, each mouse was sacrificed through cervical dislocation. Afterward, the liver and white adipose tissue (WAT), including epididymal white adipose tissue (eWAT), inguinal white adipose tissue (ingWAT), and perirenal adipose tissue, were separated and collected. The liver and adipose tissue were fixed in a 4% formaldehyde solution. The colon contents were kept in sterile tubes. The sample was snap-frozen in liquid nitrogen and then stored at −80 °C until analysis.

Table 1. Group name and treatment.

Group Name	Treatment
Group 1, control group: CON	mice were fed the control diet
Group 2, high-fat diet: HF	mice were fed a high-fat diet
Group 3, HF-IRW (Ile-Arg-Trp)	mice were fed a high-fat diet and drank water supplemented with 0.03 g/L IRW starting from the fifth week
Group 4, HF-IQW (Ile-Gln-Trp)	mice were fed a high-fat diet and drank water supplemented with 0.03 g/L IQW starting from the fifth week

The intraperitoneal glucose test (IGTT) and insulin tolerance test (ITT) were carried out nine days and seven days before the end of the experiment, respectively. After 6 h of fasting, the mice were injected intraperitoneally with 1 g/kg body weight of glucose, and blood glucose concentrations was detected through the tail vein at 0, 15, 60, 90, and 120 min. The mice were injected intraperitoneally with 0.65 U/kg body weight of insulin, and blood glucose concentrations were measured at 0, 15, 30, 60, 90, and 120 min.

4.2. Biochemical Assays

The biochemical indicators measured in serum included alanine aminotransferase (ALT), aspartate aminotransferase (AST), glucose (GLU), triglyceride (TG), high-density lipoprotein (HDL), and low-density lipoprotein (LDL), following the manufacturer's instructions (Nanjing Jiancheng Bioengineering Institute, Nanjing, China). The levels of leptin and adiponectin were detected according to the manufacturer's instructions (CUSABIO, Wuhan, China).

4.3. Detection of Triglyceride (TG), Total Cholesterol (TC) and Malondialdehyde (MDA) in Mice of Liver Tissue

The liver tissues were homogenized at 1:9 (m:v) and centrifuged at 6000 rpm for 10 min at 4 °C before transferring the supernatant for further detection. TG, TC, and MDA levels were measured following the instructions mentioned previously (Jiancheng Bioengineering Institute). The concentration of protein was determined according to the BCA Protein Assay Kit (Beyotime, Shanghai, China).

4.4. Pathological Observation of Liver and Adipose Tissue

The liver was immobilized in 4% formaldehyde for 48 h. It was cut into 8-μm thick slices, which were roasted, dewaxed, and stained with hematoxylin and eosin (HE) for 1 min. The slices were dehydrated using a gradient concentration of alcohol and sealed with neutral gum. Liver tissues of pathological status were observed using an optical microscope (Motic, Beijing, China).

4.5. Oil Red O Stain and Hepatic Lipid Determination

The frozen embedded liver tissues were cut into 6-μm thick slices, and 100 μL oil red O was added. The slices were placed in a wet box, incubated at room temperature for 20 min, and observed under a microscope (Motic, Beijing, China) after hematoxylin staining and buffered glycerin sectioning.

4.6. Quantitative Polymerase Chain Reaction (PCR) Analysis

Total RNA from the liver was extracted using TRIZOL reagent (Invitrogen, Carlsbad, CA, USA) and then treated with DNase I (Invitrogen). The reverse transcriptional program was performed at 37 °C for 15 min and at 95 °C for 5 s. The primers used in this study are shown in Table S1. β-actin served as the housekeeper gene for gene normalization. The PCR cycling program was conducted at 94 °C for 5 s and at 5 °C and 72 °C for 30 s. The target gene of the relative expression in the comparison with β-actin was calculated using the method of comparing the Ct value.

4.7. 16S rRNA Amplicon Sequencing for Colon Microbe

The process of 16S rRNA sequencing for the colon microbe was performed following a previous study [55]: genomic DNA quality control, design, and synthesis of the primer splice, PCR amplification and purification of PCR products, quantification and homogenization of PCR products, and MiSeq high-throughput sequencing (Illumina, San Diego, CA, USA).

4.8. Determination of Faecal Short-Chain Fatty Acid

The concentration of SCFAs and other related organic acids (acetic acid, butyrate, valeric acid, isovaleric acid, etc.) was analyzed using a gas chromatography (GC) system (Agilent 7890A, Agilent, Palo Alto, CA, USA). The GC detection program parameters are shown in the supplementary materials. A freeze-dried colon digesta of 100 mg was mixed with 1 mL of ultrapure water and centrifuged at 10,000 rpm for 10 min at 4 °C. The supernatant was adjusted to 1 mL. Then, 25% metaphosphoric acid was added at a 9:1 ratio: a 900 μL sample and 100 μL of 25% metaphosphoric acid (Sinopharm, Shanghai, China) were mixed and added into a 2 mL tube. After mixing, the mixture was left to sit at a standing reaction time of 3–4 h at an ambient temperature. The mixture was then centrifuged at 12,000 rpm for 15 min. The supernatant was filtered using a 45-μm microporous filter membrane (nylon series) and then added to the sample bottle for on-board detection.

4.9. Data Analysis

Data analysis was conducted using a one-way analysis of variance to check the homogeneity of variance, using Levene's test and Student's test through SPSS 21 (Chicago, IL, USA) for Windows. GraphPad Prism 7.00 (San Diego, CA, USA) software was used to conduct a Pearson correlation between the colonic microbe with SCFA and the lipid deposition-related factors in the serum. A significant difference was determined at $p < 0.05$.

5. Conclusions

This study proved that IQW and IRW exhibited a mitigative effect on HF diet-induced obesity in a rodent model. The results showed that IQW and IRW ameliorated the negative effect of the HF diet on GLU tolerance and insulin resistance, reduced hepatic lipid deposition and inflammation, improved the hypertrophy of adipose tissue, and reprogrammed gut microbiota. Moreover, IQW prevented weight gain, but IRW did not. A Pearson correlation analysis found that *Parabacteroides* was negatively correlated with GLU, LDL, and leptin, but positively correlated with propionic acid. This finding is consistent with previous studies in that the abundance of *Parabacteroides* is negatively correlated with obesity. Conversely, Firmicutes was positively correlated with LDL and HDL, but negatively correlated with adiponectin. This suggests that *Firmicutes* may be related to lipid deposition and energy regulation. However, more research is needed to confirm the potential role of these bacteria in obesity regulation, and to provide more evidence for egg protein transferrin-derived peptides IRW and IQW in alleviating the adverse effects of obesity.

Supplementary Materials: The supporting information can be downloaded at: https://www.mdpi.com/article/10.3390/ijms231911227/s1.

Author Contributions: Conceptualization, H.J. and J.F.; Data curation, Z.L.; Formal analysis, H.J.; Funding acquisition, J.F.; Software, J.F.; Writing—original draft, Z.L. and J.F.; Writing—review & editing, S.D. All authors have read and agreed to the published version of the manuscript.

Funding: This study was supported by the National Natural Science Foundation of China (nos. 31672457 and 31772642), Ministry of Agricultural of the People's Republic of China (2015-Z64 and 2016-X47), Hunan Provincial Science and Technology Department (2016NK2101, 2017NK2322, 2016WK2008, and 2016TP2005), Local Science and Technology Development Project guided by the Central Government (YDZX20184300002303), Double First-Class Construction Project of Hunan Agricultural University (SYL201802003), China Postdoctoral Science Foundation (2018M632963 and 2019T120705), Science and Technology Innovation and Entrepreneurship Project for University Students of Hunan Province (2021RC1004), Postgraduate Scientific Research Innovation Project of Hunan Province (CX20210654), and Science and Technology Innovation and Entrepreneurship Project for University Students of Hunan Province (2021RC1004).

Institutional Review Board Statement: The study was conducted in accordance with the Hunan Agricultural University's rules for the care and use of laboratory animals, with approval from the animal care committee of Hunan Agricultural University.

Informed Consent Statement: Not applicable.

Data Availability Statement: The raw sequence data in this study are uploaded in the NCBI database, the accession is PRJNA853190 (http://www.ncbi.nlm.nih.gov/bioproject/?term=PRJNA853190) and https://www.ncbi.nlm.nih.gov/Traces/study/?acc=PRJNA853190&o=acc_s%3Aa.

Conflicts of Interest: The authors declare no conflict of interest.

References

1. NCD Risk Factor Collaboration. Worldwide trends in body-mass index, underweight, overweight, and obesity from 1975 to 2016: A pooled analysis of 2416 population-based measurement studies in 128.9 million children, adolescents, and adults. *Lancet* **2017**, *390*, 2627–2642. [CrossRef]
2. Nguyen, D.M.; El-Serag, H.B. The epidemiology of obesity. *Gastroenterol. Clin. N. Am.* **2010**, *39*, 1–7. [CrossRef]

3. Bleich, S.N.; Ku, R.; Wang, Y.C. Relative contribution of energy intake and energy expenditure to childhood obesity: A review of the literature and directions for future research. *Int. J. Obes.* **2011**, *35*, 1–15. [CrossRef]
4. Jo, J.; Gavrilova, O.; Pack, S.; Jou, W.; Mullen, S.; Sumner, A.E.; Cushman, S.W.; Periwal, V. Hypertrophy and/or Hyperplasia: Dynamics of Adipose Tissue Growth. *PLoS Comput. Biol.* **2009**, *5*, e1000324. [CrossRef] [PubMed]
5. Jernas, M.; Palming, J.; Sjoholm, K.; Jennische, E.; Svensson, P.A.; Gabrielsson, B.G.; Levin, M.; Sjogren, A.; Rudemo, M.; Lystig, T.C.; et al. Separation of human adipocytes by size: Hypertrophic fat cells display distinct gene expression. *FASEB J. Off. Publ. Fed. Am. Soc. Exp. Biol.* **2006**, *20*, 1540–1542. [CrossRef] [PubMed]
6. Monteiro, R.; de Castro, P.M.; Calhau, C.; Azevedo, I. Adipocyte size and liability to cell death. *Obes. Surg.* **2006**, *16*, 804–806. [CrossRef]
7. Vanni, E.; Bugianesi, E.; Kotronen, A.; De Minicis, S.; Yki-Jarvinen, H.; Svegliati-Baroni, G. From the metabolic syndrome to NAFLD or vice versa? *Dig. Liver Dis. Off. J. Ital. Soc. Gastroenterol. Ital. Assoc. Study Liver* **2010**, *42*, 320–330. [CrossRef]
8. Karavia, E.A.; Papachristou, D.J.; Kotsikogianni, I.; Triantafyllidou, I.E.; Kypreos, K.E. Lecithin/cholesterol acyltransferase modulates diet-induced hepatic deposition of triglycerides in mice. *J. Nutr. Biochem.* **2013**, *24*, 567–577. [CrossRef]
9. The Human Microbiome Project Consortium. Structure, function and diversity of the healthy human microbiome. *Nature* **2012**, *486*, 207–214. [CrossRef]
10. Ridaura, V.K.; Faith, J.J.; Rey, F.E.; Cheng, J.; Duncan, A.E.; Kau, A.L.; Griffin, N.W.; Lombard, V.; Henrissat, B.; Bain, J.R.; et al. Gut microbiota from twins discordant for obesity modulate metabolism in mice. *Science* **2013**, *341*, 1241214. [CrossRef]
11. Turnbaugh, P.J.; Ley, R.E.; Mahowald, M.A.; Magrini, V.; Mardis, E.R.; Gordon, J.I. An obesity-associated gut microbiome with increased capacity for energy harvest. *Nature* **2006**, *444*, 1027–1031. [CrossRef] [PubMed]
12. Liang, Y.; Zhan, J.; Liu, D.; Luo, M.; Han, J.; Liu, X.; Liu, C.; Cheng, Z.; Zhou, Z.; Wang, P. Organophosphorus pesticide chlorpyrifos intake promotes obesity and insulin resistance through impacting gut and gut microbiota. *Microbiome* **2019**, *7*, 19. [CrossRef]
13. Chassaing, B.; Koren, O.; Goodrich, J.K.; Poole, A.C.; Srinivasan, S.; Ley, R.E.; Gewirtz, A.T. Dietary emulsifiers impact the mouse gut microbiota promoting colitis and metabolic syndrome. *Nature* **2015**, *519*, 92–96. [CrossRef]
14. Suez, J.; Korem, T.; Zeevi, D.; Zilberman-Schapira, G.; Thaiss, C.A.; Maza, O.; Israeli, D.; Zmora, N.; Gilad, S.; Weinberger, A.; et al. Artificial sweeteners induce glucose intolerance by altering the gut microbiota. *Nature* **2014**, *514*, 181–186. [CrossRef] [PubMed]
15. Org, E.; Blum, Y.; Kasela, S.; Mehrabian, M.; Kuusisto, J.; Kangas, A.J.; Soininen, P.; Wang, Z.; Ala-Korpela, M.; Hazen, S.L.; et al. Relationships between gut microbiota, plasma metabolites, and metabolic syndrome traits in the METSIM cohort. *Genome Biol.* **2017**, *18*, 70. [CrossRef] [PubMed]
16. Rios-Covian, D.; Ruas-Madiedo, P.; Margolles, A.; Gueimonde, M.; de Los Reyes-Gavilan, C.G.; Salazar, N. Intestinal Short Chain Fatty Acids and their Link with Diet and Human Health. *Front. Microbiol.* **2016**, *7*, 185. [CrossRef]
17. Matsumoto, M.; Inoue, R.; Tsukahara, T.; Ushida, K.; Chiji, H.; Matsubara, N.; Hara, H. Voluntary running exercise alters microbiota composition and increases n-butyrate concentration in the rat cecum. *Biosci. Biotechnol. Biochem.* **2008**, *72*, 572–576. [CrossRef]
18. Queipo-Ortuno, M.I.; Seoane, L.M.; Murri, M.; Pardo, M.; Gomez-Zumaquero, J.M.; Cardona, F.; Casanueva, F.; Tinahones, F.J. Gut microbiota composition in male rat models under different nutritional status and physical activity and its association with serum leptin and ghrelin levels. *PLoS ONE* **2013**, *8*, e65465. [CrossRef]
19. De Filippo, C.; Cavalieri, D.; Di Paola, M.; Ramazzotti, M.; Poullet, J.B.; Massart, S.; Collini, S.; Pieraccini, G.; Lionetti, P. Impact of diet in shaping gut microbiota revealed by a comparative study in children from Europe and rural Africa. *Proc. Natl. Acad. Sci. USA* **2010**, *107*, 14691–14696. [CrossRef]
20. Cardoso, I.; Bovet, P.; Viswanathan, B.; Luke, A.; Marques-Vidal, P. Nutrition transition in a middle-income country: 22-year trends in the Seychelles. *Eur. J. Clin. Nutr.* **2013**, *67*, 135–140. [CrossRef]
21. Tremaroli, V.; Backhed, F. Functional interactions between the gut microbiota and host metabolism. *Nature* **2012**, *489*, 242–249. [CrossRef]
22. Petriz, B.A.; Castro, A.P.; Almeida, J.A.; Gomes, C.P.; Fernandes, G.R.; Kruger, R.H.; Pereira, R.W.; Franco, O.L. Exercise induction of gut microbiota modifications in obese, non-obese and hypertensive rats. *BMC Genom.* **2014**, *15*, 511. [CrossRef] [PubMed]
23. Liao, W.; Fan, H.; Davidge, S.T.; Wu, J. Egg White-Derived Antihypertensive Peptide IRW (Ile-Arg-Trp) Reduces Blood Pressure in Spontaneously Hypertensive Rats via the ACE2/Ang (1-7)/Mas Receptor Axis. *Mol. Nutr. Food Res.* **2019**, *63*, e1900063. [CrossRef] [PubMed]
24. Beermann, C.; Euler, M.; Herzberg, J.; Stahl, B. Anti-oxidative capacity of enzymatically released peptides from soybean protein isolate. *Eur. Food Res. Technol.* **2009**, *229*, 637–644. [CrossRef]
25. Ascencio, C.; Torres, N.; Isoard-Acosta, F.; Gomez-Perez, F.J.; Hernandez-Pando, R.; Tovar, A.R. Soy protein affects serum insulin and hepatic SREBP-1 mRNA and reduces fatty liver in rats. *J. Nutr.* **2004**, *134*, 522–529. [CrossRef]
26. Ma, Y.; Ding, S.; Liu, G.; Fang, J.; Yan, W.; Duraipandiyan, V.; Al-Dhabi, N.A.; Esmail, G.A.; Jiang, H. Egg Protein Transferrin-Derived Peptides IRW and IQW Regulate Citrobacter rodentium-Induced, Inflammation-Related Microbial and Metabolomic Profiles. *Front. Microbiol.* **2019**, *10*, 643. [CrossRef] [PubMed]
27. Hori, G.; Wang, M.F.; Chan, Y.C.; Komatsu, T.; Wong, Y.; Chen, T.H.; Yamamoto, K.; Nagaoka, S.; Yamamoto, S. Soy protein hydrolyzate with bound phospholipids reduces serum cholesterol levels in hypercholesterolemic adult male volunteers. *Biosci. Biotechnol. Biochem.* **2001**, *65*, 72–78. [CrossRef]

28. Majumder, K.; Chakrabarti, S.; Davidge, S.T.; Wu, J. Structure and activity study of egg protein ovotransferrin derived peptides (IRW and IQW) on endothelial inflammatory response and oxidative stress. *J. Agric. Food Chem.* **2013**, *61*, 2120–2129. [CrossRef] [PubMed]
29. Wu, J. A Novel Angiotensin Converting Enzyme 2 (ACE2) Activating Peptide: A Reflection of 10 Years of Research on a Small Peptide Ile-Arg-Trp (IRW). *J. Agric. Food Chem.* **2020**, *68*, 14402–14408. [CrossRef]
30. Majumder, K.; Chakrabarti, S.; Morton, J.S.; Panahi, S.; Kaufman, S.; Davidge, S.T.; Wu, J. Egg-derived ACE-inhibitory peptides IQW and LKP reduce blood pressure in spontaneously hypertensive rats. *J. Funct. Foods* **2015**, *13*, 50–60. [CrossRef]
31. Singh, B.P.; Vij, S.; Hati, S. Functional significance of bioactive peptides derived from soybean. *Peptides* **2014**, *54*, 171–179. [CrossRef] [PubMed]
32. Zuconelli, C.R.; Brock, R.; Adjobo-Hermans, M.J.W. Linear Peptides in Intracellular Applications. *Curr. Med. Chem.* **2017**, *24*, 1862–1873. [CrossRef] [PubMed]
33. Hartmann, R.; Meisel, H. Food-derived peptides with biological activity: From research to food applications. *Curr. Opin. Biotechnol.* **2007**, *18*, 163–169. [CrossRef]
34. Sawada, Y.; Sakamoto, Y.; Toh, M.; Ohara, N.; Hatanaka, Y.; Naka, A.; Kishimoto, Y.; Kondo, K.; Iida, K. Milk-derived peptide Val-Pro-Pro (VPP) inhibits obesity-induced adipose inflammation via an angiotensin-converting enzyme (ACE) dependent cascade. *Mol. Nutr. Food Res.* **2015**, *59*, 2502–2510. [CrossRef]
35. Trayhurn, P.; Beattie, J.H. Physiological role of adipose tissue: White adipose tissue as an endocrine and secretory organ. *Proc. Nutr. Soc.* **2001**, *60*, 329–339. [CrossRef] [PubMed]
36. Bloor, I.D.; Symonds, M.E. Sexual dimorphism in white and brown adipose tissue with obesity and inflammation. *Horm. Behav.* **2014**, *66*, 95–103. [CrossRef] [PubMed]
37. Nogueiras, R.; Perez-Tilve, D.; Veyrat-Durebex, C.; Morgan, D.A.; Varela, L.; Haynes, W.G.; Patterson, J.T.; Disse, E.; Pfluger, P.T.; Lopez, M.; et al. Direct control of peripheral lipid deposition by CNS GLP-1 receptor signaling is mediated by the sympathetic nervous system and blunted in diet-induced obesity. *J. Neurosci. Off. J. Soc. Neurosci.* **2009**, *29*, 5916–5925. [CrossRef]
38. Yadav, A.; Kataria, M.A.; Saini, V.; Yadav, A. Role of leptin and adiponectin in insulin resistance. *Clin. Chim. Acta* **2013**, *417*, 80–84. [CrossRef]
39. Mantzoros, C.S.; Magkos, F.; Brinkoetter, M.; Sienkiewicz, E.; Dardeno, T.A.; Kim, S.Y.; Hamnvik, O.P.; Koniaris, A. Leptin in human physiology and pathophysiology. *Am. J. Physiol. Endocrinol. Metab.* **2011**, *301*, E567–E584. [CrossRef]
40. Eriksson, B.; Lof, M.; Olausson, H.; Forsum, E. Body fat, insulin resistance, energy expenditure and serum concentrations of leptin, adiponectin and resistin before, during and after pregnancy in healthy Swedish women. *Br. J. Nutr.* **2010**, *103*, 50–57. [CrossRef]
41. Brochu-Gaudreau, K.; Rehfeldt, C.; Blouin, R.; Bordignon, V.; Murphy, B.D.; Palin, M.F. Adiponectin action from head to toe. *Endocrine* **2010**, *37*, 11–32. [CrossRef] [PubMed]
42. Wainwright, P.; Byrne, C.D. Bidirectional Relationships and Disconnects between NAFLD and Features of the Metabolic Syndrome. *Int. J. Mol. Sci.* **2016**, *17*, 367. [CrossRef] [PubMed]
43. Bedogni, G.; Miglioli, L.; Masutti, F.; Tiribelli, C.; Marchesini, G.; Bellentani, S. Prevalence of and risk factors for nonalcoholic fatty liver disease: The Dionysos nutrition and liver study. *Hepatology* **2005**, *42*, 44–52. [CrossRef] [PubMed]
44. Sae-Tan, S.; Grove, K.A.; Kennett, M.J.; Lambert, J.D. (-)-Epigallocatechin-3-gallate increases the expression of genes related to fat oxidation in the skeletal muscle of high fat-fed mice. *Food Funct.* **2011**, *2*, 111–116. [CrossRef] [PubMed]
45. Jung, U.J.; Lee, M.K.; Jeong, K.S.; Choi, M.S. The hypoglycemic effects of hesperidin and naringin are partly mediated by hepatic glucose-regulating enzymes in C57BL/KsJ-db/db mice. *J. Nutr.* **2004**, *134*, 2499–2503. [CrossRef]
46. Waltner-Law, M.E.; Wang, X.L.; Law, B.K.; Hall, R.K.; Nawano, M.; Granner, D.K. Epigallocatechin gallate, a constituent of green tea, represses hepatic glucose production. *J. Biol. Chem.* **2002**, *277*, 34933–34940. [CrossRef]
47. Chang, S.C.; Yang, W.V. Hyperglycemia, tumorigenesis, and chronic inflammation. *Crit. Rev. Oncol. Hematol.* **2016**, *108*, 146–153. [CrossRef]
48. Mantovani, A.; Allavena, P.; Sica, A.; Balkwill, F. Cancer-related inflammation. *Nature* **2008**, *454*, 436–444. [CrossRef]
49. Cani, P.D.; Bibiloni, R.; Knauf, C.; Waget, A.; Neyrinck, A.M.; Delzenne, N.M.; Burcelin, R. Changes in gut microbiota control metabolic endotoxemia-induced inflammation in high-fat diet-induced obesity and diabetes in mice. *Diabetes* **2008**, *57*, 1470–1481. [CrossRef]
50. Ley, R.E.; Turnbaugh, P.J.; Klein, S.; Gordon, J.I. Microbial ecology: Human gut microbes associated with obesity. *Nature* **2006**, *444*, 1022–1023. [CrossRef]
51. Cani, P.D.; Delzenne, N.M. Interplay between obesity and associated metabolic disorders: New insights into the gut microbiota. *Curr. Opin. Pharmacol.* **2009**, *9*, 737–743. [CrossRef] [PubMed]
52. Duncan, S.H.; Lobley, G.E.; Holtrop, G.; Ince, J.; Johnstone, A.M.; Louis, P.; Flint, H.J. Human colonic microbiota associated with diet, obesity and weight loss. *Int. J. Obes.* **2008**, *32*, 1720–1724. [CrossRef] [PubMed]
53. Wang, K.; Liao, M.; Zhou, N.; Bao, L.; Ma, K.; Zheng, Z.; Wang, Y.; Liu, C.; Wang, W.; Wang, J.; et al. Parabacteroides distasonis Alleviates Obesity and Metabolic Dysfunctions via Production of Succinate and Secondary Bile Acids. *Cell Rep.* **2019**, *26*, 222–235.e5. [CrossRef] [PubMed]

54. Ding, N.; Zhang, X.; Zhang, X.D.; Jing, J.; Liu, S.S.; Mu, Y.P.; Peng, L.L.; Yan, Y.J.; Xiao, G.M.; Bi, X.Y.; et al. Impairment of spermatogenesis and sperm motility by the high-fat diet-induced dysbiosis of gut microbes. *Gut* **2020**, *69*, 1608–1619. [CrossRef] [PubMed]
55. Ding, S.; Ma, Y.; Liu, G.; Yan, W.; Jiang, H.; Fang, J. Lactobacillus brevis Alleviates DSS-Induced Colitis by Reprograming Intestinal Microbiota and Influencing Serum Metabolome in Murine Model. *Front. Physiol.* **2019**, *10*, 1152. [CrossRef] [PubMed]

International Journal of *Molecular Sciences*

MDPI

Review

The Impact of Plant Phytochemicals on the Gut Microbiota of Humans for a Balanced Life

Sarusha Santhiravel [1,2,*], Alaa El-Din A. Bekhit [3], Eresha Mendis [4], Joe L. Jacobs [5,6], Frank R. Dunshea [7,8], Niranjan Rajapakse [4] and Eric N. Ponnampalam [9,*]

1 Postgraduate Institute of Agriculture, University of Peradeniya, Peradeniya 20400, Sri Lanka
2 Department of Biochemistry, Memorial University of Newfoundland, St. John's, NL A1C 5S7, Canada
3 Department of Food Sciences, University of Otago, P.O. Box 56, Dunedin 9054, New Zealand; aladin.bekhit@otago.ac.nz
4 Department of Food Science and Technology, Faculty of Agriculture, University of Peradeniya, Peradeniya 20400, Sri Lanka; ereshamendis@agri.pdn.ac.lk (E.M.); niranjanp@agri.pdn.ac.lk (N.R.)
5 Animal Production Sciences, Agriculture Victoria Research, Department of Jobs, Precincts and Regions, Ellinbank, VIC 3821, Australia; joe.jacobs@agriculture.vic.gov.au
6 Centre for Agricultural Innovation, School of Agriculture and Food, Faculty of Veterinary and Agricultural Sciences, The University of Melbourne, Parkville, VIC 3010, Australia
7 School of Agriculture and Food, Faculty of Veterinary and Agricultural Sciences, The University of Melbourne, Parkville, VIC 3010, Australia; fdunshea@unimelb.edu.au
8 Faculty of Biological Sciences, University of Leeds, Leeds LS2 9JT, UK
9 Animal Production Sciences, Agriculture Victoria Research, Department of Jobs, Precincts and Regions, Bundoora, VIC 3083, Australia
* Correspondence: eric.ponnampalam@agriculture.vic.gov.au (E.N.P.); ssanthiravel@mun.ca (S.S.)

Citation: Santhiravel, S.; Bekhit, A.E.-D.A.; Mendis, E.; Jacobs, J.L.; Dunshea, F.R.; Rajapakse, N.; Ponnampalam, E.N. The Impact of Plant Phytochemicals on the Gut Microbiota of Humans for a Balanced Life. *Int. J. Mol. Sci.* **2022**, *23*, 8124. https://doi.org/10.3390/ijms23158124

Academic Editor: Virginia Garcia-Cañas

Received: 4 July 2022
Accepted: 21 July 2022
Published: 23 July 2022

Abstract: The gastrointestinal tract of humans is a complex microbial ecosystem known as gut microbiota. The microbiota is involved in several critical physiological processes such as digestion, absorption, and related physiological functions and plays a crucial role in determining the host's health. The habitual consumption of specific dietary components can impact beyond their nutritional benefits, altering gut microbiota diversity and function and could manipulate health. Phytochemicals are non-nutrient biologically active plant components that can modify the composition of gut microflora through selective stimulation of proliferation or inhibition of certain microbial communities in the intestine. Plants secrete these components, and they accumulate in the cell wall and cell sap compartments (body) for their development and survival. These compounds have low bioavailability and long time-retention in the intestine due to their poor absorption, resulting in beneficial impacts on gut microbiota population. Feeding diets containing phytochemicals to humans and animals may offer a path to improve the gut microbiome resulting in improved performance and/or health and wellbeing. This review discusses the effects of phytochemicals on the modulation of the gut microbiota environment and the resultant benefits to humans; however, the effect of phytochemicals on the gut microbiota of animals is also covered, in brief.

Keywords: plant foods; phytochemicals; gut microbiota; digestive process; metabolic diseases; health and wellness

1. Introduction

Around 400 B.C., Hippocrates said, "death sits in the bowels" and "bad digestion is the root of all evil", indicating the significant contribution of the human intestine to improved health [1]. Taking humans as an example, gut microbiota (GM), also called gut flora, is the name given to the microbial loads inhabiting our gastrointestinal (GI) tract, and around 100 trillion micro-organisms comprising, primarily bacteria, inhabit the human colon [2]. The human intestine is also inhabited by a lesser number of eukaryotic organisms, such as fungi, protozoa, and viruses. Anaerobic bacteria are the predominant group of the microbial community present in the human colon, which creates one of the most densely populated and diversified bacterial ecosystems in nature [3].

Although several hundred micro-organism species reside in the human colon, many relevant studies and current findings collected from the Human Microbiome Project explain that among individuals, the microbial population is greatly varied in composition [4,5]. The colonization of bacteria begins in utero, and lifelong changes take place in the composition of the GM, of which the significant alterations in number and diversity occur during the breastfeeding period and at the commencement of solid food consumption. Micro-organisms differ in number, type, and function throughout the GI tract. Still, the GM is densely populated in the large intestine where the microbiota is involved in the fermentation of undigested food components, especially fibers, some carbohydrates, and oxidized proteins, together with more associated metabolic functions [1].

In the context of herbivores, ruminants (cattle, sheep, goats, alpacas, deer, etc.) cannot directly digest plant material due to the lack of enzymes capable of breaking down high-fibrous cell wall contents, such as cellulose and hemicellulose. Rather, the GI tract of ruminants is inhabited by a large population of bacteria, protozoans, and fungi in their four-chambered stomach (rumen, reticulum, omasum, and abomasum) capable of digesting the high-fibrous plant materials. Partially or fully digested food passes through the last compartment of the stomach to the small intestine and large intestine, and further digestion and absorption of nutrients takes place by the host animal. The anatomical and functional attributes of the small intestine of ruminants is similar to humans and other animals, and ranges in length by approximately 12–30 times the animal's body length [6].

Gut microbiota have been identified as capable sources of modern curative medicines [7]. Various details of the GM and its impacts on human health, including during childhood [8] and certain diseases, such as inflammatory bowel diseases, cardiometabolic disorders, cancer, and neuropsychiatric diseases, are considered in a large number of recent studies and reviews [9,10]. Signals that stimulate the generation of cytokines are provided by the GM, resulting in the alteration in the general development of the host immune system functions, facilitated by maturation of the immune cells [11]. The relationship between the microbiota and the host has a reciprocal nature, and it has been said that "feed your microbiota and get fed by it" [12,13].

Dysbiosis is the impairment of GM and/or their functions due to several factors, such as poor diet, insufficient exercise, stress, age, drugs, and xenobiotics [1,14]. The inter-relationship among dysbiosis and diseases, namely those related to the GI tract, such as ulcerative colitis, inflammatory bowel disease, colorectal cancer, and Crohn's disease as well as some disorders associated with extra-intestinal metabolism, such as obesity, diabetes, cardiovascular disease, and related micro- and macrovascular complications are supported by considerable evidence [15,16]. Correspondingly, an imbalance in the microbial community can be caused by a pathological state. For example, metabolic disorders may be promoted by dysfunction of the innate immune system through the modulation of the GM [17,18].

The diet normally provides nutrients and energy for growth, development, and maintenance of life. At the same time, some food components of varying chemical structures and functionalities offer health benefits that extend beyond their nutritional value when consumed regularly, leading to improved health. These components are known as "nutraceuticals", and foods containing these bioactive molecules are coined as "functional foods" [19,20]. Functional foods are categorized as non-nutrient (probiotics and phytochemicals) and macro- and micro-nutrients (fatty acids and vitamins) [21]. In humans, herbivores, and mice, the composition of the GM can be modified by diet [22]. Long-term dietary habits considerably influence human gut microbiota. Diet and changes in gut pH can affect several physiological aspects of the intestinal environment, particularly the absorption of nutraceuticals, micronutrients, and vitamins, consequently changing the equilibrium of the intestinal ecology [23]. Many investigators examined the effects of nutraceuticals on GM and their potential to remove pathogenic bacteria (or pathobionts) without harming the favorable bacteria (symbionts).

Phytochemicals are non-nutrient plant compounds that are biologically active and produced by the primary and secondary metabolism of plants. Phytochemicals are natural functional ingredients widely present in fruits, vegetables, seeds and nuts, whole grain products, legumes, dark chocolate, and tea, whose regular dietary intake was suggested to reduce the occurrence of many chronic illnesses [24,25]. Relatively few types of phytochemicals have been identified and isolated from plants but there are tens of thousands of phytochemicals [26]. Phytochemicals benefit the human and animal body by altering the intestinal microflora by selectively inducing the growth of some bacterial populations in the gut, referred to as "probiotics" [27]. These consist of endosymbionts, including yeast, bifidobacterium, lactic acid bacteria, and bacilli involved in human and animal GI metabolism [28].

Phytochemicals may be categorized as polyphenols, alkaloids, terpenoids (carotenoid terpenoids and non-carotenoid terpenoid), organosulfur compounds, and nitrogen-containing compounds based on their biosynthetic origins [29–31]. Classification of dietary phytochemicals is shown in Figure 1. Polyphenols constitute the largest group among these phytochemicals. There is increasing evidence demonstrating the ability of phytochemicals to decrease inflammation, slow the growth rate of cancer cells, reduce the formation of carcinogenic compounds, regulate gene expression and hormones' intracellular signaling, enhance the immune system, alleviate DNA damage, decrease oxidative destruction of cells, and activate insulin receptors [32–34].

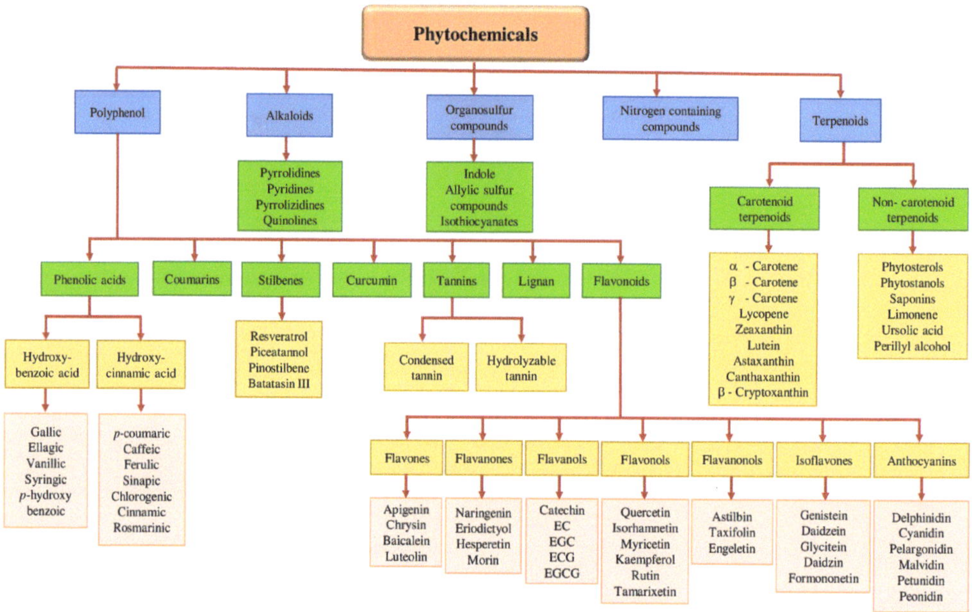

Figure 1. Classification of Dietary Phytochemicals (self-generated) [EC: epicatechin; EGC: epigallocatechin; ECG: epicatechin gallate; EGCG: epigallocatechin 3-gallate].

Compared to the micro- and macro-nutrients, the bioavailability of phytochemicals is very low within the human body due to their complex chemical structure and to their being metabolized as a xenobiotic [35]. The poor absorption of phytochemicals leads to extended retention times in the intestine where they may play a beneficial role by influencing the intestinal ecology [36]. During the last two decades, the influence of dietary phytochemicals on GI microbial population and the fundamental mechanisms assumed for their favorable effects on extra-intestinal and GI disorders have been illustrated [37–39]. However, there is a need to comprehensively review recent studies related to the impact of phytochemicals on gut microbiota of humans and their mechanism of action, including their relationship with metabolic diseases. This review mainly covers the effects of phytochemicals on the modulation of the gut microbiota environment and the resultant benefits to human health to establish clear directions for their potential use in human diets; however, the effect of phytochemicals on gut microbiota modulation of animals is also covered, in brief.

2. Human Gut Morphology and Composition of Healthy Microbiota

The surface area of the GI tract of humans is about 260–300 m^2, which represents major interactions among many parts of the human body, interior antigens, and exterior environmental agents [40]. The intestinal microflora is a complex community consisting of 10–100 trillion microbes represented by about 1000 species. A large number of external micro-organisms and approximately sixty tonnes of food that pass through the GI tract during the average lifetime threaten the intestine's integrity [41]. A few decades ago, the human GM was deemed to be the 'forgotten organ' and was simply considered to be a part of an excretion system [42]; now it is recognized as an individual organ for metabolism in the human body [43].

Humans and animals show a complex and mutual relationship with their symbiotic collective community of intestinal micro-organisms, namely bacteria, eukaryotes (mainly yeasts), viruses (mainly phages), archaea, and other microbial species, which have evolved through thousands of years of joint development [39]. Even though many microbial species exist, the human GI tract harbors approximately 1014 cells, chiefly anaerobes, such as *Bacteroidetes*, *Firmicutes*, Proteobacteria, and Actinobacteria [9,44]. Nearly 90% of the microbes present in the GI tract of adults are under the phyla *Firmicutes* (Gram-positive) (for example, *Lactobacillales*, *Clostridiales*, and *Ruminococcus* species) and *Bacteroidetes* (Gram-negative) (including *Bacteroides* and *Prevotella* species). Other phyla, such as Proteobacteria (Gram-negative), Actinobacteria (Gram-positive, including *Bifidobacterium*), Fusobacteria, and Verrucomicrobia (Gram-negative, including *Akkermansia muciniphila*) as well as some facultative anaerobic bacteria represent a very low proportion [2,45–48].

The gut microbial population is profoundly influenced by the dietary habits associated with the consumption of different types and amounts of phytochemicals [38,49,50]. Figure 2 depicts some of the known effects of phytochemicals on intestinal microflora. The gut's microbial composition changes along the entire length of the GI tract, depending on the structure and function of the regions of the digestive system. From the proximal part of the intestine to the distal portion, a significant increase in the microbial population density and enhanced population of anaerobes occurs [44,51]. Because of the acidic environment imparted by the gastric juices (chyme) secreted by the stomach, pancreas, and biliary, the composition of the GM in the initial part of the proximal gut, mainly in the duodenum, is similar to that of the stomach [52,53].

From the duodenum to the ileum, the bacterial population increases in number and diversity toward the distal part of the gut as a result of the gradual increase in pH. This distal portion is dominated by the phyla *Bacteroidetes*, Proteobacteria (mainly *Escherichia coli*), *Firmicutes* (mainly *Lactobacillus* and *Clostridium* species), and Gram-negative facultative anaerobes [41,47]. A more favorable nutritional environment with pH in the range of 5.7–6.8 exists in the large intestine, mainly in the colon, and this induces the proliferation of the microbial population. This helps produce a more concentrated, complex, and diverse community of intestinal microflora, containing obligate anaerobes that can live at very low oxygen concentrations. The lumen of the intestine and the mucus layer of the inner lining of the intestinal tract also have different compositions and physiology of microflora. In addition, the distribution of aerobic and anaerobic microbial species also varies between them [54,55].

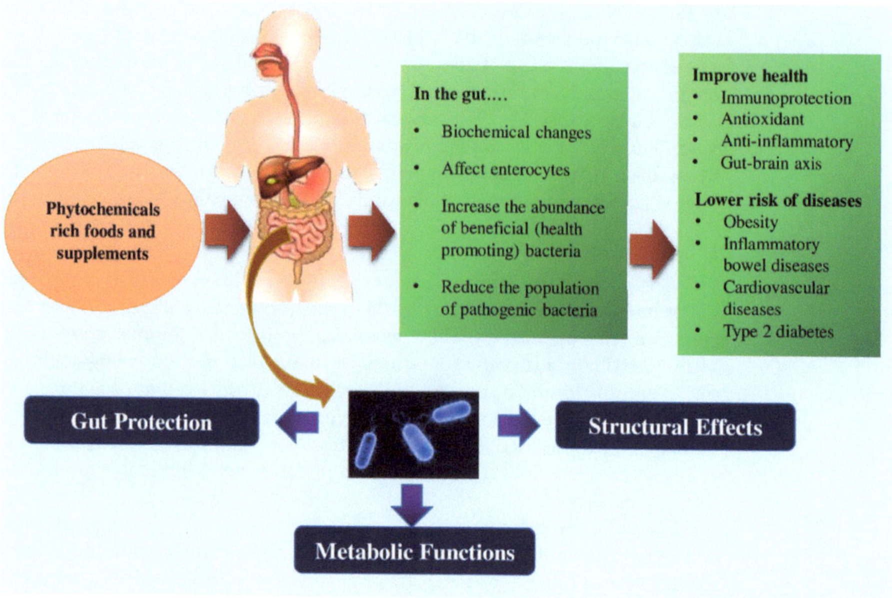

Figure 2. The influence of phytochemicals on GM and key roles of GM in humans (Self-generated).

The colonization of microbes in the GI tract varies in humans and animal species. For humans, the colonization starts in the first two years of life. During gestation, the gut of infants is considered sterile, and it is not until after birth that the micro-organisms appear in the gut of infants. The first colonization of the microbes and subsequent proliferation throughout the intestine is crucial in achieving favorable outcomes by influencing the host's immune system [56]. As the human grows and the digestive tract develops, at around 2–3 years of age, the phylogenetic diversity of intestinal microflora is established, and a stable and complex microbial ecosystem is generated [57].

The healthy intestinal ecology influences several vital functions (Figure 2) in the host body, including the production of bile acid, satiety, lipogenesis, digestion process, absorption of dietary nutrients, and innate immunity [58–60]. The GM contributes several key metabolic functions (co-metabolism) to the host body, including its ability to protect against harmful bacteria. This is carried out by maintaining and regulating the integrity and permeability of the intestinal barrier, thereby influencing the homeostasis of the host. In addition, their capacity to carry out bile biotransformation while establishing immunity of the host and fermenting carbohydrates, proteins, and lipids and synthesizing a large number of vitamins, essential and non-essential amino acids, and short-chain fatty acids (SCFAs) [61,62]. Furthermore, some dietary components such as complex oligosaccharides that are not easily digested, including a few indigestible polysaccharides (e.g., cellulose, hemicellulose, pectins, unabsorbed sugars, resistant starches, alcohols, and gums) can be metabolized by GM. This supports the bacteria in obtaining energy and nutrients for their growth and proliferation, resulting in the host regaining absorbable components from the digesta [62,63].

3. Gut Microbial Metabolism of Phytochemicals

The metabolic pathway of dietary phytochemicals in the human body is shown in Figure 3. Several reactions, including hydration, oxidation, hydroxylation, decarboxylation, methylation, dehydrogenation, glycosylation, and isomerization are involved and cause various modifications to the parent phytochemicals. The interaction of high molecular-weight phytochemicals to intestinal tissues is influenced by the metabolism of phytochemicals in food and their degradation/modification by intestinal microflora [64].

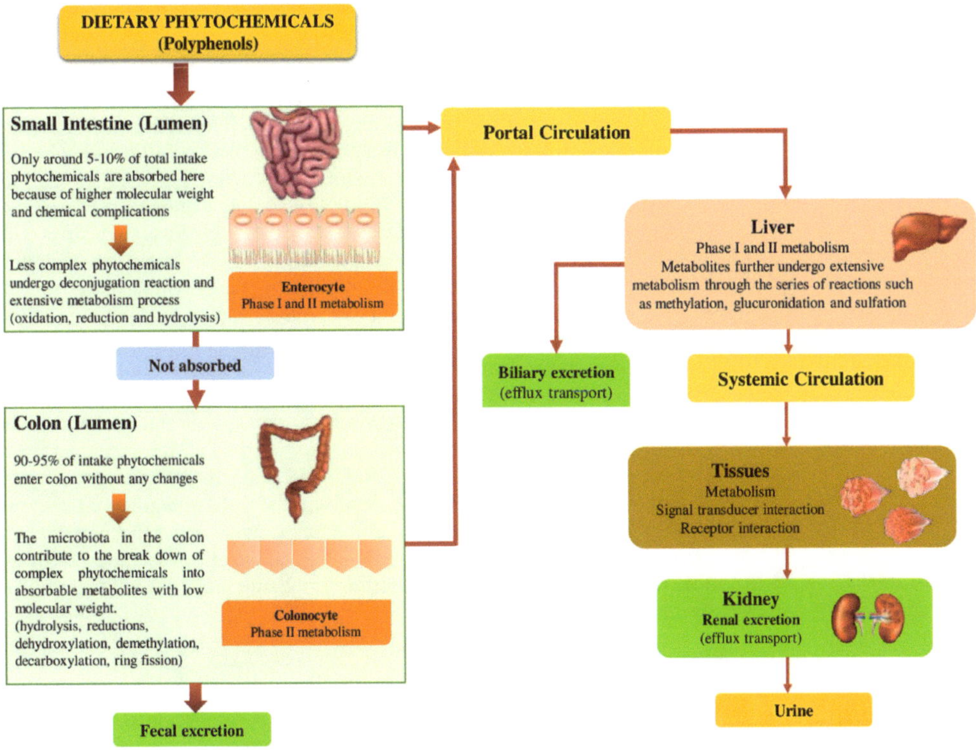

Figure 3. Simple illustration of the metabolic pathway of dietary phytochemicals in human body (self-generated).

Considering polyphenols as an example, following absorption by the small intestine, a sequence of conjugated metabolites (water-soluble derivatives of sulfate, glucuronide, and methyl compounds) is produced in the hepatocytes and enterocytes, called Phase I. Phase I is a complex process that involves hydrolysis, oxidation, and reduction reactions. Phase II is the biological conversion of polyphenol compounds to hydrophilic metabolites through conjugation by either membrane bound or soluble cytosolic enzymes, which is less complex [65].

Intestinal bacteria produce metabolites, which can play several crucial functions in the body, from the unabsorbed polyphenols (around 90–95% of the total ingestion of polyphenol) by enzymatically acting on the backbone of those polyphenols remaining in the large intestine [1]. The polyphenols are metabolized by intestinal micro-organisms through glycosidic bond-splitting and heterocyclic backbone breakdown [66]. Upon absorption, the resulting metabolites of polyphenols will enter the liver via the portal vein, producing active metabolites (sulfation, methylation, and glucuronidation) by undergoing considerable degradation reactions. Further, the target tissues and cells will be exposed to these active metabolites after they are released into systematic circulation, where they can play significant physiological roles. Finally, the remaining unutilized metabolites that are in excess for the body will be excreted in urine [53].

4. Effect of Different Phytochemical Compounds on the Modulation of Gut Microbiota

4.1. Effect of Polyphenols

Polyphenols are chemical compounds widely distributed in plants such as vegetables, fruits, cereals, coffee, tea, and wine [67]. Dietary polyphenols are a broad group of natural heterogeneous components consisting of phenyl moieties, that are hydroxylated. Commonly polyphenols are categorized as flavonoids and non-flavonoids according to their complexity and chemical structure, including the type and number of substituting groups and the amounts of phenolic rings [68]. Examples of non-flavonoids include phenolic acids, stilbenes, curcumin, tannins, lignans, and coumarin [34,69].

Even though the most commonly found modifications are polymerization or esterification, many polyphenols are present in plants in the glycosylated form. Generally, polyphenols are received as xenobiotics in the host body after ingestion. Therefore, their bioavailability is relatively lower than macro- and micro-nutrients. Moreover, the small intestine may absorb low molecular weight compounds such as monomers and dimers due to their low complexity [70]. The compounds with a complex structure, such as oligomeric and polymeric structures, may enter the large intestine almost without any modification, degraded by microflora and absorbed by the host animal [39].

The actual reason behind the health benefits obtained from the intake of foods rich in polyphenols is the metabolites of polyphenol produced by GM, rather than the original components present in foods [64]. Recent studies indicated that the prebiotic properties and antimicrobial effects of dietary phenolic compounds and the antagonistic effect produced by aromatic metabolites against harmful gut microbiota, may alter the intestinal microbial ecosystem [1,39,71,72]. The evidence regarding the effect of each type of polyphenol, including flavonoids, phenolic acid, curcumin, stilbenes, lignans, and tannins, on GM is discussed below.

4.1.1. Flavonoids

Flavonoids are the largest subgroup of polyphenols having more than 6000 identified compounds that have been isolated from plants [73], with the most notable being those containing pigments of flowers and plants, and play a crucial role by acting as free radical scavengers. Flavonoids can be classified into seven subgroups: flavones, flavanones, flavanols, flavonols, flavononols, isoflavones, and anthocyanins [34]. Each subgroup constitutes unique compounds that exhibit various biological activities.

Flavones

Many human and animal studies investigated the effects of specific flavones, including chrysin, apigenin, baicalein, and luteolin (Figure 4), on the intestinal microflora. Apigenin is particularly present at high levels in parsley (215 mg per 100 g) and celery (19 mg per 100 g). It is also found in rutabagas, tea, oranges, wheat sprouts, onions, cilantro, and chamomile [74]. The influence of apigenin on the alteration to the intestinal ecosystem is not well understood yet [75,76]. Wang et al. (2017) examined the effects of pure apigenin on the levels of both the single strain and community of human intestinal bacteria [74]. Growth profiles of the anaerobic bacteria were measured to study the influence of apigenin on the single strains of intestinal bacteria, including *Enterococcus caccae*, *Lactobacillus rhamnosus* GG, *Bifidobacterium catenulatum*, and *Bacteroides galacturonicus* as well as a fecal inoculum, were cultured by reproducing the human large intestine in vitro. The findings from that study highlighted that the growth of *E. caccae* and the microbial community cultured in vitro were effectively suppressed by apigenin compared to other examined species. It was noted that apigenin lowered the ratio between *Firmicutes* and *Bacteroidetes*.

Chrysin is widely distributed in honey and propolis [77]. Andrade et al. (2019) have examined the effect of fructose on gut microbiota and the capacity of the chrysin to influence the observed putative variations in an in vivo study [78]. They found that chrysin did not modify the composition of the intestinal ecology, but the abundance of *E. coli* and

Lactobacillus were considerably enhanced by fructose, while the ratio between *Firmicutes* and *Bacteroidetes* was increased in rats treated with fructose and chrysin. This study was the first to report that chrysin can interfere with the impacts of fructose on the intestinal ecosystem, which may be responsible for the features of metabolic syndrome induced by fructose.

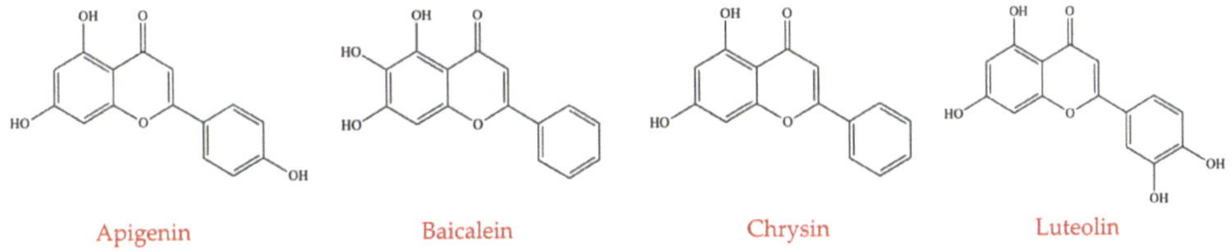

Figure 4. Chemical structures of flavones (apigenin, baicalein, chrysin, and luteolin) (self-generated).

Baicalein has several pharmacological effects, including anti-inflammatory, antioxidant, and antitumor properties. Gao et al. (2018) reported that treatment with baicalein significantly modified the population density of six bacterial genera in senescence-accelerated mouse prone 8 (SAMP8) mice by analyzing their intestinal microflora [79]. In another study, Zhang et al. (2018) treated male Wistar rats with baicalein for four weeks and examined the ability of baicalein to manipulate the intestinal microflora. It was observed that the ratio of *Bacteroidetes* to *Firmicutes* was significantly decreased, the relative abundance of *Bacteroidales*, Bacteroidaceae, Porphyromonadaceae, and Verrucomicrobiaceae was significantly increased, while that of Streptococcaceae, Desulfarculaceae Deferribacteraceae, and Ruminococcaceae was reduced in baicalein-treated rats [80].

Flavanones

Citrus fruits are the richest source of dietary flavanones for humans, while aromatic herbs (mint) and tomatoes contain flavanones, but in lower percentages [81]. Although flavanones are found in limited amounts in the diet, they are also one of the main dietary flavonoids since citrus fruits and juices are frequently consumed worldwide. Naringenin and hesperetin (Figure 5) are the most abundant flavanones present in grapefruit and sweet oranges, respectively, and are the primary reason for their specific bitter taste [82].

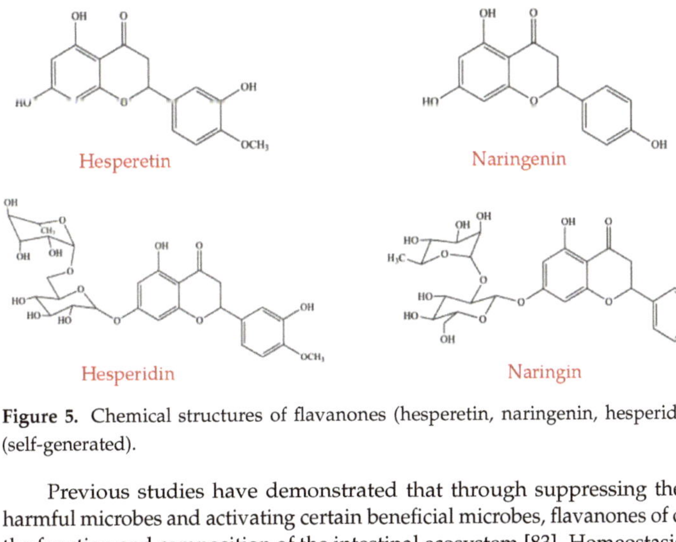

Figure 5. Chemical structures of flavanones (hesperetin, naringenin, hesperidin, and naringin) (self-generated).

Previous studies have demonstrated that through suppressing the growth of the harmful microbes and activating certain beneficial microbes, flavanones of citrus could alter the function and composition of the intestinal ecosystem [83]. Homeostasis of the intestine could be maintained by ingesting citrus flavanones, which might explain the mechanism of their beneficial health effects [84].

The primary target of the studies examining the influence of citrus flavanones or citrus fruit-based food products on the modulation of gut microflora is their capacity to promote the growth of beneficial microbes (such as the species of *Lactobacillus* and *Bifidobacterium*), to decrease the population of harmful bacteria, and to trigger generation of short-chain fatty acids (SCFAs). The findings of an in vitro study [85] stated that, even though the parent compounds failed to influence microbial species, the incubation of both naringenin and hesperetin (aglycones of citrus flavanone) decreased the population of several bacterial species after 24 h of incubation, which included the abundance of *Enterococcus caccae*, *Bifidobacterium catenulatum*, *Ruminococcus gauvreauii*, *Bacteroides galacturonicus*, and *E. coli*. Only naringenin suppressed the growth of *Lactobacillus* spp., and it showed the inhibitory action at a minimum level of 250 μg/mL [85]. These findings are in agreement with the results reported by Parkar et al. (2008) who also reported that naringenin can suppress the proliferation of *Salmonella typhimurium*, *E. coli*, *Lactobacillus rhamnosus*, and *Staphylococcus aureus* at a minimum inhibition concentration (MIC) of 62.5 μg/mL for *Staphylococcus aureus* and at 125 μg/mL for the other three bacteria [86].

Hesperidin (Figure 5), which is formed by the conjugation of hesperetin with rutinose, is the principal flavonoid widely distributed in citrus fruits [87]. Once the hesperidin enters the colon, the connected rutinose groups are split by GM, producing hesperetin for further absorption in the large intestine [88]. A recent study reported that oral supplementation of hesperidin at two dose levels of 100 and 200 mg/kg for four weeks (three times per week) significantly altered the intestinal microbiota composition in rats [89]. The population of *Staphylococcus* was increased and the proportion of *Clostridium coccoides* and *Eubacterium rectale* was decreased by both hesperidin dosages, while the abundance of *Lactobacillus* was enhanced by the high dose treatment of hesperidin.

Previous studies conducted in humans examined the influence of hesperidin administration on the composition of the intestinal ecosystem and the production of SCFAs [84,90,91]. The ratio between butyrate and total SCFA was increased by the daily oral administration of 500 mg of a citrus extract (containing hesperidin of >80% and naringin of >4%) for 12 weeks in a randomized, placebo-controlled study of healthy individuals showing features of metabolic disorders, while there was no change observed in the absolute concentrations of SCFAs in the fecal samples [84,92].

Lima et al. (2019) showed that continuous consumption of orange juice for two months could notably enhance the proportion of total fecal anaerobes and *Lactobacillus* spp. in healthy volunteers [90]. However, the study did not state the amount of flavanone. Additionally, a considerable decline in the concentration of ammonium and a rise in the ratio of acetate to total SCFA were observed compared to the control treatment. In healthy volunteers, modulation of the composition of the microbial community was observed with the daily ingestion of two orange juices that had different flavanone concentrations for seven days [91]. The findings showed that the most significant increase was observed in the population belonging to the functional taxonomic unit of clostridia from the families of Veillonellaceae, Ruminococcaceae, Odoribacteraceae, Tissierellaceae, and Mogibacteriaceae [91].

Briefly, the findings from several human, animal, and in vitro experiments indicate that supplementation of citrus flavanone can shift the intestinal ecology composition or proliferation of a particular taxon. Even though many studies have repeatedly depicted the suppression of the growth of Enterobacteriaceae by citrus flavanone, the results of some other research vary. Unfortunately, there are no studies involving the analysis of fecal metabolites of humans to investigate the influence of flavanones on the intestinal microflora [84].

Flavanols

The flavanol subgroup is primarily composed of catechins, which are present in higher concentrations in the fruit's skin as opposed to the pulp of the fruit. Catechins are found at high concentrations in green tea and are responsible for several health benefits of green tea. Although there are several compounds included in this group, only epicatechin, epigallocatechin, epicatechin gallate, and epigallocatechin gallate (Figure 6) are reported to affect the intestinal community [31].

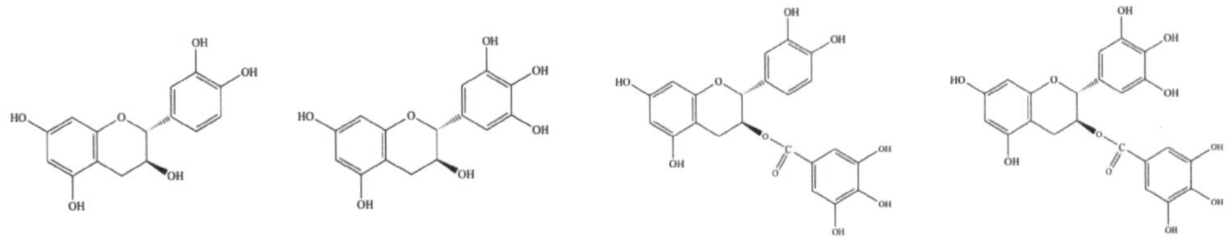

Epicatechin Epigallocatechin Epicatechin gallate Epigallocatechin gallate

Figure 6. Chemical structures of flavanols (epicatechin, epigallocatechin, epicatechin gallate, and epigallocatechin gallate) (self-generated).

Cueva et al. (2013) examined in vitro fermentation of bacterial loads by flavan-3-ol using two purified fractions from grape seed extract (GSE): GSE-M (70% monomers and 28% procyanidins) and GSE-O (21% monomers and 78% procyanidins) and observed, using a fluorescent in situ hybridization technique, the appearance of the metabolites of phenols. During the fermentation, both grape seed flavanol fractions increased the proportion of *Enterococcus* and *Lactobacillus* and reduced the proportion of *C. histolyticum*. Overall, the results revealed that flavan-3-ol can stimulate changes that could influence their potential bioactivity and bioavailability by modifying the composition of colonic microflora and intrinsic catabolic activity [93].

Tzosunis et al. (2008) used a batch-culture fermentation system where pH is controlled modeling the distal portion of the large intestine colon to evaluate the bidirectional metabolic relationship among intestinal microflora and (+)-catechin and (−)-epicatechin. Remarkably, the proliferation of certain bacterial loads was affected by (+)-catechin, where the growth of *Bifidobacterium* spp., *E. coli*, and a group of *C. coccoides* and *E. rectale* was considerably enhanced, at the same time, the proportion of the *C. histolyticum* bacteria was decreased, leading to the beneficial effects in the body. Despite this, (−)-epicatechin significantly promoted only the proliferation of the group of *C. coccoides* and *E. rectale*, proving their poor activity. Both (+)-catechin and (−)-epicatechin showed noticeable potential prebiotic properties at a minimum concentration of 150 mg/L. The considerable change in the specific populations of gut microbiota resulted from the incubation of (+)-catechin could be related to the transformation of (+)-catechin to (−)-epicatechin by bacteria since the same metabolites of 5-(3′, 4′-dihydroxyphenyl)-γ-valerolactone, 5-phenyl-γ-valerolactone and phenyl-propionic acid are produced by the transformation of both (−)-epicatechin and (+)-catechin. In conclusion, these results suggest that the intake of diets rich in flavanol may contribute to intestinal health by acting as prebiotics [94].

The ability of (−)-epigallocatechin gallate (EGCG) to influence the modification of intestinal microflora and the output of fermentation in the large intestine was investigated by Unno et al. (2014) who found that the growth of specific microbial species such as *Clostridium* spp., *Bacteroides*, *Bifidobacterium*, and *Prevotella* in the colon was affected by dietary EGCG in rats [95]. Further, Cheng et al. (2017) investigated the influence of (−)-epigallocatechin 3-*O*-(3-*O*-methyl) gallate (EGCG3″Me) in a human flora-associated mice model with obesity induced by a high-fat diet. After 8 weeks of supplementation of EGCG3″Me, the proportion of *Bacteroidetes* was dramatically increased, while the population of *Firmicutes* had declined. In this study, it was found that EGCG3″Me can act as a prebiotic and potentially have therapeutic effects in modulating the colonic ecosystem, which assists in gut dysbiosis inhibition [96]. In a study by Lee et al. (2006), the growth of the *E. coli*, *C. coccoides*, and *E. rectale* was promoted, while the growth *C. histolyticum* was suppressed by (+)-catechin. Moreover, (+)-catechin did not affect the population of beneficial microflora, namely *Lactobacillus* spp. and *Bifidobacterium* spp. [97]. Preclinical data of an in vitro study in a bacterial medium revealed that epicatechin gallate sensitizes methicillin-resistant β-lactam antibiotics *S. aureus* to become sensitive to β-lactam antibiotics [98].

Flavonols

The flavonols subgroup constitutes different compounds, including quercetin, kaempferol, myricetin, rutin, and isorhamnetin (Figure 7). Generally, only quercetin and kaempferol appear to have been the focus of many studies investigating their ability to manipulate the GM. Fruits and vegetables contain quercetin in high amounts, and the approximate daily intake of quercetin in a typical Western diet was estimated to range from 0 to 30 mg/day, depending on the intake of fruits and vegetables [99]. Foods such as berries, apples, onions, and kale are considered the richest sources of quercetin [100]. Recent studies reported that the abundance of beneficial *Lactobacillus*, *Bacteroides*, Clostridia, and *Bifidobacterium* was significantly increased, and the proportion of *Enterococcus* and *Fusobacterium* was decreased by the administration of quercetin, which modified the composition of the intestinal ecosystem [101,102]. In an earlier study, single-molecule RNA sequencing was used along with Helicos technology to investigate the influence of quercetin supplementation on the commensal colonic bacteria, including *Enterococcus caccae*, *Ruminococcus gauvreauii*, and *Bifidobacterium catenulatum*, by analyzing the gene expression profiles and examining the cell morphology and growth patterns between groups treated with and without quercetin [103]. It was found that phenotypically, quercetin moderately suppressed the growth of *E. caccae*; mildly inhibited the growth of *B. catenulatum*; and did not affect the growth of *R. gauvreauii* [103].

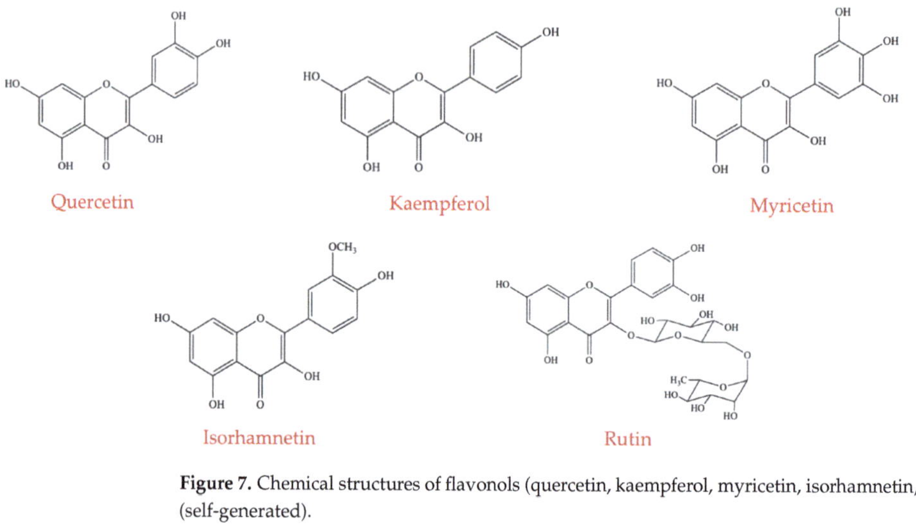

Figure 7. Chemical structures of flavonols (quercetin, kaempferol, myricetin, isorhamnetin, and rutin) (self-generated).

Etxeberria et al. (2015) examined the consequences of the supplementation of quercetin and *trans*-resveratrol on a high-fat sucrose diet (HFS)-induced dysbiosis of gut microbiota. In this study, Wistar rats were randomly divided into four groups, namely those receiving an HFS diet enriched with or without quercetin (30 mg/kg BW/day), *trans*-resveratrol (15 mg/kg BW/day), or a mixture of the two components (quercetin and *trans*-resveratrol) at the same dosages used in the other treatments. This study revealed that the administration of quercetin resulted in a significant effect on the composition of GM at various levels of taxonomy, reducing the ratio between *Firmicutes* and *Bacteroidetes* and suppressing the multiplication of gut microbes linked to diet-induced obesity, including *E. cylindroides*, Erysipelotrichaceae, and *Bacillus*. In summary, dysbiosis of gut microbiota induced by the HFS diet was effectively attenuated by the supplementation of quercetin. However, there was no significant influence on the GM composition by the supplementation of *trans*-resveratrol alone or in combination with quercetin. Moreover, the individual effect of quercetin was reduced when combined with *trans*-resveratrol [104].

Numerous medicinal and edible plants and herbs contain kaempferol, which is regarded as an important compound in the flavonol subgroup. Kaempferol has several pharmacological effects, including its ability to act as an antioxidant and anti-inflammatory agent [105]. A study observed that the colonic ecosystem and its metabolism were regulated by a high concentration of kaempferol in the large intestine [106]. Moreover, Kawabata et al. (2013) conducted an in vitro study to investigate the growth of *B. adolescentis* treated with flavonols by incubating the *B. adolescentis* obtained from the human colon with various flavonols, including quercetin, kaempferol, fisetin, myricetin, and galangin under anaerobic conditions. Quercetin and fisetin did not affect or mildly affected the growth rate (inhibited by 20% after 6 h of treatment). In comparison, galangin showed about 30% to 70% suppression in the growth rate of *B. adolescentis* when incubated for 1 to 6 h. This study suggested that, except for galangin, other tested flavonols exhibited no or mild anti-bacterial properties against *B. adolescentis*, which is considered beneficial for gut health [107].

An in vitro experiment was carried out by Duda Chodak (2012) to evaluate the effects of rutin and quercetin on certain colonic microbial species. Rutin was used at rates of 20, 100, and 250 μg/mL, and quercetin was examined at 4, 20, and 50 μg/mL against six inoculate bacteria species (*B. galacturonicus*, *B. catenulatum*, *Ruminococcus gauvreauii*, *Lactobacillus* sp., *E. coli*, and *E. caccae*). The obtained results revealed that, although the inhibitory action of rutin against bacterial proliferation was weak, quercetin had a significant inhibitory activity, which depended on the concentration used against the examined bacterial species [85].

Flavanonols

The flavanonols subgroup constitutes astilbin, engeletin, and taxifolin (Figure 8). Generally, intestinal microflora metabolizes these compounds and the flavanonols have anti-inflammatory properties [108,109]. There are no studies related to the effects of these compounds on the intestinal microbiota.

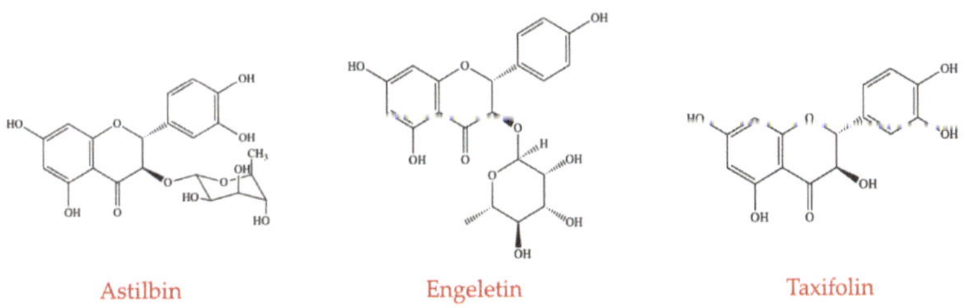

Figure 8. Chemical structures of flavanonols (astilbin, engeletin, and taxifolin) (self-generated).

Isoflavones

Isoflavones, one of the flavonoids subgroups, contain different compounds, including daidzin, daidzein, formononetin, glycitein, and genistein (Figure 9). The chemical structures of isoflavones resemble the structure of 17-β-estradiol, an estrogen steroid hormone. Isoflavones exhibit estrogenic activity and are naturally present in many plants, among which soy is considered one of the most abundant sources [110]. Generally, isoflavones present as conjugates of isoflavone glycoside (such as genistin, daidzin, and glycitin) in unfermented soy foods and soy milk, which have low estrogenic effects and bioavailability. Isoflavone aglycones (genistein, daidzein, and glycitein) should be liberated from the corresponding glycosides to be absorbed and attain a complete activity [111]. The cellular β-glucosidases and β-glucosidases from the components of intestinal microflora are responsible for the release of isoflavone aglycones [112]. Cellular enzymes and other constituents of the colonic microflora can further hydrolyze and metabolize isoflavone aglycones producing more active compounds (for example, equol from daidzein) or inactive metabolites [113].

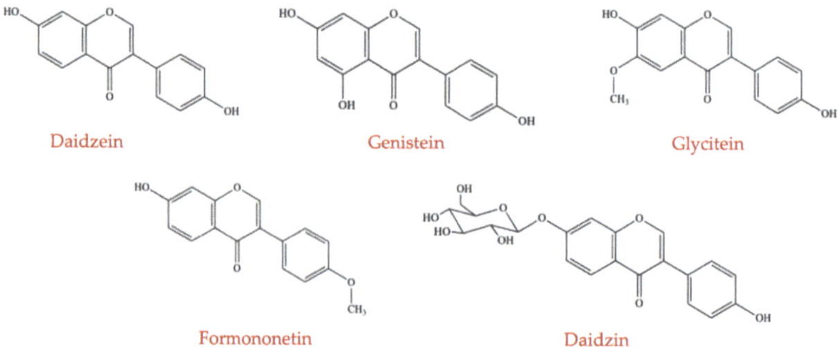

Figure 9. Chemical structures of isoflavones (daidzein, genistein, glycitein, formononetin, and daidzin) (self-generated).

Soy and soy products are the major sources of dietary isoflavones. Factors that affect the total quantities of genistein and daidzein (including glycosides) present in soy products are the type of cultivar, preparation method, and extent of ripeness [114]. Frequent consumption of soy and soy products in Asian populations resulted in the greater prevalence of equol-producing microbes within the colon [115]. Another study revealed that the intake of dietary daidzein significantly increased the population of two equol-generating bacteria, *Slackia isoflavoniconvertens* and *Asaccharobacter celatus* [116]. Moreover, numerous studies conducted in humans and rodents showed that consumption of soy or soy food products affects intestinal ecology [114,117–120]. It was also stated that the total amount and/or relative percentage of the particular microbial ecosystem in the intestine might be altered by genistein in food or supplemental forms [121].

A study showed that after one week of soy isoflavone supplementation, colonic microflora composition was significantly modified in 17 postmenopausal women receiving a soy bar that contained 1 g of saponin and 160 mg of soy isoflavones (including daidzein, genistein, and glycitein) [122]. It was found that after the administration of soy isoflavones, all subjects showed a significant increase in the population of bifidobacteria in the colonic microflora while reducing that of lactobacilli, and in equol-generating subjects, the growth of bifidobacteria and eubacteria were greatly promoted than in non-producers [122]. Another study showed similar results where soy isoflavones (100 mg per day) were given to 39 postmenopausal women for 2 months. In this study, isoflavone supplementation enhanced the proportion of *Faecalibacterecterium, Eubacterium, Clostridium, Lactobacillus, Enterococcus*, and *Bifidobacterium*. At the same time, the levels of *Clostridium* and *Eubacterium* were significantly increased in equol producers ($n = 12$) [123]. When cows were used as a model, supplementation with soy isoflavones also increased the population of *Firmicutes* [124]. However, no modification in the composition of gut microbiota was observed in an experiment involving 16 menopausal women receiving soy isoflavone (80 mg per day) [125] suggesting that soy isoflavones may have a subject-dependent and/or dose-dependent influence on the intestinal ecology.

Although different studies revealed varying results, animal and human studies suggest that the ingestion of soy-based foods promoted lactobacilli and bifidobacteria growth, and the ratio of *Firmicutes* to *Bacteroidetes* can be changed [126]. That said, a reciprocal relationship between the soy isoflavones and colonic microflora in studies in healthy volunteers is difficult to establish. Therefore, further research is required to explain the effect of soy isoflavones on the colonic microflora structure involving gnotobiotic animal models transplanted with microflora obtained from various donors [127].

Anthocyanins

Anthocyanins belong to the flavonoids group of polyphenols. They are plant pigments imparting deep red/purple/blue colors in plant-derived food products [128]. Although anthocyanins are naturally present in the form of aglycones and glycosides of flavylium (2-phenyl benzopyrylium) salts, they are different from the structure of the salts [129]. It is reported that approximately 700 anthocyanin compounds have been identified and isolated from plants, but only 6 anthocyanidins are widely studied, namely delphinidin, cyanidin, pelargonidin, malvidin, petunidin, and peonidin (Figure 10) [130].

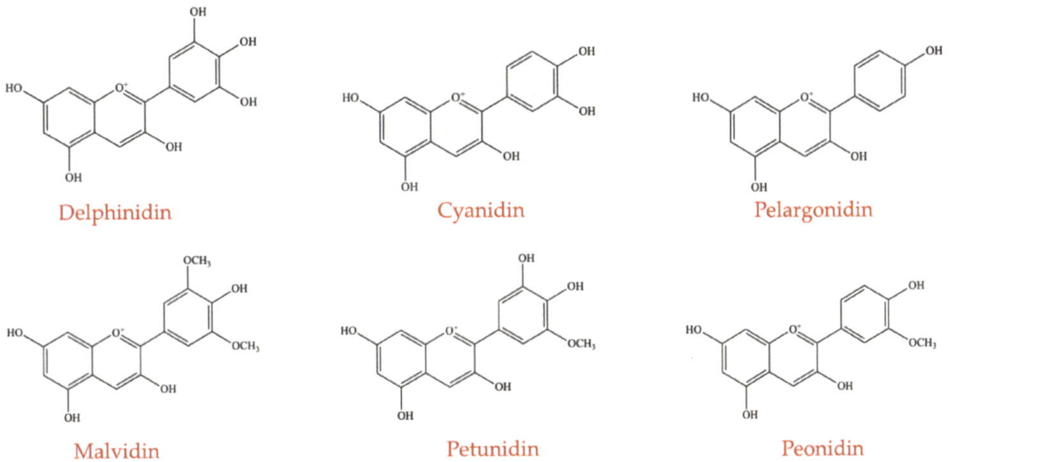

Figure 10. Chemical structures of anthocyanins (delphinidin, cyanidin, pelargonidin, malvidin, petunidin, and peonidin) (self-generated).

The majority of ingested anthocyanins are biologically converted to their metabolites by the gut microbiota and are absorbed in the large intestine since they are not utilized in the upper part of the GI tract. Intestinal microbiota can metabolize anthocyanins, and in turn, anthocyanins and/or their metabolites can alter the composition of the gut ecosystem by altering the abundance of particular bacteria [131]. Under in vitro conditions and in humans, anthocyanins can increase the prevalence of favorable bacteria, including *Lactobacillus-Enterococcus* spp. and *Bifidobaterium* spp. [132,133]. Similarly, Sun et al. (2018) revealed that the growth of *Bifidobacterium infantis*, *Bifidobacterium bifidum*, *Lactobacillus acidophilus*, and *Bifidobacterium adolescentis* could be stimulated and the proliferation of harmful *S. typhimurium* and *S. aureus* were suppressed by purple sweet potato anthocyanins and monomers of peonidin-derived anthocyanin in in vitro microbial cultivations [134]. In a study by Chen et al. (2018) [135], it was observed that the prevalence of beneficial *Faecalibacterium prausnitzii*, *Eubacterium rectale*, and *Lactobacillus* was increased and the abundance of *Desulfovibrio* spp. and *Enterococcus* spp. was decreased by the supplement of black raspberry anthocyanins. Similar findings were reported by Zhu et al. (2018), who reported that anthocyanin of black rice and cyanidine-3-*O*-glucoside considerably stimulated the growth of lactobacilli and bifidobacteria genus [136].

In an earlier study, malvidin-3-*O*-glucoside was incubated with fecal slurry under in vitro conditions. The results indicated that malvidin-3-*O*-glucoside increased the population of total bacteria, such as *Lactobacillus* spp. and *Bifidobaterium* spp., while the abundance of *Bacteroides* spp. was unaffected [132]. This stimulating effect of malvidin-3-*O*-glucoside on the proliferation of beneficial microbes can be improved by mixing malvidin-3-*O*-

glucoside with other anthocyanins. In an in vivo study that analyzed the fresh fecal samples of eight healthy volunteers (25–30 years), Zhang et al. (2016) reported that anthocyanins of purple sweet potato increased the population of *Lactobacillus-Enterococcus* spp. and *Bifidobacterium*, while *Clostridium histolyticum* and *Bacteroides-Prevotella* were decreased, although total bacteria count was not affected by the treatment [137].

Red wine extract was incubated with bacteria isolated from human feces, and the results indicated that the abundance of *C. histolyticum* was decreased with no other observed changes [138]. While this does not reflect the pathway expected during digestion, another randomized, crossover-controlled intervention study has also shown a similar observation concerning the abundance of *C. histolyticum* after consuming red wine in humans [139]. It is likely that anthocyanins were not the only compounds responsible for the above effect found with red wine consumption since red wine contains a complex mixture of polyphenols such as flavanols, flavonols, anthocyanins, phenolic acids, etc. [138]. Further, Vendrame et al. (2011) showed that the population of *Bifidobaterium* spp. was significantly enhanced in human volunteers after consuming a blueberry drink for 6 weeks [140].

A study conducted on polygenic obese mouse models fed with diets supplemented by six types of berries containing different profiles of anthocyanin for 12 weeks showed that the proportions of *Actinobacteria* and obligate anaerobic bacteria were significantly increased in the intestine by the treatment [141]. Later, the influence of a mixture of prebiotics and anthocyanin on the colonic ecosystem and colonic inflammation was determined using uncomplicated obese male and female volunteers in an open-label study for 8 weeks. The frequent intake of the mixture of prebiotics and anthocyanins resulted in a favorable alteration in the intestinal community of microflora [142].

Overall, in vitro animal and human intervention studies suggest that anthocyanins can promote the proliferation of beneficial bacteria, including *Bifidobacterium* spp. and *Lactobacillus* spp. These species contribute beneficial roles in the colon, such as the antimicrobial action against harmful bacteria via competing for adhesion sites and growth substrate, and the production of SCFAs. Furthermore, a decrease in the abundance of pathogenic bacteria, such as *C. histolyticum*, has been observed, which is responsible for inflammatory bowel disease and tumor-inducing actions [132,143]. Further studies are required to understand the overall impact of these compounds and define the mechanism of action that anthocyanins exert on the community of intestinal microbiota. This can be attained by performing well-designed clinical trials in humans and animals with a balanced experiment design of the subjects' age and genetic backgrounds. This should cover different forms and dosages of anthocyanins and establish a consistent approach to control diets [128,144].

4.1.2. Curcumin

Curcumin (Figure 11) belongs to the subgroup of polyphenols and the rhizome part of the *Curcuma longa* is considered the richest source of curcumin, which is used for cooking and conventional medicine [31]. Typically, it is extracted from the turmeric rhizome by solvent extraction and refinement of the extract is by crystallization. Numerous animal studies have reported that the intestinal microbial community may be influenced by curcumin. For instance, a deficiency of estrogen causes detrimental changes in the colonic microbiota, and administration of curcumin can partly re-establish the composition of the typical GM in ovariectomized rats [145]. Ohno et al. (2017) revealed that the prevalence of bacteria-generating butyrate and level of fecal butyrate can be increased by curcumin nanoparticles at a dosage of 0.2% (w/w). Further, curcumin nanoparticles stimulate the NF-κB in epithelial cells of the intestine and the inhibition of mucosal mRNA expression of inflammatory mediators [146].

In a colitis mouse model, curcumin enhanced the relative proportion of *Lactobacillales*, while reducing that of *Coriobacterales* [147]. Curcumin may alter intestinal barrier function by remaining in the colonic mucosa in higher concentrations, thereby minimizing inflammation caused by circulating lipopolysaccharide producing bacteria. Feng et al. (2017) reported that the administration of curcumin restores the gut barrier function in high-fat-diet-fed rats and alters the intestinal microflora composition and diversity [148]. Higher proportions of anti-inflammatory lactobacilli and bifidobacteria and fewer loads of pro-inflammatory *enterococci* and *enterobacteria* were observed in animals supplemented with curcumin [149].

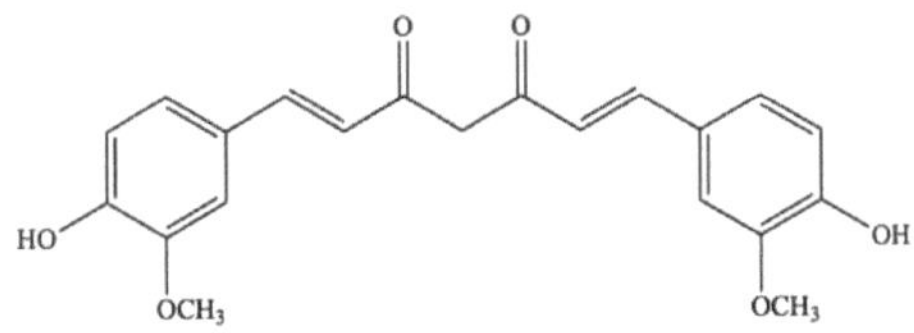

Figure 11. Chemical structure of curcumin (self-generated).

In a double-blind, randomized, placebo-controlled pilot study, the influence of supplementation of dietary curcumin and turmeric on the human intestinal ecosystem was evaluated [150]. It was found that, with time, colonic microflora showed a notable and individualized modification. The relative prevalence of 71 and 56 taxa were significantly reduced by the supplementation of turmeric and curcumin, respectively. These studies reveal that proliferation, growth, or the existence of beneficial microbes in the intestinal community might be promoted by curcumin.

4.1.3. Phenolic Acids

Phenolic acids (aromatic acids) are the second major subgroup of polyphenols, consisting of a phenyl ring and a carboxylic group [31], and are produced by the shikimate pathway. Although phenolic acids are present in numerous food products, they are widely distributed in berries, wine, and whole grains. Phenolic acids are divided into two main groups, namely hydroxybenzoic acids and hydroxycinnamic acids [151]. Phenolic acids have enormous health benefits.

Hydroxybenzoic Acid

Hydroxybenzoic acid is a phenolic derivative of benzoic acid, which can be obtained both naturally and synthetically. It contributes to several derivatives such as gallic acid, protocatechuic acid, syringic acid, vanillic acid, and *p*-hydroxy-benzoic acid (Figure 12) [151]. Protocatechuic acid (3, 4-dihydroxybenzoic acid) is abundant in human diets through foods such as white grapes, bran, brown rice, gooseberries, olive oil, onions, plums, and almonds [152]. It has been reported using a rodent model that altered gut microbial communities can be repaired by an n-butanol fraction of *Trianthema portulacastrum* rich in protocatechuic acid that lowered the relative abundance of inflammatory microbes such as *Helicobacter*, *Mucispirillum*, and Lachnospiraceae [153]. Wang et al. (2019) examined the influence of protocatechuic acid on the gut health of Chinese yellow-feathered broilers. The results revealed that microflora diversity was altered by dietary protocatechuic acid. In comparison to control group broilers, protocatechuic acid-treated broilers had more *Firmicutes* and Actinobacteria that are beneficial to gut health and fewer Proteobacteria and *Bacteroidetes* that have pro-inflammatory effects [154].

Hydroxycinnamic Acid

Hydroxycinnamic acids (HCAs) are the most abundant phenolic acids in plants [155]. The major factor contributing to this is that HCAs are attached to the cell walls of plants [156]. Many enzymes needed for the hydrolysis and metabolism of the complex structure of this polyphenolic compound are not present in the human genome [157]. However, these glycans can be fermented by the anaerobic intestinal microflora, where different microorganisms liberate HCAs in the human intestine, thus considerably impacting human health [59,158]. Hydroxycinnamic acids contain a wide range of compounds, such as caffeic acid, *p*-coumaric acid, sinapic acid, chlorogenic acid, and ferulic acid (Figure 13) which are all recognized to manipulate the intestinal ecosystem [159].

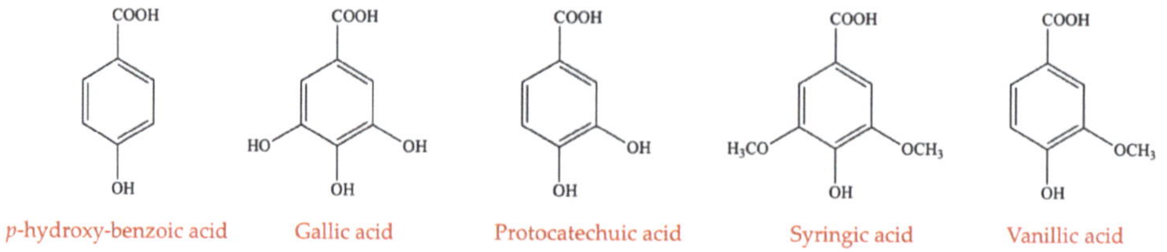

Figure 12. Chemical structures of hydroxybenzoic acids (*p*-hydroxybenzoic acid, gallic acid, protocatechuic acid, syringic acid, and vanillic acid) (self-generated).

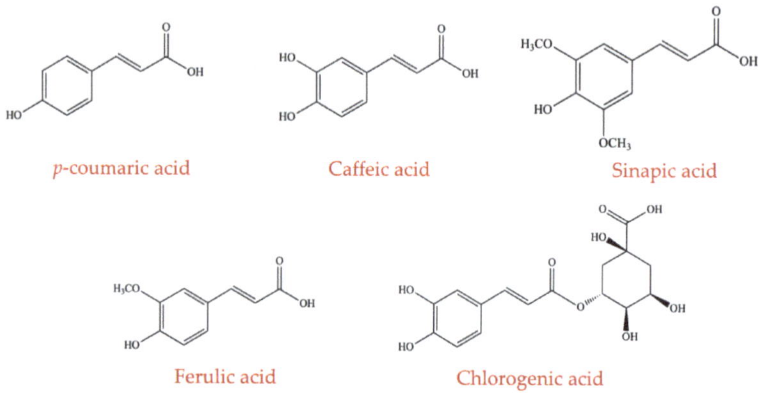

Figure 13. Chemical structures of hydroxycinnamic acids (*p*-coumaric acid, caffeic acid, sinapic acid, ferulic acid, and chlorogenic acid) (self-generated).

Oral administration of 1 mmol/L of caffeic acid for 15 days enhanced the abundance of mucin-deteriorating *Akkermansia* and could restore the abundance of gut microbiota and suppress the rise in the ratio between *Firmicutes* and *Bacteroidetes* in female C57BL/6 mice model with colitis induced by dextran sulfate sodium (DSS) [160]. In an in vitro study analyzing the fecal microbiota of humans, Mills et al. (2015) showed that after 10 h of exposure to chlorogenic acid at a concentration of 80.8 mg, colonic microbiota was selectively modulated via increasing the abundance of *Bifidobacterium* spp., *E. rectale*, and *C. coccoides* [161]. Oral administration of 150 mg of chlorogenic acid for 6 weeks in ICR male mice with dysbiosis induced by a high-fat diet enhanced the proliferation of Bacteroidaceae and Lactobacillaceae, while suppressing that of Erysipelotrichaceae, Lachnospiraceae, Ruminococcaceae, and Desulfovibrionaceae, indicating restoration of normal intestinal microbiota [162].

Ma et al. (2019) used male apolipoprotein E (ApoE−/−) mice with non-alcoholic fatty liver disease induced by a high-fat diet to investigate the influence of oral administration of ferulic acid (30 mg/kg). This study suggested that the colonic microflora composition was modified by decreasing the indole-3-acetic acid secretion and changing the ratio between *Firmicutes* and *Bacteroidetes* [163]. Yang et al. (2019) investigated the influence of oral administration of 200 mg/kg sinapic acid for 8 weeks in 30-week-old male Wistar rats fed a high-fat diet (45% fat) [164]. Gut microbiota diversity was improved by enhancing the population of *Dorea* and *Blautia* of the Lachaospiraceae family while decreasing that of Desulfovibrionaceae and *Bacteroides*, which generally contribute to human diseases and inflammation [165,166].

4.1.4. Stilbenes

Stilbenes are widely distributed in red grapes, certain berries, peanuts, and many other plants. Stilbenes are generally produced by the phenylpropanoid pathway, and one of their important features is the presence of aromatic rings connected to an ethane bridge [167]. Stilbenes comprise different compounds such as resveratrol, piceatannol (Figure 14), pinostilbene, batatasin III, oxyresveratrol, and thunalbene. Jaimes et al. (2019) investigated the influence of six stilbenoids on the composition of the intestinal microflora, and the results revealed that the analyzed stilbenoids modify the GM as observed in a human gut model of fecal fermentation at physiological levels of 10 μg/mL. The ratio between *Firmicutes* and *Bacteroidetes* was significantly lowered, and the responses from different strains from the Lachnospiraceae family and the relative proportion of strains from the *Clostridium* genus showed a consistent reduction. Among the stilbenoids groups studied, resveratrol and piceatannol highly contributed to the observed responses, followed by thunalbene and batatasin III [168].

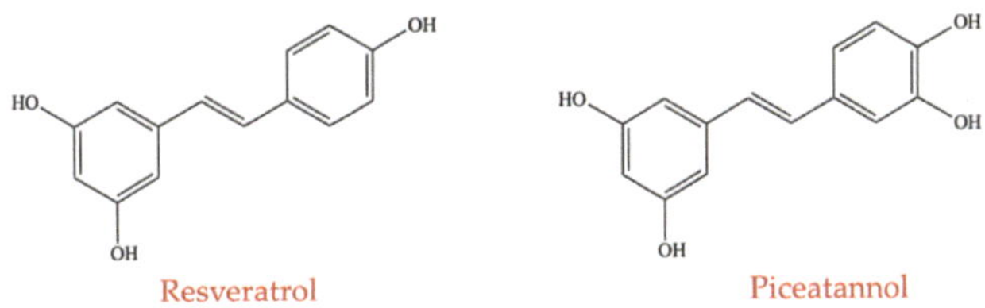

Figure 14. Chemical structures of resveratrol and piceatannol (self-generated).

Resveratrol

Resveratrol, also known as 3, 5, 4′-trihydroxystilbene, is a widely distributed polyphenol subgroup present in different plants. Although grapes, red wine, and peanuts contain considerable amounts of resveratrol, *Polygonum cuspidatum* is the richest natural plant source of the compound [169]. Resveratrol is also available commercially as tablets and is used as a nutritional supplement [170]. In recent years, studies on resveratrol have significantly increased due to its wide range of biological activities such as antioxidant, anti-diabetes, anti-obesity, and improvement in colonic micro-organisms [171,172]. Numerous studies revealed that resveratrol can reduce body fat and body weight while improving obesity and glucose homeostasis parameters by stimulating alterations in the colonic microflora. Other studies have shown that treatment with 0.4% resveratrol could enhance the population of Parabacteroides and Bacteroides while reducing those of *Akkermansia*, Lachnospiraceae, *Moryella*, and Turicibacteraceae in C57Bl/6N mice [173,174].

It has been suggested that resveratrol could improve atherosclerosis by suppressing the generation of trimethylamine in the gut, which is a trimethylamine-N-oxide (TMAO) precursor, by simulating intestinal microflora in mice. Fecal excretion and de-conjugation of bile acid are also increased by resveratrol via enhancement of the abundance of colonic bacteria with bile salt hydrolase activity, such as *Bifidobacterium* and *Lactobacillus* [175]. Moreover, the ratio of *Firmicutes* to Proteobacteria was increased by resveratrol (50 mg/L) examined in Sprague Dawley rats [176]. Most et al. (2017) conducted a study supplementing a combination of resveratrol and epigallocatechin-3-gallate in humans for 12 weeks. It was observed that the prevalence of *Bacteroidetes* was significantly decreased and the proportion of *Faecalibacterium prausnitzii* tended to reduce in overweight men [177]. Moreover, resveratrol promoted the multiplication of *Lactobacillus* and *Bifidobacterium* and prevented the virulence factors of *Proteus mirabilis* [178].

Piceatannol

Several groups of plants, particularly white tea and grapes, contain piceatannol, which is a hydroxylated analog of resveratrol [31]. Setoguchi et al. (2014) indicated that metabolic stability of piceatannol is greater in comparison to resveratrol [179]. Piceatannol is distributed at nearly equal concentrations as resveratrol in edible plants, fruits, and red wines [180]. Using C57BL/6 mice fed with a high-fat diet, the influence of piceatannol on gut microbiota was evaluated. It was observed that gut microbiota composition was significantly altered in piceatannol-treated animals and the modulation of intestinal microflora stimulated by the high-fat diet was restored by piceatannol, via significantly reducing the abundance of *Firmicutes* that are unfavorable and increasing those of *Bacteroidetes* that are favorable to enhanced gut health [181].

4.1.5. Lignans

Dietary lignans are a group of phytoestrogens found as aglycones or glycosides in plants [182]. Different parts of about 70 diverse species of plants, such as roots, stems, rhizomes, leaves, fruits, and seeds contain a significant amount of lignans. Specifically, oilseeds, mainly flaxseeds and grains with bran, are the richest sources [183]. Although estrogen-like activities are not found in the dietary lignans themselves, the colonic microbial ecosystem metabolizes them to produce enterolignans (or mammalian lignans), including enterolactone (EL) and enterodiol (ED). The prevalence of *Ruminococcus* species, such as *R. lactaris* and *R. bromii* [184], and the abundance of *Methanobrevibacter* [185] and *Lactobacillus-Enterococcus* [186] are associated with the production of EL.

Recently, Corona et al. (2020) evaluated how the composition of the colonic microbiota of both younger and premenopausal females is affected by lignans present in oilseed mix and it was observed that oilseeds rich in lignans have a strong impact on the fecal microbiota, generating a varying profile of enterolignan. Further research is required to investigate the long-term impacts of diets rich in lignan on the intestinal microflora and to discover ways to increase the abundance of bacterial species producing enterolactone. Plants and plant products containing lignans consumed by animals and humans have many advantages, not only improving the GM population and gut health but also providing a range of health benefits. Lignans can improve the antioxidant status of both tissue systems and the whole body, prevent cancer by limiting the proliferation of cells via anticarcinogenic effects, improve the immune status of individuals by providing defense against infectious diseases, and act as pro-inflammatory compounds in reducing arthritis and obesity [187].

4.1.6. Tannins

Tannins are a subgroup of polyphenols which comprise several components of different molecular weights that are widely distributed in nature [188]. Proteins can be precipitated by tannins. Depending on the molecular structure, tannins are divided into hydrolyzable tannins (HTs) and condensed tannins (CTs). An example of hydrolyzable tannins is ellagitannins which can release ellagic acid by hydrolyzing in vivo and can produce urolithin through gut microbial metabolism. In general, the antimicrobial action of tannins has been determined in vitro, and the impact of these tannins on the abundance of the colonic ecosystem in vivo has not been sufficiently illustrated.

Bialonska et al. (2009) used a liquid culturing method to investigate the influence of a 0.01% extract of commercial pomegranate and its major components at the level of 0.05% on the proliferation of different species of bacteria present in the human intestinal ecosystem in an in vitro study. The findings of this study demonstrated that byproducts of pomegranate and punicalagins suppressed the growth of pathogenic *S. aureus* and Clostridia, while ellagitannins normally did not affect the abundance of probiotic bifidobacteria and lactobacilli [189]. In another study, Bialonska et al. (2010) used a healthy individual's fecal samples inoculated in a batch-culture fermentation system simulating the environment of the intestinal ecosystem to confirm the maintenance of the above trend. At the end of this study, pomegranate extract resulted in an accelerated increase in the total number of bacteria, promoting the growth of *Enterococcus*, *Bifidobacterium* spp., and *Lactobacillus*. At the same time, the abundance of the *C. histolyticum* group was not affected [190]. It was observed that pomegranate ellagitanins and urolithin A, which is their major metabolite produced by the microbiota, modified the composition of the colonic ecosystem in rats by increasing the prevalence of *Bifidobacterium* and *Lactobacillus* [191].

Moreover, Li et al. (2015) monitored alterations in the composition of GM in 20 healthy individuals consuming a pomegranate extract (POM) of 1 g for four weeks in an in vivo study. Individuals were classified into three separate groups depending on the amount of urolithin A in the feces and urine. In this experiment, it was observed that the abundance of beneficial Actinobacteria was significantly increased while those of *Firmicutes* in individuals producing urolithin A were reduced. Furthermore, in the fecal samples of individuals who produce urolithin A, the proportion of *Akkermansia muciniphila* of phyla Verrucomicrobia was greater than non-producers. After four weeks, the prevalence of *Lactobacillus, Escherichia, Butyrivibrio, Prevotella, Veillonella, Enterobacter,* and *Serratia* genera was enhanced while those of *Collinsella* in urolithin A producers was reduced. In some participants, POM extract ingestion led to the production of metabolites, which may stimulate beneficial health effects next to modification of the intestinal microflora [192].

Samanta et al. (2004) pioneered research conducted using an in vivo model to investigate the influence of tannic acid on the communities of culturable organisms. It was observed that tannic acid could alter the ecological balance of the intestinal microbiota of the rat. This study proved the toxic and anti-nutrient properties of tannic acid from the observations of enormous microbial growth and the reduction in the bodyweight of the experimental animal after tannic acid supplementation for 21 days [193]. Smith and Mackie (2004) stated that the tannin diet could reduce the abundance of low G + C Gram-positive bacteria. In contrast, the proportion of Enterobacteriaceae and *Prevotella, Bacteroides,* and *Porphyromonas* increased in the samples of rats [194].

Condensed Tannins (Proanthocyanidins)

Condensed tannins or proanthocyanidins are flavonoid oligomers that comprise one of two large groups of tannins. They are widely abundant in plants, including ferns, gymnosperm, and angiosperms and contribute to the purple, red, or blue colors of flowers, fruits, and sometimes leaves. In general, condensed tannins show antioxidant activity in their monomeric, oligomeric, or polymeric forms. Smith and Mackie (2004) investigated the effect of condensed tannins on the fecal microbial population of rats and found that dietary supplementation with condensed tannins extracted from *Acacia angustissima* modified the gut microbial population, resulting in a shift in the predominant bacteria towards tannin-resistant Gram-negative Enterobacteriaceae and Bacteroides species and a decrease in the abundance of the Gram-positive *C. leptum* group [194]. In another study, rats were fed proanthocyanidin-rich cocoa preparation, and there was a significant reduction in the abundance of *Clostridium, Bacteroides,* and *Staphylococcus* genera in the feces [195]. Further, apple proanthocyanidins showed a significant increase in the population of Actinobacteria when incubated with colonic microbiota in a batch culture model [196]. However, Tao et al. (2019) stated that the differences in the animal model and types and sources of proanthocyanidins affect the mutual relationship between proanthocyanidins and gut microbiota [197].

In terms of ruminant animal productivity and health, condensed tannins have many advantages and disadvantages, and these aspects will not be extensively addressed in this paper. In brief, condensed tannins can increase animal growth performance by increasing nitrogen retention in the diet, improve animal health via reducing the worm load in the intestine, and reduce enteric methane emission through their ability to lower the activity of methanogenic microbes in the rumen.

4.2. *Effect of Organosulfur Compounds*

The organosulfur group comprises several compounds, including indoles, isothiocyanates, and allylic sulfur compounds. Generally, many cultures have used garlic (*Allium sativum* L.) in cooking and traditional medicine since ancient times. Garlic is the richest source of organosulfur compounds, containing about 1.1–3.5%. During the storage of intact garlic, S-allyl-l-cysteine sulfoxides (alliin) are produced by the hydrolyzation and oxidation of γ-glutamyl-S-allyl-l-cysteines (G-SAC), which is the chief organosulfur compound (OSC) in garlic [198]. Alliinase is released from garlic during chopping or slicing or crushing or chewing, by which alliin is activated to allicin and other thiosulfates [199]. Bioactive OSCs are responsible for several beneficial health impacts, mainly producing defense compounds, which have wide-ranging antimicrobial properties [200].

According to Zhai et al. (2018), the abundance of the family Lachnospiraceae was decreased by the alliin derived from garlic [201]. Chen et al. (2019) studied intact garlic's impact and mode of action on the intestinal microflora using C57BL/6N male mice fed with or without whole garlic in a normal diet or a high-fat diet. It was found that ingestion of whole garlic, which constituted fructan, alliin, and other organosulfur derivatives, including allicin, S-allylcysteines and G-SAC could increase the α-diversity of the GM. This particularly enhanced the relative abundance of the Lachnospiraceae family and decreasing the population of *Prevotella* genus resulting in the attenuation of high-fat diet-induced dyslipidemia and disturbances of GM [202].

4.3. *Effect of Carotenoids*

Carotenoids are pigments that contribute to the red, orange, and yellow colors of several vegetables, fruits, and food products derived from them. Generally, carotenoids are classified into two groups: 1. carotenes (such as lycopene, α-carotene, and β-carotene) and 2. xanthophylls (such as zeaxanthin, lutein, and meso-zeaxanthin isomer) [203]. Humans must obtain carotenoids from the diet since they cannot synthesize carotenoids from endogenous precursors. The carotenoids in the blood have poor bioavailability (10–40%) as they are fat-soluble bioactive molecules and gut microbiota ferment the carotenoids once the enter the colon [204]. However, the biological roles of carotenoids in the intestinal ecosystem and their gut microbial utilization are yet to be fully understood [205]. Carotenoids act as antioxidants at low concentrations, while at high dosages, they have been shown in clinical trials to cause toxic effects based [206].

4.3.1. Astaxanthin

Astaxanthin (Figure 15) is an oxycarotenoid pigment, and is widely distributed in certain marine animals, including shrimp and salmon as well as in particular microalgae [207]. In *Helicobacter pylori*-infected mice, the abundance of total bacteria was notably decreased, and gastrointestinal inflammation was reduced by the supplementation of astaxanthin [208]. In another study, astaxanthin-treated stressed rats showed a notable reduction in the abundance of microbial community and inflammation scores [209]. Later, Yonei et al. (2013) used a real-time PCR assay to evaluate the influence of astaxanthin on the alterations of the colonic microbiota expression in mice fed a high-fat diet (fat 35%). This study identified that the abundance of the Bacteroides genus, *C. leptum*, and *C. coccoides* species were increased. In contrast, the prevalence of the *Streptococcus* genus (lactobacilli) was reduced due to high-fat diet consumption compared with a control diet containing 3.9% of fat. However, the above modifications were inhibited in the mice receiving astaxanthin-supplemented diets [210].

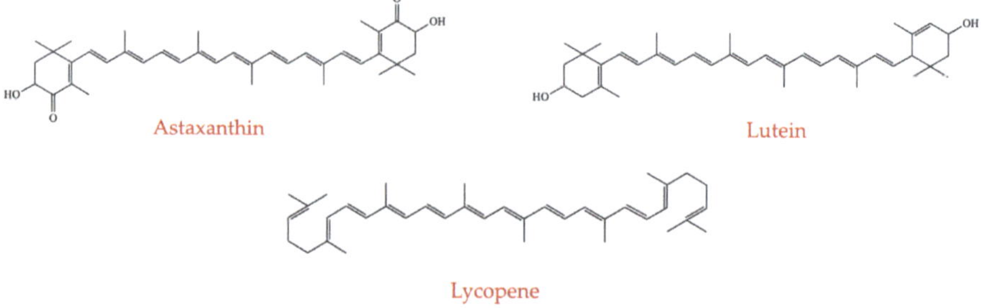

Figure 15. Chemical structures of carotenoids (astaxanthin, lutein, and lycopene) (self-generated).

Liu et al. (2018) used C57BL/6J mice as a biomedical model for humans to evaluate the intervention of astaxanthin on the gut microflora and to examine its ability to protect against liver injury induced by alcohol. The results showed that astaxanthin treatment notably increased the species from *Akkermansia* and *Verrucomicrobia* and reduced species from the genera *Parabacteroides*, *Bilophila*, and *Butyricimonas* and the phyla Proteobacteria and *Bacteroidetes* in comparison to the mice group fed with ethanol. Furthermore, astaxanthin supplementation considerably relieves inflammation and reduces the excessive accumulation of lipid and serum markers of liver injury [211]. The findings of an in vivo pilot study stated that intestinal microbiota could be altered by 0.04% (w/w) dietary astaxanthin at the phylum level by both genotype and gender [204]. Supplementation of astaxanthin selectively decreased the prevalence of colonic *Bacteroides* and Proteobacteria in female wild-type and BCO2 knockout C57BL/6J mice. Moreover, the population of *Bifidobacterium* and Actinobacteria significantly increased by astaxanthin only in male wild-type mice, which would result in a favorable effect on gut health.

4.3.2. Lutein

Lutein (Figure 15) belongs to the xanthophyll group, and it is an oxygenated carotenoid pigment. Humans and mammals obtain lutein from their diet [212]. In a human phytotherapy study, two products which contained blackcurrant extract powder, lutein, and lactoferrin were identified to have potential prebiotic actions by remarkably enhancing the abundance of lactobacilli and bifidobacteria, while decreasing other bacterial loads, including *Clostridium* spp. and *Bacteroides* spp. [213]. Animals maintained under grazing or rangeland consume a larger amount of lutein over the lifespan than those grown indoors on formulated rations or feedlot diets. This lutein can be stored in adipose tissue and fat droplets within the muscle tissue of farm animals. It may act as a scavenger of free radical molecules or substances that reduce oxidative damage in the body [212].

4.3.3. Lycopene

Lycopene (Figure 15), one of the primary carotenoids, is a red pigment widely distributed in watermelon, tomatoes, and some other fruits. Wiese et al. (2019) investigated the influence of dark chocolate and lycopene on intestinal microflora, skin, liver metabolism, blood, and oxygenation of skeletal muscle tissue. It was found that lycopene compounds showed alterations in the composition of colonic microflora by promoting the abundance of beneficial *B. longum* and *B. adolescentis*, and was dependent on the dosage level [214].

5. Mechanism of Action of Phytochemical

5.1. *Effect on the Gut Microbiome*

In general, dietary phytochemicals (polyphenols) are received as xenobiotics in humans once consumed. Their absorption in the small intestine and their biological availability is relatively low due to their structural complexity and polymerization [215]. Numerous studies revealed that dietary phytochemicals entering the intestinal ecosystem and their metabolites can alter the composition of microbial ecology beneficially by acting as prebiotics as well as antimicrobial agents against harmful gut microbiota [49,72,216]. The potential benefits of phytochemicals associated with GM are summarized in Figure 16.

The structure of phytochemicals, the type of microbial strain, and the dosage level determine the effect of phytochemicals on bacterial growth, multiplication, and metabolism [217]. To exemplify, Gram-negative bacteria show greater resistance than Gram-positive bacteria to phytochemicals which may be due to the variations in the composition of their cell wall [218]. Numerous studies demonstrated several mechanisms of the effect of phytochemicals on bacteria cells, their growth, and propagation. Phytochemicals may alter the functions of the bacterial cell membranes and, thus, suppress the cell growth by binding to the membranes in a dose-dependent manner [219]. Through the production of hydrogen peroxide and via changing the permeability of the bacterial cell membranes, phytochemicals, including catechins, affect several species of bacteria such as *Klebsiella pneumonie*, *E. coli*, *Salmonella choleraesis*, *Bordetella bronchiseptica*, *Bacillus subtilis*, *Serratia marcescens*, *Pseudomonas aeruginosa*, and *S. aureus* [1,220].

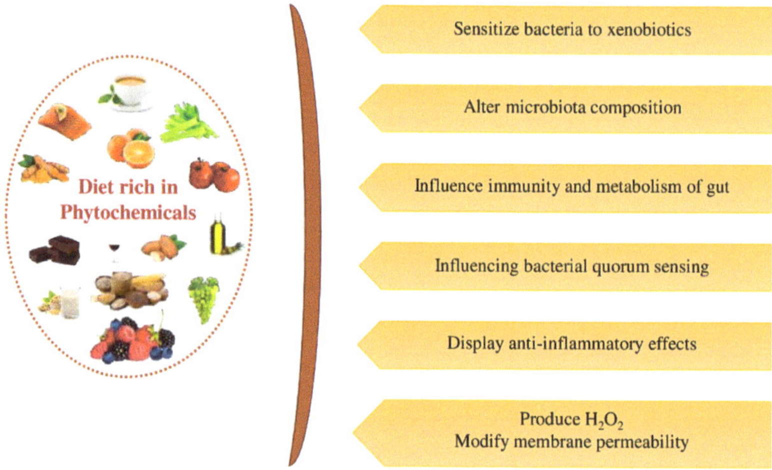

Figure 16. The potential benefits of phytochemicals associated with gut microbiota (self-generated).

Sirk et al. (2009) and Sirk et al. (2011) also stated that hydrogen bond formation between the hydroxyl groups and lipid bilayers of bacterial cell membranes may govern the anticancer, antimicrobial, and some other health benefits of theaflavins and catechins. The absorption of catechins and its ability to form a hydrogen bond with the groups of lipid heads are notably influenced by their molecular structure and collective conditions. The configuration of the theaflavins and catechins is affected by the molecular structure when binding to the surface of the lipid bilayer [221,222].

Stapleton et al. (2007) revealed that (−)-epicatechin gallate (ECg), which is a component of green tea, enhances the aggregation of staphylococcal cells, sensitizes methicillin-resistant *S. aureus* strains to β-lactam antibiotics, and increases the thickness of cell-walls. Variations in the physical properties of the bilayer, which is associated with ECg, can cause alterations in the structure of teichoic acid in the cell wall that leads to the modification of the features of the cell-surface required to support the phenotype, which is resistant to β-lactam [98].

Bacterial quorum sensing can also be influenced by phytochemicals via synthesizing, liberating, and detecting autoinducers, which are the small signaling molecules (for example, oligopeptides in Gram-positive bacteria and acylated homoserine lactones in Gram-negative bacteria) [223,224]. For instance, Hubert et al. (2003) stated that polyphenols influence the synthesis of small signaling molecules by bacteria such as *Burkholderia cepacia*, *Pseudomonas putida*, and *E. coli*, that stimulate an exponential increase in the abundance of bacteria [225]. In another study, Monagas et al. (2003) revealed that isolated or synthesized flavan-3-ols metabolites by Phase II-conjugation reactions have influence, rather than their antioxidant actions, by interfering with the signaling pathways involved in the process of disease development [226].

Additionally, the production of toxin VacA (Vacuolating toxin A), which is a primary virulence factor of *Helicobacter pylori*, was greatly suppressed by the polyphenols of green tea and red wine [227]. The destruction of bacteria cell membranes, suppression of urease activity, and disturbance of bacteria multiplication are some of the modes of inhibitory action of dietary phytochemicals on *H. pylori*. Through these mechanisms, cells become more sensitive to foreign substances such as antibiotics, resulting in the disturbance of the proton motive force via functions related to the cell membrane and the loss of H^+-ATPase [228].

Furthermore, the suppressive action on the synthesis of DNA and RNA is possibly due to the influence of the flavonoids' B-ring on the hydrogen bonding of the flavonoids with nucleic acid bases [229]. Plaper et al. (2003) stated that quercetin impedes the ATPase activity of the enzyme by binding to the GyrB subunit of *E. coli* DNA gyrase [230]. To confirm previous conclusions, Gradisar et al. (2007) revealed that catechins can attach to the ATP binding site of the gyrase B subunit and thus suppress the DNA gyrase of bacteria [231].

In vitro and animal experiments have shown that polyphenol compounds might suppress the generation of water-insoluble glucans, and this might be the reason for the anticaries action of cocoa powder [232,233]. Moreover, onion extract, a rich source of flavonoids, has been revealed to affect important bacteria which cause adult periodontitis, such as *Prevotella intermedia* and *Porphyromonas gingivalis*, and also on *Streptococcus sobrinus* and *Streptococcus mutans*, which also contribute to harmful effects on gut health [234].

A further study suggested that sensitive gut bacterial populations, mostly aerobic microbes, might be affected by iron deficiency in the gut due to the formation of polyphenol-metal ion complexes [235]. Iron is necessary for aerobic micro-organisms for different functions, such as forming heme groups and decreasing DNA ribonucleotide precursors. However, Freestone et al. (2007) suggested that the abundance of enteropathogenic bacteria may be increased by dietary catechols, which supply iron under conditions when the iron is limited and allow the growth of intestinal bacteria [236]. Further research in animals and humans is necessary to understand different mechanisms and mode of actions of phytochemicals on particular gut microflora proliferation and functions since most of them are not fully clarified at this time.

5.2. Studies Performed on the Gut Microbiome of Animals

Health and medical studies aimed at human intervention trials are normally conducted in phases, with animal studies a first step prior to human trials. Biological effects and mechanisms of action that occur in the human body could be better understood by working with animals as a biomedical model for humans. The main focus of animal studies has been the evaluation of the safety of the compound and understanding the metabolism of phytochemicals, particularly their influence on microbial action, digestibility and metabolic diseases in humans, and animals that are used for human purposes such as food, companion, clothing, travel, recreation, etc. The interaction between the composition and diversity of the gut microflora and the metabolites of phytochemicals that are derived from the host has been investigated in only a small number of animal studies. Culture-independent comparison studies have revealed that even though the distal portion of the intestinal ecosystem of humans and mice inhabits similar phyla of bacteria, many species and genera of bacteria present in mice are not found in humans. Therefore, when generalizing the findings of the animal studies to humans, caution should be taken [237,238]. Studies carried out in animals to evaluate the impacts of phytochemicals on the alteration in the composition of the gut microbiome are summarized in Table 1.

Table 1. Studies carried out in animals to evaluate the impacts of phytochemicals on the alteration in the gut microbiome.

Phytochemical	Animal	Effect on Microbiome and Related Mechanism	Reference
Tea polyphenols	Pigs	Enhanced the prevalence of lactobacilli, while reducing that of Bacteroidaceae, *C. perfringens*, and total bacteria	[239]
Coffee and caffeic acid	Rats with colon cancer	Supplementation specifically suppressed neoplastic cell transformation and colon cancer metastasis in mice via inhibition of TOPK (T-LAK cell-originated protein kinase) and MEK1	[240]
Green tea extracts	Calves	Decreased the abundance of *C. perfringens*, *Bifidobacterium* spp., and *Lactobacillus* spp.	[241]

Table 1. *Cont.*

Phytochemical	Animal	Effect on Microbiome and Related Mechanism	Reference
Grape pomace extracts	Lamb	Suppressed the growth of pathogenic bacteria *E. coli* and *Enterobacteriacae* while inducing the growth of facultative probiotic bacteria	[242]
Seaweed extract	White sheep ewes	Lactic acid bacteria count in ewes and lambs was decreased and the growth of *Enterococcus* sp. was inhibited	[243]
Red wine extract rich in proanthocyanidin	Rats with colon cancer	Supplemented rats showed a significantly greater abundance of *Bifidobacterium* spp., *Bacteroides*, and *Lactobacillus* and a reduced prevalence of *Clostridium* spp.	[244]
Quercetin	High-fat-diet fed rats	Down-regulated *Eubacterium cylindroides*, Erysipelotrichaceae, and *Bacillus*. Decreased body weight. Reduced the abundance of *Bacillus* genus, *Firmicutes*, and Erysipelotrichi class.	[104]
Proanthocyanidins extracted from ***Acacia angustissima***	Rats	Increased the prevalence of *Porphyromonas* group, *Bacteroides fragilis* group, Enterobacteriaceae, and Bacteroides *Prevotella* and reduced the abundance of *C. leptum* group	[194]
Resveratrol	Rats with DSS-induced colitis	Promoted the cell counts of *Bifidobacterium* spp. and *Lactobacillus* in feces	[245]
Polyphenols present in Chinese propolis, Brazilian propolis	Rats with DSS-induced colitis	Altered the composition of intestinal microflora, including a decrease in *Bacteroides* spp.	[246]
Lowbush wild blueberries	Rats	Increased the population of *Slackia* spp., *Thermonospora* spp., and *Corynebacteria* spp., while reducing that of *Enterococcus* spp. and *Lactobacillus* spp.	[247]
Resveratrol	Rats with colon cancer	Decreased functions of host intestinal mucosal and fecal enzymes, including β-galactosidase, α-glucoronidase, α-glucosidase, nitroreductase, and mucinase	[248]
Polyphenols from fungi	Rats with DSS-induced colitis	Modified the composition of colonic microflora by decreasing the ratio of *Bacteroidetes* to *Firmicutes* and restoring the abundance of *Lactobacillus* spp.	[249]
Grape pomace concentrate (GPC), grape seed extract (GSE)	Broiler chicks	Increased the population of *Lactobacillus* spp., *E. coli*, and *Enterococcus* spp.	[250]
Polyphenols present in *Prunella vulgaris* honey	Rats with DSS induced colitis	Modified the composition of colonic microflora, by increasing the ratio of *Bacteroidetes* to *Firmicutes* and restoring the abundance of *Lactobacillus* spp.	[251]

Although studies have been widely carried out in rats, some experiments used larger animals, including pigs, cattle, chickens, or sheep. Animal experiments conducted on calves [241] and pigs [239] demonstrated that the gut microbial composition was improved by the supplementation of tea polyphenols. In a study using pigs, the supplementation of tea polyphenols notably enhanced the growth of lactobacilli while decreasing the abundance of Bacteroidaceae and total bacteria. Furthermore, there was a tendency to reduce clostridia that was lecithinase-positive such as *C. perfringens* [239], though, *Lactobacillus* spp. and *Bifidobacterium* spp. showed a slow rate of the decrease, and *C. perfringens* showed a faster reduction rate in calves administrated with the extract of green tea [241].

Kafantaris et al. (2017) investigated the effect of grape pomace, a by-product of the wine-making process rich in polyphenols, on the gut microbiota of lamb. Twenty-four lambs were divided into two experimental groups, one receiving control feed and the other receiving a diet supplemented with grape pomace for 55 days. The effect was examined by analyzing the population of fecal microflora of the lambs. It was found that the experimental feed suppressed the growth of pathogenic bacteria *E. coli* and *Enterobacteriacae* while inducing the growth of facultative probiotic bacteria [242]. In another study, the influence of commercial seaweed extract, collected in northern Norway, on the intestinal microflora of Norwegian white sheep ewes was evaluated in vivo and in vitro. The lactic acid bacteria count in the ewes was decreased and the growth of *Enterococcus* sp. was inhibited by the bioactive compounds present in the extracts of red and brown seaweeds [243].

Ruminants (cattle, sheep, goats, alpacas, deer, etc.) cannot directly digest plant material due to the lack of enzymes capable of breaking down cell walls (cellulose and hemicellulose). Rather, the GI of ruminants is inhabited by a large population of bacteria, protozoans, and fungi in their four-chambered stomach (rumen, reticulum, omasum, and abomasum) capable of digesting a diet with a large proportion (80–85%) of high-fibrous plant materials (roughage diets), whereas monogastric animals consume smaller proportion (15–20%) of roughage diets. The micro-organisms present play a major role in the degradation of undigestible fibrous materials enabling the use of nutrients by themselves as well as providing a medium for digestion and absorption in the small (duodenum, jejunum, and ileum) and large (cecum, colon, and rectum) intestines of the host animals. The rumen and reticulum are home to the micro-organisms involved in the fermentation and breakdown of plant materials, producing volatile fatty acids and releasing other nutrients that both microbes and host animals use for energy and stimulating several metabolic reactions. Partially or fully digested food passes through the last compartment of the stomach (abomasum) to the small intestine and large intestine, and further digestion and absorption of nutrients take place by the host animal. The anatomical and functional attributes of the small intestine of ruminants is similar to non-ruminants and ranges in length by approximately 12–30 times the animal's body length [6].

The microbial community (bacteria, protozoa, archaea, fungi, and bacteriophage) in the rumen is responsible for the fermentation of complex plant materials consumed by ruminants [252,253]. However, methane gas, a greenhouse gas with global warming potential, is generated as a metabolic by-product of enteric fermentation by rumen methanogens [254]. Studies conducted in ruminants show that dietary supplementation with phytochemicals could reduce methane gas emission by modulating the rumen microbiota. For instance, over 21 seaweeds have been proven to decrease methane emission in vitro [255] which is believed to be, at least partially, due to the bio-actives and/or phytochemicals present in seaweed. The production of methane gas by ruminants accounts for nearly 81% of greenhouse gas emissions from the livestock sector, 90% of which comes from rumen microbial methanogenesis [256–258].

A complex community of ciliate protozoan, anaerobic fungi, and bacteria produce H_2 and CO_2 in the rumen. These gases are converted to methane by a population of methanogenic archaea [259,260]. Feeding ruminants with diets containing phytochemicals could reduce the population of methanogenic archaea, ciliate protozoan, anaerobic fungi, and bacteria in the rumen, thereby reducing methane gas emissions. For example, saponin compounds, which are defaunation agents for protozoa, could reduce methane production by decreasing the abundance of protozoa and their associated methanogens in rumen fluid [261]. In the future, climate variability and the sustainability of animal production are challenging areas in managing the ecosystems in relation to methane mitigation. Selecting multispecies pastures and fodders containing different types and levels of phytonutrients, or by supplementing agricultural by-products consisting of bioactives/phytochemicals to ruminants could potentially improve milk and meat productivity while reducing the enteric methane emission [262]. It is feasible to manipulate the activity of methanogenic bacteria and protozoa in the rumen (gut microbiota) resulting in lower enteric methane emission and improved productivity. Some examples are, the inclusion of seaweeds, oilseed by-products, oils, and by-products of the wine industry into ruminant diets has merits in reducing the activity and/or proliferation of methanogenic archaea in the rumen and diverting this dietary energy loss as methane emission into body gain [263–265].

The findings of studies using rats demonstrated that a significant modification in the intestinal microflora was observed in carcinogen-treated F344 rats supplemented with wine polyphenols in comparison to rats fed control diets [244]. Even though the feces of polyphenol-supplemented rats and control rats had the equivalent total number of bacteria and ratio between aerobic and anaerobic micro-organisms, the abundance of bifidobacteria and lactobacilli was enhanced and those of Clostridia, *Propionibacterium*, and *Bacteroides* was reduced. The authors suggested that the influence of wine phytochemicals on the function of the gut and carcinogenesis may be associated with alterations to the intestinal ecosystem, decreases in oxidative damage, and changes in gene expression.

In another study, fresh stool samples of rats that were administrated with apple juice as a replacement for drinking water, exhibited a higher abundance of bifidobacteria and lactobacilli, which varied from the rats fed a control diet by one log10 colony-forming units [266]. Molan et al. (2010) further proved the prebiotic properties of extracts of wild blackcurrant in rats following the findings of previous in vitro experiments. After regular supplementation of those extracts to rats, a notable growth in bifidobacteria and lactobacilli was recorded [267]. In an earlier study, the influence of the supplementation of extracts of grape pomace on the modification of the gut microbiome of broiler chicks was investigated [250]. The results demonstrated that the prevalence of *Enterococcus*, *E. coli*, and *Lactobacillus* species was higher in the gut of birds fed with grape extracts than in other birds. In conclusion, it was stated that the intestinal morphology and gut microflora were altered by the compounds rich in polyphenols from grapes, which improved the extent of the biodiversity of gut microbiota in broiler chicks.

Overall, numerous studies suggested that the application of the appropriate animal model provides a robust technique to explore the gut microbial diversity and associated metabolic functions, even though the intestinal microflora of animals and humans is not entirely the same. In general, the presence/absence and abundance of a gene, profiles of the gut microbial ecosystem, and the range of functions of microbes can be determined by metagenomic studies. However, they can influence only an observed phenotype because the presence of a gene does not indicate its functionality or that expression is there [268].

6. Gut Microbiota and Metabolic Diseases (MD)

6.1. *Influence of Gut Microbiota on some Metabolic Diseases*

Socio-demographic and environmental factors significantly contribute to the status of people with metabolic disorders, including diabetes mellitus, whereas human genetics show a lower influence [269]. Colonic microflora also shows a constructive interaction with metabolic diseases. In addition to its digestive functions, the intestinal ecosystem maintains the optimum condition of human health via contributing non-human genome encoded enzymes. This includes the generation of vitamins and the breakdown of polyphenols and polysaccharides [59]. The intestinal microflora is also responsible for the etiologies of metabolic disorders, such as obesity, hypertension, cardiovascular disease, diabetes mellitus, and inflammation.

It is reported that intestinal microflora controls most of the features of human physiology, including regulation of colonic function, immune system modification, exogenous toxins removal from the body, and defense mechanism against several pathogens. Several epidemiological and experimental data revealed that energy homeostasis and maladaptation are significantly influenced by intestinal microbiota, which is interrelated with insulin resistance and obesity [270,271].

Furthermore, Clarke et al. (2010) stated that metabolic disorders may be regulated by intestinal microbiota via affecting the immune system of the host and modifying the inflammatory signaling pathway [272]. It was found that the incidence of various features of metabolic disorders can be induced and suppressed by the lack of Toll-like receptors functions, namely TLR4 and TLR5 [273,274]. These toll-like receptors can recognize the pattern of the cell membrane, and they significantly contribute to the non-specific defense mechanisms of the host [274]. In an earlier study, Clarke et al. (2010) indicated that intestinal microflora also can produce other pro-inflammatory compounds, including flagellin, peptidoglycan, and lipoproteins, etc., which can bind to toll-like receptors [272]. Overall, the colonic microbiota plays a major role in the incidence of metabolic disorders via altering the signaling pathways associated with the initiation and progression of inflammations.

6.2. Impact of Phytochemicals on Metabolic Syndrome by Modulating Gut Microbiota

It was observed that chronic low-grade inflammation is generated due to the complex interaction between a person's diet and the intestinal microbiota [275,276]. The key reason for indicating that metabolic syndrome is inter-related with chronic diseases is the reciprocal association between the diet and the gut microbiota.

Several researchers focused on phytochemicals because of their significant influence on human health. Even though the presence of dietary fibers, vitamins, minerals, etc. are responsible for the beneficial health effects of fruits, many studies revealed that the phytochemical content of the fruits also favorably influences human health by decreasing the complications related to obesity and other disorders [171,277,278]. Numerous experimental and epidemiological data demonstrate that phytochemicals can protect against diabetes and its related complications by modifying intestinal microbiota [279,280]. Figure 17 shows a simple illustration of the impact of phytochemicals and a diet rich in phytochemicals on metabolic diseases by modifying the intestinal microflora.

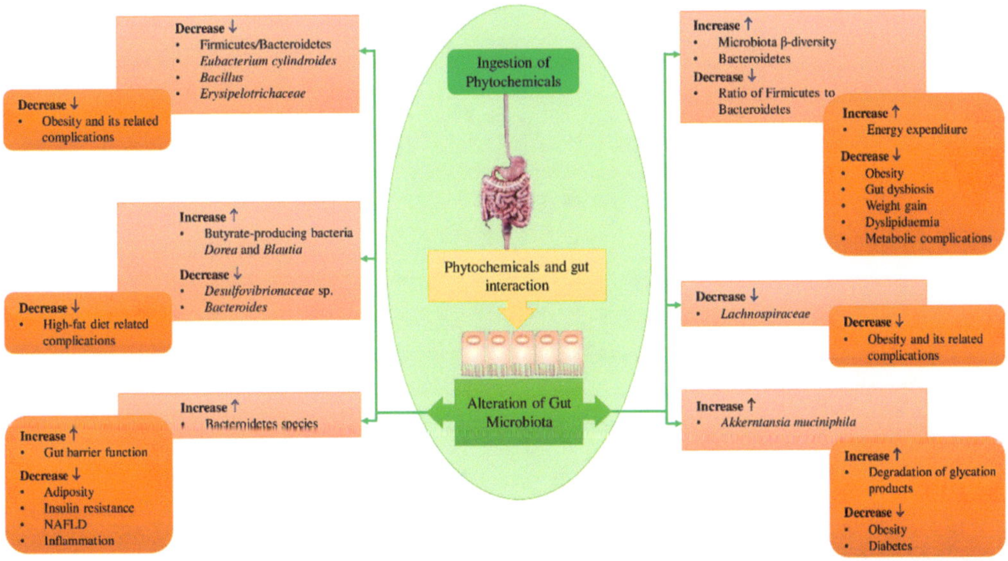

Figure 17. Impact of phytochemicals and diet rich in phytochemicals on metabolic diseases by modifying intestinal microflora (self-generated).

Gut dysbiosis can be inhibited with the supplementation of phytochemicals, which enhances the prevalence of several species of beneficial bacteria and β-diversity of microflora while reducing the abundance of opportunistic harmful bacteria. Alterations in colonic microflora can lead to enhanced intestinal barrier function, increased breakdown of glycan, and expenditure of energy. It can alleviate the incidence of adiposity, inflammation, dyslipidemia, weight gain, and insulin resistance. Abovementioned beneficial alterations in the host body consequently resulted in decreased metabolic diseases and its related complications [53].

7. Conclusions and Future Perspectives

In recent years, the interaction of dietary phytochemicals with the intestinal microflora has garnered a greater interest because of their impact on gastroenterology, fermentation, digestion, disease prevention, and human health. Several preclinical and clinical studies revealed that phytochemicals could act as antimicrobial agents and exhibit prebiotic effects on harmful intestinal microbiota. Despite numerous studies on this topic, the clarification and understanding of the specific mechanism of each phytochemical are not well understood. Complexities during the implementation, complications in interpreting the in vitro findings into in vivo applications, and ethical and economic issues are some constraints for in vivo experiments. The review indicates that phytochemicals play a crucial role in improving human (and animal) health, particularly gut health, via promoting the abundance of favorable bacteria and protozoa and suppressing harmful bacteria, exhibiting prebiotic activities. The bioactivity and metabolic function of some phytochemicals in the body are examined so far. Therefore, further studies should focus on examining the therapeutic potential of remaining phytochemicals and explaining the complex mechanisms and identifying which phytochemical will specifically influence which micro-organism in modulating gut microbiota and maintaining good health. Furthermore, the interaction of dietary phytochemicals with other nutrients, such as minerals, vitamins and essential fatty acids, in the microbiome and their effects in terms of fermentation, digestion, bio-accessibility of nutrients in the gut, and the maintenance of health in humans and animals using in vivo studies warrants further research. Due to inadequate human trials and animal model studies, another challenge is to utilize these natural phytochemicals in the pharmaceutical industry to develop drugs targeting improved gut health. In future, more animal model studies and human trials should be conducted to pave the path for drug development using phytochemicals alone or diets containing phytochemicals and other nutrients such as vitamins, essential fatty acids, etc.

Author Contributions: Conceptualization, S.S. and E.N.P.; writing—original draft preparation, S.S., E.N.P. and A.E.-D.A.B.; writing—review and editing, J.L.J., F.R.D., E.M. and N.R. All authors have read and agreed to the published version of the manuscript.

Funding: This research received no external funding.

Institutional Review Board Statement: Not applicable.

Informed Consent Statement: Not applicable.

Data Availability Statement: This is a review article. This study did not report any data.

Acknowledgments: The authors are grateful to the support of their corresponding organisations.

Conflicts of Interest: The authors declare no conflict of interest.

References

1. Kumar Singh, A.; Cabral, C.; Kumar, R.; Ganguly, R.; Kumar Rana, H.; Gupta, A.; Rosaria Lauro, M.; Carbone, C.; Reis, F.; Pandey, A.K. Beneficial Effects of Dietary Polyphenols on Gut Microbiota and Strategies to Improve Delivery Efficiency. *Nutrients* **2019**, *11*, 2216. [CrossRef] [PubMed]
2. Dekaboruah, E.; Suryavanshi, M.V.; Chettri, D.; Verma, A.K. Human microbiome: An academic update on human body site specific surveillance and its possible role. *Arch. Microbiol.* **2020**, *202*, 2147–2167. [CrossRef] [PubMed]
3. FadlAlla; Sadeek, E.A. Gut Microbiota, Probiotics and Phytochemicals. *Int. J. Curr. Adv. Res.* **2015**, *4*, 330–340.
4. Weissman, J.L.; Dogra, S.; Javadi, K.; Samantha, B.; Flint, R.; Davati, C.; Beattie, J.; Dixit, K.; Peesay, T.; Awan, S.; et al. Exploring the functional composition of the human microbiome using a hand-curated microbial trait database. *BMC Bioinform.* **2021**, *22*, 306. [CrossRef]
5. Manor, O.; Dai, C.L.; Kornilov, S.A.; Smith, B.; Price, N.S.; Lovejoy, J.C.; Gibbons, S.M.; Magis, A.T. Health and disease markers correlate with gut microbiome composition across thousands of people. *Nat. Commun.* **2020**, *11*, 5206–5218. [CrossRef]
6. Swanson, K.C. Small intestinal anatomy, physiology, and digestion in ruminants. In *Reference Module in Food Science*; Elsevier: Amsterdam, The Netherlands, 2019.
7. Bassetti, M.; Bandera, A.; Gori, A. Therapeutic Potential of the Gut Microbiota in the Management of Sepsis. *Crit. Care* **2020**, *24*, 105–112. [CrossRef]

8. Sprockett, D.; Fukami, T.; Relman, D.A. Role of priority effects in the early-life assembly of the gut microbiota. *Nat. Rev. Gastroenterol. Hepatol.* **2018**, *15*, 197–205. [CrossRef]
9. Zafar, H.; Saier, M.H., Jr. Gut *Bacteroides* species in health and disease. *Gut Microbes* **2021**, *13*, 1848158. [CrossRef]
10. Chen, Y.; Zhou, J.; Wang, L. Role and Mechanism of Gut Microbiota in Human Disease. *Front. Cell. Infect. Microbiol.* **2021**, *11*, 625913. [CrossRef]
11. Smith, C.C.; Snowberg, L.K.; Caporaso, J.G.; Knight, R.; Bolnick, D.I. Dietary input of microbes and host genetic variation shape among-population differences in stickleback gut microbiota. *ISME J.* **2015**, *9*, 2515–2526. [CrossRef]
12. Hasan, N.; Yang, H. Factors affecting the composition of the gut microbiota, and its modulation. *PeerJ* **2019**, *7*, e7502. [CrossRef] [PubMed]
13. Münger, E.; Montiel-Castro, A.J.; Langhans, W.; Pacheco-López, G. Reciprocal Interactions Between Gut Microbiota and Host Social Behavior. *Front. Integr. Neurosci.* **2018**, *12*, 21. [CrossRef] [PubMed]
14. Martinez, J.E.; Kahana, D.D.; Ghuman, S.; Wilson, H.P.; Wilson, J.; Kim, S.C.J.; Lagishetty, V.; Jacobs, J.P.; Sinha-Hikim, A.P.; Friedman, T.C. Unhealthy Lifestyle and Gut Dysbiosis: A Better Understanding of the Effects of Poor Diet and Nicotine on the Intestinal Microbiome. *Front. Endocrinol. (Lausanne)* **2021**, *12*, 667066. [CrossRef] [PubMed]
15. Kasprzak-Drozd, K.; Oniszczuk, T.; Stasiak, M.; Oniszczuk, A. Beneficial Effects of Phenolic Compounds on Gut Microbiota and Metabolic Syndrome. *Int. J. Mol. Sci.* **2021**, *22*, 3715. [CrossRef] [PubMed]
16. Sultan, S.; El-Mowafy, M.; Elgaml, A.; Ahmed, T.; Hassan, H.; Mottawea, W. Metabolic Influences of Gut Microbiota Dysbiosis on Inflammatory Bowel Disease. *Front. Physiol.* **2021**, *12*, 715506. [CrossRef]
17. Liébana-García, R.; Olivares, M.; Bullich-Vilarrubias, C.; López-Almela, I.; Romaní-Pérez, M.; Sanz, Y. The gut microbiota as a versatile immunomodulator in obesity and associated metabolic disorders. *Best Pract. Res. Clin. Endocrinol. Metab.* **2021**, *35*, 101542. [CrossRef]
18. Wang, P.X.; Deng, X.R.; Zhang, C.H.; Yuan, H.J. Gut microbiota and metabolic syndrome. *Chin. Med. J.* **2020**, *133*, 808–816. [CrossRef]
19. Arora, J.; Ramawat, K.G. Bioactive molecules, nutraceuticals, and functional foods in indian vegetarian diet and during postpartum healthcare. In *Bioactive Molecules in Food. Reference Series in Phytochemistry*; Mérillon, J.M., Ramawat, K., Eds.; Springer: Cham, Switzerland, 2018; pp. 1–30.
20. Cencic, A.; Chingwaru, W. The role of functional foods, nutraceuticals, and food supplements in intestinal health. *Nutrients* **2010**, *2*, 611–625. [CrossRef]
21. Abuajah, C.I. Functional Components and Medicinal Properties of Food. In *Bioactive Molecules in Food. Reference Series in Phytochemistry*; Mérillon, J.M., Ramawat, K., Eds.; Springer: Cham, Switzerland, 2017; ISBN 978-3-319-54528-8. [CrossRef]
22. Hugenholtz, F.; de Vos, W.M. Mouse models for human intestinal microbiota research: A critical evaluation. *Cell. Mol. Life Sci.* **2018**, *75*, 149–160. [CrossRef]
23. Lobionda, S.; Sittipo, P.; Kwon, H.Y.; Lee, Y.K. The Role of Gut Microbiota in Intestinal Inflammation with Respect to Diet and Extrinsic Stressors. *Microorganisms* **2019**, *7*, 271. [CrossRef]
24. Vasmehjani, A.A.; Darabi, Z.; Nadjarzadeh, A.; Mirzaei, M.; Hosseinzadeh, M. The relation between dietary phytochemical index and metabolic syndrome and its components in a large sample of Iranian adults: A population-based study. *BMC Public Health* **2021**, *21*, 1587. [CrossRef] [PubMed]
25. Mendoza, N.; Silva, E.M.E. Introduction to phytochemicals: Secondary metabolites from plants with active principles for pharmacological importance. In *Phytochemicals: Source of Antioxidants and Role in Disease Prevention*; Asao, T., Asaduzzaman, M., Eds.; IntechOpen: London, UK, 2018; pp. 25–47.
26. Xiao, J.; Bai, W. Bioactive phytochemicals. *Crit. Rev. Food Sci. Nutr.* **2019**, *59*, 827–829. [CrossRef] [PubMed]
27. Peng, M.; Tabashsum, Z.; Anderson, M.; Truong, A.; Houser, A.K.; Padilla, J.; Akmel, A.; Bhatti, J.; Rahaman, S.O.; Biswas, D. Effectiveness of probiotics, prebiotics, and prebiotic-like components in common functional foods. *Compr. Rev. Food Sci. Food Saf.* **2020**, *19*, 1908–1933. [CrossRef]
28. Adjei-Fremah, S.; Ekwemalor, K.; Worku, M.; Ibrahim, S. Probiotics and Ruminant Health. In *Probiotics—Current Knowledge and Future Prospects*; Enany, S., Ed.; InTech: London, UK, 2018; pp. 133–150; ISBN 978-1-78923-386-5.
29. Lara, M.V.; Bonghi, C.; Famiani, F.; Vizzotto, G.; Walker, R.P.; Drincovich, M.F. Stone Fruit as Biofactories of Phytochemicals with Potential Roles in Human Nutrition and Health. *Front. Plant Sci.* **2020**, *11*, 562252. [CrossRef] [PubMed]
30. Liu, R.H. Potential synergy of phytochemicals in cancer prevention: Mechanism of action. *J. Nutr.* **2004**, *134*, 3479S–3485S. [CrossRef] [PubMed]
31. Carrera-Quintanar, L.; LópezRoa, R.I.; Quintero-Fabián, S.; Sánchez-Sánchez, M.A.; Vizmanos, B.; Ortuño-Sahagún, D. Phytochemicals that influence gut microbiota as prophylactics and for the treatment of obesity and inflammatory diseases. *Mediat. Inflamm.* **2018**, *2018*, 9734845. [CrossRef] [PubMed]
32. Talib, W.H.; Alsayed, A.R.; Abuawad, A.; Daoud, S.; Mahmod, A.I. Melatonin in Cancer Treatment: Current Knowledge and Future Opportunities. *Molecules* **2021**, *26*, 2506. [CrossRef]
33. Zhu, D.; Zhang, Y.; Wang, S. Histone citrullination: A new target for tumors. *Mol. Cancer* **2021**, *20*, 90–107. [CrossRef]
34. Wu, M.S.; Aquino, L.B.B.; Barbaza, M.Y.U.; Hsieh, C.L.; De Castro-Cruz, K.A.; Yang, L.L.; Tsai, P.W. Anti-Inflammatory and Anticancer Properties of Bioactive Compounds from *Sesamum indicum* L.—A Review. *Molecules* **2019**, *24*, 4426. [CrossRef]

35. Selby-Pham, S.N.B.; Miller, R.B.; Howell, K.; Dunshea, F.; Bennett, L.E. Physicochemical properties of dietary phytochemicals can predict their passive absorption in the human small intestine. *Sci. Rep.* **2017**, *7*, 1931. [CrossRef]
36. Mukonowenzou, N.C.; Adeshina, K.A.; Donaldson, J.; Ibrahim, K.G.; Usman, D.; Erlwanger, K.H. Medicinal Plants, Phytochemicals, and Their Impacts on the Maturation of the Gastrointestinal Tract. *Front. Physiol.* **2021**, *12*, 1123–1131. [CrossRef] [PubMed]
37. Beane, K.E.; Redding, M.C.; Wang, X.; Pan, J.H.; Le, B.; Cicalo, C.; Jeon, S.; Kim, Y.J.; Lee, J.H.; Shin, E.C.; et al. Effects of dietary fibers, micronutrients, and phytonutrients on gut microbiome: A review. *Appl. Biol. Chem.* **2021**, *64*, 36–54. [CrossRef]
38. Dingeo, G.; Brito, A.; Samouda, H.; Iddir, M.; La Frano, M.R.; Bohn, T. Phytochemicals as modifiers of gut microbial communities. *Food Funct.* **2020**, *11*, 8444–8471. [CrossRef] [PubMed]
39. Kawabata, K.; Yoshioka, Y.; Terao, J. Role of Intestinal Microbiota in the Bioavailability and Physiological Functions of Dietary Polyphenols. *Molecules* **2019**, *24*, 370. [CrossRef] [PubMed]
40. Helander, H.F.; Fändriks, L. Surface area of the digestive tract—Revisited. *Scand. J. Gastroenterol.* **2014**, *49*, 681–689. [CrossRef]
41. Thursby, E.; Juge, N. Introduction to the human gut microbiota. *Biochem. J.* **2017**, *474*, 1823–1836. [CrossRef]
42. O'Hara, A.M.; Shanahan, F. The gut flora as a forgotten organ. *EMBO Rep.* **2006**, *7*, 688–693. [CrossRef]
43. Riccio, P.; Rossano, R. The human gut microbiota is neither an organ nor a commensal. *FEBS Lett.* **2020**, *594*, 3262–3271. [CrossRef]
44. Hillman, E.T.; Lu, H.; Yao, T.; Nakatsu, C.H. Microbial Ecology along the Gastrointestinal Tract. *Microbes Environ.* **2017**, *32*, 300–313. [CrossRef]
45. Ghosh, S.; Pramanik, S. Structural diversity, functional aspects and future therapeutic applications of human gut microbiome. *Arch. Microbiol.* **2021**, *203*, 5281–5308. [CrossRef]
46. Magne, F.; Gotteland, M.; Gauthier, L.; Zazueta, A.; Pesoa, S.; Navarrete, P.; Balamurugan, R. The Firmicutes/Bacteroidetes Ratio: A Relevant Marker of Gut Dysbiosis in Obese Patients? *Nutrients* **2020**, *12*, 1474. [CrossRef] [PubMed]
47. Rinninella, E.; Raoul, P.; Cintoni, M.; Franceschi, F.; Miggiano, G.; Gasbarrini, A.; Mele, M.C. What is the Healthy Gut Microbiota Composition? A Changing Ecosystem across Age, Environment, Diet, and Diseases. *Microorganisms* **2019**, *7*, 14. [CrossRef] [PubMed]
48. Derrien, M.; Alvarez, A.S.; de Vos, W.M. The Gut Microbiota in the First Decade of Life. *Trends Microb.* **2019**, *27*, 997–1010. [CrossRef]
49. Catalkaya, G.; Venema, K.; Lucini, L.; Rocchetti, G.; Delmas, D.; Daglia, M.; Capanoglu, E. Interaction of dietary polyphenols and gut microbiota: Microbial metabolism of polyphenols, influence on the gut microbiota, and implications on host health. *Food Front.* **2020**, *1*, 109–133. [CrossRef]
50. Yin, R.; Kuo, H.C.; Hudlikar, R.; Sargsyan, D.; Li, S.; Wang, L.; Wu, R.; Kong, A.N. Gut microbiota, dietary phytochemicals and benefits to human health. *Curr. Pharmacol. Rep.* **2019**, *5*, 332–344. [CrossRef] [PubMed]
51. Kastl, A.J., Jr.; Terry, N.A.; Wu, G.D.; Albenberg, L.G. The structure and function of the human small intestinal microbiota: Current understanding and future directions. *Cell. Mol. Gastroenterol. Hepatol.* **2020**, *9*, 33–45. [CrossRef] [PubMed]
52. Vuik, F.; Dicksved, J.; Lam, S.Y.; Fuhler, G.M.; van der Laan, L.; van de Winkel, A.; Konstantinov, S.R.; Spaander, M.; Peppelenbosch, M.P.; Engstrand, L.; et al. Composition of the mucosa-associated microbiota along the entire gastrointestinal tract of human individuals. *United Eur. Gastroenterol. J.* **2019**, *7*, 897–907. [CrossRef] [PubMed]
53. Gowd, V.; Karima, N.; Shishira, M.R.I.; Xiea, L.; Chen, W. Dietary polyphenols to combat the metabolic diseases via altering gut microbiota. *Trends Food Sci. Technol.* **2019**, *93*, 81–93. [CrossRef]
54. Li, H.; Limenitakis, J.P.; Fuhrer, T.; Geuking, M.B.; Lawson, M.A.; Wyss, M.; Brugiroux, S.; Keller, I.; MacPherson, J.A.; Rupp, S. The outer mucus layer hosts a distinct intestinal microbial niche. *Nat. Commun.* **2015**, *6*, 8292–8305. [CrossRef]
55. Sekirov, I.; Russell, S.L.; Antunes, L.C.M.; Finlay, B.B. Gut Microbiota in Health and Disease. *Physiol. Rev.* **2010**, *90*, 859–904. [CrossRef]
56. Zheng, D.; Liwinski, T.; Elinav, E. Interaction between microbiota and immunity in health and disease. *Cell Res.* **2020**, *30*, 492–506. [CrossRef] [PubMed]
57. Milani, C.; Duranti, S.; Bottacini, F.; Casey, E.; Turroni, F.; Mahony, J.; Belzer, C.; Delgado Palacio, S.; Arboleya Montes, S.; Mancabelli, L.; et al. The First Microbial Colonizers of the Human Gut: Composition, Activities, and Health Implications of the Infant Gut Microbiota. *Microbiol. Mol. Biol. Rev.* **2017**, *81*, e00036-17. [CrossRef] [PubMed]
58. Yadav, S.; Jha, R. Strategies to modulate the intestinal microbiota and their effects on nutrient utilization, performance, and health of poultry. *J. Anim. Sci. Biotechnol.* **2019**, *10*, 2–13. [CrossRef] [PubMed]
59. Rowland, I.; Gibson, G.; Heinken, A.; Scott, K.; Swann, J.; Thiele, I. Gut microbiota functions: Metabolism of nutrients and other food components. *Eur. J. Nutr.* **2018**, *57*, 1–24. [CrossRef]
60. Dahiya, D.K.; Renuka, P.M.; Shandilya, U.K. Gut microbiota modulation and its relationship with obesity using prebiotic fibers and probiotics: A review. *Front. Microb.* **2017**, *8*, 563–580. [CrossRef] [PubMed]
61. Kho, Z.Y.; Lal, S.K. The Human Gut Microbiome—A Potential Controller of Wellness and Disease. *Front. Microbiol.* **2018**, *9*, 1835–1858. [CrossRef]
62. Krishnan, S.; Alden, N.; Lee, K. Pathways and functions of gut microbiota metabolism impacting host physiology. *Curr. Opin. Biotechnol.* **2015**, *36*, 137–145. [CrossRef]
63. Bull, M.J.; Plummer, N.T. Part 1: The Human Gut Microbiome in Health and Disease. *Integr. Med. A Clin. J.* **2014**, *13*, 17–22.

64. Cardona, F.; Andrés-Lacueva, C.; Tulipani, S.; Tinahones, F.J.; Queipo-Ortuño, M.I. Benefits of polyphenols on gut microbiota and implications in human health. *J. Nutr. Biochem.* **2013**, *24*, 1415–1422. [CrossRef]
65. Liu, Z.; Hu, M. Natural polyphenol disposition via coupled metabolic pathways. *Expert Opin. Drug Metab. Toxicol.* **2007**, *3*, 389–406. [CrossRef]
66. Makarewicz, M.; Drożdż, I.; Tarko, T.; Duda-Chodak, A. The Interactions between Polyphenols and Microorganisms, Especially Gut Microbiota. *Antioxidants* **2021**, *10*, 188. [CrossRef] [PubMed]
67. Bessa, C.; Francisco, T.; Dias, R.; Mateus, N.; de Freitas, V.; Pérez-Gregorio, R. Use of Polyphenols as Modulators of Food Allergies. From Chemistry to Biological Implications. *Front. Sustain. Food Syst.* **2021**, *5*, 623611. [CrossRef]
68. Singla, R.K.; Dubey, A.K.; Garg, A.; Sharma, R.K.; Fiorino, M.; Ameen, S.M.; Haddad, M.A.; Al-Hiary, M. Natural Polyphenols: Chemical Classification, Definition of Classes, Subcategories, and Structures. *J. AOAC Int.* **2019**, *102*, 1397–1400. [CrossRef] [PubMed]
69. Shahidi, F.; Yeo, J. Bioactivities of Phenolics by Focusing on Suppression of Chronic Diseases: A Review. *Int. J. Mol. Sci.* **2018**, *19*, 1573. [CrossRef]
70. Ray, S.K.; Mukherjee, S. Evolving Interplay Between Dietary Polyphenols and Gut Microbiotan—An Emerging Importance in Healthcare. *Front. Nutr.* **2021**, *8*, 634944. [CrossRef]
71. Plamada, D.; Vodnar, D.C. Polyphenols-Gut Microbiota Interrelationship: A Transition to a New Generation of Prebiotics. *Nutrients* **2021**, *14*, 137. [CrossRef]
72. Rodríguez-Daza, M.C.; Pulido-Mateos, E.C.; Lupien-Meilleur, J.; Guyonnet, D.; Desjardins, Y.; Roy, D. Polyphenol-Mediated Gut Microbiota Modulation: Toward Prebiotics and Further. *Front. Nutr.* **2021**, *8*, 689456. [CrossRef]
73. Dias, M.C.; Pinto, D.C.G.A.; Silva, A.M.S. Plant Flavonoids: Chemical Characteristics and Biological Activity. *Molecules* **2021**, *26*, 5377. [CrossRef]
74. Wang, M.; Firrman, J.; Zhang, L.; Arango-Argoty, G.; Tomasula, P.; Liu, L.; Xiao, W.; Yam, K. Apigenin Impacts the Growth of the Gut Microbiota and Alters the Gene Expression of *Enterococcus*. *Molecules* **2017**, *22*, 1292. [CrossRef]
75. Bian, S.; Wan, H.; Liao, X.; Wang, W. Inhibitory Effects of Apigenin on Tumor Carcinogenesis by Altering the Gut Microbiota. *Mediat. Inflamm.* **2020**, *2020*, 7141970. [CrossRef]
76. Wang, M.; Firrman, J.; Liu, L.; Yam, K. A Review on Flavonoid Apigenin: Dietary Intake, ADME, Antimicrobial Effects, and Interactions with Human Gut Microbiota. *BioMed Res. Int.* **2019**, *2019*, 7010467. [CrossRef] [PubMed]
77. Stompor-Gorący, M.; Bajek-Bil, A.; Machaczka, M. Chrysin: Perspectives on Contemporary Status and Future Possibilities as Pro-Health Agent. *Nutrients* **2021**, *13*, 2038. [CrossRef] [PubMed]
78. Andrade, N.; Marques, C.; Andrade, S.; Silva, C.; Rodrigues, I.; Guardão, L.; Guimarães, J.T.; Keating, E.; Calhau, C.; Martel, F. Effect of chrysin on changes in intestinal environment and microbiome induced by fructose-feeding in rats. *Food Funct.* **2019**, *10*, 4566–4576. [CrossRef] [PubMed]
79. Gao, L.; Li, J.; Zhou, Y.; Huang, X.; Qin, X.; Du, G. The effects of baicalein on cortical pro-inflammatory cytokines and the intestinal microbiome in senescence accelerated mouse prone 8. *ACS Chem. Neurosci.* **2018**, *9*, 1714–1724. [CrossRef] [PubMed]
80. Zhang, B.; Sun, W.; Yu, N.; Sun, J.; Yu, X.; Li, X.; Xing, Y.; Yan, D.; Ding, Q.; Xiu, Z.; et al. Anti-diabetic effect of baicalein is associated with the modulation of gut microbiota in streptozotocin and high-fat-diet induced diabetic rats. *J. Funct. Foods* **2018**, *46*, 256–267. [CrossRef]
81. Chávez-González, M.L.; Sepúlveda, L.; Verma, D.K.; Luna-García, H.A.; Rodríguez-Durán, L.V.; Ilina, A.; Aguilar, C.N. Conventional and Emerging Extraction Processes of Flavonoids. *Processes* **2020**, *8*, 434. [CrossRef]
82. Sharma, K.; Mahato, N.; Lee, Y.R. Extraction, characterization and biological activity of citrus flavonoids. *Rev. Chem. Eng.* **2019**, *35*, 265–284. [CrossRef]
83. Duque, A.L.R.F.; Monteiro, M.; Adorno, M.A.T.; Sakamoto, I.K.; Sivieri, K. An exploratory study on the influence of orange juice on gut microbiota using a dynamic colonic model. *Food Res. Int.* **2016**, *84*, 160–169. [CrossRef]
84. Stevens, Y.; Rymenant, E.V.; Grootaert, C.; Camp, J.V.; Possemiers, S.; Masclee, A.; Jonkers, D. The Intestinal Fate of Citrus Flavanones and Their Effects on Gastrointestinal Health. *Nutrients* **2019**, *11*, 1464. [CrossRef]
85. Duda-Chodak, A. The inhibitory effect of polyphenols on human gut microbiota. *J. Physiol. Pharmacol.* **2012**, *63*, 497–503.
86. Parkar, S.G.; Stevenson, D.E.; Skinner, M.A. The potential influence of fruit polyphenols on colonic microflora and human gut health. *Int. J. Food Microbiol.* **2008**, *124*, 295–298. [CrossRef] [PubMed]
87. Parhiz, H.; Roohbakhsh, A.; Soltani, F.; Rezaee, R.; Iranshahi, M. Antioxidant and anti-inflammatory properties of the citrus flavonoids hesperidin and hesperetin: An updated review of their molecular mechanisms and experimental models. *Pharmacol. Res.* **2015**, *29*, 323–331. [CrossRef] [PubMed]
88. Garg, A.; Garg, S.; Zaneveld, L.J.; Singla, A.K. Chemistry and pharmacology of the citrus bioflavonoid hesperidin. *Pharmacol. Res.* **2001**, *15*, 655–669. [CrossRef] [PubMed]
89. Estruel-Amades, S.; Massot-Cladera, M.; Perez-Cano, F.J.; Franch, A.; Castell, M.; Camps-Bossacoma, M. Hesperidin effects on gut microbiota and gut-associated lymphoid tissue in healthy rats. *Nutrients* **2019**, *11*, 324. [CrossRef]
90. Lima, A.C.D.; Cecatti, C.; Fidelix, M.P.; Adorno, M.A.T.; Sakamoto, I.K.; Cesar, T.B.; Sivieri, K. Effect of daily consumption of orange juice on the levels of blood glucose, lipids, and gut microbiota metabolites: Controlled clinical trials. *J. Med. Food* **2019**, *22*, 202–210. [CrossRef]

91. Brasili, E.; Hassimotto, N.M.A.; Del Chierico, F.; Marini, F.; Quagliariello, A.; Sciubba, F.; Miccheli, A.; Putignani, L.; Lajolo, F. Daily consumption of orange juice from *Citrus sinensis* L. Osbeck cv. Cara Cara and cv. Bahia differently affects gut microbiota profiling as unveiled by an integrated meta-omics approach. *J. Agric. Food Chem.* **2019**, *67*, 1381–1391. [CrossRef]
92. Salden, B.N.; Troost, F.J.; Possemiers, S.; Stevens, Y.; Masclee, A.A.; Maastricht University, Maastricht, The Netherlands. Unpublished work. 2019.
93. Cueva, C.; Sánchez-Patán, F.; Monagas, M.; Walton, G.E.; Gibson, G.R.; Martín-Álvarez, P.J.; Bartolome, B.; Moreno-Arribas, M.V. In vitro fermentation of grape seed flavan-3-ol fractions by human faecal microbiota: Changes in microbial groups and phenolic metabolites. *FEMS Microbiol. Ecol.* **2013**, *83*, 792–805. [CrossRef]
94. Tzounis, X.; Vulevic, J.; Kuhnle, G.G.C.; George, T.; Leonczak, J.; Gibson, G.R.; Kwik-Uribe, C.; Spencer, J.P.M. Flavanol monomer-induced changes to the human faecal microflora. *Br. J. Nutr.* **2008**, *99*, 782–792. [CrossRef]
95. Unno, T.; Sakuma, M.; Mitsuhashi, S. Effect of Dietary Supplementation of (-)-Epigallocatechin Gallate on Gut Microbiota and Biomarkers of Colonic Fermentation in Rats. *J. Nutr. Sci. Vitaminol.* **2014**, *60*, 213–219. [CrossRef]
96. Cheng, M.; Zhang, X.; Miao, Y.; Cao, J.; Wu, Z.; Weng, P. The modulatory effect of (-)-epigallocatechin 3-O-(3-O-methyl) gallate (EGCG3"Me) on intestinal microbiota of high fat diet-induced obesity mice model. *Food Res. Int.* **2017**, *92*, 9–16. [CrossRef]
97. Lee, H.C.; Jenner, A.M.; Low, C.S.; Lee, Y.K. Effect of tea phenolics and their aromatic fecal bacterial metabolites on intestinal microbiota. *Microbiol. Res.* **2006**, *157*, 876–884. [CrossRef] [PubMed]
98. Stapleton, P.D.; Shah, S.; Ehlert, K.; Hara, Y.; Taylor, P.W. The beta-lactam-resistance modifier (-)-epicatechin gallate alters the architecture of the cell wall of *Staphylococcus aureus*. *Microbiology* **2007**, *153*, 2093–2103. [CrossRef] [PubMed]
99. Egert, S.; Wolffram, S.; Bosy-Westphal, A.; Boesch-Saadatmandi, C.; Wagner, A.E.; Frank, J.; Rimbach, G.; Mueller, M.J. Daily Quercetin Supplementation Dose-Dependently Increases Plasma Quercetin Concentrations in Healthy Humans. *J. Nutr.* **2008**, *138*, 1615–1621. [CrossRef] [PubMed]
100. Costa, L.G.; Garrick, J.M.; Roque, P.J.; Pellacani, C. Mechanisms of neuro- protection by quercetin: Counteracting oxidative stress and more. *Oxid. Med. Cell. Longev.* **2016**, *7*, 2986796. [CrossRef]
101. Lin, R.; Piao, M.; Song, Y. Dietary Quercetin Increases Colonic Microbial Diversity and Attenuates Colitis Severity in *Citrobacter rodentium*-Infected Mice. *Front. Microbiol.* **2019**, *10*, 1092–1100. [CrossRef] [PubMed]
102. Nie, J.; Zhang, L.; Zhao, G.; Du, X. Quercetin reduces atherosclerotic lesions by altering the gut microbiota and reducing atherogenic lipid metabolites. *J. Appl. Microbiol.* **2019**, *127*, 1824–1834. [CrossRef]
103. Firrman, J.; Liu, L.; Zhang, L.; Argoty, G.A.; Wang, M.; Tomasula, P.; Kobori, M.; Pontious, S.; Xiao, W. The effect of quercetin on genetic expression of the commensal gut microbes *Bifidobacterium catenulatum*, *Enterococcus caccae* and *Ruminococcus gauvreauii*. *Anaerobe* **2016**, *42*, 130–141. [CrossRef]
104. Etxeberria, U.; Ariasc, N.; Boquéd, N.; Macarullac, M.T.; Portilloc, M.P.; Martíneza, J.A.; Milagroa, F.I. Reshaping faecal gut microbiota composition by the intake of trans-resveratrol and quercetin in high-fat sucrose diet-fed rats. *J. Nutr. Biochem.* **2015**, *26*, 651–660. [CrossRef]
105. Alam, W.; Khan, H.; Shah, M.A.; Cauli, O.; Saso, L. Kaempferol as a Dietary Anti-Inflammatory Agent: Current Therapeutic Standing. *Molecules* **2020**, *25*, 4073. [CrossRef]
106. Aa, L.; Fei, F.; Qi, Q. Rebalancing of the gut flora and microbial metabolism is responsible for the anti-arthritis effect of kaempferol. *Acta Pharm. Sin.* **2019**, *41*, 73–81. [CrossRef]
107. Kawabata, K.; Sugiyama, Y.; Sakano, T.; Ohigashi, H. Flavonols enhanced production of anti-inflammatory substance(s) by *Bifidobacterium adolescentis*: Prebiotic actions of galangin, quercetin, and fisetin. *Biofactors* **2013**, *39*, 422–429. [CrossRef] [PubMed]
108. Jeong, J.W.; Lee, H.H.; Han, M.H.; Kim, G.Y.; Kim, W.J.; Choi, Y.H. Anti-inflammatory effects of genistein via suppression of the toll-like receptor 4-mediated signaling pathway in lipopolysaccharide-stimulated BV2 microglia. *Chem. Biol. Interact.* **2014**, *212*, 30–39. [CrossRef] [PubMed]
109. Lu, C.L.; Zhu, Y.F.; Hu, M.M. Optimization of astilbin extraction from the rhizome of Smilax glabra, and evaluation of its anti-inflammatory effect and probable underlying mechanism in lipopolysaccharide-induced RAW264.7 macrophages. *Molecules* **2015**, *20*, 625. [CrossRef] [PubMed]
110. Pabich, M.; Materska, M. Biological Effect of Soy Isoflavones in the Prevention of Civilization Diseases. *Nutrients* **2019**, *11*, 1660. [CrossRef]
111. Izaguirre, J.K.; Barañano, L.; Castañón, S.; Alkorta, I.; Quirós, L.M.; Garbisu, C. Optimization of the Bioactivation of Isoflavones in Soymilk by Lactic Acid Bacteria. *Processes* **2021**, *9*, 963. [CrossRef]
112. Landete, J.M.; Arqués, J.; Medina, M.; Gaya, P.; de Las Rivas, B.; Muñoz, R. Bioactivation of phytoestrogens: Intestinal bacteria and health. *Crit. Rev. Food Sci. Nutr.* **2016**, *56*, 1826–1843. [CrossRef]
113. Kim, M.; Han, J. Isoflavone metabolism by human intestinal bacteria. *Planta Med.* **2016**, *81* (Suppl. 1), S1–S381. [CrossRef]
114. Kolátorová, L.; Lapcik, O.; Starka, L. Phytoestrogens and the intestinal microbiome. *Physiol. Res.* **2018**, *67*, 401–408. [CrossRef]
115. Yoshikata, R.; Myint, K.Z.; Ohta, H.; Ishigaki, Y. Inter-relationship between diet, lifestyle habits, gut microflora, and the equol-producer phenotype: Baseline findings from a placebo-controlled intervention trial. *Menopause* **2018**, *26*, 273–285. [CrossRef]
116. Iino, C.; Shimoyama, T.; Iino, K.; Yokoyama, Y.; Chinda, D.; Sakuraba, H. Daidzein intake is associated with equol producing status through an increase in the intestinal bacteria responsible for equol production. *Nutrients* **2019**, *11*, 433. [CrossRef]
117. Swallah, M.S.; Fan, H.; Wang, S.; Yu, H.; Piao, C. Prebiotic Impacts of Soybean Residue (Okara) on Eubiosis/Dysbiosis Condition of the Gut and the Possible Effects on Liver and Kidney Functions. *Molecules* **2021**, *26*, 326. [CrossRef] [PubMed]

118. Basson, A.R.; Ahmed, S.; Almutairi, R.; Seo, B.; Cominelli, F. Regulation of Intestinal Inflammation by Soybean and Soy-Derived Compounds. *Foods* **2021**, *10*, 774. [CrossRef] [PubMed]
119. Paul, B.; Royston, K.J.; Li, Y.; Stoll, M.L.; Skibola, C.F.; Wilson, L.S. Impact of genistein on the gut microbiome of humanized mice and its role in breast tumor inhibition. *PLoS ONE* **2017**, *12*, e0189756. [CrossRef]
120. Cross, T.L.; Zidon, T.M.; Welly, R.J.; Park, Y.M.; Britton, S.L.; Koch, L.G. Soy improves cardiometabolic health and cecal microbiota in female low-fit rats. *Sci. Rep.* **2017**, *7*, 9261–9276. [CrossRef] [PubMed]
121. Vázquez, L.; Flórez, A.B.; Guadamuro, L.; Mayo, B. Effect of Soy Isoflavones on Growth of Representative Bacterial Species from the Human Gut. *Nutrients* **2017**, *9*, 727. [CrossRef]
122. Nakatsu, C.H.; Armstrong, A.; Clavijo, A.P.; Martin, B.R.; Barnes, S.; Weaver, C.M. Fecal bacterial community changes associated with isoflavone metabolites in postmenopausal women after soy bar consumption. *PLoS ONE* **2014**, *9*, e108924. [CrossRef]
123. Clavel, T.; Fallani, M.; Lepage, P.; Levenez, F.; Mathey, J.; Rochet, V.; Sérézat, M.; Sutren, M.; Henderson, G.; Bennetau-Pelissero, C. Isoflavones and functional foods alter the dominant intestinal microbiota in postmenopausal women. *J. Nutr.* **2005**, *135*, 2786–2792. [CrossRef]
124. Kasparovska, J.; Pecinkova, M.; Dadakova, K.; Krizova, L.; Hadrova, S.; Lexa, M.; Lochman, J.; Kasparovsky, T. Effects of isoflavone-enriched feed on the rumen microbiota in dairy cows. *PLoS ONE* **2016**, *11*, e0154642. [CrossRef]
125. Guadamuro, L.; Delgado, S.; Redruello, B.; Florez, A.B.; Suarez, A.; Martínez-Camblor, P.; Mayo, B. Equol status and changes' in fecal microbiota in menopausal women receiving long-term treatment for menopause symptoms with a soy-isoflavone concentrate. *Front. Microbiol.* **2015**, *6*, 777. [CrossRef]
126. Huang, H.; Krishnan, H.B.; Pham, Q.; Yu, L.L.; Wang, T.T.Y. Soy and Gut Microbiota: Interaction and Implication for Human Health. *J. Agric. Food Chem.* **2016**, *64*, 8695–8709. [CrossRef]
127. Cady, N.; Peterson, S.R.; Freedman, S.N.; Mangalam, A.K. Beyond Metabolism: The Complex Interplay Between Dietary Phytoestrogens, Gut Bacteria, and Cells of Nervous and Immune Systems. *Front. Neurol.* **2020**, *11*, 150. [CrossRef] [PubMed]
128. Igwe, E.O.; Charlton, K.E.; Probst, Y.C.; Kent, K.; Netzel, M.E. A systematic literature review of the effect of anthocyanins on gut microbiota populations. *J. Hum. Nutr. Diet.* **2018**, *32*, 53–62. [CrossRef] [PubMed]
129. Mazza, G.; Miniati, E. *Anthocyanins in Fruits, Vegetables, and Grains*; CRC Press: Boca Raton, FL, USA, 1993; pp. 6–23.
130. Andersen, Ø.; Jordheim, M. Basic anthocyanin chemistry and dietary sources. In *Anthocyanins in Health and Disease*, 1st ed.; Wallace, T., Giusti, M., Eds.; CRC Press: New York, NY, USA, 2014; Volume 1, pp. 13–89. ISBN 9781138198685.
131. Faria, A.; Fernandes, I.; Norberto, S.; Mateus, N.; Calhau, C. Interplay between Anthocyanins and Gut Microbiota. *J. Agric. Food Chem.* **2014**, *62*, 6898–6902. [CrossRef] [PubMed]
132. Hidalgo, M.; Oruna-Concha, M.J.; Kolida, S.; Walton, G.E.; Kallithraka, S.; Spencer, J.P.E.; Gibson, G.R.; de Pascual-Teresa, S. Metabolism of anthocyanins by human gut microflora and their influence on gut bacterial growth. *J. Agric. Food Chem.* **2012**, *60*, 3882–3890. [CrossRef]
133. Boto-Ordonez, M.; Urpi-Sarda, M.; Queipo-Ortu, M.I.; Tulipani, S.; Tinahones, F.J.; Andres-Lacueva, C. High levels of bifidobacteria are associated with increased levels of anthocyanin microbial metabolites: A randomized clinical trial. *Food Funct.* **2014**, *5*, 1932–1938. [CrossRef]
134. Sun, H.; Zhang, P.; Zhu, Y.; Lou, Q.; He, S. Antioxidant and prebiotic activity of five peonidin-based anthocyanins extracted from purple sweet potato (*Ipomoea batatas* (L.) lam). *Sci. Rep.* **2018**, *8*, 5018–5030. [CrossRef]
135. Chen, L.; Jiang, B.; Zhong, C.; Guo, J.; Zhang, L.; Mu, T.; Zhang, Q.; Bi, X. Chemoprevention of colorectal cancer by black raspberry anthocyanins involved the modulation of gut microbiota and SFRP2 demethylation. *Carcinogenesis* **2018**, *39*, 471–481. [CrossRef]
136. Zhu, Y.; Sun, H.; He, S.; Lou, Q.; Yu, M.; Tang, M.; Tu, L. Metabolism and prebiotics activity of anthocyanins from black rice (*Oryza sativa* L.) in vitro. *PLoS ONE* **2018**, *13*, e0195754. [CrossRef]
137. Zhang, X.; Yang, Y.; Wu, Z. The modulatory effect of anthocyanins from purple sweet potato on human intestinal microbiota in vitro. *J. Agric. Food Chem.* **2016**, *64*, 2582–2590. [CrossRef]
138. Sanchez-Patan, F.; Cueva, C.; Monagas, M.; Walton, G.E.; Gibson, G.R.; Quintanilla-Lopez, J.E.; Lebron-Aguilar, R.; MartinAlvarez, P.J.; Moreno-Arribas, M.V.; Bartolome, B. In vitro fermentation of a red wine extract by human gut microbiota: Changes in microbial groups and formation of phenolic metabolites. *J. Agric. Food Chem.* **2012**, *60*, 2136–2147. [CrossRef]
139. Queipo-Ortuno, M.I.; Boto-Ordonez, M.; Murri, M.; GomezZumaquero, J.M.; Clemente-Postigo, M.; Estruch, R.; Cardona Diaz, F.; Andres-Lacueva, C.; Tinahones, F.J. Influence of red wine polyphenols and ethanol on the gut microbiota ecology and biochemical biomarkers. *Am. J. Clin. Nutr.* **2012**, *95*, 1323–1334. [CrossRef] [PubMed]
140. Vendrame, S.; Guglielmetti, S.; Riso, P.; Arioli, S.; Klimis-Zacas, D.; Porrini, M. Six-week consumption of a wild blueberry powder drink increases bifidobacteria in the human gut. *J. Agric. Food Chem.* **2011**, *59*, 12815–12820. [CrossRef] [PubMed]
141. Overall, J.; Bonney, S.A.; Wilson, M. Metabolic effects of berries with structurally diverse anthocyanins. *Int. J. Mol. Sci.* **2017**, *18*, 422. [CrossRef] [PubMed]
142. Hester, S.N.; Mastaloudis, A.; Gray, R.; Antony, J.M.; Evans, M.; Wood, S.M. Efficacy of an Anthocyanin and Prebiotic Blend on Intestinal Environment in Obese Male and Female Subjects. *J. Nutr. Metab.* **2018**, *2018*, 7497260. [CrossRef]
143. Flores, G.; Ruiz del Castillo, M.L.; Costabile, A.; Klee, A.; Bigetti Guergoletto, K.; Gibson, G.R. *In vitro* fermentation of anthocyanins encapsulated with cyclodextrins: Release, metabolism and influence on gut microbiota growth. *J. Funct. Foods.* **2015**, *16*, 50–57. [CrossRef]

144. Verediano, T.A.; Martino, H.S.D.; Paes, M.C.D.; Tako, E. Effects of Anthocyanin on Intestinal Health: A Systematic Review. *Nutrients* **2021**, *13*, 1331. [CrossRef]
145. Zhang, Z.; Chen, Y.; Xiang, L.; Wang, Z.; Xiao, G.C.; Hu, J. Effect of curcumin on the diversity of gut microbiota in ovariectomized rats. *Nutrients* **2017**, *9*, 1146. [CrossRef]
146. Ohno, M.; Nishida, A.; Sugitani, Y. Nanoparticle curcumin ameliorates experimental colitis via modulation of gut microbiota and induction of regulatory T cells. *PLoS ONE* **2017**, *12*, e0185999. [CrossRef]
147. McFadden, R.M.; Larmonier, C.B.; Shehab, K.W. The role of curcumin in modulating colonic microbiota during colitis and colon cancer prevention. *Inflamm. Bowel Dis.* **2015**, *21*, 2483–2494. [CrossRef]
148. Feng, W.; Wang, H.; Zhang, P. Modulation of gut microbiota contributes to curcumin-mediated attenuation of hepatic steatosis in rats. *Biochim. Biophys. Acta* **2017**, *1861*, 1801–1812. [CrossRef]
149. Bereswill, S.; Munoz, M.; Fischer, A. Anti-inflammatory effects of resveratrol, curcumin and simvastatin in acute small intestinal inflammation. *PLoS ONE* **2010**, *5*, e15099. [CrossRef] [PubMed]
150. Peterson, C.T.; Vaughn, A.R.; Sharma, V. Effects of turmeric and curcumin dietary supplementation on human gut microbiota: A double-blind, randomized, placebo controlled pilot study. *J. Evid.-Based Integr. Med.* **2018**, *23*, 1–8. [CrossRef] [PubMed]
151. Kumar, N.; Goel, N. Phenolic acids: Natural versatile molecules with promising therapeutic applications. *Biotechnol. Rep.* **2019**, *24*, e00370. [CrossRef] [PubMed]
152. Juurlink, B.H.; Azouz, H.J.; Aldalati, A.M.; AlTinawi, B.M.; Ganguly, P. Hydroxybenzoic acid isomers and the cardiovascular system. *Nutr. J.* **2014**, *13*, 63–73. [CrossRef]
153. Yadav, E. Protective Effect of Protocatechuic Acid Rich Fraction of *Trianthema portulacastrum* Against Collagen Induced Rheumatoid Arthritis via Gut Microbiota Modulation. *Ann. Rheum. Dis.* **2019**, *78*, 1098. [CrossRef]
154. Wang, Y.; Wang, Y.; Wang, B.; Mei, X.; Jiang, S.; Li, W. Protocatechuic acid improved growth performance, meat quality, and intestinal health of Chinese yellow-feathered broilers. *Poult. Sci.* **2019**, *98*, 3138–3149. [CrossRef]
155. El-Seedi, H.R.; El-Said, A.M.A.; Khalifa, S.A.M.; Göransson, U.; Bohlin, L.; Borg-Karlson, A.K. Biosynthesis, natural sources, dietary intake, pharmacokinetic properties, and biological activities of hydroxycinnamic acids. *J. Agric. Food Chem.* **2012**, *60*, 10877–10895. [CrossRef]
156. Calinoiu, L.F.; Vodnar, D.C. Whole grains and phenolic acids: A review on bioactivity, functionality, health benefits and bioavailability. *Nutrients* **2018**, *10*, 1615. [CrossRef]
157. Gill, S.R.; Pop, M.; DeBoy, R.T.; Eckburg, P.B.; Turnbaugh, P.J.; Samuel, B.S.; Gordon, J.I.; Relman, D.A.; Fraser-Liggett, C.M.; Nelson, K.E. Metagenomic Analysis of the Human Distal Gut Microbiome. *Science* **2006**, *312*, 1355–1359. [CrossRef]
158. Flint, H.J.; Scott, K.P.; Duncan, S.H.; Louis, P.; Forano, E. Microbial degradation of complex carbohydrates in the gut. *Gut Microbes* **2012**, *3*, 289–306. [CrossRef]
159. Leonard, W.; Zhang, P.; Ying, D.; Fang, Z. Hydroxycinnamic acids on gut microbiota and health. *Compr. Rev. Food Sci. Food Saf.* **2021**, *20*, 710–737. [CrossRef] [PubMed]
160. Zhang, Z.; Wu, X.; Cao, S.; Wang, L.; Wang, D.; Yang, H. Caffeic acid ameliorates colitis in association with increased *Akkermansia* population in the gut microbiota of mice. *Oncotarget* **2016**, *7*, 31790–31799. [CrossRef] [PubMed]
161. Mills, C.E.; Tzounis, X.; Oruna-Concha, M.J.; Mottram, D.S.; Gibson, G.R.; Spencer, J.P.E. In vitro colonic metabolism of coffee and chlorogenic acid results in selective changes in human faecal microbiota growth. *Br. J. Nutr.* **2015**, *113*, 1220–1227. [CrossRef] [PubMed]
162. Wang, Z.; Lam, K.L.; Hu, J.; Ge, S.; Zhou, A.; Zheng, B. Chlorogenic acid alleviates obesity and modulates gut microbiota in high-fat-fed mice. *Food Sci. Nutr.* **2019**, *7*, 579–588. [CrossRef] [PubMed]
163. Ma, Y.; Chen, K.; Lv, L.; Wu, S.; Guo, Z. Ferulic acid ameliorates nonalcoholic fatty liver disease and modulates the gut microbiota composition in high-fat diet fed ApoE-/- mice. *Biomed. Pharmacother.* **2019**, *113*, 108753. [CrossRef] [PubMed]
164. Yang, C.; Deng, Q.; Xu, J.; Wang, X.; Hu, C.; Tang, H. Sinapic acid and resveratrol alleviate oxidative stress with modulation of gut microbiota in high-fat diet-fed rats. *Food Res. Int.* **2019**, *116*, 1202–1211. [CrossRef]
165. Mukhopadhya, I.; Hansen, R.; El-Omar, E.M.; Hold, G.L. IBD-what role do Proteobacteria play? *Nat. Rev. Gastroenterol. Hepatol.* **2012**, *9*, 219–230. [CrossRef]
166. Rizzatti, G.; Lopetuso, L.R.; Gibiino, G.; Binda, C.; Gasbarrini, A. Proteobacteria: A common factor in human diseases. *BioMed Res. Int.* **2017**, *2017*, 9351507. [CrossRef]
167. Fresco, P.; Borges, F.; Diniz, C.; Marques, M.P.M. New insights on the anticancer properties of dietary polyphenols. *Med. Res. Rev.* **2006**, *26*, 747–766. [CrossRef]
168. Jaimes, J.D.; Jarosova, V.; Vesely, O.; Mekadim, C.; Mrazek, J.; Marsik, P.; Killer, J.; Smejkal, K.; Kloucek, P.; Havlik, J. Effect of Selected Stilbenoids on Human Fecal Microbiota. *Molecules* **2019**, *24*, 744. [CrossRef]
169. Perrone, D.; Fuggetta, M.P.; Ardito, F.; Cottarelli, A.; De Filippis, A.; Ravagnan, G.; De Maria, S.; Lo Muzio, L. Resveratrol (3,5,4′-trihydroxystilbene) and its properties in oral diseases. *Exp. Ther. Med.* **2017**, *14*, 3–9. [CrossRef] [PubMed]
170. Hüsken, A.; Baumert, A.; Milkowski, C.; Becker, H.C.; Strack, D.; Möllers, C. Resveratrol glucoside (Piceid) synthesis in seeds of transgenic oilseed rape (*Brassica napus* L.). *Theor. Appl. Genet.* **2005**, *111*, 1553–1562. [CrossRef] [PubMed]
171. Boccellino, M.; D'Angelo, S. Anti-Obesity Effects of Polyphenol Intake: Current Status and Future Possibilities. *Int. J. Mol. Sci.* **2020**, *21*, 5642. [CrossRef]

172. Springer, M.; Moco, S. Resveratrol and Its Human Metabolites-Effects on Metabolic Health and Obesity. *Nutrients* **2019**, *11*, 143. [CrossRef] [PubMed]
173. Sung, M.M.; Kim, T.T.; Denou, E.; Soltys, C.M.; Hamza, S.M.; Byrne, N.J.; Masson, G.; Park, H.; Wishart, D.S.; Madsen, K.L. Improved glucose homeostasis in obese mice treated with resveratrol is associated with alterations in the gut microbiome. *Diabetes* **2017**, *66*, 418–425. [CrossRef] [PubMed]
174. Kim, T.T.; Parajuli, N.; Sung, M.M.; Bairwa, S.C.; Levasseur, J.; Soltys, C.M.; Wishart, D.S.; Madsen, K.; Schertzer, J.D.; Dyck, J.R.B. Fecal transplant from resveratrol-fed donors improves glycaemia and cardiovascular features of the metabolic syndrome in mice. *Am. J. Physiol. Endocrinol. Metab.* **2018**, *315*, 511–519. [CrossRef] [PubMed]
175. Chen, M.L.; Yi, L.; Zhang, Y. Resveratrol attenuates trimethylamine-Noxide (TMAO)-induced atherosclerosis by regulating TMAO synthesis and bile acid metabolism via remodeling of the gut microbiota. *mBio* **2016**, *7*, 2210–2215. [CrossRef]
176. Tain, Y.L.; Lee, W.C.; Wu, K.L.H.; Leu, S.; Chan, J.Y.H. Resveratrol prevents the development of hypertension programmed by maternal plus post-weaning high-fructose consumption through modulation of oxidative stress, nutrient-sensing signals, and gut microbiota. *Mol. Nutr. Food Res.* **2018**, *62*, 1800066. [CrossRef]
177. Most, J.; Penders, J.; Lucchesi, M.; Goossens, G.H.; Blaak, E.E. Gut microbiota composition in relation to the metabolic response to 12-week combined polyphenol supplementation in overweight men and women. *Eur. J. Clin. Nutr.* **2017**, *71*, 1040–1045. [CrossRef]
178. Larrosa, M.; Luceri, C.; Vivoli, E. Polyphenol metabolites from colonic microbiota exert anti-inflammatory activity on different inflammation models. *Mol. Nutr. Food Res.* **2009**, *53*, 1044–1054. [CrossRef]
179. Setoguchi, Y.; Oritani, Y.; Ito, R.; Inagaki, H.; Maruki-Uchida, H.; Ichiyanagi, T.; Ito, T. Absorption and metabolism of piceatannol in rats. *J. Agric. Food Chem.* **2014**, *62*, 2541–2548. [CrossRef] [PubMed]
180. Buiarelli, F.; Coccioli, F.; Jasionowska, R.; Merolle, M.; Terracciano, A. Analysis of some stilbenes in Italian wines by liquid chromatography/tandem mass spectrometry. *Rapid Commun. Mass Spectrom.* **2007**, *21*, 2955–2964. [CrossRef] [PubMed]
181. Tung, Y.C.; Lin, Y.H.; Chen, H.J.; Chou, S.C.; Cheng, A.C.; Kalyanam, N.; Ho, C.T.; Pan, M.H. Piceatannol Exerts Anti-Obesity Effects in C57BL/6 Mice through Modulating Adipogenic Proteins and Gut Microbiota. *Molecules* **2016**, *21*, 1419. [CrossRef] [PubMed]
182. Peterson, J.; Dwyer, J.; Adlercreutz, H.; Scalbert, A.; Jacques, P.; McCullough, M.L. Dietary lignans: Physiology and potential for cardiovascular disease risk reduction. *Nutr. Rev.* **2010**, *68*, 571–603. [CrossRef]
183. Rodríguez-García, C.; Sánchez-Quesada, C.; Toledo, E.; Delgado-Rodríguez, M.; Gaforio, J.J. Naturally Lignan-Rich Foods: A Dietary Tool for Health Promotion? *Molecules* **2019**, *24*, 917. [CrossRef]
184. Hullar, M.A.; Lancaster, S.M.; Li, F.; Tseng, E.; Beer, K.; Atkinson, C.; Wähälä, K.; Copeland, W.K.; Randolph, T.W.; Newton, K.M. Enterolignan-producing phenotypes are associated with increased gut microbial diversity and altered composition in premenopausal women in the United States. *Cancer Epidemiol. Biomark. Prev.* **2015**, *24*, 546–554. [CrossRef]
185. Holma, R.; Kekkonen, R.A.; Hatakka, K.; Poussa, T.; Vapaatalo, H.; Adlercreutz, H.; Korpela, R. Low serum enterolactone concentration is associated with low colonic Lactobacillus Enterococcus counts in men but is not affected by a synbiotic mixture in a randomised, placebo-controlled, double-blind, cross-over intervention study. *Br. J. Nutr.* **2014**, *111*, 301–309. [CrossRef]
186. Lagkouvardos, I.; Kläring, K.; Heinzmann, S.S.; Platz, S.; Scholz, B.; Engel, K.; Schmitt-Kopplin, P.; Haller, D.; Rohn, S.; Skurk, T. Gut metabolites and bacterial community networks during a pilot intervention study with flaxseeds in healthy adult men. *Mol. Nutr. Food Res.* **2015**, *59*, 1614–1628. [CrossRef]
187. Corona, G.; Kreimes, A.; Barone, M.; Turroni, S.; Brigidi, P.; Keleszade, E.; Costabile, A. Impact of lignans in oilseed mix on gut microbiome composition and enterolignan production in younger healthy and premenopausal women: An in vitro study. *Microb. Cell Fact.* **2020**, *19*, 82–96. [CrossRef]
188. Hassanpour, S.; Maheri-Sis, N.; Eshratkhah, B.; Mehmandar, F.B. Plants and secondary metabolites (tannins): A review. *Int. J. For. Soil Eros.* **2011**, *1*, 47–53.
189. Bialonska, D.; Kasimsetty, S.G.; Schrader, K.K.; Ferreira, D. The effect of pomegranate (*Punica granatum* L.) byproducts and ellagitannins on the growth of human gut bacteria. *J. Agric. Food Chem.* **2009**, *57*, 8344–8349. [CrossRef] [PubMed]
190. Bialonska, D.; Ramnani, P.; Kasimsetty, S.G.; Muntha, K.R.; Gibson, G.R.; Ferreira, D. The influence of pomegranate by-product and punicalagins on selected groups of human intestinal microbiota. *Int. J. Food Microb.* **2010**, *140*, 175–182. [CrossRef] [PubMed]
191. Larrosa, M.; Gonzalez-Sarrıas, A.; Yanez-Gascon, M.J.; Selma, M.V.; Azorın-Ortuno, M.; Toti, S.; Tomas-Barberan, F.A.; Dolara, P.; Espın, J.C. Anti-inflammatory properties of a pomegranate extract and its metabolite urolithin-A in a colitis rat model and the effect of colon inflammation on the phenolic metabolism. *J. Nutr. Biochem.* **2010**, *21*, 717–725. [CrossRef] [PubMed]
192. Li, Z.; Henning, S.M.; Lee, R.P.; Lu, Q.Y.; Summanen, P.H.; Thames, G.; Corbett, K.; Downes, J.; Tsena, C.H.; Finegold, S.M.; et al. Pomegranate extract induces metabolite formation and changes stool microbiota in healthy volunteers. *Food Funct.* **2015**, *6*, 2487–2495. [CrossRef]
193. Samanta, S.; Giri, S.; Parua, S.; Nandi, D.K.; Pati, B.R.; Mondal, K.C. Impact of tannic acid on the gastrointestinal microflora. *Microb. Ecol. Health Dis.* **2004**, *16*, 32–34. [CrossRef]
194. Smith, A.H.; Mackie, R.I. Effect of Condensed Tannins on Bacterial Diversity and Metabolic Activity in the Rat Gastrointestinal Tract. *Appl. Environ. Microbiol.* **2004**, *70*, 1104–1115. [CrossRef]
195. Massot-Cladera, M.; Perez-Berezo, T.; Franch, A.; Castell, M.; Perez-Cano, F.J. Cocoa modulatory effect on rat faecal microbiota and colonic crosstalk. *Arch. Biochem. Biophys.* **2012**, *527*, 105–112. [CrossRef]

196. Koutsos, A.; Lima, M.; Conterno, L.; Gasperotti, M.; Bianchi, M.; Fava, F.; Vrhovsek, U.; Lovegrove, J.A.; Tuohy, K.M. Effects of commercial apple varieties on human gut microbiota composition and metabolic output using an in vitro colonic model. *Nutrients* **2017**, *9*, 533. [CrossRef]
197. Tao, W.; Zhang, Y.; Shen, X.; Cao, Y.; Shi, J.; Ye, X.; Chen, S. Rethinking the mechanism of the health benefits of proanthocyanidins: Absorption, metabolism, and interaction with gut microbiota. *Compr. Rev. Food Sci. Food Saf.* **2019**, *18*, 971–985. [CrossRef]
198. Yamaguchi, Y.; Kumagai, H. Characteristics, biosynthesis, decomposition, metabolism and functions of the garlic odour precursor, *S*-allyl-L-cysteine sulfoxide. *Exp. Ther. Med.* **2020**, *19*, 1528–1535. [CrossRef]
199. Amagase, H. Clarifying the real bioactive constituents of garlic. *J. Nutr.* **2006**, *136*, 716S–725S. [CrossRef] [PubMed]
200. Bhatwalkar, S.B.; Mondal, R.; Krishna, S.B.N.; Adam, J.K.; Govender, P.; Anupam, R. Antibacterial Properties of Organosulfur Compounds of Garlic (*Allium sativum*). *Front. Microbiol.* **2021**, *12*, 613077. [CrossRef] [PubMed]
201. Zhai, B.; Zhang, C.; Sheng, Y.; Zhao, C.; He, X.; Xu, W.; Huang, K.; Luo, Y. Hypoglycemic and hypolipidemic effect of S-allyl-cysteine sulfoxide (alliin) in DIO mice. *Sci. Rep.* **2018**, *8*, 3527. [CrossRef]
202. Chen, K.; Xie, K.; Liu, Z.; Nakasone, Y.; Sakao, K.; Hossain, A.; Hou, D.X. Preventive Effects and Mechanisms of Garlic on Dyslipidemia and Gut Microbiome Dysbiosis. *Nutrients* **2019**, *11*, 1225. [CrossRef] [PubMed]
203. Rinninella, E.; Mele, M.C.; Merendino, N.; Cintoni, M.; Anselmi, G.; Caporossi, A.; Gasbarrini, A.; Minnella, A.M. The Role of Diet, Micronutrients and the Gut Microbiota in Age-Related Macular Degeneration: New Perspectives from the Gut–Retina Axis. *Nutrients* **2018**, *10*, 1677. [CrossRef]
204. Lyu, Y.; Wu, L.; Wang, F.; Shen, X.; Lin, D. Carotenoid supplementation and retinoic acid in immunoglobulin A regulation of the gut microbiota dysbiosis. *Exp. Biol. Med.* **2018**, *243*, 613–620. [CrossRef]
205. Palczewski, G.; Widjaja-Adhi, M.A.K.; Amengual, J.; Golczak, M.; von Lintig, J. Genetic dissection in a mouse model reveals interactions between carotenoids and lipid metabolism. *J. Lipid Res.* **2016**, *57*, 1684–1695. [CrossRef]
206. Bohn, T. Bioactivity of carotenoids—Chasms of knowledge. *Int. J. Vitam. Nutr. Res.* **2017**, *10*, 5–9. [CrossRef]
207. Stachowiak, B.; Szulc, P. Astaxanthin for the Food Industry. *Molecules* **2021**, *26*, 2666. [CrossRef]
208. Bennedsen, M.; Wang, X.; Willén, R.; Wadstrom, T.; Andersen, L.P. Treatment of *H. pylori* infected mice with antioxidant astaxanthin reduces gastric inflammation, bacterial load and modulates cytokine release by splenocytes. *Immunol. Lett.* **1999**, *70*, 185–189. [CrossRef]
209. Nishikawa, Y.; Minenaka, Y.; Ichimura, M.; Tatsumi, K.; Nakamoto, T.; Urabe, K. Effects of astaxanthin and vitamin C on the prevention of gastric ulcerations in stressed rats. *J. Nutr. Sci. Vitaminol.* **2005**, *51*, 135–141. [CrossRef] [PubMed]
210. Yonei, Y.; Yagi, M.; Nakamura, M.; Parengkuan, L.; Ogura, M.; Taira, T.; Asano, S.; Liu, H.H. Effects of Astaxanthin on Intestinal Microflora in Mice Fed a High-fat Diet. *Anti-Aging Med.* **2013**, *10*, 77–91.
211. Liu, H.; Liu, M.; Fu, X.; Zhang, Z.; Zhu, L.; Zheng, X.; Liu, J. Astaxanthin Prevents Alcoholic Fatty Liver Disease by Modulating Mouse Gut Microbiota. *Nutrients* **2018**, *10*, 1298. [CrossRef] [PubMed]
212. Giordano, E.; Quadro, L. Lutein, zeaxanthin and mammalian development: Metabolism, functions and implications for health. *Arch. Biochem. Biophys.* **2018**, *647*, 33–40. [CrossRef] [PubMed]
213. Molan, A.L.; Liu, Z.; Plimmer, G. Evaluation of the Effect of Blackcurrant Products on Gut Microbiota and on Markers of Risk for Colon Cancer in Humans. *Phytother. Res.* **2014**, *28*, 416–422. [CrossRef]
214. Wiese, M.; Bashmakov, Y.; Chalyk, N.; Nielsen, D.S.; Krych, Ł.; Kot, W.; Klochkov, V.; Pristensky, D.; Bandaletova, T.; Chernyshova, M.; et al. Prebiotic Effect of Lycopene and Dark Chocolate on Gut Microbiome with Systemic Changes in Liver Metabolism, Skeletal Muscles and Skin in Moderately Obese Persons. *BioMed Res. Int.* **2019**, *2019*, 4625279. [CrossRef]
215. Appeldoorn, M.M.; Vincken, J.P.; Gruppen, H.; Hollman, P.C.H. Procyanidin Dimers A1, A2, and B2 Are Absorbed without Conjugation or Methylation from the Small Intestine of Rats. *J. Nut.* **2009**, *139*, 1469–1473. [CrossRef]
216. Aravind, S.M.; Wichienchot, S.; Tsao, R.; Ramakrishnan, S.; Chakkaravarthi, S. Role of dietary polyphenols on gut microbiota, their metabolites and health benefits. *Food Res. Int.* **2021**, *142*, 110189. [CrossRef]
217. Hervert-Hernandez, D.; Goñi, I. Dietary polyphenols and human gut microbiota: A review. *Food Rev. Int.* **2011**, *27*, 154–169. [CrossRef]
218. Puupponen-Pimiä, R.; Nohynek, L.; Hartman-Schmidlin, S.; Kähkönen, M.; Heinonen, M.; Mata-Riihinen, K. Berry phenolics selectively inhibit the growth of intestinal pathogens. *J. Appl. Microbiol.* **2005**, *98*, 991–1000. [CrossRef]
219. Lillehoj, H.; Liu, Y.; Calsamiglia, S.; Fernandez-Miyakawa, M.E.; Chi, F.; Cravens, R.L.; Oh, S.; Gay, C.G. Phytochemicals as antibiotic alternatives to promote growth and enhance host health. *Vet. Res.* **2018**, *49*, 76–94. [CrossRef] [PubMed]
220. Piekarska-Radzik, L.; Klewicka, E. Mutual influence of polyphenols and *Lactobacillus* spp. bacteria in food: A review. *Eur. Food Res. Technol.* **2021**, *247*, 9–24. [CrossRef]
221. Sirk, T.W.; Brown, E.F.; Friedman, M.; Sum, A.K. Molecular binding of catechins to biomembranes: Relationship to biological activity. *J. Agric. Food Chem.* **2009**, *57*, 6720–6728. [CrossRef]
222. Sirk, T.W.; Friedman, M.; Brown, E.F. Molecular binding of black tea theaflavins to biological membranes: Relationship to bioactivities. *J. Agric. Food Chem.* **2011**, *59*, 3780–7787. [CrossRef]
223. González, J.E.; Keshavan, N.D. Messing with bacterial quorum sensing. *Microbiol. Mol. Biol. Rev.* **2006**, *70*, 859–875. [CrossRef] [PubMed]
224. Williams, P. Quorum sensing, communication and cross-kingdom signalling in the bacterial world. *Microbiology* **2007**, *153*, 3923–3938. [CrossRef]

225. Hubert, B.; Eberl, L.; Feucht, W.; Polster, J. Influence of polyphenols on bacterial biofilm formation and quorum-sensing. *Z. Naturforsch.* **2003**, *58*, 879–884. [CrossRef]
226. Monagas, M.; Urpi-Sarda, M.; Sánchez-Patán, F.; Llorach, R.; Garrido, I.; GómezCordovés, C. Insights into the metabolism and microbial biotransformation of dietary flavan-3-ols and the bioactivity of their metabolites. *Food Funct.* **2010**, *1*, 233–253. [CrossRef]
227. Tombola, F.; Campello, S.; De Luca, L.; Ruggiero, P.; Del Giudice, G.; Papini, E. Plant polyphenols inhibit VacA, a toxin secreted by the gastric pathogen Helicobacter pylori. *FEBS Lett.* **2003**, *543*, 184–189. [CrossRef]
228. Lin, Y.T.; Vattem, D.A.; Labbe, R.G.; Shetty, K. Inhibition of Helicobacter pylori and associated urease by oregano and cranberry phytochemical synergies. *Appl. Environ. Microbiol.* **2005**, *71*, 8558–8564. [CrossRef]
229. Cushnie, T.P.; Lamb, A.J. Antimicrobial activity of flavonoids. *Int. J. Antimicrob. Agents* **2005**, *26*, 343–356. [CrossRef] [PubMed]
230. Plaper, A.; Golob, M.; Hafner, I.; Oblak, M.; Solmajer, T.; Jerala, R. Characterization of quercetin binding site on DNA gyrase. *Biochem. Biophys. Res. Commun.* **2003**, *306*, 530–536. [CrossRef]
231. Gradišar, H.; Pristovsek, P.; Plaper, A.; Jerala, R. Green tea catechins inhibit bacterial DNA gyrase by interaction with its ATP binding site. *J. Med. Chem.* **2007**, *50*, 264–271. [CrossRef] [PubMed]
232. Park, K.M.; You, J.S.; Lee, H.Y.; Baek, N.I.; Hwang, J.K. Kuwanon, G: An antibacterial agent from the root bark of *Morus alba* against oral pathogens. *J. Ethnopharmacol.* **2003**, *84*, 181–185. [CrossRef]
233. Percival, R.S.; Devine, D.A.; Duggal, M.S.; Chartron, S.; Marsh, P.D. The effect of cocoa polyphenols on the growth, metabolism, and biofilm formation by *Streptococcus mutans* and *Streptococcus sanguinis*. *Eur. J. Oral Sci.* **2006**, *114*, 343–348. [CrossRef]
234. Prabu, G.R.; Gnanamani, A.; Sadulla, S. Guaijaverin—A plant flavonoid as potential antiplaque agent against *Streptococcus mutans*. *J. Appl. Microbiol.* **2006**, *101*, 487–495. [CrossRef]
235. Smith, A.H.; Zoetendal, E.; Mackie, R.I. Bacterial mechanisms to overcome inhibitory effects of dietary tannins. *Microb. Ecol.* **2005**, *50*, 197–205. [CrossRef]
236. Freestone, P.E.F.; Walton, N.J.; Haigh, R.; Lyte, M. Influence of dietary catechols on the growth of enteropathogenic bacteria. *Int. J. Food Microbiol.* **2007**, *119*, 159–169. [CrossRef]
237. Shapiro, K. Human-Animal Studies: Remembering the Past, Celebrating the Present, Troubling the Future. *Soc. Anim.* **2020**, *28*, 797–833. [CrossRef]
238. Ley, R.E.; Backhed, F.; Turnbaugh, P.; Lozupone, C.A.; Knight, R.D.; Gordon, J.I. Obesity alters gut microbial ecology. *Proc. Natl. Acad. Sci. USA* **2005**, *102*, 11070–11075. [CrossRef]
239. Hara, H.; Orita, N.; Hatano, S. Effect of tea polyphenols on fecal flora and fecal metabolic products of pigs. *J. Vet. Med. Sci.* **1995**, *57*, 45–49. [CrossRef] [PubMed]
240. Kang, N.J.; Lee, K.W.; Kim, B.H.; Bode, A.M.; Lee, H.J.; Heo, Y.S.; Boardman, L.; Limburg, P.; Lee, H.J.; Dong, Z. Coffee phenolic phytochemicals suppress colon cancer metastasis by targeting MEK and TOPK. *Carcinogenesis* **2011**, *32*, 921–928. [CrossRef] [PubMed]
241. Ishihara, N.; Chu, D.C.; Akachi, S.; Juneja, L.R. Improvement of intestinal microflora balance and prevention of digestive and respiratory organ diseases in calves by green tea extracts. *Livest. Prod. Sci.* **2001**, *68*, 217–229. [CrossRef]
242. Kafantaris, I.; Kotsampasi, B.; Christodoulou, V.; Kokka, E.; Kouka, P.; Terzopoulou, Z.; Gerasopoulos, K.; Stagos, D.; Mitsagga, C.; Giavasis, I.; et al. Grape pomace improves antioxidant capacity and faecal microflora of lambs. *J. Ani. Physiol. Ani. Nutr.* **2017**, *101*, e108–e121. [CrossRef]
243. Novoa-Garrido, M.; Rebours, C.; Aanensen, L.; Torp, T.; Lind, V.; Steinshamn, H. Effect of seaweed on gastrointestinal microbiota isolated from Norwegian White sheep. *Acta Agric. Scand. Sect. A Anim. Sci.* **2016**, *66*, 152–160. [CrossRef]
244. Dolara, P.; Luceri, C.; De Filippo, C.; Femia, A.P.; Giovannelli, L.; Caderni, G.; Cecchini, C.; Silvi, S.; Orpianesi, C.; Cresci, A. Red wine polyphenols influence carcinogenesis, intestinal microflora, oxidative damage and gene expression profiles of colonic mucosa in F344 rats. *Mutat. Res. Fundam. Mol. Mech. Mutagen.* **2005**, *591*, 237–246. [CrossRef]
245. Larrosa, M.; Yañéz-Gascón, M.J.; Selma, M.V.; González-Sarrías, A.; Toti, S.; Cerónm, J.J.; Tomás-Barberán, F.; Dolara, P.; Espín, J.C. Effect of a Low Dose of Dietary Resveratrol on Colon Microbiota, Inflammation and Tissue Damage in a DSS-Induced Colitis Rat Model. *J. Agric. Food Chem.* **2009**, *57*, 2211–2220. [CrossRef] [PubMed]
246. Wang, K.; Jin, X.; Li, Q.; Sawaya, A.C.H.F.; Le Leu, R.K.; Conlon, M.A.; Wu, L.; Hu, F. Propolis from Different Geographic Origins Decreases Intestinal Inflammation and *Bacteroides* spp. Populations in a Model of DSS-Induced Colitis. *Mol. Nutr. Food Res.* **2018**, *62*, e1800080. [CrossRef]
247. Lacombe, A.; Li, R.W.; Klimis-Zacas, D.; Kristo, A.S.; Tadepalli, S.; Krauss, E.; Young, R.; Wu, V.C. Lowbush wild blueberries have the potential to modify gut microbiota and xenobiotic metabolism in the rat colon. *PLoS ONE* **2013**, *8*, e67497. [CrossRef]
248. Sengottuvelan, M.; Nalini, N. Dietary supplementation of resveratrol suppresses colonic tumour incidence in 1,2-dimethylhydrazine-treated rats by modulating bio transforming enzymes and aberrant crypt foci development. *Br. J. Nutr.* **2006**, *96*, 145–153. [CrossRef]
249. Jayachandran, M.; Xiao, J.; Xu, B.A. Critical Review on Health Promoting Benefits of Edible Mushrooms through Gut Microbiota. *Int. J. Mol. Sci.* **2017**, *18*, 1934. [CrossRef] [PubMed]
250. Viveros, A.; Chamorro, S.; Pizarro, M.; Arija, I.; Centeno, C.; Brenes, A. Effects of dietary polyphenol-rich grape products on intestinal microflora and gut morphology in broiler chicks. *Poult. Sci.* **2011**, *90*, 566–578. [CrossRef]

251. Wang, K.; Wan, Z.; Ou, A.; Liang, X.; Guo, X.; Zhang, Z.; Wu, L.; Xue, X. Monofloral honey from a medical plant, Prunella Vulgaris, protected against dextran sulfate sodium-induced ulcerative colitis via modulating gut microbial populations in rats. *Food Funct.* **2019**, *10*, 3828–3838. [CrossRef] [PubMed]
252. Huws, S.A.; Creevey, C.J.; Oyama, L.B.; Mizrahi, I.; Denman, S.E.; Popova, M.; Muñoz-Tamayo, R.; Forano, E.; Waters, S.M.; Hess, M.; et al. Addressing global ruminant agricultural challenges through understanding the rumen microbiome: Past, present, and future. *Front. Microbiol.* **2018**, *9*, 2161. [CrossRef] [PubMed]
253. Sasson, G.; Ben-Shabat, S.K.; Seroussi, E.; Doron-Faigenboim, A.; Shterzer, N.; Yaacoby, S.; Miller, M.E.B.; White, B.A.; Halperin, E.; Mizrahi, I. Heritable bovine rumen bacteria are phylogenetically related and correlated with the cow's capacity to harvest energy from its feed. *mBio* **2017**, *8*, e00703-17. [CrossRef] [PubMed]
254. Tapio, I.; Snelling, T.J.; Strozzi, F.; Wallace, R.J. The ruminal microbiome associated with methane emissions from ruminant livestock. *J. Anim. Sci. Biotechnol.* **2017**, *8*, 7. [CrossRef] [PubMed]
255. Abbott, D.W.; Aasen, I.M.; Beauchemin, K.A.; Grondahl, F.; Gruninger, R.; Hayes, M.; Huws, S.; Kenny, D.A.; Krizsan, S.J.; Kirwan, S.F.; et al. Seaweed and seaweed bioactives for mitigation of enteric methane: Challenges and opportunities. *Animals* **2020**, *10*, 2432. [CrossRef]
256. McAllister, T.A.; Meale, S.J.; Valle, E.; Guan, L.L.; Zhou, M.; Kelly, W.J.; Henderson, G.; Attwood, G.T.; Janssen, P.H. Ruminant nutrition symposium: Use of genomics and transcriptomics to identify strategies to lower ruminal methanogenesis. *J. Anim. Sci.* **2015**, *93*, 1431–1449. [CrossRef]
257. Moate, P.J.; Jacobs, J.L.; Hixson, J.L.; Deighton, M.H.; Hannah, M.C.; Morris, G.L.; Ribaux, B.E.; Jacobs, J.L.; Wales, W.J.; Williams, S.R.O. Effects of feeding either red or white grape marc on milk production and methane emissions from early-lactation dairy cows. *Animals* **2020**, *10*, 976. [CrossRef]
258. Moate, P.J.; Williams, S.R.O.; Grainger, C.; Hannah, M.C.; Ponnampalam, E.N.; Eckard, R.J. Influence of cold-pressed canola, brewers grains and hominy meal as dietary supplements suitable for reducing enteric methane emissions from lactating dairy cows. *Anim. Feed Sci. Technol.* **2011**, *166*, 254–264. [CrossRef]
259. Morgavi, D.P.; Forano, E.; Martin, C.; Newbold, C.J. Microbial ecosystem and methanogenesis in ruminants. *Animal* **2010**, *4*, 1024–1036. [CrossRef] [PubMed]
260. Martin, C.; Morgavi, D.P.; Doreau, M. Methane mitigation in ruminants: From microbe to the farm scale. *Animal* **2010**, *4*, 351–365. [CrossRef] [PubMed]
261. Newbold, C.J. Assessing antiprotozoal agents. In *In Vitro Screening of Plant Resources for Extra-Nutritional Attributes in Ruminants;* Vercoe, P.E., Makkar, H.P.S., Schilink, C., Eds.; Springer: Dordrecht, The Netherlands, 2010; pp. 47–53.
262. Williams, S.R.O.; Moate, P.J.; Deighton, M.H.; Hannah, M.C.; Wales, W.J.; Jacobs, J.L. Milk production and composition, and methane emissions from dairy cows fed lucerne hay with forage brassica or chicory. *Anim. Prod. Sci.* **2016**, *56*, 304–311. [CrossRef]
263. Ponnampalam, E.N.; Burnett, V.; Norng, S.; Dunshea, F.R.; Lewandowski, P.; Jacobs, J.L. Natural algae boosts health benefits of lambs. In *Ag in Focus Victoria;* Department of Primary Industries: Victoria, Australia, 2012.
264. Hopkins, D.L.; Clayton, E.H.; Lamb, T.A.; van de Ven, R.J.; Refshauge, G.; Kerr, M.J.; Bailes, K.; Lewandowski, P.; Ponnampalam, E.N. The impact of supplementing lambs with algae on growth, meat traits and oxidative status. *Meat Sci.* **2014**, *98*, 135–141. [CrossRef] [PubMed]
265. Ponnampalam, E.N.; Lewandowski, P.A.; Fahri, F.T.; Burnett, V.F.; Dunshea, F.R.; Plozza, T.; Jacobs, J.L. Forms of n-3 (ALA, C18:3n-3 or DHA, C22:6n-3) fatty acids affect carcass yield, blood lipids, muscle n-3 fatty acids and liver gene expression in lambs. *Lipids* **2015**, *50*, 1133–1143. [CrossRef]
266. Sembries, S.; Dongowski, G.; Mehrlander, K.; Will, F.; Dietrich, H. Physiological effects of extraction juices from apple, grape, and red beet pomaces in rats. *J. Agric. Food Chem.* **2006**, *54*, 10269–10280. [CrossRef]
267. Molan, A.L.; Liu, Z.; Kruger, M. The ability of blackcurrant extracts to positively modulate key markers of gastrointestinal function in rats. *World J. Microbiol. Biotechnol.* **2010**, *26*, 1735–1743. [CrossRef]
268. Saad, R.; Rizkallah, M.R.; Aziz, R.K. Gut Pharmacomicrobiomics: The tip of an iceberg of complex interactions between drugs and gut-associated microbes. *Gut Pathog.* **2012**, *4*, 16. [CrossRef]
269. Dendup, T.; Feng, X.; Clingan, S.; Astell-Burt, T. Environmental Risk Factors for Developing Type 2 Diabetes Mellitus: A Systematic Review. *Int. J. Environ. Res. Public Health.* **2018**, *15*, 78. [CrossRef]
270. Guerra-Valle, M.; Orellana-Palma, P.; Petzold, G. Plant-Based Polyphenols: Anti-Helicobacter pylori Effect and Improvement of Gut Microbiota. *Antioxidants* **2022**, *11*, 109. [CrossRef]
271. Xie, L.; Su, H.; Sun, C.; Zheng, X.; Chen, W. Recent advances in understanding the anti-obesity activity of anthocyanins and their biosynthesis in microorganisms. *Trends Food Sci. Technol.* **2018**, *72*, 13–24. [CrossRef]
272. Clarke, T.B.; Davis, K.M.; Lysenko, E.S.; Zhou, A.Y.; Yu, Y.; Weiser, J.N. Recognition of peptidoglycan from the microbiota by Nod1 enhances systemic innate immunity. *Nat. Med.* **2010**, *16*, 228–231. [CrossRef]
273. Chassaing, B.; Ley, R.E.; Gewirtz, A.T. Intestinal epithelial cell toll-like receptor 5 regulates the intestinal microbiota to prevent low-grade inflammation and metabolic syndrome in mice. *Gastroenterology* **2014**, *147*, 1363–1377. [CrossRef] [PubMed]
274. Tsukumo, D.M.L.; Carvalho-Filho, M.A.; Carvalheira, J.B.C.; Prada, P.O.; Hirabara, S.M.; Schenka, A.A. Loss-of-function mutation in toll-like receptor 4 prevents diet-induced obesity and insulin resistance. *Diabetes* **2007**, *56*, 1986–1998. [CrossRef] [PubMed]

275. Ma, W.; Nguyen, L.H.; Song, M.; Wang, D.D.; Franzosa, E.A.; Cao, Y.; Joshi, A.; Drew, D.A.; Mehta, R.; Ivey, K.L.; et al. Dietary fiber intake, the gut microbiome, and chronic systemic inflammation in a cohort of adult men. *Genome Med.* **2021**, *13*, 102–115. [CrossRef]
276. Roopchand, D.E.; Carmody, R.N.; Kuhn, P.; Moskal, K.; Rojas-Silva, P.; Turnbaugh, P.J. Dietary polyphenols promote growth of the gut bacterium *Akkermansia muciniphila* and attenuate high-fat diet-induced metabolic syndrome. *Diabetes* **2015**, *64*, 2847–2858. [CrossRef]
277. Kalt, W.; Cassidy, A.; Howard, L.R.; Krikorian, R.; Stull, A.J.; Tremblay, F.; Zamora-Ros, R. Recent Research on the Health Benefits of Blueberries and Their Anthocyanins. *Adv. Nutr.* **2020**, *11*, 224–236. [CrossRef] [PubMed]
278. Anhe, F.F.; Desjardins, Y.; Pilon, G.; Dudonne, S.; Genovese, M.I.; Lajolo, F.M. Polyphenols and type 2 diabetes: A prospective review. *PharmaNutrition* **2013**, *1*, 105–114. [CrossRef]
279. Kong, M.; Xie, K.; Lv, M.; Li, J.; Yao, J.; Yan, K.; Wu, X.; Xu, Y.; Ye, D. Anti-inflammatory phytochemicals for the treatment of diabetes and its complications: Lessons learned and future promise. *Biomed. Pharmacother.* **2021**, *133*, 110975. [CrossRef]
280. Sun, C.; Zhao, C.; Guven, E.C.; Paoli, P.; Simal-Gandara, J.; Ramkumar, K.M.; Xiao, J. Dietary polyphenols as antidiabetic agents: Advances and opportunities. *Food Front.* **2020**, *1*, 18–44. [CrossRef]

International Journal of *Molecular Sciences*

MDPI

Review

Prebiotics as a Tool for the Prevention and Treatment of Obesity and Diabetes: Classification and Ability to Modulate the Gut Microbiota

Ashwinipriyadarshini Megur, Eric Banan-Mwine Daliri, Daiva Baltriukienė and Aurelijus Burokas *

Department of Biological Models, Institute of Biochemistry, Life Sciences Center, Vilnius University, Sauletekio Ave. 7, LT-10257 Vilnius, Lithuania; ashwinipriyadarshini.megur@bchi.stud.vu.lt (A.M.); eric.daliri@gmc.vu.lt (E.B.-M.D.); daiva.baltriukiene@bchi.vu.lt (D.B.)

* Correspondence: aurelijus.burokas@gmc.vu.lt

Abstract: Diabetes and obesity are metabolic diseases that have become alarming conditions in recent decades. Their rate of increase is becoming a growing concern worldwide. Recent studies have established that the composition and dysfunction of the gut microbiota are associated with the development of diabetes. For this reason, strategies such as the use of prebiotics to improve intestinal microbial structure and function have become popular. Consumption of prebiotics for modulating the gut microbiota results in the production of microbial metabolites such as short-chain fatty acids that play essential roles in reducing blood glucose levels, mitigating insulin resistance, reducing inflammation, and promoting the secretion of glucagon-like peptide 1 in the host, and this accounts for the observed remission of metabolic diseases. Prebiotics can be either naturally extracted from non-digestible carbohydrate materials or synthetically produced. In this review, we discussed current findings on how the gut microbiota and microbial metabolites may influence host metabolism to promote health. We provided evidence from various studies that show the ability of prebiotic consumption to alter gut microbial profile, improve gut microbial metabolism and functions, and improve host physiology to alleviate diabetes and obesity. We conclude among other things that the application of systems biology coupled with bioinformatics could be essential in ascertaining the exact mechanisms behind the prebiotic–gut microbe–host interactions required for diabetes and obesity improvement.

Keywords: prebiotics; obesity; diabetes; gut microbiota; biotherapeutics; dietary fiber

Citation: Megur, A.; Daliri, E.B.-M.; Baltriukienė, D.; Burokas, A. Prebiotics as a Tool for the Prevention and Treatment of Obesity and Diabetes: Classification and Ability to Modulate the Gut Microbiota. *Int. J. Mol. Sci.* **2022**, *23*, 6097. https://doi.org/10.3390/ijms23116097

Academic Editors: Baojun Xu and Matteo Bordiga

Received: 15 April 2022
Accepted: 26 May 2022
Published: 29 May 2022

1. Introduction

The condition of obesity and diabetes has risen drastically in the last decade, leading to a public health emergency. In a recent study, 463 million people were estimated to suffer from diabetes worldwide and the number is expected to increase in the coming years [1]. Diabetes is typically preceded by insulin resistance, where insulin action in peripheral tissues including the liver, skeletal muscles, and adipose tissues are impaired. This results in reduced insulin-stimulated glucose disposal, reduced lipolysis rates, and decreased insulin-induced suppression of hepatic glucose production [2]. There is increasing evidence that disruption of the gut microbiota function and composition could contribute to the pathogenesis of metabolic diseases such as diabetes [3] and obesity [4–6]. Consequently, it is crucial to evaluate the cross talk between the gut microbial composition in the gut, the development of metabolic disorders, and the potential therapeutic strategies to prevent these metabolic syndromes.

The mammalian gastrointestinal tract (GIT) is home to trillions of microorganisms, collectively known as the gut microbiota (GM) [7]. The GM is defined as an ecological community of commensal microorganisms that live symbiotically and pathogenically in the gut [8]. Colonization of neonatal gut may start during birth [9]. GM represents a complex ecosystem, consisting of numerous diverse sets of microorganisms such as viruses, fungi, bacteria, archaea, and phages, deeply implicated in different functions of host metabolism [10]. The most abundant phyla consists of *Firmicutes* (64%), *Bacteroidetes* (23%), *Proteobacteria* (8%), and *Actinobacteria* (3%) [11]. GM makes a crucial contribution to the production of enzymes that are not encoded by the human genome, for example, the breakdown of polysaccharides, polyphenols, and the synthesis of vitamins [12]; is pivotal for human development and physiology [13]; and plays a vital role in regulatory functions in health and disease [14].

The composition of the GM differs between person-to-person and can fluctuate significantly within an individual [15]. Variation in GM composition could be caused not only by differences in the host's genome, but also by environmental factors, such as antibiotic use, lifestyle, hygiene, and diet administration [16,17]. Significant alterations in gut microbial composition (dysbiosis) can be unfavorable and can predispose an individual to disease. For instance, acute and chronic disorders such as obesity, inflammatory bowel disease, irritable bowel syndrome, diabetes, colon cancer, and antibiotic-associated diarrhea have all been associated with dysbiosis [12,18,19].

Food is considered as a substrate that greatly contributes to the growth of GM and has a significant influence on its composition [20]. In 1980, it was proposed that definite components of the diet could promote the proliferation of specific bacterial strains inhabiting in the GIT, which are associated with the benefit of the host's health [21]. The dietary intervention with prebiotics can be classified as dietary fibers; however, not all fibers can be considered as prebiotics [22]. Dietary modulation of GM with prebiotics has shown great potential as an agent to ameliorate and perpetuate a balanced microbial composition to improve health and well-being [23–26].

In this review, we discussed prebiotics, their classification, and the modulatory capacity of GM for health promotion in the host. We also discussed in vivo and in vitro studies and human clinical trials to provide better insight into the benefits of prebiotics on health. Finally, we focused on the therapeutic uses of prebiotics in the treatment/prevention of obesity and type 2 diabetes mellitus (T2DM).

2. Prebiotics

Prebiotics are a class of nutritional compounds categorized together, not necessarily by structural affinity, but by the potential to promote the growth and/or activity of specific beneficial bacteria (probiotics) in the GM. The concept of prebiotics came into recognition due to Glenn Gibson and Marcel Roberfroid in 1995 [23]. A prebiotic is known as "a non-digestible food constituent that beneficially influences the host by selectively promoting the growth and/or activity of one or a restricted number of bacteria in the colon, and thus improving the host health" [27]. In 2004, prebiotics were upgraded to include four criteria: (1) resistance to hydrolysis by mammalian enzymes, gastric acidity, and gastrointestinal absorption; (2) they should only be fermented by GM; (3) induce systemic or luminal effects that are beneficial to host health; and (4) selectively stimulate the growth and activity of GM associated with health and well-being [28]. The health benefits of prebiotics are diverse and include immune modulation through increased immune-regulatory interleukins and intestinal-specific immunoglobulins; reduction of pro-inflammatory interleukins [29,30]; and production of short-chain fatty acids (SCFAs) such as acetate, propionate, and butyrate [31] (Figure 1). SCFAs are carboxylic acids with aliphatic tails of one to six carbons that are produced by anaerobic fermentation of dietary fibers in the intestine by the GM [32]. SCFAs are an important indicator of bacterial fermentation in the colon and are known to improve the gut health by maintaining intestinal barrier integrity [33], mucus production [34], protection against inflammation, and reduction in colorectal cancer and obesity [35].

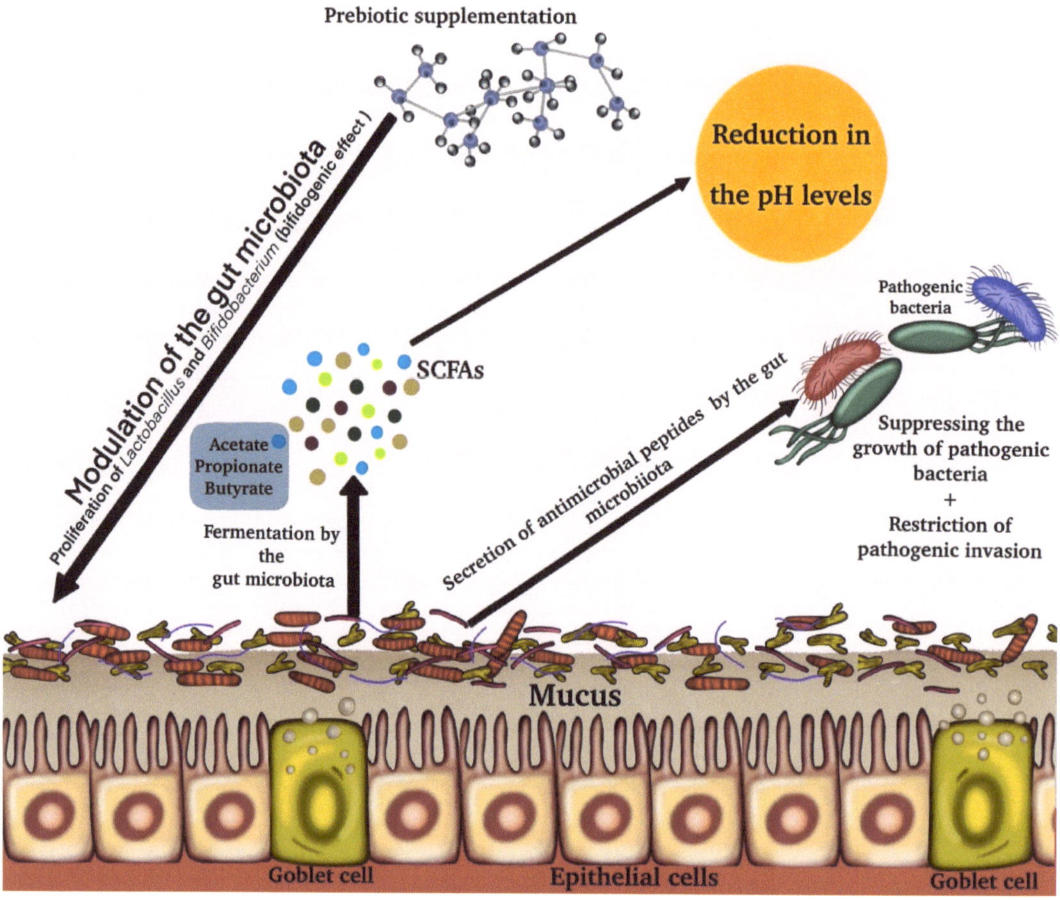

Figure 1. Mechanism of action of prebiotic supplementation. Prebiotic administration in a regular diet increases bacterial growth and functionality of specific species or genera, leading to modulation of the GM and showing a strong bifidogenic effect. The goblet cells play a key role in the production of mucus, which helps to protect the mucous membrane and form a layer in the colon that helps to reduce the inflammation caused by the bacterial interaction with intestinal epithelial cells. The modulated GM ferments prebiotics to form SCFAs (acetate, propionate, and butyrate), from which health benefits can be accrued. The production of antimicrobial agents and the reduction in the pH levels of the intestine due to prebiotic supplementation can suppress and restrict the growth of pathogenic bacteria, which can lead to positive health effects.

Among the abundant food ingredients available, some peptides and proteins, particular lipids, and non-digestible carbohydrates are components of prebiotics [36]. The chemical structures of these components are not absorbed in the upper part of the GIT or hydrolyzed by the digestive enzymes of humans. Hence, these ingredients are called colonic foods [37]. In colonic food, non-digestible carbohydrates are naturally occurring and meet all the criteria of prebiotics. These carbohydrates include non-starch polysaccharides, resistant starch, and non-digestible oligosaccharides [38]. However, not all of them are prebiotics [39]. In order to be classified as prebiotics, carbohydrates must fulfil the following criteria: (i) they are dietary fibers with a degree of polymerization (DP) between three and nine [40], and (ii) the endogenous enzymes produced in the small intestine should not hydrolyze them [41]. It should be taken into account that fermentation and fiber solubility are generally not curtailed [22].

Bacterial genera that promote health such as *Lactobacillus* and *Bifidobacterium* is proliferated by the administration of prebiotics, so that the fermented metabolites can be easily absorbed by the mammalian gut and have an influence on host physiology [42] (Figure 1). The prebiotics share several characteristics with dietary fiber, which includes partial or total resistance to digestion and fermentation by the GM. Due to its selectivity, prebiotics highlight the key condition to be demonstrated in an in-vivo experiment (including complex human or animal GM) using validated and relevant methodologies to quantify a wide variety of species that make up the GM [43].

Through characteristic and selective assimilation of prebiotics by subsequent fermentation, there is a production of SCFAs at high levels, having immunomodulation and metabolic effects on the host [44]. In this case, a reduction in the intestinal pH is also observed, creating an environment that competitively hinders the growth of pathogenic bacteria [45]. Some prebiotics prevent the adhesion of pathogenic microbiota to the GIT by mimicking an intestinal binding site [46] (Figure 1).

The application of prebiotics is well known in pharmaceuticals, and products for people with diabetes (as a natural sweetener) [47]. The large number of scientific data on prebiotics has focused on compounds associated with two major chemical groups: fructo-oligosaccharides and galacto-oligosaccharides [48]. They can be derived and/or extracted from food sources such as seeds, whole grains, legumes, chicory roots, Jerusalem artichokes, onions, garlic, and some vegetables [49], but in a recent study it was found that some aquatic plants (seaweeds and microalgae) contain prebiotics [50]. Prebiotics include a variety of forms such as fructo-oligosaccharides (FOS), galacto-oligosaccharides (GOS), human milk oligosaccharides (HMO), lactulose, lactosucrose, inulin, resistant starches (RS), arabinoxylans (AX), xylooligosaccharides (XOS), and pectin [24]. More attention has been given by researchers towards FOS as a prebiotic in improving human health [51].

3. Classification of Prebiotics

As mentioned above, there are many types of prebiotics that can be classified into different groups [52]. They differ in structure and can have a health benefit to the host through numerous different mechanisms [44]. Prebiotics also have the potential to modulate GM by selectively stimulating the growth of *Bifidobacteria* and *Lactobacilli*, by assimilation via beneficial GM and subsequent fermentation. In the fermentation process, these GM produce high levels of butyrate, isobutyrate, valerate, propionate, and acetate, which has various physiological functions in an organism [53]. The majority of prebiotics are mostly the subset of carbohydrate groups, more specifically, oligosaccharide carbohydrates. There are many relevant articles on oligosaccharide carbohydrates [54,55], but there are also few pieces of evidence showing that prebiotics are not only carbohydrates [56].

3.1. Inulin (Fructan)

Inulin-type prebiotics are members of an immense group called "fructans". Fructans constitute a group of compounds that confine all naturally occurring plant oligosaccharides and polysaccharides in which one or more fructosyl–fructose linkages form the majority of glycosidic bonds [57]. Hence, they are the primarily polymers of fructose units. Fructans can also be characterized by the DP, which refers to the number of repeated units in a polymer or oligomer chain [58]. The category of fructans consists of inulin and oligofructose (FOS) [59] (Table 1).

Inulin is a collective term that comprises all linear fructans with β (2→1) fructosyl–fructose glycosidic bonds [60]; this specific type of glycosidic bond gives inulin its distinctive physiological and structural properties. Inulin-type fructans resist enzymatic hydrolysis by small intestine digestive enzymes and human salivary enzymes because of the beta configuration bonds between fructose monomers [61]. Chemically, the linear chain of inulin is either an α-D-glucopyranosyl-[β-D-fructofuranosyl](n-1)-β-D-fructofuranoside (GpyFn) or αβ-D-fructopyranosyl-[β-D-fructofuranosyl](n-1)-β-D-fructofuranoside (FpyFn) [62].

3.2. Fructo-Oligosaccharides (Fructan)

Another type of fructans i.e., FOS, is a natural component that can be found in plants [63]. FOS are commercially prepared from chicory in a hydrolysis reaction using inulinase and may also be derived in an enzymatic synthetic reaction via the transfer of fructosyl units from sucrose molecules [64]. When presented structurally, FOS consist of a sucrose molecule linked by a chain of 3–30 fructosyl units. FOS are oligomeric linear fructans with β-(2–1) or β-(2–6) fructosyl-fructose linkages with the first monomer of the chain either being α-D- glucopyranosyl or β- D -fructopyranosyl residue [60]. The DP of inulin is up to 60 and the DP of FOS is less than 10 [65] (Table 1).

3.3. Galactooligosaccharides

GOS are the product of lactose extension and are included among non-digestible oligosaccharides. They are arranged in two subgroups: (i) with excess galactose at C_3, C_4 and C_6; and (ii) manufactured from lactose through enzymatic trans-glycosylation [66]. The mixture of the product depends on the reaction conditions and the enzymes used. β-galactosidase of various origins, such as *Aspergillus oryzae, Bacillus circulans,* and *Cryptococcus laurentii,* is used for the industrial production of GOS [67]. The general constituents of this oligosaccharide are from tri- to penta-saccharide with β (1→6), β (1→3), and β (1→4) linkages. This category of GOS is known as trans-galacto-oligosaccharide [68]. Culture studies of *Bifidobacteria* and most of *Lactobacilli* and enterobacteria, including some streptococci-metabolized trans-oligosaccharide, with *Bifidobacteria* showed robust growth [69]. There are some GOS derived from the isomers of lactose, due to influential factors such as the source of the enzyme, temperature, pH, and substrate concentration. They are also considered as prebiotics [70] (Table 1).

3.4. Human Milk Oligosaccharides

HMO are complex and non-digestible carbohydrates, recently classified as prebiotic substances. They are present in high abundance in maternal breast milk (10–15 g/L) [57,71,72]. The length of the HMO chain can vary from 3 to 15 carbohydrate units and is synthesized in the mammary gland [73]. The HMO concentration in the lactating mother is higher during the early stages and gradually decreases over time [74–76]. Structurally, HMO are composed of five monosaccharides: glucose, galactose, N-acetylglucosamine, fucose, and N-acetylneuraminic acid or sialic acid [77–79]. They are synthesized from a lactose core (galactose-β (1→4) glucose) by glycosyl transferases in the lactocyte. Some HMO are branched with a fucose or sialic acid monosaccharide residue attached to the lactose core via α1–2/3/4 and α2–3/6 linkages, due to the action of fucosyltransferases and sialyltransferases, respectively [80,81]. Among its several types, less than 50 HMO have a representative abundance in human breast milk. HMO 2′-fucosyllactose has been identified as the most abundant HMO in breast milk [82]. Breast milk due to its high levels of 2′-fucosyllactose has shown advantages for the infant because of its efficiency to promote an early high *Bifidobacteria*-dominated GM [83]. Several experiments conducted on the supplementation of HMO documented beneficial effects on the overall health of an individual, which includes modification of the GM [78,81,84], effects on immune development [78,85,86], anti-adhesive antimicrobial effects [87], and brain development [88,89] (Table 1).

3.5. Glucose-Derived Oligosaccharides

An example of glucose-derived oligosaccharide is polydextrose (PDX), which is non-digestible and widely used in the food industry [90]. PDX is a randomly bonded glucose polymer with an average DP of 12, but ranging from 2 to 120. This molecule contains the combination of α- and β-linked 1→2, 1→3, 1→4, and 1→6 glycosidic linkages [91]. PDX has been acknowledged as a soluble fiber that has beneficial effects on gut health, satiety, and postprandial glycemia [92]. Daily intake of 4–12 g of PDX has been found to have a large improvement in physiological functions without showing any adverse effect [93].

3.6. Resistant Starches

The starch that is resistant to the upper gut digestion is termed as RS [94]. RS cannot be digested by human pancreatic amylase in the small intestine, reaching the colon, promoting health benefits by producing a high level of butyrate, suggesting it to be classified as a prebiotic [39,95]. RS consumption has been related to improving the diabetes condition by reducing postprandial glycemic and insulinemic responses, and is also associated with decreased levels of cholesterol and triglycerides [96].

3.7. Pectic Oligosaccharides

Pectic oligosaccharides (POS) originate from a polysaccharide, known as pectin, which is a structural element of intracellular regions and cell walls of the plants and is vastly present in fruits and vegetable waste materials [97]. Chemically, POS are based on the extension of rhamnose or galacturonic acid, and different types of sugars (galactose, xylose, and arabinose) or ferulic acid are linked to the side chains [98]. In humans, gastric juice and saliva are not capable of degrading pectin. Furthermore, digestive enzymes like trypsin, pepsin, and rennet cannot breakdown pectin in the small intestine [99]. It has been reported that pectin undergoes slow fermentation and exhibits prebiotic effects by producing SCFAs [100]. It has been shown that pectic oligosaccharide has the potential to show bifidogenic effects [101]. Experiments conducted on pectic oligosaccharides revealed health benefits that include antiobesity, anticancer, and antioxidant properties [102].

Table 1. Summary of the structure and formula of prebiotics.

Abbreviation	Chemical Composition	DP	Chemical Formula	References
Inulin	linear chain of fructose with β(2→1) linkages	3–60	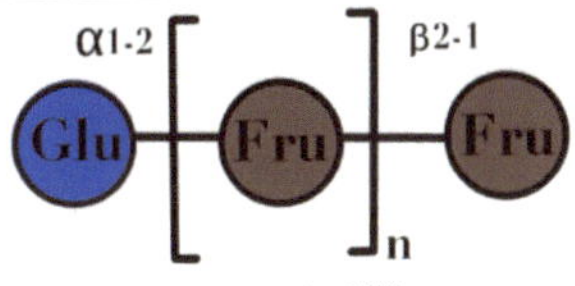 n = up to 100	[59,60]
FOS	linear chain of fructose with β(2→1) linkages	<10	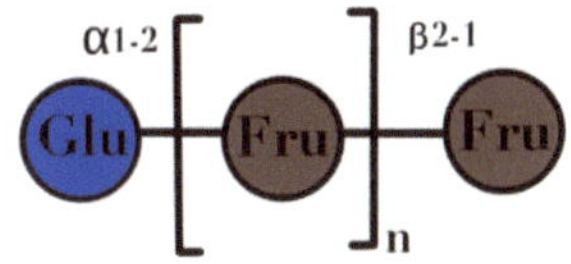 n = 1–5	[59,60]
GOS	Chain of galactosyl residues and a terminal glucose linked by β-(1–2), β-(1–3), β-(1–4), or β-(1–6) glycosidic bonds	2–8	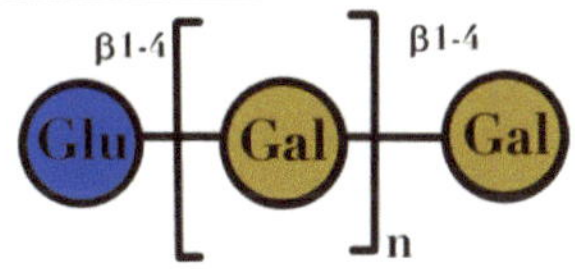 n = 1–4	[60,103]

Table 1. *Cont.*

Abbreviation	Chemical Composition	DP	Chemical Formula	References
HMO	composed of five monosaccharides: glucose, galactose, N-acetylglucosamine, fucose, and N-acetylneuraminic acid or sialic acid	<7	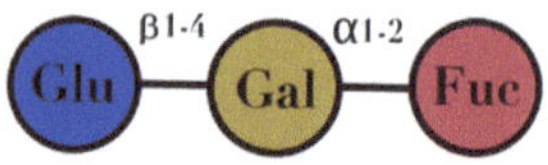 2′fucosyllactose 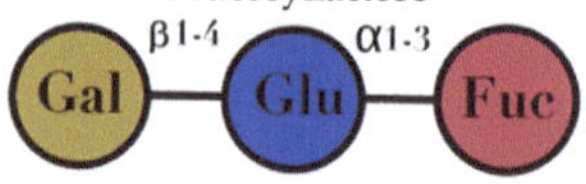 3′-fucosyllactose 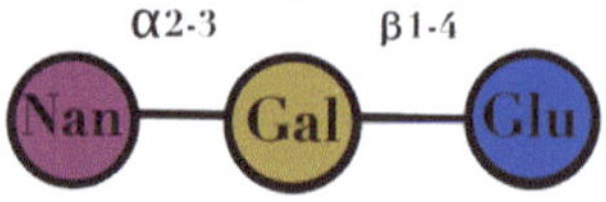 3′-sialyllactose 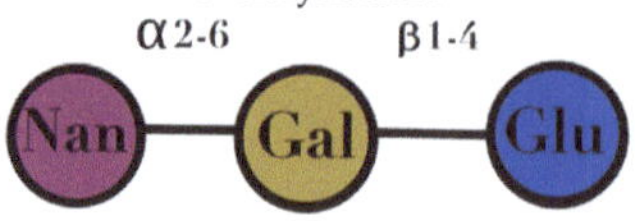 6′-sialyllactose 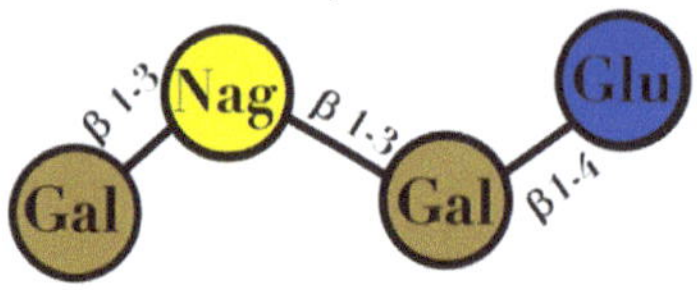 Lacto-N-tetraose 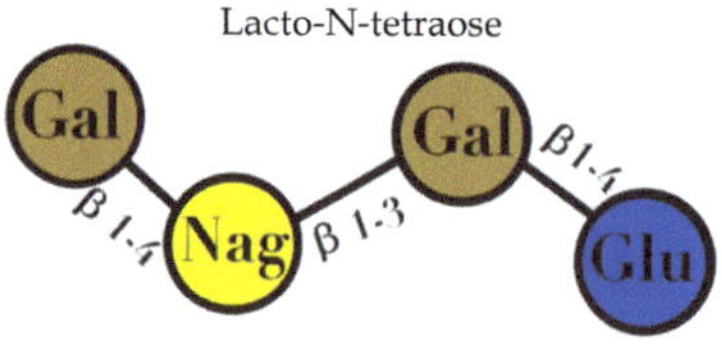 Lacto-N-neotetraose 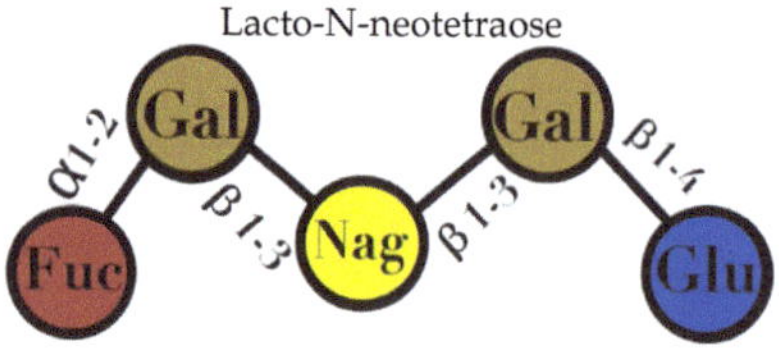 Lacto-N-fucopentaose I	[104,105]
Lactulose	consisting of galactose and fructose moieties	-	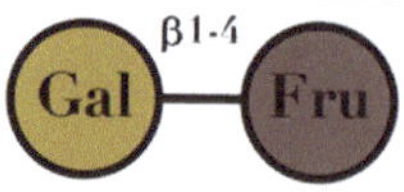	[106,107]
Lactosucrose	composed of galactose, fructose, and glucose monomers	-	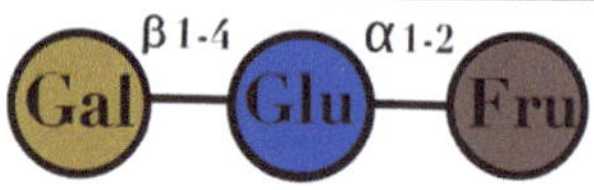	[108]

Table 1. *Cont.*

Abbreviation	Chemical Composition	DP	Chemical Formula	References
AX	β-1,4-linked D-xylopyranoside units substituted with arabinose residues on the c(o)-2 or c(o)-3 position	1–60	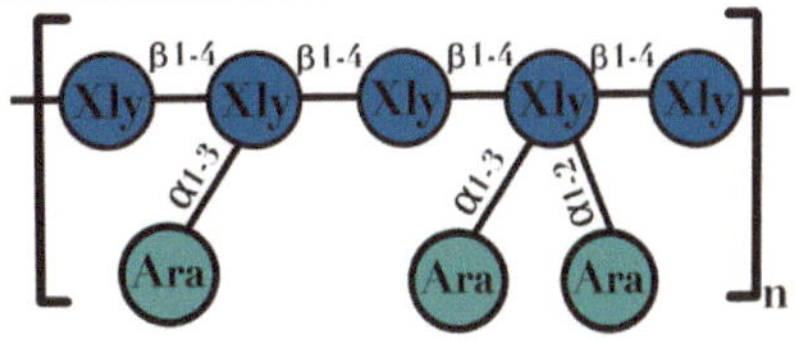	[109]
XOS	xylose moeities linked by β-(1→4) glyosidic bonds	2–4	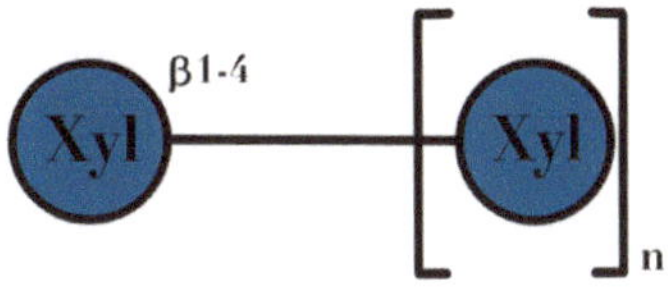 n = 2–4	[110–112]
Symbols used in Table 1: their meaning and chemical structure.			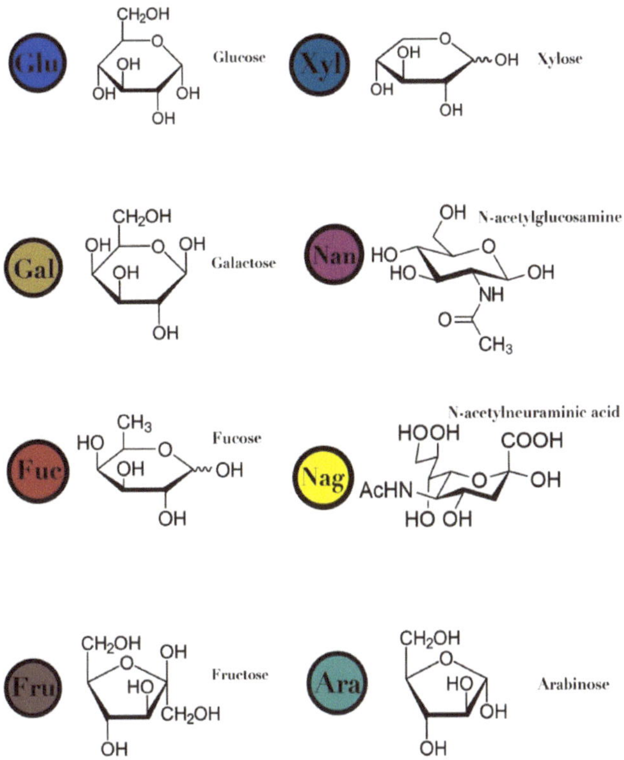	

3.8. *Lactulose*

Lactulose is a synthetically produced non-digestible ketose disaccharide that consists of galactose and fructose linked by a bond resistant to lactase [113]. Lactulose is extracted from lactose (milk sugar), chemically known as 4-O-β-d-galactopyranosyl-d-fructose, and the enzyme used for the biocatalytic production is β-galactosidase [114]. It is used medically for the treatment of constipation [115]. The human small intestinal mucosa does not have the enzymes to breakdown lactulose, and hence it reaches the large bowel unchanged [116]. Lactulose is metabolized by colonic bacteria to monosaccharides and then to methane, volatile fatty acids, and hydrogen [117]. In human studies, the lactulose have a significantly modified GM by increasing *Bifidobacterium*, *Lactobacillus*, and *Streptococcus*, and having favorable health benefits [81,118] (Table 1).

3.9. *Lactosucrose*

Lactosucrose is also known as galactosylsucrose, lactosylfructoside, and galactosucrose, and is synthetically produced trisaccharide, which is composed of galactose, fructose, and glucose monomers [119]. Raffinose, an isomer of lactulose, has a potential bifidogenic effect [120]. Lactosucrose is used as a commercial food supplement in many healthy foods and beverages with the intention of altering gastrointestinal functions and improving health [121]. Lactosucrose has shown promising effects as a bifidogenic compound modulating immune functions [122,123] (Table 1).

3.10. *Arabinoxylans*

Arabinoxylans (AX) are predominant non-cellulosic polysaccharides of cell walls in plants. AX were first identified by Hoffman and Gortner in 1927, as viscous gum in wheat flour [124]. Their structural properties, heterogeneity, and recovery depend on their location, which is strongly influenced by the other components of the cell wall [125]. AX are called as "pentosans" as they consist of pentoses xylose and arabinose. Chemically, it is heteroxylan consisting of a backbone of β-1,4-linked D-xylopyranoside units substituted with arabinose residues on the c(o)-2 or c(o)-3 position [126]. The DP of AX is between 1 and 60 [109]. AX have the potential to show high technological importance. There are several biological studies that have been reported on the behalf of AX, including antioxidant activity [127], cholesterol-lowering agents [128], blood sugar modifiers [129], and immunity enhancers [130] (Table 1).

3.11. *Xylooligosaccharides*

Xylooligosaccharides (XOS) or xylan are to be considered as the second most abundant biopolymer in the plant kingdom. These are the sugar oligomers of β-1,4-linked xylose (a pentose sugar) found naturally in food sources such as honey, bamboo shoots, fruits, vegetables, and milk [131]. On the basis of substituted groups, xylan can be categorized into three classes: (i) glucuronoxylan, (ii) neutral arabinoxylan, and (iii) glucuronoarabinoxylan [132].The DP of the XOS used in commercial food products ranges from 2 to 10 [133]. The complete utilization of XOS is based on the activities of a number of enzymes, including β-xylosidase, α-glucuronides, and acetyl esterases released by different strains of GM, and produces SCFAs [134]. XOS have shown a bifidogenic effect [135], with the support of in-vivo animal studies, and offers modification in the composition and activity of the GM [136] (Table 1).

Excess consumption of prebiotics can promote severe discomfort in an individual, therefore, optimal intake is necessary [137] (Table 2).

Table 2. Recommended intake of prebiotics.

Prebiotic	Doses Suggested	Reference
Inulin	2–12 g/day	[138]
FOS	12.5–20 g/day	[139]
GOS	2–20 g/day	[140]
HMO	10–20 g/day	[141]
PDX	4–12 g/day	[93]
RS	10–15 g/day	[142]
POS	10–20 g/day	[143]
Lactulose	10–30 g/day	[144]
Lactosucrose	Not estimated	-
AX	Not estimated	-
XOS	1–5 g/day	[145]

4. Efficacy of Prebiotics on Gut Microbiota Composition: In Vivo and In Vitro Studies

The experiments conducted on the administration of prebiotics have shown selective changes in the GM composition. Different categories of prebiotics can stimulate the growth of various indigenous bacterial communities in the GM. Collective evidence from animal model trials, human studies, and in-vitro modeling systems has concluded that they affect the composition of GM, leading to proliferation in health-promoting organisms such as *Bifidobacteria* and *Lactobacilli* [146–148]. Prebiotics have ameliorative properties such as maintaining intestinal integrity and homeostasis, production of SCFAs, and regulation of gastrointestinal transit [115]. Indeed, it has been suggested that the use of prebiotics should have ameliorative properties on gastrointestinal diseases like irritable bowel disease, Chron's disease, and ulcerative colitis [149]. Selective stimulation of GM growth and/or activity is potentially associated with health protection and well-being [24,150].

4.1. Inulin

Inulin is a non-digestible oligosaccharide that is fermented by the GM and has resistance to the degradation by the human digestive enzymes. It reaches the colon almost as an intact molecule and acts as a fermentable substrate for GM [62]. In vivo and in vitro studies on inulin concluded that it has selective stimulation of bacterial growth; this has been observed in numerous studies carried out either in defined pure culture fermentation or by using human feces [151–153]. Inulin supplementation for 19 days to a group of 10 elderly women with a dose beginning at 20 g/day from days 1 to 8 and gradually increasing to 40 g/day during days 9 to 19, showed a significant increase in *Bifidobacteria* that can be utilized during fermentation, and a decrease in the number of *Enterococci* and *Enterobacteriaceae*, while no statistically significant changes were observed in *Bacteroides*, *Clostridia*, or *Faecalibacterium prausnitzii* [154]. In another study conducted on 10 healthy volunteers with inulin supplementation of 8 g per day for 14 days, a significant increase in *Bifidobacteria* was shown. In this case, a number of *Clostridia* increased also, but the magnitude of *Clostridia* was one tenth of *Bifidobacteria*. These data supported a bifidogenic effect of inulin [155]. Importantly, inulin fermentation leads to the production of SCFAs. In an experiment conducted on rats cecum (colonic part of the GIT), it was demonstrated that inulin has significantly higher efficiency in producing SCFAs compared with other dietary fibers [156] (Table 3).

4.2. FOS

FOS have great potential as ingredients due to their prebiotic activity and low caloric value. Gibson and Roberfroid [23] showed the bifidogenic characteristics of FOS using 15 g per day as dietary supplementation. The GM was modulated and there was a significant decrease in the number of *Bacteroides, Fusobacterium,* and *Clostridium.* Therefore, it was concluded that FOS is better utilized by *Bifidobacteria,* and, on the other hand, they can cause unfavorable changes for harmful bacteria in the GIT [23].

It was verified that the addition of NeosugarR (a trade name for fructooligosaccharide) to the human diet, i.e., 15 g per day, can cause a 10-fold increase in the population of *Bifidobacteria* in the large intestine [138]. In addition to its bifidogenic property, the regular and adequate intake of FOS has beneficial effects in the case of disorders associated with obesity, diarrhea, osteoporosis, atherosclerotic, gastrointestinal disorders, cardiovascular, and T2DM diseases [157]. The fermentation of FOS by GM generates SCFAs and organic acids that decrease luminal pH, thereby enhancing the bioavailability of nutritionally important minerals [158]. It was also found that a diet supplemented with FOS promotes the production of butyrate, which influences lipid metabolism in humans [159] (Table 3).

4.3. GOS

GOS are a type of non-digestible fiber with prebiotic activity [133], which has also been demonstrated by a dynamic in-vitro colon model and the ^{13}C labeling technique with GOS consumption. The results showed an increase in *Bifidobacterium longum, B. bifidum, B. catenulatum, Lactobacillus gasseri,* and *L. salivarius,* but changes in numbers of *Enterobacteriaceae* (a family of Gram-negative bacteria that includes some harmless symbionts) and several familiar pathogens, such as *Salmonella, Yersinia pestis, Klebsiella, Escherichia coli,* and *Shigella,* were rather negligible [160]. In another study, the prebiotic activity of GOS was analyzed by pyrosequencing of fecal samples from healthy human volunteers with GOS administration. The data obtained showed a statistically significant increase in *Bifidobacteria* and *Faecalibacterium prausnitzii,* and a decrease in *Bacteroides* [161]. It was also concluded that 90% of GOS resist digestion in the upper GIT and then enter the colon, which then get intact to the tract and act as fermentation substrates for the resident microbiota [162] (Table 3). An in-vitro study showed that the fermentation of GOS by GM generates SCFAs and organic acids that decrease luminal pH, thereby enhancing the bioavailability of nutritionally important minerals [163]. Interestingly, GOS administration showed anxiolytic effects in both animals [164] and humans [165] (Table 3).

4.4. HMO

One of the multifarious functions of HMO is that they act as prebiotics and stimulate the colonization of beneficial GM [166]. In vitro studies provided strong evidence that HMO promotes the growth of selective *Bifidobacteria* [78]. *Bifidobacterium longum subsp. infantis* proliferates well on 2′-FL, as the sole source of carbohydrates [81,84,167,168]. These *Bifidobacterium longum subsp. infantis* produce SCFAs, which create an environment that favors the growth of commensal bacteria and prevents the adhesion of pathogenic bacteria [169]. Some structures of HMO are similar to the intestinal epithelial cell surface glycan receptors, which serve as decoy receptors to prevent pathogen binding and increase pathogen removal [78]. A study on HMO supplementation suggested that breast-fed infants have a higher number of *Bifidobacteria* compared to the formula-fed infants [170]. In a human study, investigation into the interaction between *Bifidobacteria* and *Eubacterium hallii* demonstrated that *E. hallii* consume acetate, lactate, and 1,2-propanediol (which are the products formed by the fermentation of HMOs by *Bifidobacteria*) and eventually lead to the production of butyrate and propionate [171] (Table 3). On the other hand, the study conducted on bioengineered 2′-FL showed inhibition of the adhesion of *Campylobacter jejuni, Salmonella enterica, E.coli,* and *Pseudomonas aeruginosa* to an intestinal human cell line [172]. Research on HMO, specifically 2′-FL, has shown that it is even more potent than standard commercial prebiotics, such as FOS, and has many different functions, including immune, GM, and cognition benefits [173] (Table 3).

4.5. PDX

In vitro studies have indicated that PDX has all the characteristics to be a prebiotic [174,175]. It has been shown that daily intake can beneficially modify the composition and activity of GM. In a study in humans, PDX favored intestinal function and improved the ease of bowel movement. Furthermore, it inhibited the absorption of glucose in the small intestine and the fermentation for the production of SCFAs in the large intestine favoring the reduction of gut pH [176]. Supplementation with PDX in healthy humans with a dose of 8 g per day for 3 weeks showed a significant increase in the number of *Ruminococcus intestinalis*, the main producer of butyrate, and slow fermentation of PDX in the colon was observed [177]. Another study carried out in healthy adult males with 21 g of PDX supplementation per day significantly suppressed the number of phylum *Firmicutes* and significantly increased the number of bacteroidetes when compared to the control group [174]. These data concluded that PDX supplementation had a positive impact on the bacterial composition of GM (Table 3).

4.6. RS

A number of studies demonstrated that RS is capable of modifying the GM composition towards the heath benefit of the host. An experiment carried out in mice for 8 weeks showed that mice fed with diets containing high amylose RS2 (one of the types of RS) were colonized by higher levels of *Bifidobacterium*, *Akkermansia*, and *Allobactum* [178]. The nutritional intervention study revealed that RS, when supplemented in the diet, can induce a 10-fold increase of gut *Bifidobacteria* [179]. On the other hand, one of the byproducts of RS is SCFAs. A study carried out on rats for 12 weeks treated with two concentrations of RS (0 and 27% weight of diet) showed an increase in propionate, butyrate, and acetate [180] (Table 3).

4.7. POS

POS is a new class of prebiotics that derives SCFAs from the GM fermentation [181]. In an in vitro study, the POS from the citrus peel and sugar beet pulp were fermented by the human fecal samples, leading to an increase in the bacterial population of eight different groups. POS from sugar beet showed the highest bifidogenic effect and utmost SCFAs concentration. On the other hand, the POS from citrus peel showed an increase in the population of *Lactobacillus* [182]. In a recent study, it was concluded that the concentration of SCFAs was higher in the POS supplementation, when compared to FOS [183] (Table 3).

4.8. Lactulose

An investigation conducted on lactulose degradation determined that human and calf β-galactosidases do not degrade it [184]. An in-vitro study performed using fecal samples on agars, and an analysis of enzymes produced and putrefactive compounds of lactulose fermentation, concluded a selective and significant increase in *Bifidobacteria*, decreasing the abundance of streptococci, bacteroides, *C. perfringens*, and *Lactobacilli* [185]. Studies carried out on humans demonstrated that lactulose selectively and significantly modifies GM by increasing *Lactobacillus*, *Bifidobacterium*, and *Streptococcus* [186] (Table 3).

4.9. Lactosucrose

Strong evidence has been observed in the administration of lactosucrose, selectively promoting the number of *Bifidobacteria* in in-vitro and in-vivo studies on animals and humans [187–189]. Lactosucrose fermentation was evaluated using *Bifidobacterium*, *Lactobacillus*, and *Streptococcus* probiotic strains in the in-vitro study, and the results led to the growth of four bacterial strains: *Lactobacillus casei*, *Lactobacillus reuteri*, *Lactobacillus acidophilus*, and *Streptococcus salivarius* [187]. Animal studies have shown a significant increase in *Lactobacilli* and *Bifidobacteria*, while restraining the levels of pathogens, such as *Clostridium perfringens*, *Staphylococci*, and *Bacteroidaceae*, after the consumption of lactosucrose [122,185]. Lactosucrose fermentation by the GM produces SCFAs and shows a consequent reduction in the pH of fecal contents [190,191]. An in-vitro study on different fish species such as *Pagrus major*, *Cyprinus carpio* L., and *Oncorhynchus mykiss* showed that the lactosucrose fermentation results in the production of SCFAs and gases, concluding that the lactosucrose can also be fermented in herbivorous, omnivorous, and carnivorous fishes [192–194] (Table 3).

4.10. AX

AX are not digested by the enzymes produced by the GIT, thus these provide the carbon source for the GM that inhibits the large bowel [195]. Many experiments have been performed on the regular supplementation of AX, resulting in an enhancement in the proliferation of the growth of health-promoting bacteria. In-vitro studies of AX, implemented in anaerobic batch cultures inoculated with human feces, demonstrated that fermentation of wheat endosperm AX resulted in the production of acetate, propionate, and butyrate [196]. In the in-vitro digestibility test carried out on pigs, it was established that only 15% of the ingested AX is recovered in the feces, while the major fraction of AX is fermented in the cecum, which represents the high fermentability of AX [197] (Table 3).

Table 3. Prebiotic efficiency in modulating the GM.

Prebiotics	Model	Strategy/Duration of Feeding	Dose Supplemented	Form	No. of Applications	Re-Calculated Dose *	Fecal Microbial Changes Relative to Control	Reference
Inulin	17 elderly women (mean age = 76.4 years body weight not reported)	8 days, (3 days adaptation) Feeding was continued for 8 days	20 g/day and increased to 40 g/day	Dissolved in drinking water	Once/day	285.7 mg/kg/day and increased to 571.4 mg/kg/day	significant ↑ in *Bifidobacteria* ↓ in Enterococci and *Enterobacteriaceae*	[154]
	10 healthy volunteers (age = between 20 and 55 years Body weight not reported)	14 days	8 g/day	Dissolved in drinking water	Twice/day	114.3 mg/kg/day	significant ↑ in *Bifidobacteria* ↑ in the number of Clostridia	[155]
	Germ-free adult male Fischer rats (age = 10 weeks and body weight = 280 g)	8 weeks	1.84 g/day of the diet	Mixed with chow	During the day	6.57 g/kg/day	significant ↑ in producing SCFAs	[156]
FOS	Male Wistar rats (age = 2 months and body weight 403.2 ± 48.1 g)	7 days	8% of the diet	Mixed with chow	During the day	3.4 g/kg/day	↑ the bioavailability of nutritionally important minerals	[158]
FOS + GOS	10 Male C57BL/6J mice (age 8 weeks old mice; mean body weight = 28 g)	10 weeks	0.3 –0.4 g/mouse/day	Dissolved in drinking water	During the day	1.1–1.43 g/kg/day	↑ *Akkermansia* abundance	[164]
GOS	18 healthy human (age and body weight not indicated)	3 weeks	2.5 g/day 5 g/day, 10 g/day	Administered in edible chews	Once/day	35.7 mg/kg/day 71.4 mg/kg/day, 142 mg/kg/day	significant ↑ in abundance of *Bifidobacteria* and *Faecalibacterium prausnitzii*, ↓ in *Bacteroides*	[161]
	Mud crab (age not reported and body weight 63.6 ± 8.8 g)	24 h	0.05 g/day	Dissolved in water	During the day	786 mg/kg/day	↑ *Bacteroidetes*	[163]
PDX/FOS	77 Children (age 5.8 ± 1.3; body weight not reported)	2 weeks	4.17 g PDX + 0.45 g FOS	Dissolved in drinking water	Once/day	PDX 200 mg/kg/day + FOS 22 mg/kg/day	↑ in number of *Bifidobacterium* and *Lactobacillus*	[176]

Table 3. *Cont.*

Prebiotics	Model	Strategy/Duration of Feeding	Dose Supplemented	Form	No. of Applications	Re-Calculated Dose *	Fecal Microbial Changes Relative to Control	Reference
PDX	20 Healthy men (Age = 27.5 6 ± 4.33; body weight = 86.26 ± 13.48 kg)	21 days	21 g/day	Mixed in bar	Once/day	243.4 mg/kg/day	↑ in number of *Faecalibacterium*, *Phascolarctobacterium*, and *Dialister*	[175]
	15 Healthy volunteers (age = 18–50 body weight not reported)	3 weeks	8 g/day	Powder	Once/day	243.5 mg/kg/day	↑ *Ruminococcus intestinalis*, *Clostridium clusters I*, *II* and *IV*, significantly ↓ levels of *Lactobacillus* and *Enterococcus* group	[177]
RS	6 Male C57BL/6J mice (18–20 month old and body weight not reported)	8 weeks	0.54 g/day	Mixed with chow	During the day	18 g/kg/day	↑ in number of *Bacteroidetes*, *Bifidobacterium* and *Akkermansia species*	[178]
	Sprague-Dawley rats (age 6 weeks and body weight not reported)	12 weeks	27% of the diet	Mixed with chow	During the day	18 g/kg/day	↑ in SCFAs	[180]
POS	Pigs' fecal inoculum (age 4 years and the mean body weight 233.0 ± 10.02 kg)	48 h	9 g/of POS to 1 mL of inoculum	Mixed with the chow	-	-	↑ in SCFAs	[183]
Lactulose	12 healthy volunteers (age = (24 to 31 years and body weight not reported)	4 weeks	20 g/day	Mixed with chow	Twice/day	285.7 mg/kg/day	↑ in number of *Bifidobacterium* and *Lactobacillus*.	[186]

Table 3. *Cont.*

Prebiotics	Model	Strategy/Duration of Feeding	Dose Supplemented	Form	No. of Applications	Re-Calculated Dose *	Fecal Microbial Changes Relative to Control	Reference
Lactosucrose	Red seabream *Pagrus major* (age and body weight not reported)	9 months	20 mg/kg/day	Mixed with chow	Once/day	20 mg/kg/day	↑ production of SCFAs	[192]
	8 Shepherd dogs (body weight = 22 to 32 kg; mean age = 13.5 months)	2 weeks	1.5 g/day	Mixed with chow	Twice/day	55.6 mg/kg/day	↓ in the levels of f *Clostridium pefringms* ↑ *Bifidobacterium*	[122]
	16 Broiler chickens (20–62 days and body weight not reported)	62 days	825 mg/day	Mixed with chow	During the day	458 mg/kg/day	↑ in the number of *Bifidobacterium* ↓ the number of *Bacteriodaceae; Staphylococci;* and total anaerobic bacteria, *C. perfringens*	[198]
	8 Cats (Mean agae + 7; body weight 3.5 kg)	2 weeks	50 mg of lactosucrose/day	Mixed with the chow	During the day	14 mg/kg/day	↑ in *Lactobacilli* and *Bifidobacterium* ↓ in *Clostridium perfringens*, clostridia, *Spirochaetaceae*, and *Enterobacteriaceae*	[185]
AX	10 human children (mean age, 3 years, 7 months body weight not reported) (in vitro)	48 h	10 g/liter	Dissolved in drinking water	-	–	↑ in number of *Lactobacillus*	[196]

Table 3. *Cont.*

Prebiotics	Model	Strategy/Duration of Feeding	Dose Supplemented	Form	No. of Applications	Re-Calculated Dose *	Fecal Microbial Changes Relative to Control	Reference
XOS	12 healthy adult women (mean age for women = 33.6 years and body weight not reported) and 11 healthy men (mean age = 30.1 and body weight not reported)	8 weeks	1.4 g/day or 2.8 g/day	Capsule	Once/day	20 or 40 mg/kg/day	↑ *Bacteroides fragilis*, ↑ *Bifidobacterium*	[199]
	13 elderly human (body weight = 58.6 ± 10.1 kg body weight not reported)	3 weeks	4 g/day	Mixed with chow	Once/day	68.3 mg/kg/day	↑ in number of *Bifidobacterium* species	[200]

↑—Increase, ↓—Decrease, * unless indicated, the average adult human weight was estimated as 70 kg and the average rat weight was estimated to be 280 g.

4.11. XOS

Animal studies have furnished evidence that oral administration of XOS remarkably increases fecal weight, bone properties, fecal moisture, and number of *Bifidobacteria*, with a parallel increase in SCFAs production in mice [201], rats [202], and humans (elderly) [200]. A recent study on a healthy human adult demonstrated that XOS intake increases *Bifidobacterium* counts without affecting the number of *Lactobacillus* [199]. The potential of *Bifidobacteria* to metabolize XOS is based on the activity of their xylan-degrading enzyme systems. Human study on the prebiotic XOS and their effects on modulating the GM in vivo is limited, particularly regarding the efficiency (Table 3).

Prebiotics are also able to remodulate the composition of the GM. Compared to a different category of prebiotics, only the fructans (inulin and FOS), GOS, and lactulose had highly selective effects on human GM modification [203]. As mentioned before, fermentation products of prebiotics such as SCFAs also have modulatory effects on the gut pH [204]. The pH alteration can have an influence on the population of acid-sensitive species, such as *Bacteroides*, and promote butyrate formation by *Firmicutes* [205].

5. Prebiotics for the Treatment of Obesity and Diabetes

Globally, the population of diabetes patient is increasing, imposing a great social and economic burden on public health [206,207]. T2DM is a chronic metabolic syndrome of abnormal lipid and glucose metabolism that leads to neuropathy, retinopathy, leg ulcers, and gangrene [208]. The factors that could have an impact on T2DM development are obesity, genetics, smoking, age, hypertension, and sedentary lifestyle [207]. In recent studies, it has been proposed that the remolding of the GM composition from obesity could lead to the pathogenesis of T2DM [209–213].

As mentioned above, the two dominant bacteria groups in human GIT are *Bacteroidetes* and *Firmicutes* [209]. A link between obesity and GM composition has been reported in humans, showing an increase in the number of *Firmicutes* and a decrease in the diversity of *Bacteroidetes* [213] (Figure 2).

Prebiotics have gained a considerable place in the management of obesity and diabetes due to their ability to modulate GM composition, thereby affecting the status of GIT and exerting anti-diabetic effects [214,215]. As prebiotics consist of different forms, their supplementation can be considered as a dietary therapy for the prevention and treatment of T2DM [216], and also in the fight against obesity by affecting food intake and appetite and metabolic activities [10] (Figure 2).

FOS have numerous desirable characteristics such as low calories, safety for diabetics, no carcinogenicity, and bifidus-stimulating functionality [65]. Due to these properties, FOS are considered a functional food ingredient that improves health status [217]. Increasing studies demonstrated the functional properties of FOS including the reduction of blood glucose levels, cholesterol levels, and lowering of blood pressure [218–220] (Table 4).

Meanwhile, inulin as a prebiotic has shown mixed results on the glycemic scale [221,222]. A study carried out on 54 middle-aged (between 35 and 65 years) healthy adults (men and women) as a double-blind, randomized, placebo-controlled parallel groups with 10 g of inulin supplementation for 8 weeks did not show any significant changes in the body weight [223]. A decrease in plasma insulin level was observed after 4 weeks of treatment and remained lower up to the 8th week, along with lower plasma triglyceride concentrations. Total cholesterol (TC) was lower in the inulin-supplemented group when compared to the placebo group. The study concluded that inulin supplementation may influence the degradation of triglyceride-rich lipoprotein particles [223]. In human trials conducted on obese women treated with inulin, greater proportions of *Bifidobacterium* and *Faecalibacterium* were observed, an effect that coincided with reduced fat mass and serum lipopolysaccharide [224]. The important role of *Bifidobacterium* in the fight against obesity has recently been demonstrated by *Bifidobacterium longum* both in preclinical obesity models and in humans [225].

Table 4. Effect of different prebiotics on the treatment of obesity and diabetes in animal and human studies.

Prebiotic Used	Tested Species	Dose	Re-Calculated Dose	Period	Outcomes	Reference
FOS	27 women with Type-2 diabetes, age = 20–65 years; 76.0 (12.2)	10 g/day	131.6 mg/kg/day	8 weeks	• ↓ Fasting plasma glucose (19.2 mg/dL; 9.50%), glycosylated hemoglobin (1.0%; 8.40%), interleukin-6 (1.3 pg/mL; 8.15%), tumor necrosis factor-α (3.0 pg/mL; 19.80%) and plasma lipopolysaccharide (6.0 EU/mL; 21.95%).	[220]
FOS	10-week-old C57BL/6J mice, body weight not reported	0.3 g/mouse/day		8 weeks	• plasma TG, LPS • ↑ plasma glucagon-like peptide-1 and colon proglucagon mRNA • ↑ colon L-cells number	[226]
GOS	6 rats alloxan-induced diabetic rats, 6 weeks old; Average weight = 90 g	100 g/kg of diet	1.11 g/kg of diet	42 days	• ↑ level of antioxidative enzymes • ↓ blood glucose, lipid profile, serum urea • ↓ fecal coliform count	[227]
	Human (women with overweight age 18–65 years and body weight not reported)	5.5 g/day of GOS	5.5 g/kg/day	12 weeks	• ↓ fasting insulin levels, triglycerides, TC, and HDL cholesterol • ↓ in fecal calprotectin	[228]
PDX	Rats (Wistar rats age not reported and body weight 43.0 ± 4.5 g)	5 g/100 g diet	5 g/100 g diet	60 days	• ↓ the of triglyceride (17%) • lowered the hepatic cholesterol • showed lower serum malondialdehyde	[229]
RS	Human (over weight and obese adults—11 men and 22 women age 18–69 years and body weight not reported)	15 g/kg/day of HAM-RS2 v. 30 g/kg/day HAM-RS21	15 g/kg/day of HAM-RS2 v. 30 g/kg/day HAM-RS21	4 weeks	• ↑ insulin sensitivity by insulin-modified intravenous Glucose Tolerance Test	[230]
	Human (diabetic adults age = 55 ± 2.4 years body weight not reported)	40 g/day of	571.4 mg/kg/day	12 weeks	• ↓ postprandial glucose by meal tolerance test • ↑ glucagon-like peptide-1 • ↓ tumor necrosis factor α	[231]
Lactulose	Human (patients with obesity age and body weight not reported)	8.2 g/day	-	2 days	• ↓ mean daytime glucose and insulin	[232]

Table 4. *Cont.*

Prebiotic Used	Tested Species	Dose	Re-Calculated Dose	Period	Outcomes	Reference
AX	Rats (wild type rats with high cholesterol diet age 7 weeks body weight not reported)	8% corn arabinoxylan	5.8 g/kg/day	20 days	• ↓ uptake of cholesterol from the diet • ↓ serum cholesterol levels • abbreviated cholesterol accumulation in the liver	[233]
	Human (T2DM); mean age = 55 years and body weight not reported)	49.2 g/day	702.9 mg/kg/day	35 days	• ↓ fasting serum glucose levels. • ↓ serum glucose and insulin level 2 h after oral glucose intake	[234]
XOS	Rats (Male Wistar rats treated with streptozotocin to induce diabetes, age = 8 weeks; body weight = 180 ± 8 g)	0.325 g/day	1.81 mg/kg/day	5 weeks	• ↓ diabetic weight loss • ↓ serum glucose, triglycerides	[235]

↑—Increase, ↓—Decrease.

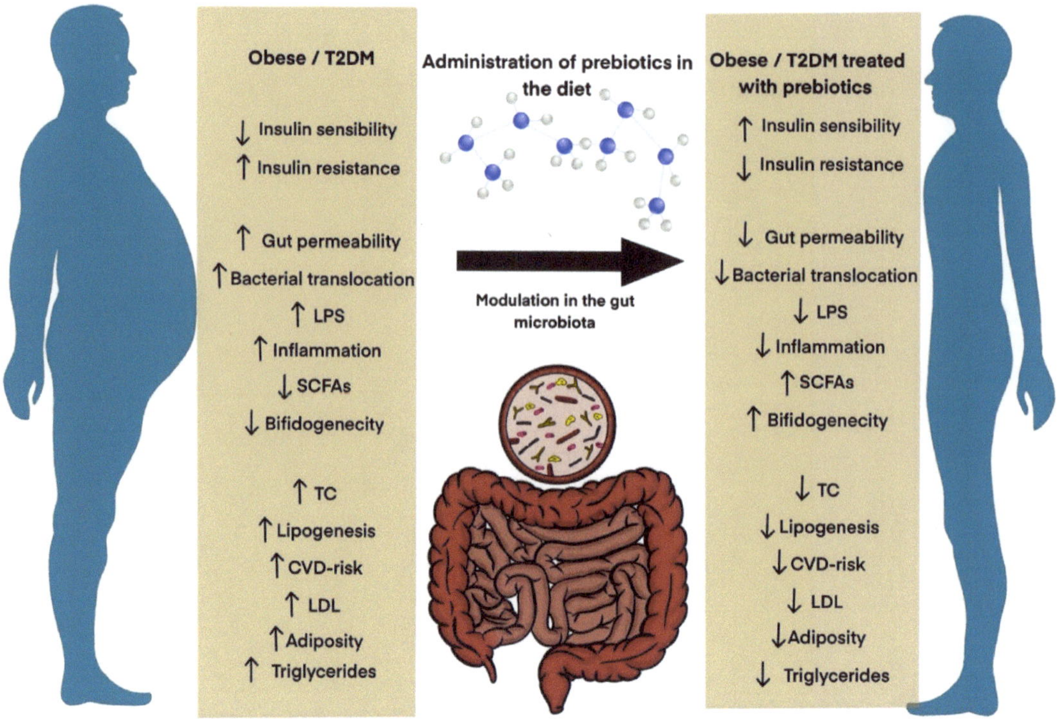

Figure 2. An overview of the improvement in the health of obese/T2DM patients treated by modulating their GM using prebiotics supplementation in a regular diet. Administration of prebiotics has the potential to modulate GM composition in patients suffering from T2DM and obesity and can be used as a therapeutic approach to cure the adverse effects of metabolic diseases. The daily intake of prebiotics in a designed diet has a major influence on GM by decreasing gut permeability, bacterial translocation, and reducing LPS-induced inflammation. However, this diet increases SCFAs and bifidogenecity in the gut, leading to lower TC levels, lipogenesis, LDL triglycerides, and adiposity, eventually resulting in lower risk of cardiovascular diseases.

XOS studies indicated that they have the potential to reduce serum cholesterol and triglycerides levels, which are the main risk factors for diabetes. The administration of XOS to wild-type rats for 28 days showed a reduction in LDL levels, TC in serum, triglycerides, and body weight [236] (Table 4).

The consumption of RS improved insulin sensitivity in subjects with metabolic syndrome and appears to have a favorable effect on insulin sensitivity [231] (Table 4).

In summary, prebiotics show efficiency not only in modulating or restructuring and stabilizing the host microbiome, but also in the regulation of many mechanisms associated with the development and metabolic consequences of obesity. Furthermore, prebiotics should be enriched in popular foods, increasing the chances of consistent consumption and improving overall health. At least, dietary prebiotic supplementation represents a safe, well-tolerated, and inexpensive therapeutic approach and should be considered as a potential therapy for the treatment and prevention of T2DM and obesity (Figure 2).

6. Conclusions and Future Perspectives

Recent studies have proven that increased inflammatory state (as seen in obesity and diabetes) has a paramount influence on glucose metabolism, and eubiosis ensures appropriate immune responses. This implies that the implementation of appropriate GM modulatory strategies could be a new and promising therapy against metabolic diseases. Meanwhile, the appropriate dosage, duration of treatment, and long-term effects of the intervention of different prebiotics remain unknown. For this reason, more clinical trials are needed before prebiotics can be rationally suggested for the prevention and/or treatment of obesity and diabetes.

Although in-vivo and in-vitro studies conducted on animals and humans revealed that many prebiotics increase the growth of *Bifidobacterium* spp., and *Lactobacillus* spp. and cause a diverse change in the *Bacteroidetes* and *Firmicutes* phyla, it is still not fully understood how these carbohydrates interact with the GM with their widely diversified structures. Further research is required to reveal the mechanisms of these carbohydrates' structures on the GM and the host. In addition, it is well established that GM ferments the prebiotics, leading to the formation of SCFAs and acidification of the colonic contents. These by-products formed by the fermentation process play an extensive role in maintaining the host's health and ameliorate diseases. Despite advances in our understanding of prebiotics, there remain numerous knowledge gaps concerning the SCFAs molecular signaling mechanisms and their association with prebiotic chemical composition and structural conformations, along with their modulatory effects at the genetic, cellular, organelle, and systemic levels.

Meanwhile, the application of systems biology coupled with bioinformatics could be a powerful strategy to unveil mechanistic insights into the action of prebiotics on the gut microorganisms and lead to an understanding of how these compounds (and their metabolites) alter both microbial and host metabolic functions at the molecular level. These insights and population-based studies could uncover new strategies to improve dietary relevance and clinical efficacy.

Author Contributions: Conceptualization, A.M.; investigation, A.M.; writing—original draft preparation, A.M., E.B.-M.D., D.B. and A.B.; writing—review and editing, A.M., A.B., E.B.-M.D. and D.B.; visualization, A.M.; supervision, A.B.; project administration, A.B. All authors have read and agreed to the published version of the manuscript.

Funding: This project received funding from the European Regional Development Fund (project No. 01.2.2-LMT-K-718-02-0014) under grant agreement with the Research Council of Lithuania (LMT LT).

Institutional Review Board Statement: Not applicable.

Informed Consent Statement: Not applicable.

Data Availability Statement: Not applicable.

Acknowledgments: We thank Virginija Bukelskienė for the correction of the first draft of the manuscript.

Conflicts of Interest: The authors declare no conflict of interest.

Abbreviations

GIT—gastrointestinal tract, GM—gut microbiota, SCFAs—short-chain fatty acids, DP—degree of polymerization, FOS—fructo-oligosaccharides, GOS—galacto-oligosaccharides, HMO—human milk oligosaccharide, PDX—polydextrose, RS—resistant starch, POS—pectin oligosaccharides, AX—arabinoxylans, XOS—Xylooligosaccharide, T2DM—Type-2 diabetes mellitus, TC—total cholesterol, LDL—low-density lipoprotein, HDL—high-density lipoprotein.

References

1. Saeedi, P.; Petersohn, I.; Salpea, P.; Malanda, B.; Karuranga, S.; Unwin, N.; Colagiuri, S.; Guariguata, L.; Motala, A.A.; Ogurtsova, K.; et al. Global and regional diabetes prevalence estimates for 2019 and projections for 2030 and 2045: Results from the International Diabetes Federation Diabetes Atlas, 9(th) edition. *Diabetes Res. Clin. Pract.* **2019**, *157*, 107843. [CrossRef] [PubMed]
2. Morigny, P.; Houssier, M.; Mouisel, E.; Langin, D. Adipocyte lipolysis and insulin resistance. *Biochimie* **2016**, *125*, 259–266. [CrossRef] [PubMed]
3. Pascale, A.; Marchesi, N.; Govoni, S.; Coppola, A.; Gazzaruso, C. The role of gut microbiota in obesity, diabetes mellitus, and effect of metformin: New insights into old diseases. *Curr. Opin. Pharmacol.* **2019**, *49*, 1–5. [CrossRef]

4. Arnoriaga-Rodríguez, M.; Mayneris-Perxachs, J.; Contreras-Rodríguez, O.; Burokas, A.; Ortega-Sanchez, J.-A.; Blasco, G.; Coll, C.; Biarnés, C.; Castells-Nobau, A.; Puig, J.; et al. Obesity-associated deficits in inhibitory control are phenocopied to mice through gut microbiota changes in one-carbon and aromatic amino acids metabolic pathways. *Gut* **2021**, *70*, 2283–2296. [CrossRef]
5. Arnoriaga-Rodríguez, M.; Mayneris-Perxachs, J.; Burokas, A.; Contreras-Rodríguez, O.; Blasco, G.; Coll, C.; Biarnés, C.; Miranda-Olivos, R.; Latorre, J.; Moreno-Navarrete, J.-M.; et al. Obesity Impairs Short-Term and Working Memory through Gut Microbial Metabolism of Aromatic Amino Acids. *Cell Metab.* **2020**, *32*, 548–560.e7. [CrossRef] [PubMed]
6. Arnoriaga-Rodríguez, M.; Mayneris-Perxachs, J.; Burokas, A.; Pérez-Brocal, V.; Moya, A.; Portero-Otin, M.; Ricart, W.; Maldonado, R.; Fernández-Real, J.-M. Gut bacterial ClpB-like gene function is associated with decreased body weight and a characteristic microbiota profile. *Microbiome* **2020**, *8*, 59. [CrossRef]
7. Milani, C.; Duranti, S.; Bottacini, E.; Casey, E.; Turroni, F.; Mahony, J.; Belzer, C.; Montes, A.S.; Mancabelli, L.; Lugli, G.A.; et al. The First Microbial Colonizers of the Human Gut: Composition, Activities, and Health Implications of the Infant Gut Microbiota. *Microbiol. Mol. Biol. Rev.* **2017**, *81*, 1–67. [CrossRef]
8. Adak, A.; Khan, M.R. An insight into gut microbiota and its functionalities. *Cell. Mol. Life Sci.* **2019**, *76*, 473–493. [CrossRef]
9. Avershina, E.; Lundgård, K.; Sekelja, M.; Dotterud, C.; Storrø, O.; Øien, T.; Johnsen, R.; Rudi, K. Transition from infant- to adult-like gut microbiota. *Environ. Microbiol.* **2016**, *18*, 2226–2236. [CrossRef]
10. Cerdó, T.; García-Santos, J.A.; Bermúdez, M.G.; Campoy, C. The role of probiotics and prebiotics in the prevention and treatment of obesity. *Nutrients* **2019**, *11*, 635. [CrossRef]
11. Peterson, J.; Garges, S.; Giovanni, M.; McInnes, P.; Wang, L.; Schloss, J.A.; Bonazzi, V.; McEwen, J.E.; Wetterstrand, K.A.; Deal, C.; et al. The NIH Human Microbiome Project. *Genome Res.* **2009**, *19*, 2317–2323. [CrossRef] [PubMed]
12. Rowland, I.; Gibson, G.; Heinken, A.; Scott, K.; Swann, J.; Thiele, I.; Tuohy, K. Gut microbiota functions: Metabolism of nutrients and other food components. *Eur. J. Nutr.* **2018**, *57*, 1–24. [CrossRef] [PubMed]
13. Dominguez-Bello, M.G.; Godoy-Vitorino, F.; Knight, R.; Blaser, M.J. Role of the microbiome in human development. *Gut* **2019**, *68*, 1108–1114. [CrossRef] [PubMed]
14. Megur, A.; Baltriukienė, D.; Bukelskienė, V.; Burokas, A. The microbiota–gut–brain axis and Alzheimer's disease: Neuroinflammation is to blame? *Nutrients* **2021**, *13*, 37. [CrossRef]
15. Sommer, F.; Bäckhed, F. The gut microbiota-masters of host development and physiology. *Nat. Rev. Microbiol.* **2013**, *11*, 227–238. [CrossRef]
16. Moschen, A.R.; Wieser, V.; Tilg, H. Dietary factors: Major regulators of the Gut's microbiota. *Gut Liver* **2012**, *6*, 411–416. [CrossRef]
17. Turnbaugh, P.J.; Ley, R.E.; Hamady, M.; Fraser-Liggett, C.M.; Knight, R.; Gordon, J.I. The Human Microbiome Project. *Nature* **2007**, *449*, 804–810. [CrossRef]
18. Marchesi, J.R.; Adams, D.H.; Fava, F.; Hermes, G.D.A.; Hirschfield, G.M.; Hold, G.; Quraishi, M.N.; Kinross, J.; Smidt, H.; Tuohy, K.M.; et al. The gut microbiota and host health: A new clinical frontier. *Gut* **2016**, *65*, 330–339. [CrossRef]
19. Mayneris-Perxachs, J.; Arnoriaga-Rodríguez, M.; Luque-Córdoba, D.; Priego-Capote, F.; Pérez-Brocal, V.; Moya, A.; Burokas, A.; Maldonado, R.; Fernández-Real, J.-M. Gut microbiota steroid sexual dimorphism and its impact on gonadal steroids: Influences of obesity and menopausal status. *Microbiome* **2020**, *8*, 136. [CrossRef]
20. De Angelis, M.; Garruti, G.; Minervini, F.; Bonfrate, L.; Portincasa, P.; Gobbetti, M. The Food-gut Human Axis: The Effects of Diet on Gut Microbiota and Metabolome. *Curr. Med. Chem.* **2019**, *26*, 3567–3583. [CrossRef]
21. Janssen, A.W.F.; Kersten, S. The role of the gut microbiota in metabolic health. *FASEB J.* **2015**, *29*, 3111–3123. [CrossRef] [PubMed]
22. Slavin, J. Fiber and prebiotics: Mechanisms and health benefits. *Nutrients* **2013**, *5*, 1417–1435. [CrossRef] [PubMed]
23. Gibson, G.R.; Roberfroid, M.B. Dietary modulation of the human colonic microbiota: Introducing the concept of prebiotics. *J. Nutr.* **1995**, *125*, 1401–1412. [CrossRef] [PubMed]
24. Davani-Davari, D.; Negahdaripour, M.; Karimzadeh, I.; Seifan, M.; Mohkam, M.; Masoumi, S.J.; Berenjian, A.; Ghasemi, Y. Prebiotics: Definition, types, sources, mechanisms, and clinical applications. *Foods* **2019**, *8*, 92. [CrossRef] [PubMed]
25. Vallianou, N.; Stratigou, T.; Christodoulatos, G.S.; Tsigalou, C.; Dalamaga, M. Probiotics, Prebiotics, Synbiotics, Postbiotics, and Obesity: Current Evidence, Controversies, and Perspectives. *Curr. Obes. Rep.* **2020**, *9*, 179–192. [CrossRef] [PubMed]
26. Da Silva, T.F.; Casarotti, S.N.; de Oliveira, G.L.V.; Penna, A.L.B. The impact of probiotics, prebiotics, and synbiotics on the biochemical, clinical, and immunological markers, as well as on the gut microbiota of obese hosts. *Crit. Rev. Food Sci. Nutr.* **2021**, *61*, 337–355. [CrossRef]
27. Salminen, S.; Collado, M.C.; Endo, A.; Hill, C.; Lebeer, S.; Quigley, E.M.M.; Sanders, M.E.; Shamir, R.; Swann, J.R.; Szajewska, H.; et al. The International Scientific Association of Probiotics and Prebiotics (ISAPP) consensus statement on the definition and scope of postbiotics. *Nat. Rev. Gastroenterol. Hepatol.* **2021**, *18*, 649–667. [CrossRef]
28. Holscher, H.D. Dietary fiber and prebiotics and the gastrointestinal microbiota. *Gut Microbes* **2017**, *8*, 172–184. [CrossRef]
29. Pluta, R.; Ułamek-Kozioł, M.; Januszewski, S.; Czuczwar, S.J. Gut microbiota and pro/prebiotics in Alzheimer's disease. *Aging* **2020**, *12*, 5539–5550. [CrossRef]
30. Shokryazdan, P.; Faseleh Jahromi, M.; Navidshad, B.; Liang, J.B. Effects of prebiotics on immune system and cytokine expression. *Med. Microbiol. Immunol.* **2017**, *206*, 1–9. [CrossRef]
31. Macfarlane, G.T.; Macfarlane, S. Fermentation in the human large intestine: Its physiologic consequences and the potential contribution of prebiotics. *J. Clin. Gastroenterol.* **2011**, *45*, 120–127. [CrossRef] [PubMed]

32. Venegas, D.P.; De La Fuente, M.K.; Landskron, G.; González, M.J.; Quera, R.; Dijkstra, G.; Harmsen, H.J.M.; Faber, K.N.; Hermoso, M.A. Short chain fatty acids (SCFAs)mediated gut epithelial and immune regulation and its relevance for inflammatory bowel diseases. *Front. Immunol.* **2019**, *10*, 277. [CrossRef] [PubMed]
33. Peng, L.; Li, Z.-R.; Green, R.S.; Holzman, I.R.; Lin, J. Butyrate enhances the intestinal barrier by facilitating tight junction assembly via activation of AMP-activated protein kinase in Caco-2 cell monolayers. *J. Nutr.* **2009**, *139*, 1619–1625. [CrossRef] [PubMed]
34. Gaudier, E.; Rival, M.; Buisine, M.-P.; Robineau, I.; Hoebler, C. Butyrate enemas upregulate Muc genes expression but decrease adherent mucus thickness in mice colon. *Physiol. Res.* **2009**, *58*, 111–119. [CrossRef]
35. O'Keefe, S.J.D. Diet, microorganisms and their metabolites, and colon cancer. *Nat. Rev. Gastroenterol. Hepatol.* **2016**, *13*, 691–706. [CrossRef]
36. Wan, M.L.Y.; Ling, K.H.; El-Nezami, H.; Wang, M.F. Influence of functional food components on gut health. *Crit. Rev. Food Sci. Nutr.* **2019**, *59*, 1927–1936. [CrossRef]
37. Bengmark, S. Colonic food: Pre- and probiotics. *Am. J. Gastroenterol.* **2000**, *95*, 9–11. [CrossRef]
38. Zdunczyk, Z. Physiological effect of low digestible oligosaccharides in diets for animals and humans. *Pol. J. Food Nutr. Sci.* **2004**, *13*, 115–130.
39. Leeman, A.; Sjöö, M.; Eliasson, A.-C.; Björck, I. Resistant starch formation in temperature treated potato starches varying in amylose/amylopectin ratio. *Carbohydr. Polym.* **2006**, *65*, 306–313. [CrossRef]
40. Dai, F.J.; Chau, C.F. Classification and regulatory perspectives of dietary fiber. *J. Food Drug Anal.* **2017**, *25*, 37–42. [CrossRef]
41. Howlett, J.F.; Betteridge, V.A.; Champ, M.; Craig, S.A.S.; Meheust, A.; Jones, J.M. The definition of dietary fiber—Discussions at the Ninth Vahouny Fiber Symposium: Building scientific agreement. *Food Nutr. Res.* **2010**, *54*, 1–5. [CrossRef] [PubMed]
42. Slizewska, K.; Chlebicz-Wójcik, A. The in vitro analysis of prebiotics to be used as a component of a synbiotic preparation. *Nutrients* **2020**, *12*, 1272. [CrossRef] [PubMed]
43. Mohanty, D.P.; Mohapatra, S.; Misra, S.; Sahu, P.S. Milk derived bioactive peptides and their impact on human health—A review. *Saudi J. Biol. Sci.* **2016**, *23*, 577–583. [CrossRef] [PubMed]
44. Bindels, L.B.; Delzenne, N.M.; Cani, P.D.; Walter, J. Towards a more comprehensive concept for prebiotics. *Nat. Rev. Gastroenterol. Hepatol.* **2015**, *12*, 303–310. [CrossRef] [PubMed]
45. Markowiak, P.; Śliżewska, K. Effects of Probiotics, Prebiotics, and Synbiotics on Human Health. *Nutrients* **2017**, *9*, 1021. [CrossRef] [PubMed]
46. Simpson, H.L.; Campbell, B.J. Review article: Dietary fibre-microbiota interactions. *Aliment. Pharmacol. Ther.* **2015**, *42*, 158–179. [CrossRef]
47. Hanau, S.; Almugadam, S.H.; Sapienza, E.; Cacciari, B.; Manfrinato, M.C.; Trentini, A.; Kennedy, J.F. Schematic overview of oligosaccharides, with survey on their major physiological effects and a focus on milk ones. *Carbohydr. Polym. Technol. Appl.* **2020**, *1*, 100013. [CrossRef]
48. Swanson, K.S.; de Vos, W.M.; Martens, E.C.; Gilbert, J.A.; Menon, R.S.; Soto-Vaca, A.; Hautvast, J.; Meyer, P.D.; Borewicz, K.; Vaughan, E.E.; et al. Effect of fructans, prebiotics and fibres on the human gut microbiome assessed by 16S rRNA-based approaches: A review. *Benef. Microbes* **2020**, *11*, 101–129. [CrossRef]
49. Singla, V.; Chakkaravarthi, S. Applications of prebiotics in food industry: A review. *Food Sci. Technol. Int.* **2017**, *23*, 649–667. [CrossRef]
50. De Jesus Raposo, M.F.; de Morais, A.M.M.B.; de Morais, R.M.S.C. Emergent Sources of Prebiotics: Seaweeds and Microalgae. *Mar. Drugs* **2016**, *14*, 27. [CrossRef]
51. Flesch, A.G.T.; Poziomyck, A.K.; Damin, D.C. The therapeutic use of symbiotics. *Braz. Arch. Dig. Surg.* **2014**, *27*, 206–209. [CrossRef] [PubMed]
52. Lordan, C.; Thapa, D.; Ross, R.P.; Cotter, P.D. Potential for enriching next-generation health-promoting gut bacteria through prebiotics and other dietary components. *Gut Microbes* **2020**, *11*, 1–20. [CrossRef] [PubMed]
53. Den Besten, G.; Van Eunen, K.; Groen, A.K.; Venema, K.; Reijngoud, D.J.; Bakker, B.M. The role of short-chain fatty acids in the interplay between diet, gut microbiota, and host energy metabolism. *J. Lipid Res.* **2013**, *54*, 2325–2340. [CrossRef] [PubMed]
54. Guarino, M.P.L.; Altomare, A.; Emerenziani, S.; Di Rosa, C.; Ribolsi, M.; Balestrieri, P.; Iovino, P.; Rocchi, G.; Cicala, M. Mechanisms of action of prebiotics and their effects on gastro-intestinal disorders in adults. *Nutrients* **2020**, *12*, 1037. [CrossRef]
55. Uauy, R. Foreword. *Br. J. Nutr.* **2009**, *101*, S1. [CrossRef]
56. Gibson, G.R.; Hutkins, R.; Sanders, M.E.; Prescott, S.L.; Reimer, R.A.; Salminen, S.J.; Scott, K.; Stanton, C.; Swanson, K.S.; Cani, P.D.; et al. Expert consensus document: The International Scientific Association for Probiotics and Prebiotics (ISAPP) consensus statement on the definition and scope of prebiotics. *Nat. Rev. Gastroenterol. Hepatol.* **2017**, *14*, 491–502. [CrossRef]
57. Cardoso, B.B.; Amorim, C.; Silvério, S.C.; Rodrigues, L.R. Novel and emerging prebiotics: Advances and opportunities. *Adv. Food Nutr. Res.* **2021**, *95*, 41–95. [CrossRef]
58. Kelly, G. Inulin-type prebiotics: A review (Part 2). *Altern. Med. Rev.* **2009**, *14*, 36–55.
59. Yeung, C.K.; Glahn, R.P.; Welch, R.M.; Miller, D.D. Prebiotics and iron bioavailability—Is there a connection? *J. Food Sci.* **2005**, *70*, 584–590. [CrossRef]
60. Ibrahim, O.O. Functional Oligosaccharide: Chemicals Structure, Manufacturing, Health Benefits, Applications and Regulations. *J. Food Chem. Nanotechnol.* **2018**, *4*, 65–76. [CrossRef]
61. Roberfroid, M. Prebiotics: The concept revisited. *J. Nutr.* **2007**, *137*, 830S–837S. [CrossRef] [PubMed]

62. Cui, S.W.; Roberts, K.T. *Dietary Fiber. Fulfilling the Promise of Added-Value Formulations*; Academic Press: Cambridge, MA, USA, 2009; ISBN 9780123741950.
63. Chatterjee, P.; Ojiambo, M. *Food Ingredients*; Elsevier Inc.: Amsterdam, The Netherlands, 2014; Volume 11, ISBN 9780128113721.
64. Korczak, R.; Slavin, J.L. Fructooligosaccharides and appetite. *Curr. Opin. Clin. Nutr. Metab. Care* **2018**, *21*, 377–380. [CrossRef] [PubMed]
65. Yun, J.W. Fructooligosaccharides—Occurrence, preparation, and application. *Enzyme Microb. Technol.* **1996**, *19*, 107–117. [CrossRef]
66. Crittenden, R.; Playne, M. Production, properties and applications of food-grade oligosaccharides. *Trends Food Sci. Technol.* **1996**, *7*, 353–361. [CrossRef]
67. Sako, T.; Tanaka, R. Prebiotics: Types. In *Encyclopedia of Dairy Sciences*, 2nd ed.; Elsevier Ltd.: Amsterdam, The Netherlands, 2011; pp. 354–364. [CrossRef]
68. Commercially Available Oligosaccharides. Available online: https://agris.fao.org/agris-search/search.do?recordID=BE9601560 (accessed on 1 April 2022).
69. Tanaka, R.; Takayama, H.; Morotomi, M.; Kuroshima, T.; Ueyama, S.; Matsumoto, K.; Kuroda, A.; Mutai, M. Effects of Administration of TOS and Bifidobacterium breve 4006 on the Human Fecal Flora. *Bifidobact. Microflora* **1983**, *2*, 17–24. [CrossRef]
70. Mahoney, R.R. Galactosyl-oligosaccharide formation during lactose hydrolysis: A review. *Food Chem.* **1998**, *63*, 147–154. [CrossRef]
71. Bao, Y.; Chen, C.; Newburg, D.S. Quantification of neutral human milk oligosaccharides by graphitic carbon high-performance liquid chromatography with tandem mass spectrometry. *Anal. Biochem.* **2013**, *433*, 28–35. [CrossRef]
72. Thongaram, T.; Hoeflinger, J.L.; Chow, J.; Miller, M.J. Human milk oligosaccharide consumption by probiotic and human-associated bifidobacteria and lactobacilli. *J. Dairy Sci.* **2017**, *100*, 7825–7833. [CrossRef]
73. Akkerman, R.; Faas, M.M.; de Vos, P. Non-digestible carbohydrates in infant formula as substitution for human milk oligosaccharide functions: Effects on microbiota and gut maturation. *Crit. Rev. Food Sci. Nutr.* **2019**, *59*, 1486–1497. [CrossRef]
74. Xu, G.; Davis, J.C.; Goonatilleke, E.; Smilowitz, J.T.; German, J.B.; Lebrilla, C.B. Absolute Quantitation of Human Milk Oligosaccharides Reveals Phenotypic Variations during Lactation. *J. Nutr.* **2017**, *147*, 117–124. [CrossRef]
75. Chaturvedi, P.; Warren, C.D.; Altaye, M.; Morrow, A.L.; Ruiz-Palacios, G.; Pickering, L.K.; Newburg, D.S. Fucosylated human milk oligosaccharides vary between individuals and over the course of lactation. *Glycobiology* **2001**, *11*, 365–372. [CrossRef] [PubMed]
76. Thurl, S.; Munzert, M.; Henker, J.; Boehm, G.; Müller-Werner, B.; Jelinek, J.; Stahl, B. Variation of human milk oligosaccharides in relation to milk groups and lactational periods. *Br. J. Nutr.* **2010**, *104*, 1261–1271. [CrossRef] [PubMed]
77. Thomson, P.; Medina, D.A.; Garrido, D. Human milk oligosaccharides and infant gut bifidobacteria: Molecular strategies for their utilization. *Food Microbiol.* **2018**, *75*, 37–46. [CrossRef] [PubMed]
78. Bode, L. Human milk oligosaccharides: Every baby needs a sugar mama. *Glycobiology* **2012**, *22*, 1147–1162. [CrossRef]
79. Cheng, Y.-J.; Yeung, C.-Y. Recent advance in infant nutrition: Human milk oligosaccharides. *Pediatr. Neonatol.* **2021**, *62*, 347–353. [CrossRef] [PubMed]
80. Morozov, V.; Hansman, G.; Hanisch, F.G.; Schroten, H.; Kunz, C. Human Milk Oligosaccharides as Promising Antivirals. *Mol. Nutr. Food Res.* **2018**, *62*, 1–46. [CrossRef]
81. Asakuma, S.; Hatakeyama, E.; Urashima, T.; Yoshida, E.; Katayama, T.; Yamamoto, K.; Kumagai, H.; Ashida, H.; Hirose, J.; Kitaoka, M. Physiology of Consumption of Human Milk Oligosaccharides by Infant Gut-associated Bifidobacteria. *J. Biol. Chem.* **2011**, *286*, 34583–34592. [CrossRef]
82. Vandenplas, Y.; Berger, B.; Carnielli, V.P.; Ksiazyk, J.; Lagström, H.; Luna, M.S.; Migacheva, N.; Mosselmans, J.M.; Picaud, J.C.; Possner, M.; et al. Human milk oligosaccharides: 2′-fucosyllactose (2′-FL) and lacto-n-neotetraose (LNnT) in infant formula. *Nutrients* **2018**, *10*, 1161. [CrossRef]
83. Corona, L.; Lussu, A.; Bosco, A.; Pintus, R.; Cesare Marincola, F.; Fanos, V.; Dessì, A. Human Milk Oligosaccharides: A Comprehensive Review towards Metabolomics. *Children* **2021**, *8*, 804. [CrossRef]
84. Marcobal, A.; Barboza, M.; Froehlich, J.W.; Block, D.E.; German, J.B.; Lebrilla, C.B.; Mills, D.A. Consumption of human milk oligosaccharides by gut-related microbes. *J. Agric. Food Chem.* **2010**, *58*, 5334–5340. [CrossRef]
85. Donovan, S.M.; Comstock, S.S. Human Milk Oligosaccharides Influence Neonatal Mucosal and Systemic Immunity. *Ann. Nutr. Metab.* **2016**, *69* (Suppl. 2), 42–51. [CrossRef] [PubMed]
86. Comstock, S.S.; Li, M.; Wang, M.; Monaco, M.H.; Kuhlenschmidt, T.B.; Kuhlenschmidt, M.S.; Donovan, S.M. Dietary Human Milk Oligosaccharides but Not Prebiotic Oligosaccharides Increase Circulating Natural Killer Cell and Mesenteric Lymph Node Memory T Cell Populations in Noninfected and Rotavirus-Infected Neonatal Piglets. *J. Nutr.* **2017**, *147*, 1041–1047. [CrossRef] [PubMed]
87. Bode, L. The functional biology of human milk oligosaccharides. *Early Hum. Dev.* **2015**, *91*, 619–622. [CrossRef]
88. Jacobi, S.K.; Yatsunenko, T.; Li, D.; Dasgupta, S.; Yu, R.K.; Berg, B.M.; Chichlowski, M.; Odle, J. Dietary Isomers of Sialyllactose Increase Ganglioside Sialic Acid Concentrations in the Corpus Callosum and Cerebellum and Modulate the Colonic Microbiota of Formula-Fed Piglets. *J. Nutr.* **2016**, *146*, 200–208. [CrossRef] [PubMed]
89. Rousseaux, A.; Brosseau, C.; Le Gall, S.; Piloquet, H.; Barbarot, S.; Bodinier, M. Human Milk Oligosaccharides: Their Effects on the Host and Their Potential as Therapeutic Agents. *Front. Immunol.* **2021**, *12*, 680911. [CrossRef] [PubMed]
90. Luo, Y.; Tu, Y.; Ren, F.; Zhang, H. Characterization and functional properties of Maillard reaction products of β-lactoglobulin and polydextrose. *Food Chem.* **2021**, *377*, 131749. [CrossRef]

91. Lahtinen, S.J.; Knoblock, K.; Drakoularakou, A.; Jacob, M.; Stowell, J.; Gibson, G.R.; Ouwehand, A.C. Effect of molecule branching and glycosidic linkage on the degradation of polydextrose by gut microbiota. *Biosci. Biotechnol. Biochem.* **2010**, *74*, 2016–2021. [CrossRef]
92. Caselato, V.; Ouwehand, A.; Flavors, I.; Andreollo, N.A. Polydextrose: Physiological Function, and Effects on Health. *Nutrients* **2016**, *8*, 553. [CrossRef]
93. Zhong, J.; Luo, B.; Xiang, M.; Liu, H.; Zhai, Z.; Wang, T.; Craig, S.A.S. Studies on the effects of polydextrose intake on physiologic funtions in Chinese people. *Am. J. Clin. Nutr.* **2000**, *72*, 1503–1509. [CrossRef]
94. Yao, N.; Paez, A.V.; White, P.J. Structure and function of starch and resistant starch from corn with different doses of mutant amylose-extender and floury-1 alleles. *J. Agric. Food Chem.* **2009**, *57*, 2040–2048. [CrossRef]
95. Fuentes-Zaragoza, E.; Sánchez-Zapata, E.; Sendra, E.; Sayas, E.; Navarro, C.; Fernández-Lõpez, J.; Pérez-Alvarez, J.A. Resistant starch as prebiotic: A review. *Starch/Staerke* **2011**, *63*, 406–415. [CrossRef]
96. Marlatt, K.L.; White, U.A.; Beyl, R.A.; Peterson, C.M.; Corby, K.; Marco, M.L.; Keenan, M.J.; Martin, R.J.; Aryana, K.J.; Ravussin, E.; et al. Role of resistant starch on diabetes risk factors in people with prediabetes: Design, conduct, and baseline results of the STARCH trial. *Contemp. Clin. Trials* **2018**, *65*, 99–108. [CrossRef] [PubMed]
97. Tingirikari, J.M.R. Microbiota-accessible pectic poly- and oligosaccharides in gut health. *Food Funct.* **2018**, *9*, 5059–5073. [CrossRef]
98. Yoo, H.D.; Kim, D.; Paek, S.H.; Oh, S.E. Plant cell wall polysaccharides as potential resources for the development of novel prebiotics. *Biomol. Ther.* **2012**, *20*, 371–379. [CrossRef] [PubMed]
99. Yang, J.; Maldonado-Gómez, M.X.; Hutkins, R.W.; Rose, D.J. Production and in vitro Fermentation of Soluble, Non-digestible, Feruloylated Oligo- and Polysaccharides from Maize and Wheat Brans. *J. Agric. Food Chem.* **2014**, *62*, 159–166. [CrossRef]
100. Moon, J.S.; Shin, S.Y.; Choi, H.S.; Joo, W.; Cho, S.K.; Li, L.; Kang, J.-H.; Kim, T.-J.; Han, N.S. In vitro digestion and fermentation properties of linear sugar-beet arabinan and its oligosaccharides. *Carbohydr. Polym.* **2015**, *131*, 50–56. [CrossRef]
101. Wilkowska, A.; Nowak, A.; Antczak-Chrobot, A.; Motyl, I.; Czyżowska, A.; Paliwoda, A. Structurally Different Pectic Oligosaccharides Produced from Apple Pomace and Their Biological Activity In Vitro. *Foods* **2019**, *8*, 365. [CrossRef]
102. Zhang, W.; Xu, P.; Zhang, H. Pectin in cancer therapy: A review. *Trends Food Sci. Technol.* **2015**, *44*, 258–271. [CrossRef]
103. Vera, C.; Córdova, A.; Aburto, C.; Guerrero, C.; Suárez, S.; Illanes, A. Synthesis and purification of galacto-oligosaccharides: State of the art. *World J. Microbiol. Biotechnol.* **2016**, *32*, 197. [CrossRef]
104. Zhang, W.; Wang, T.; Chen, X.; Pang, X.; Zhang, S.; Obaroakpo, J.U.; Shilong, J.; Lu, J.; Lv, J. Absolute quantification of twelve oligosaccharides in human milk using a targeted mass spectrometry-based approach. *Carbohydr. Polym.* **2019**, *219*, 328–333. [CrossRef]
105. German, J.; Freeman, S.; Lebrilla, C.; Mills, D. Human milk oligosaccharides: Evolution, structures and bioselectivity as substrates for intestinal bacteria. *Nestle Nutr. Work. Ser. Pediatr. Progr.* **2008**, *62*, 205–218. [CrossRef]
106. Ackerman, D.L.; Craft, K.M.; Townsend, S.D. Infant food applications of complex carbohydrates: Structure, synthesis, and function. *Carbohydr. Res.* **2017**, *437*, 16–27. [CrossRef] [PubMed]
107. Tungland, B.C.; Meyer, D. Nondigestible oligo-and polysaccharides (dietary fiber): Their physiology and role in human health and food. *Compr. Rev. Food Sci. Food Saf.* **2002**, *1*, 90–109. [CrossRef] [PubMed]
108. Xu, W.; Liu, Q.; Yu, S.; Zhang, T.; Mu, W. Synthesis of Lactosucrose Using a Recombinant Levansucrase from *Brenneria goodwinii*. *Appl. Biochem. Biotechnol.* **2018**, *186*, 292–305. [CrossRef] [PubMed]
109. Langenaeken, N.A.; De Schutter, D.P.; Courtin, C.M. Arabinoxylan from non-malted cereals can act as mouthfeel contributor in beer. *Carbohydr. Polym.* **2020**, *239*, 116257. [CrossRef] [PubMed]
110. Gérard, D.; Méline, T.; Muzard, M.; Deleu, M.; Plantier-Royon, R.; Rémond, C. Enzymatically-synthesized xylo-oligosaccharides laurate esters as surfactants of interest. *Carbohydr. Res.* **2020**, *495*, 108090. [CrossRef] [PubMed]
111. Aachary, A.A.; Prapulla, S.G. Corncob-Induced endo-1,4-β-d-Xylanase of Aspergillus oryzae MTCC 5154: Production and Characterization of Xylobiose from Glucuronoxylan. *J. Agric. Food Chem.* **2008**, *56*, 3981–3988. [CrossRef]
112. Carvalho, A.F.A.; de Oliva Neto, P.; da Silva, D.F.; Pastore, G.M. Xylo-oligosaccharides from lignocellulosic materials: Chemical structure, health benefits and production by chemical and enzymatic hydrolysis. *Food Res. Int.* **2013**, *51*, 75–85. [CrossRef]
113. Shendurse, A.M.; Khedkar, C.D. Lactose. In *Encyclopedia of Food and Health*; Caballero, B., Finglas, P.M., Toldrá, F., Eds.; Academic Press: Oxford, UK, 2016; pp. 509–516. ISBN 978-0-12-384953-3.
114. Panesar, P.S.; Bali, V. Prebiotics. In *Encyclopedia of Food and Health*; Caballero, B., Finglas, P.M., Toldrá, F., Eds.; Academic Press: Oxford, UK, 2016; pp. 464–471. ISBN 978-0-12-384953-3.
115. Sivaprakasam, S.; Prasad, P.D.; Singh, N. Benefits of short-chain fatty acids and their receptors in inflammation and carcinogenesis. *Pharmacol. Ther.* **2016**, *164*, 144–151. [CrossRef]
116. Prasad, S.; Dhiman, R.K.; Duseja, A.; Chawla, Y.K.; Sharma, A.; Agarwal, R. Lactulose improves cognitive functions and health-related quality of life in patients with cirrhosis who have minimal hepatic encephalopathy. *Hepatology* **2007**, *45*, 549–559. [CrossRef]
117. Elkington, S.G. Lactulose. *Gut* **1970**, *11*, 1043–1048. [CrossRef] [PubMed]
118. Song, M.; Zeng, J.; Jia, T.; Gao, H.; Zhang, R.; Jiang, J.; Li, G.; Su, T. Effects of sialylated lactulose on the mouse intestinal microbiome using Illumina high-throughput sequencing. *Appl. Microbiol. Biotechnol.* **2019**, *103*, 9067–9076. [CrossRef] [PubMed]
119. Mu, W.; Chen, Q.; Wang, X.; Zhang, T.; Jiang, B. Current studies on physiological functions and biological production of lactosucrose. *Appl. Microbiol. Biotechnol.* **2013**, *97*, 7073–7080. [CrossRef] [PubMed]

120. Fernando, W.M.U.; Hill, J.E.; Zello, G.A.; Tyler, R.T.; Dahl, W.J.; Van Kessel, A.G. Diets supplemented with chickpea or its main oligosaccharide component raffinose modify faecal microbial composition in healthy adults. *Benef. Microbes* **2010**, *1*, 197–207. [CrossRef]
121. Silvério, S.C.; Macedo, E.A.; Teixeira, J.A.; Rodrigues, L.R. Perspectives on the biotechnological production and potential applications of lactosucrose: A review. *J. Funct. Foods* **2015**, *19*, 74–90. [CrossRef]
122. Terada, A.; Hara, H.; Oishi, T.; Matsui, S.; Mitsuoka, T.; Nakajyo, S.; Fujimori, I.; Hara, K. Effect of Dietary Lactosucrose on Faecal Flora and Faecal Metabolites of Dogs. *Microb. Ecol. Health Dis.* **1992**, *5*, 87–92. [CrossRef]
123. Hino, K.; Kurose, M.; Sakurai, T.; Inoue, S.; Oku, K.; Chaen, H.; Fukuda, S. Effect of Dietary Lactosucrose (4^G-β-D-Galactosylsucrose) on the Intestinal Immune Functions in Mice. *J. Appl. Glycosci.* **2007**, *54*, 169–172. [CrossRef]
124. Hoffman, W.F.; Gortner, R.A. The preparation and analysis of the various proteins of wheat flour with special reference to the globulin, albumin and proteose fractions. *Cereal Chem.* **2021**, *4*, 221–229.
125. Bastos, R.; Coelho, E.; Coimbra, M.A. 8-Arabinoxylans from cereal by-products: Insights into structural features, recovery, and applications. In *Woodhead Publishing Series in Food Science, Technology and Nutrition*; Galanakis, C., Ed.; Woodhead Publishing: Sawston, UK, 2018; pp. 227–251. ISBN 978-0-08-102162-0.
126. Izydorczyk, M.S. Arabinoxylans. In *Handbook of Hydrocolloids*, 2nd ed.; Elsevier: Amsterdam, The Netherlands, 2009; pp. 653–692. [CrossRef]
127. Paz-Samaniego, R.; Méndez-Encinas, M.; Fierro-Islas, J.M.; Marquez-Escalante, J.; Rascón-Chu, A.; Martinez-Lopez, A.L.; Carvajal-Millan, E. Ferulated arabinoxylans recovered from low-value maize by-products: Gelation and antioxidant capacity. In *Ferulic Acid: Antioxidant Properties, Uses and Potential Health Benefits*; Nova Science Publishers: Hauppauge, NY, USA, 2014; pp. 151–164.
128. Mendis, M.; Simsek, S. Arabinoxylans and human health. *Food Hydrocoll.* **2014**, *42*, 239–243. [CrossRef]
129. Lu, Z.X.; Walker, K.Z.; Muir, J.G.; Mascara, T.; O'Dea, K. Arabinoxylan fiber, a byproduct of wheat flour processing, reduces the postprandial glucose response in normoglycemic subjects. *Am. J. Clin. Nutr.* **2000**, *71*, 1123–1128. [CrossRef]
130. Mendis, M.; Leclerc, E.; Simsek, S. Arabinoxylans, gut microbiota and immunity. *Carbohydr. Polym.* **2016**, *139*, 159–166. [CrossRef] [PubMed]
131. Althubiani, A.S.; Al-Ghamdi, S.B.; Samreen; Qais, F.A.; Khan, M.S.; Ahmad, I.; Malak, H.A. *Plant-Derived Prebiotics and Its Health Benefits*; Elsevier Inc.: Amsterdam, The Netherlands, 2018; ISBN 9780128146200.
132. Gibson, G.R.; Probert, H.M.; Van Loo, J.; Rastall, R.A.; Roberfroid, M.B. Dietary modulation of the human colonic microbiota: Updating the concept of prebiotics. *Nutr. Res. Rev.* **2004**, *17*, 259–275. [CrossRef] [PubMed]
133. Mitmesser, S.; Combs, M. *Prebiotics: Inulin and Other Oligosaccharides*; Elsevier Inc.: Amsterdam, The Netherlands, 2017; ISBN 9780128040621.
134. Lin, S.-H.; Chou, L.-M.; Chien, Y.-W.; Chang, J.-S.; Lin, C.-I. Prebiotic Effects of Xylooligosaccharides on the Improvement of Microbiota Balance in Human Subjects. *Gastroenterol. Res. Pract.* **2016**, *2016*, 5789232. [CrossRef] [PubMed]
135. Parajó, J.C.; Garrote, G.; Cruz, J.M.; Dominguez, H. Production of xylooligosaccharides by autohydrolysis of lignocellulosic materials. *Trends Food Sci. Technol.* **2004**, *15*, 115–120. [CrossRef]
136. Scott, K.P.; Martin, J.C.; Duncan, S.H.; Flint, H.J. Prebiotic stimulation of human colonic butyrate-producing bacteria and bifidobacteria, in vitro. *FEMS Microbiol. Ecol.* **2014**, *87*, 30–40. [CrossRef]
137. Tuohy, K.; Rouzaud, G.; Bruck, W.; Gibson, G. Modulation of the Human Gut Microflora Towards Improved Health Using Prebiotics—Assessment of Efficacy. *Curr. Pharm. Des.* **2005**, *11*, 75–90. [CrossRef]
138. Modler, H.W. Bifidogenic factors—Sources, metabolism and applications. *Int. Dairy J.* **1994**, *4*, 383–407. [CrossRef]
139. Sabater-Molina, M.; Larqué, E.; Torrella, F.; Zamora, S. Dietary fructooligosaccharides and potential benefits on health. *J. Physiol. Biochem.* **2009**, *65*, 315–328. [CrossRef]
140. Carabin, I.G.; Flamm, W.G. Evaluation of safety of inulin and oligofructose as dietary fiber. *Regul. Toxicol. Pharmacol.* **1999**, *30*, 268–282. [CrossRef]
141. Elison, E.; Vigsnaes, L.K.; Rindom Krogsgaard, L.; Rasmussen, J.; Sørensen, N.; McConnell, B.; Hennet, T.; Sommer, M.O.A.; Bytzer, P. Oral supplementation of healthy adults with 2′-O-fucosyllactose and lacto-N-neotetraose is well tolerated and shifts the intestinal microbiota. *Br. J. Nutr.* **2016**, *116*, 1356–1368. [CrossRef]
142. Chen, L.; Liu, R.; Qin, C.; Meng, Y.; Zhang, J.; Wang, Y.; Xu, G. Sources and intake of resistant starch in the Chinese diet. *Asia Pac. J. Clin. Nutr.* **2010**, *19*, 274–282. [PubMed]
143. Brouns, F.; Theuwissen, E.; Adam, A.; Bell, M.; Berger, A.; Mensink, R.P. Cholesterol-lowering properties of different pectin types in mildly hyper-cholesterolemic men and women. *Eur. J. Clin. Nutr.* **2012**, *66*, 591–599. [CrossRef] [PubMed]
144. Salminen, S.; Salminen, E. Lactulose, lactic acid bacteria, intestinal microecology and mucosal protection. *Scand. J. Gastroenterol. Suppl.* **1997**, *222*, 45–48. [CrossRef] [PubMed]
145. Barreteau, H.; Delattre, C.; Michaud, P. Production of oligosaccharides as promising new food additive generation. *Food Technol. Biotechnol.* **2006**, *44*, 323–333.
146. Denji, K.A.; Mansour, M.R.; Akrami, R.; Ghobadi, S.; Jafarpour, S.A.; Mirbeygi, S.K. Effect of dietary prebiotic mannan oligosaccharide (mos) on growth performance, intestinal microflora, body composition, haematological and blood serum biochemical parameters of rainbow trout (oncorhynchus mykiss) juveniles. *J. Fish. Aquat. Sci.* **2015**, *10*, 255–265. [CrossRef]

147. Klatt, N.R.; Canary, L.A.; Sun, X.; Vinton, C.L.; Funderburg, N.T.; Morcock, D.R.; Quiñones, M.; Deming, C.B.; Perkins, M.; Hazuda, D.J.; et al. Probiotic/prebiotic supplementation of antiretrovirals improves gastrointestinal immunity in SIV-infected macaques. *J. Clin. Investig.* **2013**, *123*, 903–907. [CrossRef]
148. Champagne, C.P. *Some Technological Challenges in the Addition of Probiotic Bacteria to Foods*; Springer: New York, NY, USA, 2009; ISBN 9780387790572.
149. Akutko, K.; Stawarski, A. Probiotics, prebiotics and synbiotics in inflammatory bowel diseases. *J. Clin. Med.* **2021**, *10*, 2466. [CrossRef]
150. Cunningham, M.; Azcarate-Peril, M.A.; Barnard, A.; Benoit, V.; Grimaldi, R.; Guyonnet, D.; Holscher, H.D.; Hunter, K.; Manurung, S.; Obis, D.; et al. Shaping the Future of Probiotics and Prebiotics. *Trends Microbiol.* **2021**, *29*, 667–685. [CrossRef]
151. Wan, X.; Guo, H.; Liang, Y.; Zhou, C.; Liu, Z.; Li, K.; Niu, F.; Zhai, X.; Wang, L. The physiological functions and pharmaceutical applications of inulin: A review. *Carbohydr. Polym.* **2020**, *246*, 116589. [CrossRef]
152. Rezende, E.S.V.; Lima, G.C.; Naves, M.M.V. Dietary fibers as beneficial microbiota modulators: A proposal classification by prebiotic categories. *Nutrition* **2021**, *89*, 111217. [CrossRef]
153. Yu, X.; Gurry, T.; Nguyen, L.T.T.; Richardson, H.S.; Alm, E.J. Prebiotics and Community Composition Influence Gas Production of the Human Gut Microbiota. *MBio* **2020**, *11*, e00217-20. [CrossRef] [PubMed]
154. Kleessen, B.; Sykura, B.; Zunft, H.J.; Blaut, M. Effects of inulin and lactose on fecal microflora, microbial activity, and bowel habit in elderly constipated persons. *Am. J. Clin. Nutr.* **1997**, *65*, 1397–1402. [CrossRef] [PubMed]
155. Tuohy, K.M.; Finlay, R.K.; Wynne, A.G.; Gibson, G.R. A Human Volunteer Study on the Prebiotic Effects of HP-Inulin—Faecal Bacteria Enumerated Using Fluorescent In Situ Hybridisation (FISH). *Anaerobe* **2001**, *7*, 113–118. [CrossRef]
156. Roland, N.; Nugon-Baudon, L.; Andrieux, C.; Szylit, O. Comparative study of the fermentative characteristics of inulin and different types of fibre in rats inoculated with a human Whole faecal flora. *Br. J. Nutr.* **1995**, *74*, 239–249. [CrossRef] [PubMed]
157. Flores-Maltos, D.A.; Mussatto, S.I.; Contreras-Esquivel, J.C.; Rodríguez-Herrera, R.; Teixeira, J.A.; Aguilar, C.N. Biotechnological production and application of fructooligosaccharides. *Crit. Rev. Biotechnol.* **2016**, *36*, 259–267. [CrossRef] [PubMed]
158. Xiao, J.; Sakaguchi, E.; Bai, G. Short-term supplementation with dietary fructooligosaccharide and dietary mannitol elevated the absorption of calcium and magnesium in adult rats. *Czech J. Anim. Sci.* **2016**, *61*, 281–289. [CrossRef]
159. Sangeetha, P.T.; Ramesh, M.N.; Prapulla, S.G. Recent trends in the microbial production, analysis and application of Fructooligosaccharides. *Trends Food Sci. Technol.* **2005**, *16*, 442–457. [CrossRef]
160. Maathuis, A.J.H.; van den Heuvel, E.G.; Schoterman, M.H.C.; Venema, K. Galacto-Oligosaccharides Have Prebiotic Activity in a Dynamic In Vitro Colon Model Using a ^{13}C-Labeling Technique. *J. Nutr.* **2012**, *142*, 1205–1212. [CrossRef]
161. Davis, L.M.G.; Martínez, I.; Walter, J.; Goin, C.; Hutkins, R.W. Barcoded pyrosequencing reveals that consumption of galactooligosaccharides results in a highly specific bifidogenic response in humans. *PLoS ONE* **2011**, *6*, e25200. [CrossRef]
162. Van Loo, J.; Cummings, J.; Delzenne, N.; Englyst, H.; Franck, A.; Hopkins, M.; Kok, N.; Macfarlane, G.; Newton, D.; Quigley, M.; et al. Functional food properties of non-digestible oligosaccharides: A consensus report from the ENDO project (DGXII AIRII-CT94-1095). *Br. J. Nutr.* **1999**, *81*, 121–132. [CrossRef]
163. Tran, N.T.; Tang, Y.; Li, Z.; Zhang, M.; Wen, X.; Ma, H.; Li, S. Galactooligosaccharides and Resistant Starch Altered Microbiota and Short-Chain Fatty Acids in an in vitro Fermentation Study Using Gut Contents of Mud Crab (*Scylla paramamosain*). *Front. Microbiol.* **2020**, *11*, 1352. [CrossRef] [PubMed]
164. Burokas, A.; Arboleya, S.; Moloney, R.D.; Peterson, V.L.; Murphy, K.; Clarke, G.; Stanton, C.; Dinan, T.G.; Cryan, J.F. Targeting the Microbiota-Gut-Brain Axis: Prebiotics Have Anxiolytic and Antidepressant-like Effects and Reverse the Impact of Chronic Stress in Mice. *Biol. Psychiatry* **2017**, *82*, 472–487. [CrossRef] [PubMed]
165. Schmidt, K.; Cowen, P.J.; Harmer, C.J.; Tzortzis, G.; Errington, S.; Burnet, P.W.J. Prebiotic intake reduces the waking cortisol response and alters emotional bias in healthy volunteers. *Psychopharmacology* **2015**, *232*, 1793–1801. [CrossRef] [PubMed]
166. Ray, C.; Kerketta, J.A.; Rao, S.; Patel, S.; Dutt, S.; Arora, K.; Pournami, F.; Bhushan, P. Human Milk Oligosaccharides: The Journey Ahead. *Int. J. Pediatr.* **2019**, *2019*, 2390240. [CrossRef]
167. LoCascio, R.G.; Ninonuevo, M.R.; Freeman, S.L.; Sela, D.A.; Grimm, R.; Lebrilla, C.B.; Mills, D.A.; German, J.B. Glycoprofiling of bifidobacterial consumption of human milk oligosaccharides demonstrates strain specific, preferential consumption of small chain glycans secreted in early human lactation. *J. Agric. Food Chem.* **2007**, *55*, 8914–8919. [CrossRef]
168. Bunesova, V.; Lacroix, C.; Schwab, C. Fucosyllactose and L-fucose utilization of infant Bifidobacterium longum and Bifidobacterium kashiwanohense. *BMC Microbiol.* **2016**, *16*, 248. [CrossRef]
169. Gibson, G.R.; Wang, X. Regulatory effects of bifidobacteria on the growth of other colonic bacteria. *J. Appl. Bacteriol.* **1994**, *77*, 412–420. [CrossRef]
170. Collado, M.C.; Cernada, M.; Baüerl, C.; Vento, M.; Pérez-Martínez, G. Microbial ecology and host-microbiota interactions during early life stages. *Gut Microbes* **2012**, *3*, 352–365. [CrossRef]
171. Schwab, C.; Ruscheweyh, H.J.; Bunesova, V.; Pham, V.T.; Beerenwinkel, N.; Lacroix, C. Trophic interactions of infant bifidobacteria and eubacterium hallii during L-fucose and fucosyllactose degradation. *Front. Microbiol.* **2017**, *8*, 95. [CrossRef]
172. Weichert, S.; Jennewein, S.; Hüfner, E.; Weiss, C.; Borkowski, J.; Putze, J.; Schroten, H. Bioengineered 2′-fucosyllactose and 3-fucosyllactose inhibit the adhesion of Pseudomonas aeruginosa and enteric pathogens to human intestinal and respiratory cell lines. *Nutr. Res.* **2013**, *33*, 831–838. [CrossRef]

173. Reverri, E.J.; Devitt, A.A.; Kajzer, J.A.; Baggs, G.E.; Borschel, M.W. Review of the clinical experiences of feeding infants formula containing the human milk oligosaccharide 2′-fucosyllactose. *Nutrients* **2018**, *10*, 1346. [CrossRef] [PubMed]
174. Holscher, H.D.; Caporaso, J.G.; Hooda, S.; Brulc, J.M.; Fahey, G.C.J.; Swanson, K.S. Fiber supplementation influences phylogenetic structure and functional capacity of the human intestinal microbiome: Follow-up of a randomized controlled trial. *Am. J. Clin. Nutr.* **2015**, *101*, 55–64. [CrossRef] [PubMed]
175. Hooda, S.; Boler, B.M.V.; Serao, M.C.R.; Brulc, J.M.; Staeger, M.A.; Boileau, T.W.; Dowd, S.E.; Fahey, G.C.J.; Swanson, K.S. 454 pyrosequencing reveals a shift in fecal microbiota of healthy adult men consuming polydextrose or soluble corn fiber. *J. Nutr.* **2012**, *142*, 1259–1265. [CrossRef] [PubMed]
176. Toporovski, M.S.; de Morais, M.B.; Abuhab, A.; Crippa Júnior, M.A. Effect of polydextrose/fructooligosaccharide mixture on constipation symptoms in children aged 4 to 8 years. *Nutrients* **2021**, *13*, 1634. [CrossRef] [PubMed]
177. Costabile, A.; Fava, F.; Röytiö, H.; Forssten, S.D.; Olli, K.; Klievink, J.; Rowland, I.R.; Ouwehand, A.C.; Rastall, R.A.; Gibson, G.R.; et al. Impact of polydextrose on the faecal microbiota: A double-blind, crossover, placebo-controlled feeding study in healthy human subjects. *Br. J. Nutr.* **2012**, *108*, 471–481. [CrossRef]
178. Tachon, S.; Zhou, J.; Keenan, M.; Martin, R.; Marco, M.L. The intestinal microbiota in aged mice is modulated by dietary resistant starch and correlated with improvements in host responses. *FEMS Microbiol. Ecol.* **2013**, *83*, 299–309. [CrossRef]
179. Brüssow, H. Microbiota and healthy ageing: Observational and nutritional intervention studies. *Microb. Biotechnol.* **2013**, *6*, 326–334. [CrossRef]
180. Charrier, J.A.; Martin, R.J.; McCutcheon, K.L.; Raggio, A.M.; Goldsmith, F.; Goita, M.; Senevirathne, R.N.; Brown, I.L.; Pelkman, C.; Zhou, J.; et al. High fat diet partially attenuates fermentation responses in rats fed resistant starch from high-amylose maize. *Obesity* **2013**, *21*, 2350–2355. [CrossRef]
181. Gullón, B.; Gómez, B.; Martínez-Sabajanes, M.; Yáñez, R.; Parajó, J.C.; Alonso, J.L. Pectic oligosaccharides: Manufacture and functional properties. *Trends Food Sci. Technol.* **2013**, *30*, 153–161. [CrossRef]
182. Gómez, B.; Gullón, B.; Yáñez, R.; Schols, H.; Alonso, J.L. Prebiotic potential of pectins and pectic oligosaccharides derived from lemon peel wastes and sugar beet pulp: A comparative evaluation. *J. Funct. Foods* **2016**, *20*, 108–121. [CrossRef]
183. Leijdekkers, A.G.M.; Aguirre, M.; Venema, K.; Bosch, G.; Gruppen, H.; Schols, H.A. In vitro fermentability of sugar beet pulp derived oligosaccharides using human and pig fecal inocula. *J. Agric. Food Chem.* **2014**, *62*, 1079–1087. [CrossRef]
184. Gibson, G.; Angus, F. *Leatherhead Ingredients Handbook: Prebiotics and Probiotics*; Springer: New York, NY, USA; Wiley–Blackwell: Hoboken, NJ, USA, 2000.
185. Terada, A.; Hara, H.; Kato, S.; Kimura, T.; Fujimori, I.; Hara, K.; Maruyama, T.; Mitsuoka, T. Effect of Lactosucrose (4^G-β-D-Galactosylsucrose) on Fecal Flora and Fecal Putrefactive Products of Cats. *J. Vet. Med. Sci.* **1993**, *55*, 291–295. [CrossRef] [PubMed]
186. Ballongue, J.; Schumann, C.; Quignon, P. Effects of lactulose and lactitol on colonic microflora and enzymatic activity. *Scand. J. Gastroenterol. Suppl.* **1997**, *222*, 41–44. [CrossRef] [PubMed]
187. García-Cayuela, T.; Díez-Municio, M.; Herrero, M.; Martínez-Cuesta, M.C.; Peláez, C.; Requena, T.; Moreno, F.J. Selective fermentation of potential prebiotic lactose-derived oligosaccharides by probiotic bacteria. *Int. Dairy J.* **2014**, *38*, 11–15. [CrossRef]
188. Li, W.; Yu, S.; Zhang, T.; Jiang, B.; Stressler, T.; Fischer, L.; Mu, W. Efficient Biosynthesis of Lactosucrose from Sucrose and Lactose by the Purified Recombinant Levansucrase from *Leuconostoc mesenteroides* B-512 FMC. *J. Agric. Food Chem.* **2015**, *63*, 9755–9763. [CrossRef]
189. Maruyama, T.; Terada, A.; Ogata, Y.; Fujita, K.; Ishigami, H.; Hara, K.; Pet, T.; Co, F.; Kaijo, T.; Shiba, B. Effective Dose of Lactosucrose on Fecal Flora and Fecal Metabolites of Humans vitro by bifidobacteria and lactobacilli, and to a limited degree by the The effects of dietary lactosucrose on the fecal flora and fecal metab- olites were studied in eight he. *Bifidobact. Microflora* **1994**, *13*, 51–63.
190. Nath, A.; Haktanirlar, G.; Varga, Á.; Molnár, M.A.; Albert, K.; Galambos, I.; Koris, A.; Vatai, G. Biological Activities of Lactose-Derived Prebiotics and Symbiotic with Probiotics on Gastrointestinal System. *Medicina* **2018**, *54*, 18. [CrossRef]
191. Díez-Municio, M.; Herrero, M.; Olano, A.; Moreno, F.J. Synthesis of novel bioactive lactose-derived oligosaccharides by microbial glycoside hydrolases. *Microb. Biotechnol.* **2014**, *7*, 315–331. [CrossRef]
192. Kihara, M.; Ohba, K.; Sakata, T. Trophic effect of dietary lactosucrose on intestinal tunica muscularis and utilization of this sugar by gut microbes in red seabream Pagrus major, a marine carnivorous teleost, under artificial rearing. *Comp. Biochem. Physiol. Part A Physiol.* **1995**, *112*, 629–634. [CrossRef]
193. Kihara, M.; Sakata, T. Production of short-chain fatty acids and gas from various oligosaccharides by gut microbes of carp (*Cyprinus carpio* L.) in micro-scale batch culture. *Comp. Biochem. Physiol. Part A Mol. Integr. Physiol.* **2002**, *132*, 333–340. [CrossRef]
194. Kihara, M.; Sakata, T. Influences of incubation temperature and various saccharides on the production of organic acids and gases by gut microbes of rainbow trout Oncorhynchus mykiss in a micro-scale batch culture. *J. Comp. Physiol. B Biochem. Syst. Environ. Physiol.* **2001**, *171*, 441–447. [CrossRef]
195. Kamel, R.; Afifi, S.M.; Kassem, I.A.A.; Elkasabgy, N.A.; Farag, M.A. Arabinoxylan and rhamnogalacturonan mucilage: Outgoing and potential trends of pharmaceutical, environmental, and medicinal merits. *Int. J. Biol. Macromol.* **2020**, *165*, 2550–2564. [CrossRef] [PubMed]

196. Hopkins, M.J.; Englyst, H.N.; Macfarlane, S.; Furrie, E.; Macfarlane, G.T.; McBain, A.J. Degradation of Cross-Linked and Non-Cross-Linked Arabinoxylans by the Intestinal Microbiota in Children. *Appl. Environ. Microbiol.* **2003**, *69*, 6354–6360. [CrossRef] [PubMed]
197. Glitsø, L.V.; Gruppen, H.; Schols, H.A.; Højsgaard, S.; Sandström, B.; Bach Knudsen, K.E. Degradation of rye arabinoxylans in the large intestine of pigs. *J. Sci. Food Agric.* **1999**, *79*, 961–969. [CrossRef]
198. Terada, A.; Hara, H.; Sakamoto, J.; Sato, N.; Takagi, S.; Mitsuoka, T.; Mino, R.; Hara, K.; Fujimori, I.; Yamada, T. Effects of Dietary Supplementation with Lactosucrose (4G-β-D-Galactosylsucrose) on Cecal Flora, Cecal Metabolites, and Performance in Broiler Chickens. *Poult. Sci.* **1994**, *73*, 1663–1672. [CrossRef]
199. Finegold, S.M.; Li, Z.; Summanen, P.H.; Downes, J.; Thames, G.; Corbett, K.; Dowd, S.; Krak, M.; Heber, D. Xylooligosaccharide increases bifidobacteria but not lactobacilli in human gut microbiota. *Food Funct.* **2014**, *5*, 436–445. [CrossRef]
200. Chung, Y.-C.; Hsu, C.-K.; Ko, C.-Y.; Chan, Y.-C. Dietary intake of xylooligosaccharides improves the intestinal microbiota, fecal moisture, and pH value in the elderly. *Nutr. Res.* **2007**, *27*, 756–761. [CrossRef]
201. Berger, K.; Burleigh, S.; Lindahl, M.; Bhattacharya, A.; Patil, P.; Stålbrand, H.; Nordberg Karlsson, E.; Hållenius, F.; Nyman, M.; Adlercreutz, P. Xylooligosaccharides Increase Bifidobacteria and Lachnospiraceae in Mice on a High-Fat Diet, with a Concomitant Increase in Short-Chain Fatty Acids, Especially Butyric Acid. *J. Agric. Food Chem.* **2021**, *69*, 3617–3625. [CrossRef]
202. Hsu, C.K.; Liao, J.W.; Chung, Y.C.; Hsieh, C.P.; Chan, Y.C. Xylooligosaccharides and fructooligosaccharides affect the intestinal microbiota and precancerous colonic lesion development in rats. *J. Nutr.* **2004**, *134*, 1523–1528. [CrossRef]
203. Roberfroid, M.B.; Van Loo, J.A.E.; Gibson, G.R. The Bifidogenic Nature of Chicory Inulin and Its Hydrolysis Products. *J. Nutr.* **1998**, *128*, 11–19. [CrossRef]
204. Walker, A.W.; Duncan, S.H.; Carol McWilliam Leitch, E.; Child, M.W.; Flint, H.J. pH and peptide supply can radically alter bacterial populations and short-chain fatty acid ratios within microbial communities from the human colon. *Appl. Environ. Microbiol.* **2005**, *71*, 3692–3700. [CrossRef] [PubMed]
205. Duncan, S.H.; Louis, P.; Thomson, J.M.; Flint, H.J. The role of pH in determining the species composition of the human colonic microbiota. *Environ. Microbiol.* **2009**, *11*, 2112–2122. [CrossRef] [PubMed]
206. Guariguata, L.; Whiting, D.R.; Hambleton, I.; Beagley, J.; Linnenkamp, U.; Shaw, J.E. Global estimates of diabetes prevalence for 2013 and projections for 2035. *Diabetes Res. Clin. Pract.* **2014**, *103*, 137–149. [CrossRef] [PubMed]
207. Sherwin, R.S.; Anderson, R.M.; Buse, J.B.; Chin, M.H.; Eddy, D.; Fradkin, J.; Ganiats, T.G.; Ginsberg, H.; Kahn, R.; Nwankwo, R.; et al. The prevention or delay of type 2 diabetes. *Diabetes Care* **2003**, *26* (Suppl. 1), S62–S69. [CrossRef] [PubMed]
208. Nathan, D.M. Long-term complications of diabetes mellitus. *N. Engl. J. Med.* **1993**, *328*, 1676–1685. [CrossRef]
209. Ley, R.E.; Turnbaugh, P.J.; Klein, S.; Gordon, J.I. Human gut microbes associated with obesity. *Nature* **2006**, *444*, 1022–1023. [CrossRef]
210. Moya-Pérez, A.; Luczynski, P.; Renes, I.B.; Wang, S.; Borre, Y.; Anthony Ryan, C.; Knol, J.; Stanton, C.; Dinan, T.G.; Cryan, J.F. Intervention strategies for cesarean section-induced alterations in the microbiota-gut-brain axis. *Nutr. Rev.* **2017**, *75*, 225–240. [CrossRef]
211. Rabot, S.; Membrez, M.; Bruneau, A.; Gérard, P.; Harach, T.; Moser, M.; Raymond, F.; Mansourian, R.; Chou, C.J. Germ-free C57BL/6J mice are resistant to high-fat-diet-induced insulin resistance and have altered cholesterol metabolism. *FASEB J. Off. Publ. Fed. Am. Soc. Exp. Biol.* **2010**, *24*, 4948–4959. [CrossRef]
212. Yde, C.C.; Jensen, H.M.; Christensen, N.; Servant, F.; Lelouvier, B.; Lahtinen, S.; Stenman, L.K.; Airaksinen, K.; Kailanto, H.-M. Polydextrose with and without *Bifidobacterium animalis* ssp. lactis 420 drives the prevalence of Akkermansia and improves liver health in a multi-compartmental obesogenic mice study. *PLoS ONE* **2021**, *16*, e0260765. [CrossRef]
213. Snelson, M.; de Pasquale, C.; Ekinci, E.I.; Coughlan, M.T. Gut microbiome, prebiotics, intestinal permeability and diabetes complications. *Best Pract. Res. Clin. Endocrinol. Metab.* **2021**, *35*, 101507. [CrossRef]
214. Tajabadi-Ebrahimi, M.; Sharifi, N.; Farrokhian, A.; Raygan, F.; Karamali, F.; Razzaghi, R.; Taheri, S.; Asemi, Z. A Randomized Controlled Clinical Trial Investigating the Effect of Synbiotic Administration on Markers of Insulin Metabolism and Lipid Profiles in Overweight Type 2 Diabetic Patients with Coronary Heart Disease. *Exp. Clin. Endocrinol. Diabetes Off. J. Ger. Soc. Endocrinol. Ger. Diabetes Assoc.* **2017**, *125*, 21–27. [CrossRef] [PubMed]
215. Li, H.Y.; Zhou, D.D.; Gan, R.Y.; Huang, S.Y.; Zhao, C.N.; Shang, A.; Xu, X.Y.; Li, H. Bin Effects and mechanisms of probiotics, prebiotics, synbiotics, and postbiotics on metabolic diseases targeting gut microbiota: A narrative review. *Nutrients* **2021**, *13*, 3211. [CrossRef] [PubMed]
216. Vezza, T.; Abad-Jiménez, Z.; Marti-Cabrera, M.; Rocha, M.; Víctor, V.M. Microbiota-mitochondria inter-talk: A potential therapeutic strategy in obesity and type 2 diabetes. *Antioxidants* **2020**, *9*, 848. [CrossRef] [PubMed]
217. Kim, Y.A.; Keogh, J.B.; Clifton, P.M. Probiotics, prebiotics, synbiotics and insulin sensitivity. *Nutr. Res. Rev.* **2018**, *31*, 35–51. [CrossRef] [PubMed]
218. Pereira, D.I.A.; Gibson, G.R. Effects of consumption of probiotics and prebiotics on serum lipid levels in humans. *Crit. Rev. Biochem. Mol. Biol.* **2002**, *37*, 259–281. [CrossRef]
219. Costa, G.T.; Guimarães, S.B.; Sampaio, H.A.C. Fructo-oligosaccharide effects on blood glucose. An overview [Efeitos dos fruto-oligossacar{í}deos no controle glic{ê}mico. revis{ã}o]. *Acta Cir. Bras.* **2012**, *27*, 279–282. [CrossRef]

220. Dehghan, P.; Pourghassem Gargari, B.; Asghari Jafar-abadi, M. Oligofructose-enriched inulin improves some inflammatory markers and metabolic endotoxemia in women with type 2 diabetes mellitus: A randomized controlled clinical trial. *Nutrition* **2014**, *30*, 418–423. [CrossRef]
221. Letexier, D.; Diraison, F.; Beylot, M. Addition of inulin to a moderately high-carbohydrate diet reduces hepatic lipogenesis and plasma triacylglycerol concentrations in humans. *Am. J. Clin. Nutr.* **2003**, *77*, 559–564. [CrossRef]
222. Aliasgharzadeh, A.; Khalili, M.; Mirtaheri, E.; Pourghassem Gargari, B.; Tavakoli, F.; Abbasalizad Farhangi, M.; Babaei, H.; Dehghan, P. A Combination of Prebiotic Inulin and Oligofructose Improve Some of Cardiovascular Disease Risk Factors in Women with Type 2 Diabetes: A Randomized Controlled Clinical Trial. *Adv. Pharm. Bull.* **2015**, *5*, 507–514. [CrossRef]
223. Jackson, K.G.; Taylor, G.R.J.; Clohessy, A.M.; Williams, C.M. The effect of the daily intake of inulin on fasting lipid, insulin and glucose concentrations in middle-aged men and women. *Br. J. Nutr.* **1999**, *82*, 23–30. [CrossRef]
224. Dewulf, E.M.; Cani, P.D.; Claus, S.P.; Fuentes, S.; Puylaert, P.G.B.; Neyrinck, A.M.; Bindels, L.B.; De Vos, W.M.; Gibson, G.R.; Thissen, J.P.; et al. Insight into the prebiotic concept: Lessons from an exploratory, double blind intervention study with inulin-type fructans in obese women. *Gut* **2013**, *62*, 1112–1121. [CrossRef] [PubMed]
225. Schellekens, H.; Torres-Fuentes, C.; van de Wouw, M.; Long-Smith, C.M.; Mitchell, A.; Strain, C.; Berding, K.; Bastiaanssen, T.F.S.; Rea, K.; Golubeva, A.V.; et al. Bifidobacterium longum counters the effects of obesity: Partial successful translation from rodent to human. *EBioMedicine* **2021**, *63*, 103176. [CrossRef] [PubMed]
226. Everard, A.; Lazarevic, V.; Derrien, M.; Girard, M.; Muccioli, G.G.; Neyrinck, A.M.; Possemiers, S.; Van Holle, A.; François, P.; de Vos, W.M.; et al. Responses of gut microbiota and glucose and lipid metabolism to prebiotics in genetic obese and diet-induced leptin-resistant mice. *Diabetes* **2011**, *60*, 2775–2786. [CrossRef] [PubMed]
227. Sangwan, V.; Tomar, S.K.; Ali, B.; Singh, R.R.B.; Singh, A.K. Hypoglycaemic effect of galactooligosaccharides in alloxan-induced diabetic rats. *J. Dairy Res.* **2015**, *82*, 70–77. [CrossRef]
228. Vulevic, J.; Juric, A.; Tzortzis, G.; Gibson, G.R. A mixture of trans-galactooligosaccharides reduces markers of metabolic syndrome and modulates the fecal microbiota and immune function of overweight adults. *J. Nutr.* **2013**, *143*, 324–331. [CrossRef]
229. Albarracín, M.; Weisstaub, A.R.; Zuleta, Á.; Mandalunis, P.; González, R.J.; Drago, S.R. Effects of extruded whole maize, polydextrose and cellulose as sources of fibre on calcium bioavailability and metabolic parameters of growing Wistar rats. *Food Funct.* **2014**, *5*, 804–810. [CrossRef]
230. Maki, K.C.; Pelkman, C.L.; Finocchiaro, E.T.; Kelley, K.M.; Lawless, A.L.; Schild, A.L.; Rains, T.M. Resistant starch from high-amylose maize increases insulin sensitivity in overweight and obese men. *J. Nutr.* **2012**, *142*, 717–723. [CrossRef]
231. Bodinham, C.L.; Smith, L.; Thomas, E.L.; Bell, J.D.; Swann, J.R.; Costabile, A.; Russell-Jones, D.; Umpleby, A.M.; Robertson, M.D. Efficacy of increased resistant starch consumption in human type 2 diabetes. *Endocr. Connect.* **2014**, *3*, 75–84. [CrossRef]
232. Bianchi, G.; Ronchi, M.; Marchesini, G. Effect of lactulose on carbohydrate metabolism and diabetes mellitus. *Scand. J. Gastroenterol. Suppl.* **1997**, *222*, 62–64. [CrossRef]
233. Lopez, H.W.; Levrat, M.A.; Guy, C.; Messager, A.; Demigné, C.; Rémésy, C. Effects of soluble corn bran arabinoxylans on cecal digestion, lipid metabolism, and mineral balance (Ca, Mg) in rats. *J. Nutr. Biochem.* **1999**, *10*, 500–509. [CrossRef]
234. Lu, Z.X.; Walker, K.Z.; Muir, J.G.; O'Dea, K. Arabinoxylan fibre improves metabolic control in people with Type II diabetes. *Eur. J. Clin. Nutr.* **2004**, *58*, 621–628. [CrossRef] [PubMed]
235. Imaizumi, K.; Nakatsu, Y.; Sato, M.; Sedarnawati, Y.; Sugano, M. Effects of xylooligosaccharides on blood glucose, serum and liver lipids and-cecum short-chain fatty acids in diabetic rats. *Agric. Biol. Chem.* **1991**, *55*, 199–205. [CrossRef]
236. Kim, S.; Rhee, I.; Joo, G.; Ha, H. Effects of dietary xylooligosaccharide on lipid levels of serum in rats fed high cholesterol diet. *J. Korean Soc. Food Sci. Nutr.* **1998**, *27*, 945–951.

International Journal of *Molecular Sciences*

MDPI

Review

An Insight into Anti-Inflammatory Activities and Inflammation Related Diseases of Anthocyanins: A Review of Both In Vivo and In Vitro Investigations

Zilong Ma [1,2,3,†], Bin Du [3,†], Jun Li [2], Yuedong Yang [3] and Fengmei Zhu [2,*]

1 College of Food Science and Technology, Hebei Agricultural University, Baoding 071001, China; mzl2582@163.com
2 College of Food Science and Technology, Hebei Normal University of Science and Technology, Qinhuangdao 066004, China; spgcx@163.com
3 Hebei Key Laboratory of Natural Products Activity Components and Function, Hebei Normal University of Science and Technology, Qinhuangdao 066004, China; dubin@hevttc.edu.cn (B.D.); kycyyd@126.com (Y.Y.)
* Correspondence: zhufengmei@hevttc.edu.cn; Tel.: +86-335-2079074
† These two authors made an equal contribution to this work.

Abstract: Anthocyanin is a type of flavonoid pigment widely present in fruits and vegetables. It can not only be used as natural pigment, but also has a variety of health functions, for instance, antioxidant, anti-inflammatory, anti-tumor, and neuroprotective activities. Persistent proinflammatory status is a major factor in the development, progression, and complications of chronic diseases. Not surprisingly, there are thus many food ingredients that can potentially affect inflammation related diseases and many studies have shown that anthocyanins play an important role in inflammatory pathways. In this paper, the inflammation related diseases (such as, obesity, diabetes, cardiovascular disease, and cancer) of anthocyanins are introduced, and the anti-inflammatory effect of anthocyanins is emphatically introduced. Moreover, the anti-inflammatory mechanism of anthocyanins is elaborated from the aspects of NF-κB, toll like receptor, MAPKs, NO, and ROS and the main efficacy of anthocyanins in inflammation and related diseases is determined. In conclusion, this review aims to get a clear insight into the role of anthocyanins in inflammation related diseases.

Keywords: anthocyanins; inflammation; NF-κB; MAPKs; inflammation related diseases; in vitro; in vivo

Citation: Ma, Z.; Du, B.; Li, J.; Yang, Y.; Zhu, F. An Insight into Anti-Inflammatory Activities and Inflammation Related Diseases of Anthocyanins: A Review of Both In Vivo and In Vitro Investigations. *Int. J. Mol. Sci.* **2021**, *22*, 11076. https://doi.org/10.3390/ijms222011076

Academic Editors: Matteo Bordiga and Maurizio Battino

Received: 9 September 2021
Accepted: 5 October 2021
Published: 14 October 2021

1. Introduction

Inflammation is a normal physiological response, which is the protective response of the innate immune system to pathogens and injuries, usually temporary, and is one of the body's oldest defense mechanisms [1]. While acute, localized inflammation is a lifesaving mechanism to protect the body from pathogens, repeated stimulation or ineffective regulation can lead to chronic inflammation, damage the body, and induce a variety of diseases [2]. Inflammation is controlled by many factors, and NF-κB signaling pathway is one of the main influencing factors of inflammation. The activation of NF-κB can stimulate the expression of many genes and produce various cytokines, such as TNF-α, IL-6, IL-1β, MCP-1, adipokines, cell adhesion molecules, sVCAM-1 and sICAM-1, and acute phase protein (CRP) [3]. Other important inflammatory factors include PRRS, such as TLRs, and kinases such as MAPK and JNK. When these kinase cascades and nuclear transcription factors are stimulated by external stimuli such as endotoxin, viruses, ROS, cellular redox status, fatty acids, cytokines, growth factors, and carcinogens, inflammation is induced [4].

Chronic, low-grade, systemic proinflammatory state is the risk factor of insulin resistance, metabolic syndrome, atherosclerosis, type II diabetes, cardiovascular disease, and other diseases [5]. Studies have shown that a chronic inflammatory environment is a risk factor for cancer. Chronic inflammation is closely related to tumorigenesis, including cell transformation, promotion, survival, proliferation, invasion, angiogenesis, and metastasis [6]. Atherosclerotic thrombosis is often accompanied by inflammation [7]. Inflammation also regulates the production of acute phase proteins, such as CRP, a subclinical inflammatory marker associated with atherosclerosis. The role of many proinflammatory cytokines in the progression of atherosclerosis has been verified in many studies [8]. Activation of NF-κB signaling pathway produces a large number of pro-inflammatory factors, such as pcam-1, which is involved in the increase of monocyte adhesion and vascular inflammation. TNF-α is related to the pathogenesis of endothelial dysfunction [9]. Many previous studies have shown that inhibition of TNF-α expression can effectively reduce endothelial dysfunction [10,11]. In addition, obesity is closely related to inflammation, which is a manifestation of chronic and low-grade inflammation. Hotamisligil et al. first reported that the fat mass increases with the increase of TNF-α expression [12]. In obesity, the morphological changes of adipocytes lead to the change of secretory response, which is conducive to the inflammatory state. The expanded adipose tissue aggregates macrophages to produce pro-inflammatory proteins such as TNF-α, IL-6, and MCP-1, which further promote the inflammatory state [1]. In conclusion, many diseases are prevented and treated by reducing the inflammatory cascade.

Anthocyanins widely exist in plant cell fluid and are secondary metabolites of plants, such as berries, soybean seed, purple potato, purple cabbage, and black carrot [13]. They have different colors in different environments. Research showed that anthocyanins showed different colors at different pH values, being blue in alkaline environments and purplish red in acid environments. The characteristics of anthocyanins make plants show different colors in different environments. Therefore, anthocyanins of various colors can be produced in the fruits, flowers, and leaves of plants [14]. The change of temperature also has a significant effect on their color [15]. Anthocyanins are widely used as natural pigments because of their brilliant colors. Anthocyanins are a type of polyphenol pigment which belongs to flavonoids, and has many health functions, such as anti-cancer [16], anti-inflammatory [17], neuroprotective [18], eye protection, and so on [19].

Anthocyanins are flavonoid compounds formed by glycosidic bonds between anthocyans and sugars. They are polyhydroxy and methoxy derivatives of 2-benzopyrene or xanthate ions. Their basic structure is C6-C3-C6, namely two aromatic rings and one oxygen-containing heterocycle [20]. Due to the different substituents on the C6-C3-C6 nucleus, the ability to form resonance structure and the environmental factors, various anthocyanins are formed, showing a variety of different colors. At present, there are 22 types of anthocyanins in nature, and 6 types of common anthocyanins in food [21]. The chemical structure of anthocyanins is shown in Figure 1. The types of R1 and R2 substituents are shown in Table 1.

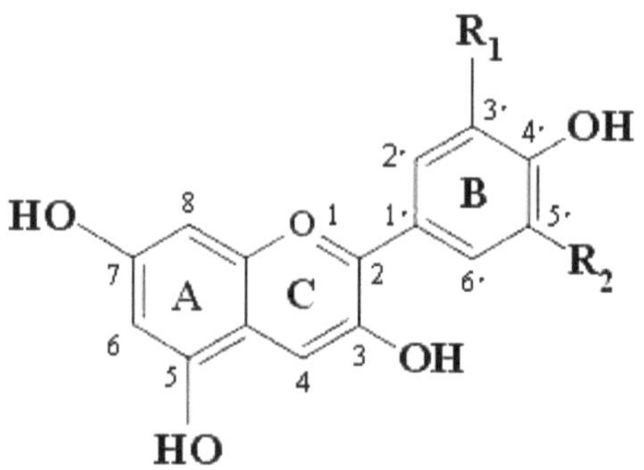

Figure 1. Basic structure of anthocyanins.

Table 1. R1 and R2 Substituents of Six Basic Anthocyanins.

Anthocyanins	R1	R2
Pelargonidin	H	H
Cyanidin	OH	H
Delphinidin	OH	OH
Peonodin	OCH_3	H
Petunidin	OCH_3	OH
Malvidin	OCH_3	OCH_3

Because of its strong polarity, it was initially thought that anthocyanins could not be directly absorbed by cells into the animal or human circulatory system. However, in vivo bioavailability tests have fully confirmed that anthocyanins can be absorbed through the gastrointestinal tract in prototype form, enter the circulatory system for transfer, transformation, and then excreted through urine [22]. Among them, stomach and small intestine are the main sites of anthocyanin absorption. Under the action of gastric acid, anthocyanins in food can be fully released and dissolved, and most of them can combine with bilirubin translocation enzyme to promote it to pass through the gastric wall mucosa, thus the absorption speed is relatively fast [23]. Studies have shown that the concentration of anthocyanins reached its peak at 2 h after ingestion and disappeared at 4–6 h after ingestion [22]. Anthocyanins in human intake are mainly from berries, vegetables, and other foods. Zamora et al. [24] found that anthocyanin intake was 26.2–90.9 mg/d, through a survey of residents in ten Western European countries. It has been reported that when the anthocyanin intake level reaches 22.3–25.1 mg/d, it can reduce the risk of myocardial infarction and diabetes [25].

In view of the important role of inflammation in various diseases, this paper reviews the therapeutic effect of anthocyanins on inflammatory related diseases, and discusses the potential anti-inflammatory mechanism of anthocyanins from three aspects: Toll like receptor, MAPKs, and NF-κB and oxidative stress. In addition, the antioxidant activity of anthocyanins can effectively eliminate free radicals, reduce the stimulation of inflammation, reduce the secretion of inflammatory factors, inhibit the activation of inflammation related signal pathways, stimulate the production of anti-inflammatory factors, and effectively reduce the inflammatory reaction. Anthocyanins, as a natural pigment, are non-toxic and harmless, and play an anti-inflammatory role in various inflammatory diseases.

2. Therapeutic Effect of Anthocyanins on Inflammation Related Diseases

The properties of anthocyans are very unstable, and there are few free anthocyans [26]. Anthocyans generally react with one or more monosaccharides, disaccharides, and trisaccharides through the condensation reaction of 3, 5, 7 carbon hydroxyl groups to form glycosidic bonds in the form of anthocyanins [27]. There are three types of monosaccharides, such as rhamnose, rhamnose and so on. In addition to condensation with sugars, the hydroxyl groups on the core of anthocyanins and the hydroxyl groups connected to glycosides can combine with one or more acylation groups to form acyl anthocyanins [28].

Anthocyanins are widely distributed in higher plants, especially in dark petals, berries, vegetables, potatoes, and cereal seed coats, which make them red, purple, and even black. The amount of anthocyanins in fruits varies greatly, usually in proportion to color, and is affected by light intensity, temperature during growth, altitude, plant hormones, genes, and other factors. The increase of growth temperature will reduce the synthesis of anthocyanins [29]. With the development of anthocyanins extraction technology, a large number of studies have determined the content of anthocyanins in different plants, such as berries (25–495 mg/100 g (fresh weight)) [1], grapes (181.2–716.4 mg/100 g) [30], and pomegranate (>300 mg/100 g) [31], and vegetables, such as red cabbage (113 mg/100 g) [32], black beans (23.1 mg/100 g) [33], eggplant (85.7 mg/100 g), and red onion (23.8–38.8 mg/100 g) [34].

2.1. Therapeutic Effect of Anthocyanins on Obesity

2.1.1. In Vivo Study

Anthocyanins extracted from berries are beneficial to weight loss. Research by Diego Luna Vital et al. found that maize rich in ferulic acid and anthocyanins can prevent obesity by regulating TLRs and MAPK signaling pathways, reducing fat production and inflammation, and promoting energy consumption [35]. Some studies have also shown that berry juice and berry powder have no obvious anti-obesity effect, while anthocyanins extracted from berries have obvious anti-obesity effect [36,37]. The reason may be the instability of anthocyanins, and the denaturation in the processing of fruit juice and fruit powder, which affects the function of anthocyanins.

2.1.2. In Vitro Study

Khan Mi et al. found that anthocyanins in Cornus officinalis inhibit lipid accumulation by regulating adipogenesis and lipogenesis related genes and signaling proteins [38].

2.2. Therapeutic Effect of Anthocyanins on Diabetes and Cardiovascular Disease

2.2.1. In Vivo Study

Many studies have shown that anthocyanins have an hypoglycemic effect; thus, it has been widely confirmed that anthocyanins can reduce blood glucose concentration. Anthocyanins can regulate the relaxation and contraction of blood vessels by controlling the activity of nitric oxide synthase and potassium channel [39]. Anthocyanins inhibit collagen, hyaluronic acid, and elastin and other important components of the inner wall of blood vessels, thereby protecting collagen, hyaluronic acid, and elastin from being degraded [40]. Anthocyanins can rapidly increase the oxidative function of mitochondria in muscle cells and brown adipose tissue after supplementing energy to increase the body's energy metabolism rate [41]. They can also inhibit the activity of hydrolytic enzymes that play a key role in the process of carbohydrate digestion, slow down the hydrolysis process of food, and avoid a sharp increase in blood glucose after a meal [42]. Anthocyanins can promote the expression of glucose vector and accelerate glucose consumption. They can be used to prepare diabetes drugs [43]. The Anahita aboonabi study found that anthocyanins supplementation has a positive effect on cardiovascular metabolic risk factors and the inflammatory cascade reaction in a metabolic syndrome population, which may play a role in the prevention or treatment of atherosclerosis [44]. This has been proved by clinical trials. Four weeks of anthocyanin supplementation significantly decreased cardiometabolic risk factors including the average serum fasting blood glucose (FBG) (by 13.3%) and lipid profiles by significant reductions in triglyceride (by 24.9%) and LDL-C (by 33.1%). These results support the hypothesis that anthocyanin supplementation exerts anti-atherogenicity effects by improving cardiometabolic risk factors and reducing thrombogenicity in the MetS population [9]. Reducing the level of alterable atherosclerotic risk factors is an important goal to prevent cardiovascular disease in the metabolic syndrome population. There are established relationships among metabolic syndrome, oxidative stress, chronic inflammation, and cardiovascular disease [45].

2.2.2. In Vitro Study

Daily intake of anthocyanins also significantly improved cardiovascular disease and coronary heart disease [46]. In another study, the authors induced an obese mouse model with a high-fat diet and fed berries containing methylanthocyanins. The results showed that the metabolic damage of the mouse model was significantly improved. Blueberries and Concord grapes (57% and 33% anthocyanins as malvidin, petunidin, or peonidin, respectively) improved the body composition through individual significant effects on energy expenditure and increased activity. Methylated anthocyanins counteract mitochondrial dysfunction associated with metabolic stress by enhancing mitochondrial respiration and eliminating mitochondrial proton gradients (proton leakage) in the adipose tissue. It is proven that methyl anthocyanins can significantly improve the metabolic damage caused by a long-term high calorie diet [47]. Therefore, anthocyanins have good antioxidant and anti-inflammatory effects, which can improve these diseases.

2.3. Therapeutic Effect of Anthocyanins on Cancer

2.3.1. In Vivo Study

Anthocyanins in blackberry, raspberry, and other berries can promote the apoptosis of cervical cancer, rectal cancer, hepatocellular carcinoma, prostate cancer, and esophageal cancer in mice [7]. Liu et al. investigated the anticancer activity of the bilberry anthocyanin combo containing macromolecules by modulating the gut microbiome and inhibiting PD-L1. The results showed that bilberry anthocyanins combo improved the proportion of butyrate in feces and increased intratumoral CD8+ T cell infiltrations. The application of the bilberry anthocyanin combo changed the species diversity of gut microbiome decreased by LCP–chitosan and attained the best control of tumor growth in colon cancer [47].

2.3.2. In Vitro Study

In addition, the anticancer and anti-inflammatory effects of anthocyanins have been widely concerned, and scholars at home and abroad have carried out a large number of studies. Feng et al. studied the effect and mechanism of anthocyanin-3-rutin in black raspberry varieties [46]. The results showed that anthocyanin-3-rutin can induce HL-60 cell apoptosis. Anthocyanin-3-rutin treatment also activated p38, MAPK, and JNK reactive oxygen species (ROS) dependence. It activates Bim through the mitochondrial pathway, and the up-regulation of Bim can promote cell apoptosis. Anthocyanins have no cytotoxicity and have potential application prospects in the treatment of leukemia [46]. The anticancer properties and mechanism of anthocyanins are a hot topic.

In conclusion, anthocyanins have a variety of health effects, as shown in Table 2. They can improve metabolic syndrome and obesity, and have anti-cancer, anti-inflammatory, and vision protection effects. They constitute a good natural health care substance.

Table 2. Health Effects of Anthocyanins.

Scheme	Dose and Duration of the Intervention	Participants	Study Design	Health Effects	References
Wild Norwegian bilberries and blackcurrant	Two capsules twice a day 4 weeks	35 male and female subjects (MetS + healthy) age = 25–75	Randomized, control design Intervention group (n = 20)-two capsule twice a day Control group (n = 15)-two capsule twice a day	Lowering inflammation and improving glucose and lipid metabolism	[42]
Fruit juice (Apples, strawberries, blueberries, grapes)	750 mL fruit juice taken in three equal portions 55 days	62 healthy male volunteers age = 20–50	Randomized, control design Intervention group (n = 30)-750 mL fruit juice is taken in three equal portions Control group (n = 27)–750 mL placebo is taken in three equal portions	Improve DNA integrity and might influence lipid metabolism in humans	[46]

Table 2. *Cont.*

Scheme	Dose and Duration of the Intervention	Participants	Study Design	Health Effects	References
Blueberries	150 g or 75 g fresh blueberries per day 21 days	115 male and female subjects (MetS) age = 50–75	A double-blind, placebo-controlled, parallel study	Improved endothelial Function, Improving metabolic syndrome	[47]
Tart cherry juice	240 mL of tart cherry juice twice a day 2 weeks	11 healthy male or female subjects with chronic insomnia age $\geq$ 50	A randomized, double-blind, placebo controlled clinical trial	improving insomnia	[48]
Fresh ripe berries of cornelian cherry	total anthocyanin 320 mg/d 12 weeks	80 patients with NAFLD age = 25–65	A double-blind randomized clinical trial	Improving NAFLD	[49]
Blood orange juice	50 mg anthocyanins/d and 500 mL blonde orange juice 4 weeks	41 participants (aged 25–84) with a waist circumference > 94 cm (men) and > 80 cm (women)	A randomized controlled trial	Lowering cholesterol	[50]
Black currant	Black currant anthocyanins 50 mg/d 2 years	38 patients with OAG	A randomized, placebo-controlled, double-masked trial	Increase eye blood flow and improve glaucoma	[51]
Black currant	Black currant capsules 300 mg	11 male patients with Parkinson's disease	Plasma and cerebrospinal fluid were collected from 11 male patients before and after 28 day supplementation of black currant capsules.	Treat neurological conditions with IGF-1 deficiency.	[52]
Bilberry and black currant	Purified anthocyanin 320 mg/d 12 weeks	21 patients with NAFLD	A randomized, double-blind, placebo-controlled pilot trial	Improving NAFLD	[53]
Black soybeans	anthocyanin-rich black soybean testa extracts 2.5 g/d 8 weeks	63 participants defined as overweight or obese by their body mass index (BMI > 23) or waist circumference (WC > 90 cm for males, >85 cm for females)	A randomized, double-blinded, and placebo-controlled clinical trial	Improve blood lipid status, Prevention of abdominal obesity caused by high fiber and low cholesterol diet	[54]

3. Anti-Inflammatory Mechanism of Anthocyanins

3.1. Nuclear Factor-κB Pathway (NF-κB)

NF-κB is an important protein complex, which mainly controls DNA transcription and cytokine production. It is a central orchestrator of the inflammatory response [55]. It exists in the cytoplasm of various types of cells in an inactive form. Under normal condition, the binding of p65/p50 heterodimer and its inhibitor protein IκB is inactive. When it is stimulated by external factors, such as ROS, ultraviolet rays, hyperglycemia, and other factors, NF-κB will degrade. The p65/p50 heterodimer is separated from IκB and is in a free state. The free p65/p50 heterodimer transfers to the nucleus and binds to the common DNA sequence to activate the expression of proinflammatory genes, as shown in Figure 2 [56]. Activation of the NF-κB pathway upregulates the expression of proinflammatory cytokines (such as TNF-α, IL-1α, IL-1β and IL-10), chemokines (IL-8), adhesion molecules (ICAMs and VCAM-1), iNOS, COX-2, and cytosolic phospholipase 2 [42].

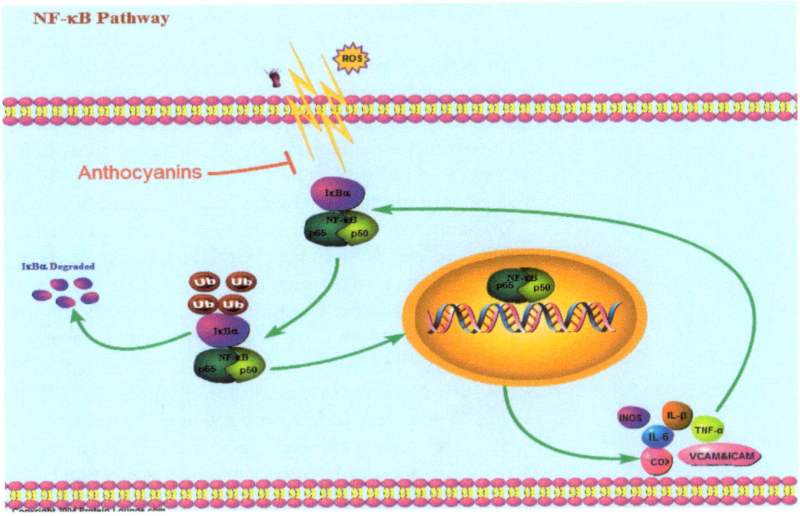

Figure 2. Anthocyanins reduce external stimulation and inhibit the activation of NF-κB signaling pathway. (Ub: ubiquitination).

Duarte et al. found that in LPS induced macrophage model, strawberry anthocyanins significantly inhibited the translocation of p65 subunit from cytoplasm to nucleus, thus inhibiting the activation of NF-κB signaling pathway [57]. Lee et al. (2017) found that the p-Coumaroyl anthocyanins mixture (contains petanin, peonanin, malvanin, and pelanin) extracted from a dark purple-fleshed potato cultivar Jayoung displayed an inhibitory effect on the transcriptional activity and translocation of NF-κB in RAW264.7 macrophages [58]. Another in vitro study reported that a pure sour cherry anthocyanins extract addition to human Caco-2 cells receded the translocation of a p65 subunit from the cytosol to nuclei [59]. Roth et al. treated colon patients with blueberry anthocyanins revealed decreased serum levels of TNF-α, IFN-γ, and activated NF-κB subunit p65 and increased serum levels of IL-10 and IL-22 [60]. IL-10 and IL-22 are anti-inflammatory cytokines involved in wound healing and production of defensins and mucins against bacterial invasion, which can effectively reduce inflammation [61]. Aboonabi et al. found that anthocyanin supplements inhibited NF-κB transactivation and decreased plasma concentrations of pro-inflammatory chemokines, cytokines, and inflammatory mediators and also increased PPAR-γ gene expression. Many studies have shown that anthocyanins inhibit inflammation by inhibiting the activation of NF-κB [62].

3.2. TLRs and MAPKs

Toll like receptors (TLRs) are innate immune receptors, which are widely distributed and can recognize the specific structures shared by some pathogens or their products, namely pathogen related molecular patterns (PAMPs) [63]. TLR4 is a receptor that mainly mediates endotoxin reactions, such as LPS, and TLR4/CD14 is an important signaling pathway related to inflammatory response. When the TLR4 receptor binds to a ligand lipopolysaccharide, the protein adapter MyD88 activates the NF-κB signaling pathway, thus promoting the expression of a large number of inflammatory factors, such as TNF-α,

IL-6, IL-1β, and COX-2, leading to a persistent inflammatory state [64]. Activator protein 1 (AP-1) is a transcription regulator, which is closely related to the activation of TLR4. AP-1 is assembled through the dimerization of a characteristic bZIP domain (basic region leucine zipper) in the Fos and Jun subunits. Moreover, AP-1 functions are heavily dependent on the specific Fos and Jun subunits, contributing to AP-1 dimers. It is mainly responsible for the control of cell differentiation, proliferation, and apoptosis in the inflammatory state [57]. Both in vivo and in vitro evidence show that anthocyanins can suppress the expression level of COX-2 as well as the transactivation of AP-1, which is a transcription factor that regulates COX-2 gene expression [65,66]. Cui et al. found that in the model of cerebral ischemia–reperfusion injury in mice, feeding *Myrica rubra* anthocyanins can significantly reduce the expression of TLR4 and TNF-α [67]. Anthocyanin pretreatment can also directly regulate ROS level, and the activity of inflammation related downstream pathways, including NO production and SOD activity [68]. In the study of Karunarathne et al., an immunohistochemistry assay revealed that anthocyanins inhibited LPS-induced TLR4 dimerization or expression on the cell surface, which consequently decreased MyD88 recruitment and IRAK4 phosphorylation, resulting in the inhibition of NF-κB activity [69].

MAPKS is also a signal pathway related to inflammation. MAPKs are a family of enzymes that respond to inflammatory stimuli by regulating cell differentiation, mitosis, and apoptosis. MAPK has no catalytic activity in its base form and needs phosphorylation to become active. Three important MAPKs are ERK, which mainly controls cell division, c-JNKs which control transcription, and p38 MAPKs which respond to inflammatory factors. When these MAPKs are activated by inflammatory factors and external environmental factors, they are associated with inflammatory related diseases. The activation of MAPKs directly regulates the activation of AP-1 and synergizes with the NFκB pathway, resulting in gene expression by simulating the promoter gene of many mediators, such as the cytokines IL-6 and TNF-α [70]. Studies have shown that anthocyanins inhibited MLK3 activation and its downstream JNK and p38 MAPK signaling cascades [71]. The protective effect of anthocyanins can be explained by the regulation of oxidative-stress and the suppression of cell apoptosis through the activation of Nrf-2 by interaction with the MAPK and NF-κB signaling pathways [72]. Wongwichai et al. found that anthocyanins and metabolites from purple rice significantly inhibited IκBα degradation, the level of p-p65, and the ERK/MAPK pathway [73]. Many studies have shown that anthocyanins rich plant foods play a protective role through different cell transduction pathways, including inflammatory transcription factors, SAPK/JNK and p38MAPK cascades, JAK/STAT signaling, NF-κB/perk/MAPK, Wnt signaling pathway, and the Nrf2 cell protection pathway [74].

In conclusion, as shown in Figure 3, anthocyanins play an anti-inflammatory role by inhibiting TLR4 protein expression and activating MAPKs signaling pathway.

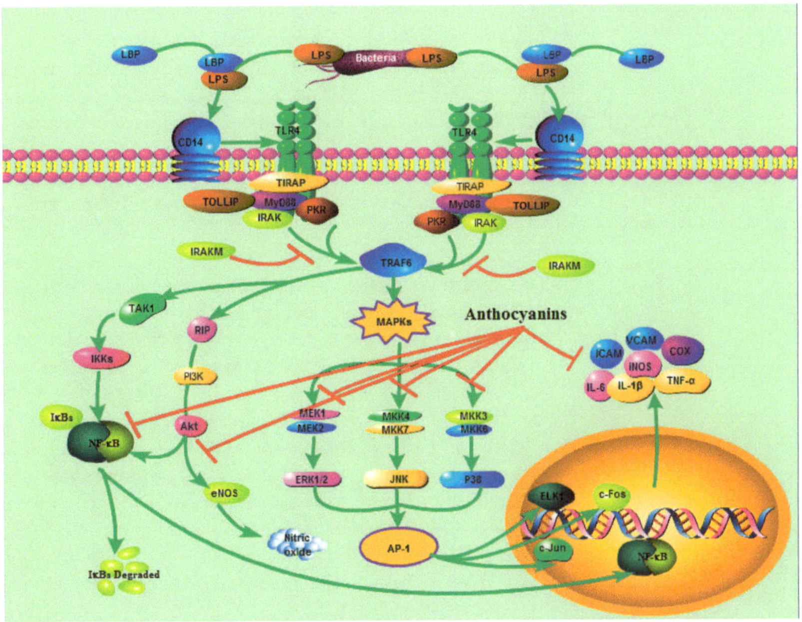

Figure 3. Anthocyanins inhibit TLR4 protein expression and MAPKs signaling pathway. The structure of TLR4 is divided into three domains: extracellular domain, transmembrane domain, and intracellular domain. Extracellular LPS binds to CD14, and since CD14 does not have a transmembrane domain, it binds to the extracellular domain of TLR4 to a transmembrane-mediated endotoxin. When the signal was transferred into the cell, the MyD88 adaptor protein and toll–irak complex began to be recruited. The intracellular TIR region of TLR4 binds to the carboxyl end of MyD88, and the amino terminal of MyD88 binds to IRAK again to activate IRAK (IRAK-M, as a negative regulator, can inhibit the phosphorylation of IRAK and interrupt signal transduction). Activated IRAK reactivates TRAF-6 and further activates the NF-κB, MAPKs, and PI3K-Akt signaling pathways, promoting the secretion of NO and inflammatory factors (Ding et al., 2018; Monica et al., 2016).

3.3. Nitric Oxide (NO)

Nitric oxide (NO) is an important messenger molecule. In pro-oxidative conditions, nitric oxide reacts with O^{2-} to form the peroxynitrite anion $ONOO^-$, a highly reactive molecule that damages DNA and lipids and promotes inflammation [75]. NO is mainly produced by nitric oxide synthase (NOS) through a series of oxidation reactions, and its three subtypes are neuronal NOS (nNOS), endothelial NOS (eNOS), and inducible NOS (iNOS) [76]. In activated macrophages, iNOS induces and kills bacteria or tumor cells by producing peroxynitrite. However, iNOS can also be induced in endothelial cells and smooth muscle cells, leading to pathological release of nitric oxide, which is the feature of endothelial dysfunction. A large amount of peroxynitrite production maintains the proinflammatory state, excessive vasoconstriction, and thrombosis. Endothelial nitric oxide synthase (eNOS), conversely, is constitutive and mainly activated by shear stress [77]. Larissa et al. found that strawberry anthocyanins (Pelargonidin-3-O-glucoside) could reduce the concentration of pleural effusion, inhibit the expression of iNOS, and reduce the level of NO in a dose-dependent manner in the mouse model of pleurisy [57].

Moreover, the stimulus–secretion coupling of high glucose-induced the synthesis and the release of NO could interact with VEGF [78]. VEGF robustly activated the PI3K-Akt pathway. Akt, the serine/threonine kinase, can activate eNOS to produce NO, thus promoting inflammation [79]. Huang et al. found that blueberry anthocyanins inhibited eNOS activity and changed NO level by inhibiting Akt expression [80]. Nizamutdinova et al. found that anthocyanins (ANT) extracted from Oryza sativa L stimulate wound healing while suppressing superfluous inflammation by inducing vascular endothelial growth factor (VEGF) production in fibroblasts and keratinocytes [81]. Winter et al. found that in LPS induced BV2 microglial inflammation model, pretreatment with protocatechuic acid can significantly reduce the production of NO [82].

In conclusion, a number of studies have shown that anthocyanins can inhibit the expression and activity of iNOS, thereby reducing the harmful pro-inflammatory effect of excessive production of nitric oxide under oxidative stress.

3.4. Reactive Oxygen (ROS)

Oxidative stress has been implicated in the damage of various cellular portions involving lipids, proteins, and nucleic acids through oxidation by ROS such as H_2O_2, OH^-, and superoxide anion radical (O^{2-}) [83]. The oxidative process involves the pathogenesis of many diseases. In particular, ROS produced by cell redox disorder is involved in the pathogenesis of various inflammatory diseases including skin injury [84]. In the process of inflammation, the cells involved in the inflammatory process are recruited to the injured site, absorb oxygen, and release ROS. In addition, cytokines and chemokines secreted by inflammatory cells can further stimulate inflammatory cells and produce more ROS. As a result, NF-κB and AP-1 are activated, leading to an increased secretion of cytokines. The vicious cycle will aggravate the inflammatory transition and lead to various chronic diseases [1].

As we all know, anthocyanins have good antioxidant activity, which can remove the excess oxidation free radicals in human body. Huang et al. (2018) found that blueberry anthocyanins can significantly inhibit the increase of ROS in endothelial cells induced by high glucose [80]. At the same time, they can inhibit the decrease of antioxidant enzymes activity. Mallow anthocyanins (Mv), the main component of blueberry extracts, can down regulate the expression of Nox4, which is the main catalytic component of NADPH oxidase and an important source of ROS. Mv inhibited the expression of the NOX4 protein by 45.96%. It is proved that blueberry anthocyanins have a good antioxidant effect. Palungwachira et al. found that the intracellular levels of ROS control the level of phosphor-IκBα by activating a kinase or inactivating a phosphatase that is specific to this protein [84]. Therefore, anthocyanins mediated low levels of ROS, by suppressing IκBα phosphorylation, may abolish the specific proteolysis of phosphorylated IκBα that induces NF-κB activation. González-Reyes et al. have found that anthocyanins have a good antioxidation effect in AD models and can significantly reduce the production of ROS [85]. This was also confirmed in studies by Ma et al. [86], anthocyanin-rich berry extracts reduced H_2O_2-induced ROS production and LPS induced No production in BV-2 microglia. Ryo Furuuchi et al., through the study of the diet induced obesity mice model, found that taking borsenberry polyphenols and anthocyanins can inhibit the production of ROS in the aorta [87]. In conclusion, the antioxidant activity of anthocyanins can effectively reduce the inflammatory reaction.

3.5. Prostaglandin E2 (PGE2)

Prostaglandin E2 (PGE2) is an important physiologically active lipid, mainly derived from membrane phospholipids. Arachidonic acid (AA) is released by phospholipase A2 (PLA2) from membrane phospholipids, and PGE is synthesized from AA via cyclooxygenase (COX-1 and COX-2) and PGE synthase [88]. PGE2 is the most abundant prostaglandin detected in various tissues. It plays a variety of physiological and pathological roles through the expression of 4 PGE receptor subtypes (EP1-4) on the cell surface. Prostaglandin E2 is produced in large quantities in inflammatory areas. PGE2 induces mast cell activation through EP3 receptor signaling pathway, thus enhancing vascular permeability and leading to acute inflammation. PGE2 also promotes Th1 cell differentiation, Th17 cell proliferation, and Th22 cell production of IL-22 through EP2 and EP4 receptors in vitro. In most cases,

PGE2 aggravates chronic inflammation and various autoimmune diseases mainly through EP4 receptor.

PGE2 is mainly synthesized by COX with AA as substrate. Therefore, non-steroidal anti-inflammatory drugs, such as aspirin, play a strong anti-inflammatory effect mainly by inhibiting COX activity [89]. As mentioned above, the activation of NF-κB signaling pathway leads to the increase of COX-2 expression, while anthocyanins can inhibit the activation of NF-κB signaling pathway and COX activity, reduce the production of PGE2 and play an anti-inflammatory effect. He et al. found that anthocyanins can reduce the inflammation induced by UVB radiation by scavenging ROS and inhibiting the expression of COX-2 [90]. Park et al. fed asthmatic mice with anthocyanins, and found that anthocyanins can reduce the development of asthma by down regulating Th2 cytokines, proinflammatory cytokines, and COX-2 [91]. Van de Velde et al. extracted anthocyanins from strawberry and blackberry, and found that the extract had 20% inhibitory effect on COX-2 gene expression by LPS stimulated raw 264.7 macrophage model, and the results showed that the anti-inflammatory effect of anthocyanins was particularly significant [92]. In addition, studies have shown that PGE2-EP3 can also activate PI3K signal, and the inhibition of PI3K can significantly inhibit the production of IL-6 induced by PGE2. Ali et al. demonstrated that anthocyanins can regulate PI3K Akt signaling pathway in AD mice [93]. Zhao et al. found that anthocyanins could down regulate the expression of PI3K protein and inhibit the expression and phosphorylation of Akt [94]. In conclusion, a number of studies have shown that anthocyanins inhibit COX activity, reduce PGE2 production, and reduce inflammation by inhibiting NF-κB and PI3K Akt signaling pathways. The mechanisms of action of anthocyanins are as shown. The underlying molecular mechanisms of action of anthocyanins are listed in Table 3.

Table 3. The Mechanisms of Action of Anthocyanins.

Source of Anthocyanins	Major Anthocyanins and Dose	Model	Biological Effects	References
Strawberry	Pelargonidin-3-O-glucoside Dose: 100–400 mg/kg	Mouse model of pleurisy	Decreased: ADA and MPO Inhibited: IkB-α, JNKMAPK	[57]
Sour cherry	cyanidin-3-rutinoside, cyanidin-3-O-glucoside, and cyanidin-3-O-glucosyl-rutinoside Dose: 50 μg/mL	HUVECs were treated with 100 ng/mL LPS	Decreased: ROS, TNF-α, IL-6, tPA, PGI2, COX-2	[95]
Mahaleb Cherry	Cyanidin 3-(6-(rhamnosyl)glucoside), Cyanidin 3-glucoside, Cyanidin 3-(6-(rhamnosyl)-2-(xylosyl)glucoside), Cyanidin 3-(2-(xylosyl)glucoside) Dose: 60 μg/mL, 50 μg/mL	TEAC, ORAC and model of vascular inflammation	Decreased: ROS, VCAM-1 and ICAM-1	[17]
Black currant	Delphinidin 3-(6-(rhamnosyl)glucoside), Cyanidin 3-(6-(rhamnosyl)glucoside) Dose: 60 μg/mL, 50 μg/mL	TEAC, ORAC and model of vascular inflammation	Decreased: ROS, VCAM-1 and ICAM-1	
Black Carrot	Cyanidin 3-(6-(6-(feruloyl)glucosyl)-2-(xylosyl)galactoside), Cyanidin 3-(6-(6-(sinapoyl)glucosyl)-2-(xylosyl)galactoside)	TEAC, ORAC and model of vascular inflammation	Decreased: ROS, VCAM-1 and ICAM-1	
"Sun Black" T omato	Petunidin 3-(6-(4-(E-p-coumaroyl)rhamnosyl)glucoside)-5-glucoside (petanin), Malvidin 3-(6-(4-(E-p-coumaroyl)rhamnosyl)glucoside)-5-glucoside Dose: 60 μg/mL, 50 μg/mL	TEAC, ORAC and model of vascular inflammation	Decreased: ROS, VCAM-1 and ICAM-1	

Table 3. *Cont.*

Source of Anthocyanins	Major Anthocyanins and Dose	Model	Biological Effects	References
Blueberries	malvidin, malvidin-3-glucoside, malvidin-3-galactoside Dose: 10 μg/mL	HRCECs	Decreased: ROS, VEGF, ICAM-1 Inhibited: Akt, NF-κB Increased: CAT, SOD	[80]
Portuguese blueberries	malvidin-3-galactoside, petunidin-3-arabinoside Dose: 100 mg/kg	TNBS induced colitis in rats	Decreased: iNOS, COX2, MPO, GPX	[96]
Black currant	delphinidin-3-rutinoside, cyanidin-3-rutinoside, delphinidin-3-glucoside Dose: 50 μg/mL	RAW 264.7 macrophages and human THP-1 monocytes	Decreased: IL-1β, iNOS, CXCL9, TNFα Increased: ARG1, CHIL3	[97]
Raspberries	Cyanidin-3-O-sophoroside, Cyanidin-3-O-glucosylrutinoside, Cyanidin-3-O-glucoside, Cyanidin-3-O-rutinoside Dose: 125 μg/mL	HL-60-Human Caucasian promyelocytic leukemia, J45.01-Human acute T cell leukemia	Decreased: LOX, COX-2	[98]
Black rice	cyanidin-3-O-glucoside, peonidin-3-O-glucoside Dose: 25 μg/mL	Rat primary dermal fibroblasts	Decreased: NF-κB p50 and p65 mRNA Increased: Induce Collagen, Type I Alpha 2 mRNA	[84]
Purple rice	Cyanidin-3-O-glucoside, peonidin-3-O-glucoside Dose: 50 μg/mL	Porcine cartilage explant	Decreased: s-GAG, HA, MMP-1, 3 and 13, Inhibited: NF-κB, ERK	[73]
Purple maize	Cyanidin-3-O-glucoside, pelargonidin-3-O-glucoside, peonidin-3-O-glucoside	RAW264.7 macrophages, 3T3-L1 adipocytes	Decreased: PGE2, NO, MCP, iNOS, COX-2, ROS Inhibited: PPARγ, DPP-IV	[99]

4. Conclusions

Anthocyanins are naturally harmless and play a role in many diseases. Anthocyanins can regulate the vasodilation and contraction of blood vessels by controlling the activity of nitric oxide synthase and potassium channels, inhibit the degradation of important components of the inner wall of blood vessels, and improve cardiovascular diseases. Anthocyanins rapidly increase the oxidative function of mitochondria in muscle cells and brown adipose tissue after supplementing energy to increase the body's energy metabolism rate, which can prevent obesity. Anthocyanins can also inhibit the activity of carbon hydrolase and exert the effect of reducing blood glucose. In addition, anthocyanins can also activate Bim through the mitochondrial pathway, promote cell apoptosis, and exert certain anti-cancer effects. As a body's self-defense mechanism, inflammation plays a very complicated role in various diseases. This article reviews the role of anthocyanins in inflammation. Anthocyanins exert anti-inflammatory effects by inhibiting the release of pro-inflammatory factors, reducing the expression of TLR4, and inhibiting the activation of the NF-κB pathway and MAPKs signaling pathway. Meanwhile, they reduce the production of NO, ROS, and PGE2 and avoid repeated stimulation. However, the stability and utilization of anthocyanins are poor, which limits the application of these compounds. Therefore, future research should focus on improving the stability and bioavailability of anthocyanins, so as to better use the anti-inflammatory effect of anthocyanins.

Funding: This research was funded by Open Project Funding by Food Processing Discipline Group of Hebei Agricultural University.

Acknowledgments: We thank Yasai Sun from Hebei Agricultural University provided the funding support and supervised this manuscript.

Conflicts of Interest: The authors declare no conflict of interest.

Abbreviations

Nuclear factor-kappa B	NF-κB
Tumor necrosis factor-α	TNF-α
Interleukin-6	IL-6
Interleukin-1β	IL-1β
Monocyte chemoattractant protein-1	MCP-1
soluble vascular cell adhesion molecule-1	sICAM-1
C-reactive protein	CRP
Pattern recognition receptors	PRRS
Toll like receptors	TLRs
Mitogen activated protein kinases	MAPK
c-Jun N-terminal kinases	JNK
Reactive oxygen species	ROS
Cyclooxygenase-2	COX-2
Vascular endothelial growth factor	VEGF
Inducible nitric oxide synthase	iNOS
Hydrogen peroxide	H_2O_2
Hydroxyl radical	OH^-
Nicotinamide adenine dinucleotide phosphate	NADPH
Alzheimer's disease	AD
Lipopolysaccharide binding protein	LBP
Lipopolysaccharide	LPS
Myeloid differentiation factor 88	MyD88
IL-1 receptor associated Kinase	IRAK
TNF-receptor association factor 6	TRAF-6
TGF-activated kinase 1	TAK1
Toll-interacting protein	TOLLIP
TIR	Toll/IL-1 receptor
RNA-activated protein kinase	PKR
Tissue-type plasminogen activator	tP A
Prostacyclin	PGI2
Human retinal capillary endothelial cells	HRCECs
Catalase	CAT
Superoxide dismutase	SOD
Glutathione peroxidase	GPX
C-X-C motif ligand 9	CXCL9
Arginase	ARG1
Chitinase-like 3	CHIL3
Lipoxygenase	LOX
Sulfated glycosaminoglycan	s-GAG
Hyaluronic acid	HA
Matrix metalloproteinases	MMP
Dipeptidyl peptidase-4	DPP-IV
Proliferator-activated receptor γ	PPARγ

References

1. Joseph, S.V.; Edirisinghe, I.; Burton-Freeman, B.M. Berries: Anti-inflammatory Effects in Humans. *J. Agric. Food Chem.* **2014**, *62*, 3886–3903. [CrossRef] [PubMed]
2. Vendrame, S.; Klimis-Zacas, D. Anti-inflammatory effect of anthocyanins via modulation of nuclear factor- B and mitogen-activated protein kinase signaling cascades. *Nutr. Rev.* **2015**, *73*, 348–358. [CrossRef] [PubMed]
3. Mitchell, S.; Vargas, J.; Hoffmann, A. Signaling via the NFκB system. *Wiley Interdiscip. Rev. Syst. Biol. Med.* **2016**, *8*, 227–241. [CrossRef] [PubMed]
4. Stewart, A.; Beart, P.M. Inflammation: Maladies, models, mechanisms and molecules. *Br. J. Pharmacol.* **2016**, *173*, 631–634. [CrossRef] [PubMed]
5. Ellulu, M.S. Obesity, cardiovascular disease, and role of vitamin C on inflammation: A review of facts and underlying mechanisms. *Inflammopharmacology* **2017**, *25*, 313–328. [CrossRef] [PubMed]

6. Singh, N.; Baby, D.; Rajguru, J.P.; Patil, P.B.; Thakkannavar, S.S.; Pujari, V.B. Inflammation and cancer. *Ann. Afr. Med.* **2019**, *18*, 121–126. [CrossRef] [PubMed]
7. Liuzzo, G.; Pedicino, D.; Flego, D.; Crea, F. Inflammation and Atherothrombosis. *Clin. Immunol.* **2018**, 935–946.e1. [CrossRef]
8. Nandkeolyar, S.; Naqvi, A.; Fan, W.; Sharma, A.; Rana, J.S.; Rozanski, A.; Shaw, L.; Friedman, J.D.; Hayes, S.; Dey, D.; et al. Utility of novel serum biomarkers to predict subclinical atherosclerosis: A sub-analysis of the EISNER study. *Atherosclerosis* **2019**, *282*, 80–84. [CrossRef] [PubMed]
9. Aboonabi, A.; Meyer, R.R.; Gaiz, A.; Singh, I. Anthocyanins in berries exhibited anti-atherogenicity and antiplatelet activities in a metabolic syndrome population. *Nutr. Res.* **2020**, *76*, 82–93. [CrossRef] [PubMed]
10. Zhang, W.-J.; Frei, B. α-Lipoic acid inhibits TNF-a-induced NF-κB activation and adhesion molecule expression in human aortic endothelial cells. *FASEB J.* **2001**, *15*, 2423–2432. [CrossRef] [PubMed]
11. Kim, H.J.; Tsoy, I.; Park, J.M.; Chung, J.I.; Shin, S.C.; Chang, K.C. Anthocyanins from soybean seed coat inhibit the expression of TNF-α-induced genes associated with ischemia/reperfusion in endothelial cell by NF-κB-dependent pathway and reduce rat myocardial damages incurred by ischemia and reperfusion in vivo. *FEBS Lett.* **2006**, *580*, 1391–1397. [CrossRef] [PubMed]
12. Hotamisligil, G.S.; Shargill, N.S.; Spiegelman, B.M. Adipose Expression of Tumor Necrosis Factor-Alpha: Direct Role in Obesity-Linked Insulin Resistance. *Science* **1993**, *259*, 87–91. [CrossRef] [PubMed]
13. Belwal, T.; Singh, G.; Jeandet, P.; Pandey, A.; Giri, L.; Ramola, S.; Bhatt, I.D.; Venskutonis, P.R.; Georgiev, M.; Clément, C.; et al. Anthocyanins, multi-functional natural products of industrial relevance: Recent biotechnological advances. *Biotechnol. Adv.* **2020**, *43*, 107600. [CrossRef]
14. Sendri, N.; Devidas, S.B.; Katoch, S.; Patial, V.; Bhandari, P. Copigmentation and UPLC-ESI-MS/MS of anthocyanin in Ipomoea nil as potential source of food colorant. *Nat. Prod. Res.* **2020**, 1–6. [CrossRef]
15. Martinsen, B.K.; Aaby, K.; Skrede, G. Effect of temperature on stability of anthocyanins, ascorbic acid and color in strawberry and raspberry jams. *Food Chem.* **2020**, *316*, 126297. [CrossRef] [PubMed]
16. Moraes, L.F.D.S.; Sun, X.; Peluzio, M.D.C.G.; Zhu, M.-J. Anthocyanins/anthocyanidins and colorectal cancer: What is behind the scenes? *Crit. Rev. Food Sci. Nutr.* **2017**, *59*, 59–71. [CrossRef] [PubMed]
17. Blando, F.; Calabriso, N.; Berland, H.; Maiorano, G.; Gerardi, C.; Carluccio, M.A.; Andersen, M. Radical Scavenging and Anti-Inflammatory Activities of Representative Anthocyanin Groupings from Pigment-Rich Fruits and Vegetables. *Int. J. Mol. Sci.* **2018**, *19*, 169. [CrossRef] [PubMed]
18. Zhang, J.; Wu, J.; Liu, F.; Tong, L.; Chen, Z.; Chen, J.; He, H.; Xu, R.; Ma, Y.; Huang, C. Neuroprotective effects of anthocyanins and its major component cyanidin-3-O-glucoside (C3G) in the central nervous system: An outlined review. *Eur. J. Pharmacol.* **2019**, *858*, 172500. [CrossRef] [PubMed]
19. Wallace, T.C.; Monica, G.M. Anthocyanins. *Adv. Nutr.* **2015**, *6*, 620–622. [CrossRef] [PubMed]
20. Clifford, M.N. Anthocyanins—Nature, occurrence and dietary burden. *J. Sci. Food Agric.* **2000**, *80*, 1118–1125. [CrossRef]
21. Saigo, T.; Wang, T.; Watanabe, M.; Tohge, T. Diversity of anthocyanin and proanthocyanin biosynthesis in land plants. *Curr. Opin. Plant Biol.* **2020**, *55*, 93–99. [CrossRef] [PubMed]
22. Czank, C.; Cassidy, A.; Zhang, Q.; Morrison, D.; Preston, T.; Kroon, P.; Botting, N.P.; Kay, C. Human metabolism and elimination of the anthocyanin, cyanidin-3-glucoside: A 13C-tracer study. *Am. J. Clin. Nutr.* **2013**, *97*, 995–1003. [CrossRef] [PubMed]
23. He, J.; Giusti, M.M. Anthocyanins: Natural Colorants with Health-Promoting Properties. *Annu. Rev. Food Sci. Technol.* **2010**, *1*, 163–187. [CrossRef] [PubMed]
24. Zamora-Ros, R.; Knaze, V.; Luján-Barroso, L.; Slimani, N.; Romieu, I.; Touillaud, M.; Kaaks, R.; Teucher, B.; Mattiello, A.; Grioni, S.; et al. Estimation of the intake of anthocyanidins and their food sources in the European Prospective Investigation into Cancer and Nutrition (EPIC) study. *Br. J. Nutr.* **2011**, *106*, 1090–1099. [CrossRef]
25. Cassidy, A.; Mukamal, K.J.; Liu, L.; Franz, M.; Eliassen, A.H.; Rimm, E.B. High Anthocyanin Intake Is Associated With a Reduced Risk of Myocardial Infarction in Young and Middle-Aged Women. *Circulation* **2013**, *127*, 188–196. [CrossRef]
26. Tarone, A.G.; Cazarin, C.B.B.; Junior, M.R.M. Anthocyanins: New techniques and challenges in microencapsulation. *Food Res. Int.* **2020**, *133*, 109092. [CrossRef]
27. Ersus, S.; Yurdagel, U. Microencapsulation of anthocyanin pigments of black carrot (*Daucus carota* L.) by spray drier. *J. Food Eng.* **2007**, *80*, 805–812. [CrossRef]
28. Riaz, M.; Zia-Ul-Haq, M.; Saad, B. Introduction to Anthocyanins. In *Techniques for Nanoencapsulation of Food Ingredients*; Springer: New York, NY, USA, 2016; pp. 21–33.
29. Turturică, M.; Rapeanu, A.M.; Bahrim, G. Anthocyanins: Naturally occuring fruit pigments with functional properties. *Ann. Univ. Dunarea de Jos Galati. Fascicle VI-Food Technol.* **2015**, *39*, 9–24.
30. Zhang, K.; Liu, Z.; Guan, L.; Zheng, T.; Jiu, S.; Zhu, X.; Jia, H.; Fang, J. Changes of Anthocyanin Component Biosynthesis in 'Summer Black' Grape Berries after the Red Flesh Mutation Occurred. *J. Agric. Food Chem.* **2018**, *66*, 9209–9218. [CrossRef]
31. Pirzadeh, M.; Caporaso, N.; Rauf, A.; Shariati, M.A.; Yessimbekov, Z.; Khan, M.U.; Imran, M.; Mubarak, M.S. Pomegranate as a source of bioactive constituents: A review on their characterization, properties and applications. *Crit. Rev. Food Sci. Nutr.* **2020**, *61*, 982–999. [CrossRef]
32. Strauch, R.C.; Mengist, M.F.; Pan, K.; Yousef, G.G.; Iorizzo, M.; Brown, A.F.; Lila, M.A. Variation in anthocyanin profiles of 27 genotypes of red cabbage over two growing seasons. *Food Chem.* **2019**, *301*, 125289. [CrossRef]

33. Yoon, B.I.; Bae, W.J.; Choi, Y.S.; Kim, S.J.; Ha, U.S.; Hong, S.-H.; Sohn, D.W.; Kim, S.W. Anti-inflammatory and Antimicrobial Effects of Anthocyanin Extracted from Black Soybean on Chronic Bacterial Prostatitis Rat Model. *Chin. J. Integr. Med.* **2013**, *24*, 621–626. [CrossRef]
34. Frond, A.D.; Iuhas, C.I.; Stirbu, I.; Leopold, L.; Socaci, S.; Andreea, S.; Ayvaz, H.; Andreea, S.; Mihai, S.; Diaconeasa, Z.; et al. Phytochemical Characterization of Five Edible Purple-Reddish Vegetables: Anthocyanins, Flavonoids, and Phenolic Acid Derivatives. *Molecules* **2019**, *24*, 1536. [CrossRef]
35. Luna-Vital, D.A.; Luzardo-Ocampo, I.; Cuellar-Nuñez, M.L.; Loarca-Piña, G.; de Mejia, E.G. Maize extract rich in ferulic acid and anthocyanins prevents high-fat-induced obesity in mice by modulating SIRT1, AMPK and IL-6 associated metabolic and inflammatory pathways. *J. Nutr. Biochem.* **2020**, *79*, 108343. [CrossRef]
36. DeFuria, J.; Bennett, G.; Strissel, K.J.; Perfield, J.W.; Milbury, P.E.; Greenberg, A.S.; Obin, M.S. Dietary Blueberry Attenuates Whole-Body Insulin Resistance in High Fat-Fed Mice by Reducing Adipocyte Death and Its Inflammatory Sequelae. *J. Nutr.* **2009**, *139*, 1510–1516. [CrossRef]
37. Prior, R.L.; Wu, X.; Gu, L.; Hager, T.J.; Hager, A.; Howard, L.R. Whole Berries versus Berry Anthocyanins: Interactions with Dietary Fat Levels in the C57BL/6J Mouse Model of Obesity. *J. Agric. Food Chem.* **2008**, *56*, 647–653. [CrossRef]
38. Khan, M.I.; Shin, J.H.; Shin, T.S.; Kim, M.Y.; Cho, N.J.; Kim, J.D. Anthocyanins from Cornus kousa ethanolic extract attenuate obesity in association with anti-angiogenic activities in 3T3-L1 cells by down-regulating adipogeneses and lipogenesis. *PLoS ONE* **2018**, *13*, e0208556. [CrossRef]
39. Vilhena, R.; Figueiredo, I.; Baviera, A.; Silva, D.; Marson, B.; Oliveira, J.; Peccinini, R.; Borges, I.; Pontarolo, R. Antidiabetic activity of Musa x paradisiaca extracts in streptozotocin-induced diabetic rats and chemical characterization by HPLC-DAD-MS. *J. Ethnopharmacol.* **2020**, *254*, 112666. [CrossRef]
40. Fang, J.; Huang, J. Accumulation of plasma levels of anthocyanins following multiple saskatoon berry supplements. *Xenobiotica* **2019**, *50*, 454–457. [CrossRef]
41. Skates, E.; Overall, J.; DeZego, K.; Wilson, M.; Esposito, D.; Lila, M.A.; Komarnytsky, S. Berries containing anthocyanins with enhanced methylation profiles are more effective at ameliorating high fat diet-induced metabolic damage. *Food Chem. Toxicol.* **2018**, *111*, 445–453. [CrossRef]
42. Aboonabi, A.; Aboonabi, A. Anthocyanins reduce inflammation and improve glucose and lipid metabolism associated with inhibiting nuclear factor-kappaB activation and increasing PPAR-γ gene expression in metabolic syndrome subjects. *Free Radic. Biol. Med.* **2020**, *150*, 30–39. [CrossRef]
43. Holvoet, P. Relations between metabolic syndrome, oxidative stress and inflammation and cardiovascular disease. *Verh.-K. Acad. Geneeskd. Belg.* **2008**, *70*, 193–219.
44. Bakuradze, T.; Tausend, A.; Galan, J.; Groh, I.A.M.; Berry, D.; Tur, J.A.; Marko, D.; Richling, E. Antioxidative activity and health benefits of anthocyanin-rich fruit juice in healthy volunteers. *Free Radic. Res.* **2019**, *53*, 1045–1055. [CrossRef]
45. Curtis, P.J.; Van Der Velpen, V.; Berends, L.; Jennings, A.; Feelisch, M.; Umpleby, A.M.; Evans, M.; Fernandez, B.O.; Meiss, M.S.; Minnion, M.; et al. Blueberries improve biomarkers of cardiometabolic function in participants with metabolic syndrome—results from a 6-month, double-blind, randomized controlled trial. *Am. J. Clin. Nutr.* **2019**, *109*, 1535–1545. [CrossRef]
46. Feng, R.; Ni, H.-M.; Wang, S.Y.; Tourkova, I.L.; Shurin, M.; Harada, H.; Yin, X.-M. Cyanidin-3-rutinoside, a Natural Polyphenol Antioxidant, Selectively Kills Leukemic Cells by Induction of Oxidative Stress. *J. Biol. Chem.* **2007**, *282*, 13468–13476. [CrossRef]
47. Liu, X.; Wang, L.; Jing, N.; Jiang, G.; Liu, Z. Biostimulating Gut Microbiome with Bilberry Anthocyanin Combo to Enhance Anti-PD-L1 Efficiency against Murine Colon Cancer. *Microorganisms* **2020**, *8*, 175. [CrossRef]
48. Losso, J.N.; Finley, J.W.; Karki, N.; Liu, A.G.; Prudente, A.; Tipton, R.; Yu, Y.; Greenway, F.L. Pilot Study of the Tart Cherry Juice for the Treatment of Insomnia and Investigation of Mechanisms. *Am. J. Ther.* **2018**, *25*, e194–e201. [CrossRef]
49. Sangsefidi, Z.S.; Hosseinzadeh, M.; Ranjbar, A.M.; Akhondi-Meybodi, M.; Fallahzadeh, H.; Mozaffari-Khosravi, H. The effect of total anthocyanin-base standardized (*Cornus mas* L.) fruit extract on liver function, tumor necrosis factor α, malondealdehyde, and adiponectin in patients with non-alcoholic fatty liver: A study protocol for a double-blind randomized clinical trial. *Nutr. J.* **2019**, *18*, 1–7. [CrossRef]
50. Hollands, W.J.; Armah, C.N.; Doleman, J.F.; Perez-Moral, N.; Winterbone, M.S.; Kroon, P.A. 4-Week consumption of anthocyanin-rich blood orange juice does not affect LDL-cholesterol or other biomarkers of CVD risk and glycaemia compared with standard orange juice: A randomised controlled trial. *Br. J. Nutr.* **2018**, *119*, 415–421. [CrossRef]
51. Ohguro, H.; Ohguro, I.; Katai, M.; Tanaka, S. Two-Year Randomized, Placebo-Controlled Study of Black Currant Anthocyanins on Visual Field in Glaucoma. *Ophthalmologica* **2012**, *228*, 26–35. [CrossRef]
52. Fan, D.; Alamri, Y.; Liu, K.; Macaskill, M.; Harris, P.; Brimble, M.; Dalrymple-Alford, J.; Prickett, T.; Menzies, O.; Laurenson, A.; et al. Supplementation of Blackcurrant Anthocyanins Increased Cyclic Glycine-Proline in the Cerebrospinal Fluid of Parkinson Patients: Potential Treatment to Improve Insulin-Like Growth Factor-1 Function. *Nutrients* **2018**, *10*, 714. [CrossRef] [PubMed]
53. Zhang, P.-W.; Chen, F.-X.; Li, D.; Ling, W.-H.; Guo, H.-H. A CONSORT-Compliant, Randomized, Double-Blind, Placebo-Controlled Pilot Trial of Purified Anthocyanin in Patients With Nonalcoholic Fatty Liver Disease. *Medicine* **2015**, *94*, e758. [CrossRef] [PubMed]
54. Lee, M.; Sorn, S.R.; Park, Y.; Park, H.-K. Anthocyanin Rich-Black Soybean Testa Improved Visceral Fat and Plasma Lipid Profiles in Overweight/Obese Korean Adults: A Randomized Controlled Trial. *J. Med. Food* **2016**, *19*, 995–1003. [CrossRef] [PubMed]

55. Jiang, N.; Chen, X.-L.; Yang, H.-W.; Ma, Y.-R. Effects of nuclear factor κB expression on retinal neovascularization and apoptosis in a diabetic retinopathy rat model. *Int. J. Ophthalmol.* **2015**, *8*, 448–452. [CrossRef]
56. Tilborghs, S.; Corthouts, J.; Verhoeven, Y.; Arias, D.; Rolfo, C.; Trinh, X.B.; Van Dam, P.A. The role of Nuclear Factor-kappa B signaling in human cervical cancer. *Crit. Rev. Oncol.* **2017**, *120*, 141–150. [CrossRef]
57. Duarte, L.J.; Chaves, V.C.; Nascimento, M.V.P.D.S.; Calvete, E.; Li, M.; Ciraolo, E.; Ghigo, A.; Hirsch, E.; Simões, C.M.O.; Reginatto, F.H.; et al. Molecular mechanism of action of Pelargonidin-3- O -glucoside, the main anthocyanin responsible for the anti-inflammatory effect of strawberry fruits. *Food Chem.* **2017**, *247*, 56–65. [CrossRef]
58. Lee, H.-H.; Lee, S.-G.; Shin, J.-S.; Lee, H.-Y.; Yoon, K.; Ji, Y.W.; Jang, D.S.; Lee, K.-T. p-Coumaroyl Anthocyanin Mixture Isolated from Tuber Epidermis of Solanum tuberosum Attenuates Reactive Oxygen Species and Pro-inflammatory Mediators by Suppressing NF-κB and STAT1/3 Signaling in LPS-Induced RAW264.7 Macrophages. *Biol. Pharm. Bull.* **2017**, *40*, 1894–1902. [CrossRef]
59. Le Phuong Nguyen, T.; Fenyvesi, F.; Remenyik, J.; Homoki, J.R.; Gogolák, P.; Bácskay, I.; Fehér, P.; Ujhelyi, Z.; Vasvári, G.; Vecsernyés, M.; et al. Protective Effect of Pure Sour Cherry Anthocyanin Extract on Cytokine-Induced Inflammatory Caco-2 Monolayers. *Nutrients* **2018**, *10*, 861. [CrossRef]
60. Roth, S.; Spalinger, M.R.; Gottier, C.; Biedermann, L.; Zeitz, J.; Lang, S.; Weber, A.; Rogler, G.; Scharl, M. Bilberry-Derived Anthocyanins Modulate Cytokine Expression in the Intestine of Patients with Ulcerative Colitis. *PLoS ONE* **2016**, *11*, e0154817. [CrossRef]
61. Ouyang, W.; O'Garra, A. IL-10 Family Cytokines IL-10 and IL-22: From Basic Science to Clinical Translation. *Immunity* **2019**, *50*, 871–891. [CrossRef]
62. Mirza, A.Z.; AlThagafi, I.I.; Shamshad, H. Role of PPAR receptor in different diseases and their ligands: Physiological importance and clinical implications. *Eur. J. Med. Chem.* **2019**, *166*, 502–513. [CrossRef]
63. Lanki, M.E.; Seppänen, H.; Mustonen, H.K.; Böckelman, C.; Juuti, A.T.; Hagström, J.K.; Haglund, C.H. Toll-like receptor 2 and Toll-like receptor 4 predict favorable prognosis in local pancreatic cancer. *Tumor Biol.* **2018**, *40*. [CrossRef]
64. Liu, T.; Zhang, L.; Joo, D.; Sun, S.-C. NF-κB signaling in inflammation. *Signal Transduct. Target. Ther.* **2017**, *2*, 17023. [CrossRef]
65. Jung, S.K.; Lim, T.-G.; Seo, S.G.; Lee, H.J.; Hwang, Y.-S.; Choung, M.-G.; Lee, K.W. Cyanidin-3-O-(2″-xylosyl)-glucoside, an anthocyanin from Siberian ginseng (*Acanthopanax senticosus*) fruits, inhibits UVB-induced COX-2 expression and AP-1 transactivation. *Food Sci. Biotechnol.* **2013**, *22*, 507–513. [CrossRef]
66. Li, L.; Wang, L.; Wu, Z.; Yao, L.; Wu, Y.; Huang, L.; Liu, K.; Zhou, X.; Gou, D. Anthocyanin-rich fractions from red raspberries attenuate inflammation in both RAW264.7 macrophages and a mouse model of colitis. *Sci. Rep.* **2014**, *4*, srep06234. [CrossRef]
67. Cui, H.-X.; Chen, J.-H.; Li, J.-W.; Cheng, F.-R.; Yuan, K. Protection of Anthocyanin from *Myrica rubra* against Cerebral Ischemia-Reperfusion Injury via Modulation of the TLR4/NF-κB and NLRP3 Pathways. *Molecules* **2018**, *23*, 1788. [CrossRef]
68. Thummayot, S.; Tocharus, C.; Jumnongprakhon, P.; Suksamrarn, A.; Tocharus, J. Cyanidin attenuates Aβ25-35-induced neuroinflammation by suppressing NF-κB activity downstream of TLR4/NOX4 in human neuroblastoma cells. *Acta Pharmacol. Sin.* **2018**, *39*, 1439–1452. [CrossRef]
69. Karunarathne, W.A.H.M.; Lee, K.T.; Choi, Y.H.; Jin, C.-Y.; Kim, G.-Y. Anthocyanins isolated from *Hibiscus syriacus* L. attenuate lipopolysaccharide-induced inflammation and endotoxic shock by inhibiting the TLR4/MD2-mediated NF-κB signaling pathway. *Phytomedicine* **2020**, *76*, 153237. [CrossRef]
70. Mansell, A.; Jenkins, B.J. Dangerous liaisons between interleukin-6 cytokine and toll-like receptor families: A potent combination in inflammation and cancer. *Cytokine Growth Factor Rev.* **2013**, *24*, 249–256. [CrossRef]
71. Zhang, Y.; Meng, Q.; Yin, J.; Zhang, Z.; Bao, H.; Wang, X. Anthocyanins attenuate neuroinflammation through the suppression of MLK3 activation in a mouse model of perioperative neurocognitive disorders. *Brain Res.* **2019**, *1726*, 146504. [CrossRef]
72. Li, K.; Zhang, M.; Chen, H.; Peng, J.; Jiang, F.; Shi, X.; Bai, Y.; Jian, M.; Jia, Y. Anthocyanins from black peanut skin protect against UV-B induced keratinocyte cell and skin oxidative damage through activating Nrf 2 signaling. *Food Funct.* **2019**, *10*, 6815–6828. [CrossRef]
73. Wongwichai, T.; Teeyakasem, P.; Pruksakorn, D.; Kongtawelert, P.; Pothacharoen, P. Anthocyanins and metabolites from purple rice inhibit IL-1β-induced matrix metalloproteinases expression in human articular chondrocytes through the NF-κB and ERK/MAPK pathway. *Biomed. Pharmacother.* **2019**, *112*, 108610. [CrossRef]
74. Farzaei, M.H.; Elsenduny, F.; Momtaz, S.; Parvizi, F.; Iranpanah, A.; Tewari, D.; Naseri, R.; Abdolghaffari, A.H.; Rezaei, N. An update on dietary consideration in inflammatory bowel disease: Anthocyanins and more. *Expert Rev. Gastroenterol. Hepatol.* **2018**, *12*, 1007–1024. [CrossRef]
75. Loscalzo, J. The Identification of Nitric Oxide as Endothelium Derived Relaxing Factor. *Circ. Res.* **2013**, *113*, 100–103. [CrossRef]
76. Baudouin, E. The language of nitric oxide signalling. *Plant. Biol.* **2010**, *13*, 233–242. [CrossRef]
77. Eqian, J.; Efulton, D. Post-translational regulation of endothelial nitric oxide synthase in vascular endothelium. *Front. Physiol.* **2013**, *4*, 347. [CrossRef]
78. Gardlik, R.; Fusekova, I. Pharmacologic Therapy for Diabetic Retinopathy. *Semin. Ophthalmol.* **2013**, *30*, 252–263. [CrossRef]
79. Gupta, K.; Sirohi, V.K.; Kumari, S.; Shukla, V.; Manohar, M.; Popli, P.; Dwivedi, A. Sorcin is involved during embryo implantation via activating VEGF/PI3K/Akt pathway in mice. *J. Mol. Endocrinol.* **2018**, *60*, 119–132. [CrossRef]
80. Huang, W.; Yan, Z.; Li, D.; Ma, Y.; Zhou, J.; Sui, Z. Antioxidant and Anti-Inflammatory Effects of Blueberry Anthocyanins on High Glucose-Induced Human Retinal Capillary Endothelial Cells. *Oxid. Med. Cell. Longev.* **2018**, *2018*, 1862462. [CrossRef]

81. Nizamutdinova, I.T.; Kim, Y.M.; Chung, J.I.; Shin, S.C.; Jeong, Y.-K.; Seo, H.G.; Lee, J.H.; Chang, K.C.; Kim, H.J. Anthocyanins from black soybean seed coats stimulate wound healing in fibroblasts and keratinocytes and prevent inflammation in endothelial cells. *Food Chem. Toxicol.* **2009**, *47*, 2806–2812. [CrossRef]
82. Winter, A.N.; Brenner, M.C.; Punessen, N.; Snodgrass, M.; Byars, C.; Arora, Y.; Linseman, D.A. Comparison of the Neuroprotective and Anti-Inflammatory Effects of the Anthocyanin Metabolites, Protocatechuic Acid and 4-Hydroxybenzoic Acid. *Oxid. Med. Cell. Longev.* **2017**, *2017*, 6297080. [CrossRef]
83. Nita, M.; Grzybowski, A. The Role of the Reactive Oxygen Species and Oxidative Stress in the Pathomechanism of the Age-Related Ocular Diseases and Other Pathologies of the Anterior and Posterior Eye Segments in Adults. *Oxid. Med. Cell. Longev.* **2016**, *2016*, 3164734. [CrossRef]
84. Palungwachira, P.; Tancharoen, S.; Phruksaniyom, C.; Klungsaeng, S.; Srichan, R.; Kikuchi, K.; Nararatwanchai, T. Antioxidant and Anti-Inflammatory Properties of Anthocyanins Extracted from *Oryza sativa* L. in Primary Dermal Fibroblasts. *Oxid. Med. Cell. Longev.* **2019**, *2019*, 2089817. [CrossRef] [PubMed]
85. González-Reyes, R.E.; Nava-Mesa, M.O.; Vargas-Sánchez, K.; Ariza-Salamanca, D.; Mora-Muñoz, L. Involvement of Astrocytes in Alzheimer's Disease from a Neuroinflammatory and Oxidative Stress Perspective. *Front. Mol. Neurosci.* **2017**, *10*, 427. [CrossRef] [PubMed]
86. Ma, H.; Johnson, S.L.; Liu, W.; DaSilva, N.A.; Meschwitz, S.; Dain, J.A.; Seeram, N.P. Evaluation of Polyphenol Anthocyanin-Enriched Extracts of Blackberry, Black Raspberry, Blueberry, Cranberry, Red Raspberry, and Strawberry for Free Radical Scavenging, Reactive Carbonyl Species Trapping, Anti-Glycation, Anti-β-Amyloid Aggregation, and Microglial Neuroprotective Effects. *Int. J. Mol. Sci.* **2018**, *19*, 461. [CrossRef]
87. Furuuchi, R.; Shimizu, I.; Yoshida, Y.; Hayashi, Y.; Ikegami, R.; Suda, M.; Katsuumi, G.; Wakasugi, T.; Nakao, M.; Minamino, T. Boysenberry polyphenol inhibits endothelial dysfunction and improves vascular health. *PLoS ONE* **2018**, *13*, e0202051. [CrossRef] [PubMed]
88. Woodward, D.F.; Jones, R.L.; Narumiya, S. International Union of Basic and Clinical Pharmacology. LXXXIII: Classification of Prostanoid Receptors, Updating 15 Years of Progress. *Pharmacol. Rev.* **2011**, *63*, 471–538. [CrossRef]
89. Kawahara, K.; Hohjoh, H.; Inazumi, T.; Tsuchiya, S.; Sugimoto, Y. Prostaglandin E2-induced inflammation: Relevance of prostaglandin E receptors. *Biochim. Biophys. Acta (BBA)—Mol. Cell Biol. Lipids* **2014**, *1851*, 414–421. [CrossRef]
90. He, Y.; Hu, Y.; Jiang, X.; Chen, T.; Ma, Y.; Wu, S.; Sun, J.; Jiao, R.; Li, X.; Deng, L.; et al. Cyanidin-3-O-glucoside inhibits the UVB-induced ROS/COX-2 pathway in HaCaT cells. *J. Photochem. Photobiol. B Biol.* **2017**, *177*, 24–31. [CrossRef]
91. Park, S.-J.; Shin, W.-H.; Seo, J.-W.; Kim, E.-J. Anthocyanins inhibit airway inflammation and hyperresponsiveness in a murine asthma model. *Food Chem. Toxicol.* **2007**, *45*, 1459–1467. [CrossRef]
92. Van De Velde, F.; Esposito, D.; Grace, M.; Pirovani, M.E.; Lila, M.A. Anti-inflammatory and wound healing properties of polyphenolic extracts from strawberry and blackberry fruits. *Food Res. Int.* **2018**, *121*, 453–462. [CrossRef]
93. Ali, T.; Kim, T.; Rehman, S.U.; Khan, M.S.; Amin, F.U.; Khan, M.; Ikram, M.; Kim, M.O. Natural Dietary Supplementation of Anthocyanins via PI3K/Akt/Nrf2/HO-1 Pathways Mitigate Oxidative Stress, Neurodegeneration, and Memory Impairment in a Mouse Model of Alzheimer's Disease. *Mol. Neurobiol.* **2017**, *55*, 6076–6093. [CrossRef]
94. Zhao, X.; Feng, P.; He, W.; Du, X.; Chen, C.; Suo, L.; Liang, M.; Zhang, N.; Na, A.; Zhang, Y. The Prevention and Inhibition Effect of Anthocyanins on Colorectal Cancer. *Curr. Pharm. Des.* **2020**, *25*, 4919–4927. [CrossRef] [PubMed]
95. Biro, A.; Markovich, A.; Homoki, J.R.; Szőllősi, E.; Hegedűs, C.; Tarapcsák, S.; Lukács, J.; Stündl, L.; Remenyik, J. Anthocyanin-Rich Sour Cherry Extract Attenuates the Lipopolysaccharide-Induced Endothelial Inflammatory Response. *Molecules* **2019**, *24*, 3427. [CrossRef] [PubMed]
96. Pereira, S.R.; Pereira, R.; Figueiredo, I.V.; Freitas, V.; Dinis, T.C.P.; Almeida, L.M. Comparison of anti-inflammatory activities of an anthocyanin-rich fraction from Portuguese blueberries (*Vaccinium corymbosum* L.) and 5-aminosalicylic acid in a TNBS-induced colitis rat model. *PLoS ONE* **2017**, *12*, e0174116. [CrossRef] [PubMed]
97. Lee, Y.; Lee, J.-Y. Blackcurrant (*Ribes nigrum*) Extract Exerts an Anti-Inflammatory Action by Modulating Macrophage Phenotypes. *Nutrients* **2019**, *11*, 975. [CrossRef] [PubMed]
98. Szymanowska, U.; Baraniak, B.; Bogucka-Kocka, A. Antioxidant, Anti-Inflammatory, and Postulated Cytotoxic Activity of Phenolic and Anthocyanin-Rich Fractions from Polana Raspberry (*Rubus idaeus* L.) Fruit and Juice—In Vitro Study. *Molecules* **2018**, *23*, 1812. [CrossRef] [PubMed]
99. Zhang, Q.; de Mejia, E.G.; Luna-Vital, D.A.; Tao, T.; Chandrasekaran, S.; Chatham, L.; Juvik, J.; Singh, V.; Kumar, D. Relationship of phenolic composition of selected purple maize (*Zea mays* L.) genotypes with their anti-inflammatory, anti-adipogenic and anti-diabetic potential. *Food Chem.* **2019**, *289*, 739–750. [CrossRef] [PubMed]

MDPI AG
Grosspeteranlage 5
4052 Basel
Switzerland
Tel.: +41 61 683 77 34

International Journal of Molecular Sciences Editorial Office
E-mail: ijms@mdpi.com
www.mdpi.com/journal/ijms

www.ingramcontent.com/pod-product-compliance
Lightning Source LLC
Chambersburg PA
CBHW041417210926
51850CB00013B/3299
* 9 7 8 3 7 2 5 8 3 2 5 4 5 *